ALSO BY OLIVER SACKS

Migraine
Awakenings
A Leg to Stand On
The Man Who Mistook His Wife for a Hat
Seeing Voices
An Anthropologist on Mars
The Island of the Colorblind
Uncle Tungsten
Oaxaca Journal
Musicophilia
The Mind's Eye
Hallucinations
On the Move
Gratitude
The River of Consciousness
Everything in Its Place

LETTERS

LETTERS

Oliver Sacks

Edited by Kate Edgar

ALFRED A. KNOPF | NEW YORK | TORONTO | 2024

THIS IS A BORZOI BOOK PUBLISHED BY ALFRED A. KNOPF
AND ALFRED A. KNOPF CANADA

www.aaknopf.com

Knopf, Borzoi Books, and the colophon are registered trademarks of Penguin Random House LLC.

LIBRARY OF CONGRESS CATALOGING-IN-PUBLICATION DATA
Names: Sacks, Oliver, 1933–2015, author. | Edgar, Kate (Editor), editor.
Title: Letters / Oliver Sacks ; edited by Kate Edgar.
Description: New York : Alfred A. Knopf, 2024. |
Includes bibliographical references and index.
Identifiers: LCCN 2023052530 (print) | LCCN 2023052531 (ebook) |
ISBN 9780451492913 (hardcover) | ISBN 9780451492920 (ebook)
Subjects: LCSH: Sacks, Oliver, 1933–2015—Correspondence. |
Sacks, Oliver, 1933–2015—Friends and associates. |
Neurologists—England—Biography. | LCGFT: Personal correspondence.
Classification: LCC RC339.52.S23 A4 2024 (print) |
LCC RC339.52.S23 (ebook) | DDC 616.80092 [B]—dc23/eng/20240222
LC record available at https://lccn.loc.gov/2023052530
LC ebook record available at https://lccn.loc.gov/2023052531

LIBRARY AND ARCHIVES CANADA CATALOGUING IN PUBLICATION
Title: Letters / Oliver Sacks.
Names: Sacks, Oliver, 1933–2015, author.
Identifiers: Canadiana (print) 20240322576 | Canadiana (ebook) 20240322592 |
ISBN 9780345811417 (hardcover) | ISBN 9780345811431 (EPUB)
Subjects: LCSH: Sacks, Oliver, 1933–2015—Correspondence. |
LCSH: Sacks, Oliver, 1933–2015—Friends and associates. |
LCSH: Neurologists—England—Biography. | LCGFT: Personal correspondence.
Classification: LCC RC339.52.S23 A3 2024 | DDC 616.80092—dc23

Front-of-jacket photograph © Rosalie Winard
Spine-of-jacket photograph by Kate Edgar
Jacket design by Chip Kidd

Manufactured in the United States of America
First Edition

oliversacks.com

1st Printing

In memory of Dan Frank

Contents

Preface

KATE EDGAR

OLIVER SACKS ADORED LETTERS. In London, where he grew up in the 1930s and 1940s, letters and postcards were the way people kept in touch. Relatively few households yet had access to a telephone, but there were two mail deliveries a day, so one could, if need be, reply by return post the same day.

By the time he was six years old, Oliver was away at boarding school, and I cannot help imagining that a letter from home must have been especially prized. Even as an adult, he loved collecting the post each day to see what it would bring him.

He always felt that one must reply to a letter—instantly, if possible. (This seemed to be a combination of good manners and his irrepressible urge to communicate.) He was even known to compose a few lines to send to the electric company along with his monthly check. He saved envelopes, especially those from important people or from exotic locations, and interesting postage stamps (we, his office staff, would later do the same, putting them aside for a particular patient or two who collected them).

Throughout his life, Oliver kept most of his incoming correspondence, and took pains to preserve his own replies—with carbon sets, rough or retyped drafts, or, later, photocopies. This, of course, was not always possible in the early 1960s, when this book begins. Many of the letters in the first chapters are reproduced from the lightweight, folding airmail missives he sent home to London after arriving in North America. It would have been difficult to copy these at the time, apart from photographing them, but fortunately, his parents saved the letters, and they were later returned to Oliver.

His sheer output of letters was prodigious. The correspondence files in his archive include something like two hundred thousand pages—some seventy bankers boxes full.

OLIVER'S LITERARY STYLE, WHILE VIVID AND LYRICAL, was rarely concise, and it was complex in both structure and content. In person, he often spoke in paragraphs with long asides, but finally circling back to the topic at hand; when he was working on an essay or a book, it was much the same. He had difficulty, however, editing his own work. Thus, when one editor or another asked him to clarify something or boil it down, he would simply crank a new piece of paper into his typewriter and start over. Voilà, a new draft. Eventually, the editor would have a pile of drafts, to say nothing of a sheaf of follow-up letters with new footnotes and addenda. It was difficult to choose the best among these, since most versions contained wonderful passages, but each headed in a different direction.

When I began working with him as a young editor, circa 1983, it seemed clear to me that the only solution was to cut and paste among the many drafts (in those pre-computer days, we did this the old-fashioned way, with scissors and tape), stitching together his various trains of thought. And so—informed by his wonderful stories and his philosophical ruminations, or often simply repeating things he had just said to me—I tried to do so. In time we worked out a dialogic process of editing.

This was new for me. I was used to getting a more or less complete manuscript from an author, reading it through a couple of times, and then making and revising my comments before returning them for the author's consideration. Oliver, on the other hand, wanted me actually sitting by his side as he tore each finished page out of the typewriter: "Here! What do you think?" I began referring to this as "combat editing."

I would arrive home after a day with Oliver, exhausted from the nonstop effort of trying to keep up with his restless intellect for eight hours. But it was also exhilarating work, and when he phoned me an hour or two later with new thoughts, I was ready to dive back in. What started for me as a freelance job, occupying a day or two a week, soon became a full-time vocation—and then some.

IN THOSE DAYS, Oliver told me a great deal about his youth and early days as a doctor and a writer. He had published two books already, both based on the patients he was seeing as a neurologist. *Migraine* (1970) and *Awakenings* (1973) already showed his patient-centered approach to his work, his curiosity, and his deep erudition. But though *Awakenings* had been a literary success, he had received little or no recognition from his medical peers—quite the opposite. Partly perhaps because of this, he was struggling to finish *A Leg to Stand On,* a book he had worked on for almost a decade. He could hardly have imagined at that point that he would one day become a hero to aspiring young doctors around the world or ignite an entire genre of what he called "clinical tales."

The letters in this volume are full of contradictions; they are ferocious, tender, observant. They betray a fair degree of self-absorption—the kind many of us might evince in adolescence—but more usually show a concern and generosity, especially for people at the margins of society: young people, old people, incarcerated people, and, of course, patients living with unusual syndromes or maladies. These letters are revelatory, even for me, despite my decades with Oliver. They speak of books conceived but never written, of books written but then lost or destroyed, of passionate love affairs, and of his wrestling as a young man with the career choices ahead of him. At one point, in 1961,* the man who would one day become the personification of a compassionate physician writes: "I am discovering, in a far more intense degree than ever before [. . .] an extreme aversion to patients, sickness, hospitals and particularly doctors. [. . .] The truth is, you know, I should never have become a doctor."

By 1970,† he is hitting his stride, and he writes: "I think that I am a good (and, very rarely, at magic moments, a great) teacher: not because I communicate facts, but because I somehow convey a sort of passion for the patient and the subject, and a feeling of the *texture* of patients, the way their symptoms dovetail into their total being, and how this, in turn, dovetails into their total environment: in short,

* Letter to his parents, July 6, 1961.

† Letter to his parents, May 6, 1970.

a sort of wonder and delight at the way everything *fits* (and it does all fit, so beautifully, like a wonderful jigsaw puzzle)."

One can see that, from the beginning, he had high aspirations for himself as a writer, perhaps even before he found his true calling as a physician. One can hear his thoughts evolving over a span of decades, returning to the same topics again and again, searching for a new understanding that would fully mature only with the development of modern neuroscience in the 1980s. Sometimes he uses letters to try out new ideas or modes of expression; at times his letters seem more akin to journal entries or attempts at semi-fictionalized narrative. They occasionally become exercises in the analysis of his own psyche, especially as he begins his journey into psychoanalysis in the mid-1960s.

One can feel the evolution of his prose as it becomes more assured, more focused, partly in response to his vast correspondence with thousands of people, from Nobel laureates to schoolchildren. By the middle of the 1980s, after the publication of *The Man Who Mistook His Wife for a Hat*, his correspondence broadened to include legions of fans, and many of those people offered him their own stories. Oliver loved writing back and forth with them, and their letters became an extension of his medical practice. Much as Darwin corresponded with ornithologists and pigeon fanciers around the world to expand his understanding of natural selection, so Oliver wrote to one and all to gain insights into human experience and individuality.

Oliver insisted—to himself and to others—that his own observations, however seemingly exotic, guide his practice. On this point, he was adamant. (He often compared himself to a natural historian, like Humboldt, Bates, or Darwin, those nineteenth-century explorers he loved to read as a boy, and he considered observation and description to be his métier.) Long before most in his profession, he was describing the effects of music and art as therapy in the patients he spent so many hours with. He was beginning to develop a new view of conditions very few had even heard of, like Tourette's syndrome and migraine aura and prosopagnosia. He revisited long-misunderstood conditions like autism and color "blindness," introducing them to his readers with his usual sense of wonder and profound empathy. Indeed, he broke with his medical colleagues by casting such conditions not as pathologies but as, simply, different modes of *being*.

ONE THING THAT IS ABUNDANTLY clear from his letters (and Oliver himself remarks on this) is that he had a seriously prolonged adolescence. One can imagine many possible reasons. As a child, he had been exiled from his family and friends during the war, sent to a remote boarding school, where he was beaten and starved but afraid to complain. As a gay teenager and adult, he was forced to hide his sexual orientation in a homophobic culture. And as a brilliant but unconventional man, he seemed to attract envy and rivalry from nearly all his bosses, as well as outright rejection or silence from his peers. His moods oscillated between rushes of creativity and periods of withdrawal.

Oliver was not unaware of this. He had spent an awful lot of his life, he told me, in immobilizing depression, which seemed to alternate with manic creativity. Was he bipolar and somehow began to grow out of it as he approached the age of forty? He sometimes wondered if this was so, or whether perhaps he was schizophrenic, like his brother Michael—though his psychiatrist of nearly fifty years thought he was neither. Then there were the amphetamines, opiates, and hallucinogens that he used to excess throughout the 1960s. His letters from these early days are sometimes grandiose, melodramatic, perhaps written while he was high on speed.

He began to emerge from his own self-doubts in the early 1970s, especially with the sudden death of his mother and the publication of *Awakenings* on his fortieth birthday. In his correspondence, we see him forced to confront and embrace his own adulthood; his mood swings, while still evident, become more moderate. His letters become more concise, more assured. (His gradual quitting of amphetamines must have helped, as did years of analysis, greater success in the literary world, and more recognition from his medical colleagues.)

Oliver sometimes complained, as time went on and he became a popular figure, about the sheer volume of correspondence he received (and thus felt compelled to answer). It is true that writing letters sometimes seemed a distraction from whatever "bigger" writing projects he was working on, but it was also crucial, his connection to the world. Often a serendipitous letter, totally unexpected, would launch him on a whole new essay or even a book. At other times, he might remember and dig up a letter from years earlier, as a topic he had been subconsciously cultivating burst into consciousness. Letters remained, as always, a lifeline, and a constant source of inspiration. Even into old age, Oliver loved to sit down with a handful of incoming mail, a pile of notepaper, and his fountain pen, poised to reply.

Editor's Note

OS COULD TYPE amazingly fast, using his two index fingers in a machine-gun staccato. His typos were many, and he had a rather Germanic sense of capitalizing certain nouns (usually conceptual ones like Action or Will) when he wanted to emphasize them. And he used many elements of punctuation interchangeably: dashes, commas, ellipses, colons, and quotation marks (to say nothing of underlinings, double underlinings, fountain pen flourishes, and ALL CAPS) are all sprinkled liberally and inconsistently throughout his work.

For this book, his various emphatic forms are all rendered as italics. Obvious typos or missing words are, for the most part, silently corrected. I have freely repunctuated his letters for legibility but kept some of his random punctuation and a few misspellings to convey a slight flavor of his style and usage. OS often confused the spelling of people's names, even those of close friends. In most cases, I have corrected these; those I have left are explained in context. His spelling also evolved, with time, from U.K. usage to U.S. custom.

Quite a few of the letters I pared down for this volume began as epistles running to a dozen or more typed pages. (One, at least, clocked in at forty pages.) All editorial excisions, large and small, are indicated by ellipses within square brackets (unbracketed ellipses are in the originals).

I have often left out certain stock phrases or thoughts that OS included in his letters. He frequently, for instance, began a letter by apologizing for his delayed reply. (In some cases this "delay" was a matter of days or weeks; occasionally, several years. But he would pick up these conversations with an immediacy that defied whatever amount of time had passed.) Similarly, he often closed a letter with apologies for his handwriting or the length of his reply; I have left a few of these as examples.

He was very fond of quoting, sometimes in extenso, all manner of people: among his favorites were Freud, Darwin, Goethe, Dickens, Lewis Carroll, and William James. I have retained some of these quotations but omitted many more. Most of them are likely recalled from memory and are often inexact. Occasionally I have corrected minor variances silently; at other times I have footnoted the actual quotation he was thinking of.

Each correspondent is identified briefly at first mention and/or the first letter addressed to them, or occasionally in a headnote. While some of the correspondents represented here are the people he was closest to, the converse is not necessarily true. Many of his dearest friends he saw frequently in person—there was rarely a need to write more than a brief note.

In many instances, especially in the era before OS had any secretarial help, it is quite difficult to know whether a letter was ultimately sent to its intended recipient. His archive contains multiple drafts of some early letters, as he used the process of letter writing to think aloud to his imaginary audience or searched for the most devastating or witty words to rant about a perceived injustice, venting his feelings safely onto a piece of paper. In cases where I am relatively certain that a letter was not ultimately sent, I have noted this. In other cases, I leave it to the reader's best guess.

Information about the various essays by OS referred to in this book may be found in the Selected Bibliography.

LETTERS

1

A New World

1960–1962

IN JULY 1960, A FEW DAYS before his twenty-seventh birthday, Oliver Sacks left England, intending to settle for a while in Canada or the United States—partly to escape the English military draft and partly to reinvent himself in a new place, without the suffocating closeness of a huge extended family. He had spent four years studying at Oxford, followed by medical school and then two years of working as an intern in London and Birmingham. During this time he developed his interest in weight lifting and motorcycles, and pursued clandestine sexual encounters, for in postwar England, homosexuality was a criminal offense, punishable by imprisonment or (in the infamous case of Alan Turing) chemical castration. He had spent a summer on a kibbutz, hiked and traveled widely in Europe, and bought the first of many motorbikes. His mind was filled with images of the wide-open West he had seen in the photos of Ansel Adams, in cowboy films, and in the paintings of Albert Bierstadt.

Looking back at this period in his 2015 memoir *On the Move,* he wrote, "I had a peculiar, unprecedented feeling of freedom: I was no longer in London, no longer in Europe; this was the New World, and—within limits—I could do what I wished."

He wrote regular letters back home, to his parents and his favorite aunt, Auntie Len, chronicling his travels with a mixture of hyperbole, gritty romanticism, parody, and an avid eye for detail.

To Elsie Sacks, Samuel Sacks, and Helena Landau

OS'S PARENTS AND HIS AUNT*

AUGUST 2, 1960
QUALICUM BEACH, VANCOUVER ISLAND

Dear Ma and Pa, and, of course, Auntie Len,

Finding myself with a lull, and with a typewriter, I am sitting down to write you a long and overdue letter [. . .].

I last wrote you, I believe, from Toronto, though I have sent a couple of postcards since. [. . .]

From Toronto I flew to Calgary, going over the prairies at night. We touched down at Winnipeg and at Regina, where I snuffed the prairie air—there was no time for anything else. In Toronto, the air is humid, and smells of frenzy, sweat and gasoline. In the prairies it is dry and warm and aromatic, and smells of cinnamon and roasted buckwheat, as if the door of some gigantic oven had been opened. However, these are not the impressions on which to base important decisions! The sun rose slowly after we had left Regina, for we were chasing at 400 mph to the West; had we been going twice as fast, the miracle of Joshua would have been re-enacted, and the sun would have stood still in the heavens. At dawn I first perceived the limitless ocean beneath us, ripening wheat for more than a thousand miles in every direction, a sight unique to the Mid West. We veered to avoid a prairie storm, which was completely isolated and circumscribed in a cloudless sky, like some aerial jellyfish, grey and livid, hurling its long streamers on a little settlement below. At 6 a.m. we landed at Calgary [. . . which] had just finished its annual "stampede," and the streets were full of loafing cowboys in jeans and buckskins, sitting the long days out with their hats crushed over their faces. But Calgary also has 300,000 citizens. It is a boom town. Oil has brought a huge influx of prospectors, investors, engineers to it. The old West life has been overwhelmed by refineries, and factories, and by offices and skyscrapers. If you want to invest some $ in a sure thing, make

* OS writes in *Uncle Tungsten* and *On the Move* of his deep attachment to one of his mother's sisters, his Aunt Helena, known as "Len" or "Lennie." In *On the Move,* he also quoted some excerpts from this lengthy letter.

it Albertan Oil, which is on its way to altering the world's markets in oil. There are also tremendous fields of uranium ore, gold and silver, and the base metals, and you can see little packets of gold dust passed from hand to hand in the taverns, and men made of solid gold behind their tanned faces and filthy overalls. I must make a comment on drink here. You know the taverns of the cowboy films, the low swing doors, the tough guys within, smoking and quarreling, dicing, gambling and shooting. It's not true, not at least in public. Canada has the most stringent licensing laws in the world, and the most prohibitive social ones. You cannot stand in a bar, cannot move to another table, cannot talk to a stranger. You cannot sing, play cards, or darts. There is nothing of the mildness and geniality of an English pub. Drinking is not gregarious here. It is hard and solitary, and Canada enjoys the highest incidence of drunkenness and alcoholism in the world. I forget whether I mentioned some of the other aspects of social prohibition in a new country: In Quebec, for example, a woman cannot vote, cannot divorce her husband, cannot have a banking account of her own, and can be arrested for wearing short sleeves or skirts in public (and frequently are). The "old country" (this is everyone's term for it, both nostalgic and derisive) is very mellow in comparison.

Not only alcoholics, but cranks, psychotics, misfits, religious maniacs in uncomputed numbers. But this is another story.

I took the CPR* to Banff, roaming excitedly in the train's "scenic dome." We passed from the boundless flat prairies through the low spruce-covered foothills of the Rockies, climbing gently all the time. And gradually the air became cooler, and scale of the country more vertical. The hillocks grew to hills, and the hills to mountains, higher and jaggeder with each mile we progressed. We puffed punily in the floor of a valley, and snowcapped mountains soared tremendous about us. The air was so clear, that one could see peaks a hundred miles away, and the mountains besides us seemed to be rearing over our very heads. Banff lies at 5500 feet, in a hollow, with peaks of 10 to 12,000 feet surrounding it in every direction. It is a tourist Mecca, bursting with fat Americans with their fat cars and their fat pocketbooks. I stayed there a day and a night, not sleeping, but writing and writing for more than fourteen hours at a stretch, while the

* Canadian Pacific Railroad.

tawdry costly night life opened, and blossomed and fell silent around 2 a.m., and the silence of the mountain fell upon the little town, so that I felt now it is mine, a still Banff beneath the mountain and the stars which nobody can take from me. At 4 I heard a genuine cuckoo, upon an augmented fourth, and then the clatter of waterfowl in the river, and at 5 the old Indian streetcleaner, with his close-cropped white head, wheeling his barrow along the street, collecting the refuse of civilization, the beer bottles and the cigar butts, and the funny hats, like the debris of a party. By 6, the early editions were being hawked, and barelegged hikers were gathering over their maps, and the old ladies had risen to see the dawn on the mountains. By 7, the great cars were passing along the road, East to West, West to East, on journeys immoderate, impossible, to a traveller in Europe. And by 8, the hamburger and ice-cream parlours were open, the groceterias and meateterias had their shutters down, and the fat Americans in their Hawaiian shirts stood on every street corner, taking pictures. It was a fascinating cross-section of a night, which seemed to retrace the evolution of Banff from a tiny settlement to a bustling tourist centre.

On my second day, I went to Sunshine Lodge, attracted by its name. It stood at 7200 ft., a luxurious cedar cabin, hung with trophies of the chase, and boasting a log fire of dimensions never seen in England. I woke next morning, and whipped open my curtains to see the sunshine. There was a blinding snowstorm, and I could see nothing. But it had cleared by 8, and after a prodigious breakfast (melon, fruit juice, enriched cereal K, trout, pancakes with maple syrup, ham with three eggs, toast and marmalade, Cuban coffee and two cigars, six thousand calories and close to my visceral heaven!), the sun was high in the cloudless sky, and the temperature over 90°.* [. . .]

A nature paragraph specially for Auntie Len: The Lodge is set in a huge alpine meadow, which was at its peak in early July. Dominant flowers are mountain avens (which were in seed when I arrived, like huge dandelion heads, alight and floating as they catch the morning sun). Indian paintbrush, in every shade from faint cream to intense

* OS was sometimes prone to exaggeration, and this list of breakfast items might seem a good example, but he did overdo, especially at breakfast buffets, even into his seventies.

dayglo vermilion. Chalice cups, Trollius, valerians, saxifrages, contorted lousewort and stinking fleabane (two of the loveliest, despite their names!). Arctic raspberries and strawberries, which rarely fruit; the three-leaved strawberries catch and hold at their centre a flashing drop of dew. Heart-shaped arnicas, calypso orchids, columbines and cinquefoils. Glacial lilies and Alpine speedwell. The rocks are clustered with succulent stonecrops. The main shrubs are willow and juniper, bilberry and buffaloberry. Various firs and spruces up to the timber line, and above this only larches, with their first white stems and downy foliage.

The birds are unnaturally tame, or rather just naturally tame (since this is a National Park, and no aggressive acts are allowed). I walked right up to a ptarmigan, which had just about shed its white winter plumage, accompanied by five chicks. [. . .]

High up, through glasses, I saw a white mountain goat, perched on an unbelievably small pinnacle or rock, its four legs crushed together. I have seen black and brown bears galore, though no grizzlies. Elk and moose browsing in the lower pastures, especially if these are intersected with streams. [. . .] I have seen trees fatally ravaged by porcupines, and I have eaten "porky" meat at a barbecue, tho' I have seen no live ones yet.

All vegetation and animal life dies away as one climbs towards the summits, except moss campion, and various mosses and lichens. [. . .] It is possible to *run* down a mountain, and this is one of the most exciting experiences in the world. And I did run down that mountain, flew it seemed, leaping from boulder to boulder, yelling and weeping and laughing all at once, miraculously exempt from fear or injury or fatigue. One of those experiences which make golf, and lumbar punctures, and all the paraphernalia of one's normal, non-transcendent life, seem very dull in comparison.

I must here introduce the American family who looked after me. There were two Magoo-like men, as similar as twins, who called each other brother, though they were not brothers, I learned later, only friends. One was the Professor of Law at Philadelphia, and the other the president of the bar association of New Jersey, but I am happy I discovered what delightful companions they were before I found out what eminent lawyers they were. They took me under their wing, and we went around a good deal together. On a horse for the first time since Braefield,* I accompanied them on the pack trails to Lake Egypt and to Mt. Assiniboine.

Riding horses is a great experience; I'm sorry I missed out on it for so long. [. . .] However, I gradually got the hang of things. We ascended into a vast mountain plateau, so high that many of the cumulus clouds were beneath us. "Man has made no changes here," cried the Professor, "he has only enlarged the goat trails." It was a strange feeling, perhaps the first time I had ever had it, to know that our party were probably the only human beings in some hundreds of square miles. High on the plateau, above the trees and the insects, we seemed to be treading on the very top of the world. And then gradually we came down, our horses treading delicately in the undergrowth, to the glacial string of lakes with their strange names. Lake Egypt, Lake Sphinx, etc., and above them the Towering Pharaoh mountains, their old faces marked with gigantic hieroglyphic markings. Ignoring the cautious warnings of the others I dived into the clear waters of Egypt (you, Pop, you couldn't have resisted either), and out of their cold, and clearness and calm, was distilled the intensest pleasure. To float on your back in an alpine lake, looking around you at peaks the majority of which are as yet unnamed and may well remain so, for why name peaks where nobody could live?

Another one of the exhilarating things about Canada is that one lives in an epoch of naming. Everything in England was named and done with half a thousand years ago, but here names are vivid and contemporary, Kicking Horse Canyon and Sorefoot Lake, and tell you of adventures which have happened within the span of a man's memory.

The professor was a wonderful companion. On a strictly practical level, he taught me to recognize glacial cirques and different sorts of moraine, to decipher the trail of moose and bear, and the telltale ravages of porcupines; to survey the terrain closely for marshy and treacherous terrain, to predict the clouds (beware the sinister lens shaped clouds which portend violent storms), and to fix landmarks in my mind so that I could not get lost. But his range was enormous,

---

* A boarding school in the English Midlands to which OS and his brother Michael were evacuated during the Second World War.

in fact complete. We spoke of law and sociology, and economics, and politics and advertising, and business. I have never known a man so profoundly in touch with every aspect of his environment, physical and human, and yet enriched by a mocking insight of his own mind and motives which balanced and rendered intensely personal everything he said. His elder "brother," whom they called Marshall (at first I thought he was a sort of emeritus Marshall, and the idea stuck), was a burly old man of nearly seventy-five, in full possession of his magnificent intellectual powers and wit, who smoked cigars before breakfast, and sang in the shower in a tremendous bass voice, and out-ate all of us, and pinched the waitress's bottom, and yarned endlessly of his travels and adventures, mixing fastidious accuracy with grotesque distortions, till we were all pulped with helpless laughter. Old Marshall had virtually opened the Rockies to the Tourist trade thirty years before, and still knew every path and landmark far better than our guides.

I went on to Lake Louise by myself for a while, going along the trails to Lake Agnes and "little beehive" (a fire lookout, commanding a view a hundred miles long in either direction along the mountain valley), and then up to the Plain of the Six Glaciers, which boasted a fairytale teahouse, so high and light and airy it might have come directly from Shangri-La.* Coming down from the Plain, I overtook a bearded man limping heavily, and supported by his tiny wife. And coming up at exactly the same time, the three of us were joined by a sleek Golders Green figure† ascending from the Lake.

"I'm a doctor," I said, "can I help?"

"I'm also a doctor," the other fellow said, "and I can also help."

Thus, by a fantastic coincidence, the only injured man in a thousand square miles met in the same moment the only two doctors in a thousand square miles. He'd been caught in an avalanche, and was lucky to escape with his wife. He suffered only a bruised back and fractured scaphoid (we agreed) on his left wrist. Other doc's name was Elman (yes, yiddishe boy!), a graduate of university in Nova

* James Hilton's hugely popular 1933 novel *Lost Horizon* was set in a remote paradisal valley he called Shangri-La. (*Lost Horizon* would become one of the first mass-market paperbacks published in the United States, in 1939.)

† Golders Green, a neighborhood in London near where OS grew up. Unclear what the term "Golders Green figure" means, though it may connote Jewishness.

Scotia. We met over drinks later in the evening and chatted about this and that. He wants to do obstetrics and go to Hawaii; good luck to him. He was employed, by the way, in a curious double capacity. Two young doctors alternate between Banff Springs Hotel and Chateau Lake Louise, the two most elaborate hotels in the Rockies, and patronized almost exclusively by elderly rich hypochondriacs. The young doctors are chosen not merely for their professional skill, but for their soothing appearance, and their good looks, so that they may act as part-time gigolos to the lonely old ladies, and this subsidiary capacity is often more lucrative than the purely medical one.

Later last week I was invited by the Parks (the Philadelphia lawyers) to join them at the lodge on Lake Bow. It was called Num-Ti-Jah, the Indian term for a black sable, this being their name for the venerable Jimmie Simpson who owns the place. JS deserves a book to himself, and will get it one day I am sure. He is eighty-five, although he runs and swims like a boy of twenty. Coming from a patrician family in Lincolnshire, he was sent here in his teens, as were so many second sons of families, until the succession was secured by the eldest brother siring a son. He quickly made his way to the West (this was in the early [eighteen-]nineties), and became famous as a trapper, climber, explorer and geologist. He blazed the trail from Banff to Jasper, which only now is being consolidated to a highway. He shot (by accident) the largest sheep in the World, which now resides in the NY Natural History Museum. And he must be one of the greatest raconteurs in the world. His voice is not unlike Jonathan's imitation of Moore or Russell,* and his wit too has something of the same quality; and it is a strange experience to hear his fantastic tales, apocalyptic† a few, but mostly true, about grizzly hunts, and gunfights, appalling climbs etc. in this lucid English voice. He is quite unpredictable, sometimes keeping to himself for days on end, and at other times becoming uncontrollably voluble. I was woken at 6 a.m. on my first day there by the sound of his tales, and went down on tiptoe to join the Parks who were listening to him. At first I tried to remember

* Jonathan Miller, one of OS's closest friends from their school days, was apt to imitate many famous (or otherwise) figures, including British philosophers Bertrand Russell and G. E. Moore, whom he likely encountered during his studies at Cambridge.

† Presumably he means "apocryphal."

his stories for future reference, but there were so many, and so varied, that it was impossible, and I just surrendered to the magic of his personality. He is the very last of the Wild West men, and was a personal friend of all the famous ones, including the most famous of all, Bill Peyto, after whom a mountain and a lake are named.

(In parenthesis, I must tell you of Peyto's cabin, which old Marshall showed us on the return journey from Egypt. Not a score of people know where it is, or even know that it exists, for it is listed officially as having been burnt down by order. Peyto was a nomad and misanthrope; a wit; a great hunter and observer of wild life; and the father of uncountable bastards. His cabin is built in the most inaccessible part of the forest, and in his lifetime none but he knew how to find it. In 1936, he had been feeling ill for some time. He scrawled on his door "Back in an hour" and rode down to Banff. He never returned. The scrawled message is still faintly visible, and inside his darkened and rotting hut, we saw his cooking utensils and ancient preserves, his mineral specimens (he operated a small talc mine), fragments of a journal, the *Illustrated London News,* piled high, from 1890 to 1926, an empty ink bottle from which the contents had evaporated, and all the eerie Marie Celeste like atmosphere of his vacated home. It was a very moving experience.)

I went with the Parks to the Columbia icefields, one of the few icefields accessible without being an accomplished climber. This was the real thing, gunmetal grey and limitless in size, not like the fairy grotto stuff of the Swiss glaciers. We rolled up three miles of it in a snowmobile (I sent you a postcard of these), and were told that we now had 1100 feet of ice beneath us. I saw a pothole into which a stream was pouring, and this was 800 feet deep. You saw the deepening blueness passing to black, you heard the rush, but never the impact of water. My first thought, foolishly, was of the marmalade pot lined tunnel through which Alice fell.*

Finally taking my leave of the kind Parks, and promising to meet them in Philadelphia, I tried a little hitchhiking. I got as far as Radium Hot Springs, a sort of New-World Bad Auenstein for those with gout and disseminated lupus, and a little later found myself

* Another one of OS's literary touchstones was Lewis Carroll's *Alice's Adventures in Wonderland* (1865).

conscripted for firefighting. British Columbia has had no rain for more than thirty days, and there are forest fires raging everywhere (you have probably read about them). A sort of Martial Law exists, and the forest commission can conscript anyone they feel is suitable. I was quite glad of the experience, and spent a day in the forests with other bewildered conscripts, dragging hoses to and fro, and trying to be useful. However, it was only for one fire they wanted me, and when at last we shared a beer over its smoking dwindling ruin, I felt a real glow of confraternal pride that it had been vanquished. British Columbia at this time of the year seems bewitched. The sky is low and purple, even at midday, from the smoke of innumerable fires, and the air has a terrible stultifying heat and stillness. People seem to move and crawl with the tedium of a slow-motion film, and a sense of imminence is never absent. In all the churches prayers are said for rain, and god knows what strange rites are practised in private to make it come. Every night lightning will strike somewhere, and more acres of valuable timber conflagrate like tinder. Or sometimes there is just an instantaneous apparently sourceless combustion rising like some multifocal cancer in a doomed area. [. . .]

Yesterday I arrived in Vancouver, which is like Toronto, which is like every other city in North America (with the exception of Montreal, Quebec, Victoria, San Francisco, New Orleans, Boston and New York, which alone have a character of their own). Horrified by its rushing traffic, I made my way over to Vancouver Island. I must tell you of a little Leacock-like* episode in Vancouver. I went into a glistening barber saloon and "hair clinic," where eighteen determined young men shaved and clipped their anonymous clients in eighteen jewelled and gadgeted chairs. He snapped, "What style, sir?"—and I snapped back "Manhattan, please," and when he said he hadn't heard of it, and what was it please, I said, humbly, "short back and sides." And then, after the cutting, without consulting me, he singed my hair, and frictioned it, and shampooed it, and vibro-massaged my scalp and neck, while I was trying to say no, no—and then perfunctorily brushed me down (I was wearing shorts and T shirt, indescribably filthy) and presented me with a bill for $4.50 which I paid numbly, with all the fight gone out of me.

* Stephen Leacock (1869–1944), a famous Canadian humorist.

Vancouver Island is different in tempo and nature from all the rest of N. America. The straits act as a valve, allowing free access to the mainland, but discouraging visits to the Island. The traffic is slower, and there is less of the tremendous pressure of supermarkets, and high-power advertising, and of the sharp and restless Motel life which *Lolita* has rendered so familiar to us. I came to Qualicum beach, attracted by the resemblance of its name to Colchicum, the autumn crocus, and Thudichum, the great chemist and polymath. (Clang associations, I hope not suggestive of early schizophrenia!)* And I am staying at the Sunset Inn, which also attracted me by its name.

It is sunset now, and the setting sun is lighting up the hollyhocks, and the croquet hoops in the back garden, the tired happy men playing golf across the way. Inside they have a Broadwood piano, with a pile of Beethoven and Mozart sonatas atop it. A few clouds, illuminated, lie still above the atolls here and there. The Pacific Ocean is warm (about 75°) and enervating after the glacial lakes. I went fishing today with an ophthalmologist here, fellow called North, once at Marys and the National,† now in practice in Victoria. He calls Vancouver Island a "little bit of heaven which got left somehow," and I think he's right in a way. It has forests and mountains and streams and lakes and the ocean. It has the highest standard of living perhaps anywhere in the world, and it is closeted away from the frenzy and fury which are almost synonymous with the American Way of Life. It attracts the elderly of the whole continent, but serene as it is, I don't think it is for me. By the way, I caught six salmon; one just lets the line trail, and they bite, bite, sweet silvery beauties, which I shall have for breakfast tomorrow.

I'll descend to California in two or three days, probably by Greyhound bus, as I gather they are particularly hard on hitchhikers, and sometimes shoot them on sight. [. . .]

I hope and expect to find some letters from you when I look into

* Clang associations, or associating words by sound rather than meaning, are often simply playful but occasionally may suggest a thought disorder such as schizophrenia. OS was fond of pairs or lists of word permutations; these were often based on their etymological roots but frequently on their sounds (if he could combine both factors, so much the better).

† Two London hospitals.

Cook's[*] in San Francisco, though I imagine there may be a considerable lag. [. . .]

I hope, Auntie Len, that you also will write to me, and tell me of your intentions and movements now you are home and entering upon your Indian Summer.[†] [. . .]

Please give my regards all round to family and friends, and especially Michael.[‡]

If you get a chance, I wonder if you could show this letter to Jonathan,[§] and perhaps through him, to any other of my friends. I am so much on the move that I do not know when I shall next have the chance to type out a mammoth letter like this.

Look after yourselves.
Love,
OLIVER

IN EARLY AUGUST, OS ARRIVED in San Francisco and took a room at the YMCA in the Embarcadero district, where he would stay for the next two months. It had a well-equipped gym and was known as a place for gay men to meet. He later recounted to friends that he enjoyed quite a lot of sexual encounters there; he also spent time in gay bars, where he often went by his middle name, Wolf. Naturally he did not describe those encounters to his parents (nor did he speak of his increasing experimentation with various illicit drugs). Still, there were plenty of other adventures for him to report on.

* Travel agencies like Thomas Cook and American Express, in those days, also served as poste restante locations.

† Len had recently retired from her career as headmistress of the Jewish Fresh Air Home and School near Manchester and had returned to London to enjoy her "Indian summer."

‡ OS's brother Michael Sacks, four years his elder. Though OS did write separately to Michael during this period, none of those letters survive.

§ Miller.

To Elsie and Samuel Sacks

AUGUST 24, 1960
C/O THOMAS COOK, 175 POST ST., SAN FRANCISCO

Dear Ma and Pa,

[. . .] I have now been in and around San Francisco for two and a half weeks, have seen a fair amount of the city and country, visited hospitals and universities, made various enquiries and contacts. [. . .] And after living here, I am now almost persuaded that the States in general, and California in particular, is likely to be my ultimate home, irrespective of my immediate course of action. Canada and the States are alike in providing space, affluence and professional opportunity of an order which would be almost inconceivable to me in England. You know as well as I how tight and tedious the professional ladder is in England, in neurology above all subjects: the long, wasted years as a peripatetic registrar etc. The only comparatively easy road to consultant status* in England lies in psychiatry, and though I could easily use this (in England, or Canada, or the States) with the assurance of professional success, yet there is something in my temperament and training which inclines me to a more tangible subject, one in which I might less suspect myself of phoniness or indifferent standards, and one again which allowed of some experimental work of a laboratory kind. Perhaps I am deluding myself here. Perhaps I do have some therapeutic urge and ability, altho I cannot now perceive these as strong qualities in myself. In any case, the decision does not have to be made forthwith.

In comparison with Canada, the States is a country of densely packed intellectual centres. [. . .] In so under populated a country as Canada, neurology hardly exists as a subject, whereas here, in California, there are enormous clinical-cum-experimental neurology set ups in all the major universities and a number of non-affiliated hospitals. In the States, a prodigious amount of money is directed to research, partly as a consequence of tax evasion. Fat industrial profits

* In England, a registrar is a young doctor with specialty training, somewhat equivalent to a resident in the United States; consultant status is roughly analogous to the more senior position of "attending physician."

are propelled towards all deserving (and non-deserving) projects, in order to keep them out of the Fort Knox coffers.

And then again, California combines in itself the natural advantages and beauties of a whole continent. Climbing, skiing, desert, ocean, forest, vineyards—all lie within a day's journey. San Francisco itself has unique natural advantages, as you probably know. The temperature gradient between the ocean and the burning interior propels a mist in and out of the city twice a day, so maintaining SF at an almost constant, and almost perfect, temperature, the year round. The city has all the cultural and intellectual assets of a huge centre like London, and yet itself has a population of less than a million, which is not expanding. It has a rich and fantastic history, which would grace a much older city. It has in itself, and is within easy range of, fantastic natural beauty of every sort.

I have put my head in at the U. of California Medical Centre, which is a triad of gigantic white buildings overlooking the Golden Gate Park, with an incomparable vista of San Francisco from the upper storeys (Neur. is very high up!), and its distant bridges, ocean, and hills. They have two neurologists there. [. . . Also] three neurological residents, all of them weightlifters! (I always felt the two disciplines went together.) The Med. School faculty buildings were only rebuilt this year, and are a sort of Walter Mitty* fantasy of what such buildings should be like. The interns are by no means overworked, having indeed only an eight-hour day, with most of their weekends off (this would cause a revolt if mentioned in the columns of the *Lancet*!).†

I have also looked in at the Mt. Zion hospital. [. . .] The staff are largely Jewish, tho' the hospital is immensely popular among all sections of the population. Two thirds of the hospital's total work is research (it has ca. 500 beds), which is as high a fraction as in any university medical school. I had a long talk with Feinstein,‡ the assistant head in the neurosurgery (and neurology) dept. who is a brilliant if somewhat obsessional type, and watched him do some stereotactic operations (MZ is the foremost centre for these in California).

---

* Mitty, the protagonist in James Thurber's story "The Secret Life of Walter Mitty," is a meek and ordinary man prone to heroic fantasies.

† One of the main English medical journals.

‡ Bert Feinstein, a prominent neurosurgeon. (His future wife, Dianne Feinstein, would become mayor of San Francisco and, later, a United States senator.)

He has a massive experimental setup behind him, with quantities of electronic engineers, etc., and seems to be turning out a lot of fine work. Stereotactic operations, by the way, allow one to put a lesion in a human or animal brain anywhere with a high degree of precision, and so is as valuable an experimental tool as a therapeutic one. In a way this is the best possible set up, for one is always involved with patients and therapeutic perspectives, which of course provide an endless series of experimental challenges. And as Feinstein puts it—a neurosurgical patient is a preparation which can talk.* He raised the possibility of my doing what was on paper my internship at MZ, but in fact something nearer neurology and neurophysiology, and receiving a rather more respectable income than an intern.† This might be very worthwhile indeed if it worked out.

Finally, I have been to Stanford Medical School, which has just moved from SF to a stunning building on the gigantic campus at Palo Alto (a pleasant township of 40,000 people, in marvellous countryside, about forty fast miles from SF). At present they have no neurological beds as such, except some purloined from the general medical side (a consequence of the recent move), but next year they will be associated with the local Veterans Administration (V.A.) hospital, which will bring them a total of 140 neurological beds, thus making Stanford the biggest neurological centre in the West. Stanford by the way has a very fine academic standing, better than UC, though has also the reputation of being very "smoochy" and snobbish, at least at the level of student selection. [. . .]

There is a gigantic amount of red tape to exasperate and obstruct the immigrant doctor. One must submit innumerable documents, take a preliminary exam (only held quarterly, with another two-month delay for results), before one can accept an internship. On the other hand, one can be employed in a nonclinical capacity while awaiting the chance to take the exam and get results. [. . .] I will go in person to Sacramento tomorrow to the Medical Board and try to get things in motion. If, and when, I have the machinery going, and

* That is, compared to a tissue culture or "preparation" such as one might use for histological studies.

† Even though OS had already completed his internship in England, California required graduates of foreign medical schools to repeat a year of internship in the United States.

a job lined up, then I shall start on my travels in the States if there is any further time in hand.

A final possibility is that of entering the [armed] forces here as a volunteer: minimum period two years. This cuts a lot of the red tape about citizenship etc. provides an excellent income (ca $6000 with all the perks), may be a way of doing one's internship, and simultaneously of receiving some specialist training. If all this could be done at a military hospital in Western California [. . .] then indeed there would be much in its favour. However, unlike the Canadian Forces, which are small and gentlemanly and to be trusted, the American forces are a gigantic and unwieldly organization, and no sort of bargaining may be possible with them.

Well, these are the prospects. Please tell me what you think of them. [. . .]

It's suppertime now, and the innumerable possibilities of SF culinary mastery lie before me. Deep sea bass down on Fisherman's wharf, Japanese food, Italian, Chinese, haute cuisine, or a 3 lb. steak washed down with a gallon of light beer. For a belly-oriented type like me, SF is second to very few places in the world. As I leave the Y,* I shall throw a glance in at the barber next door, who never seems to have any customers, but sits in his barber's chair and plays the violin all day long. I shall be careful not to trip over the insensible "winos" littering the pavement, and must turn a stern face to the appeals of the alcoholic beggars who swarm the waterfront. In ten yards I can overhear as many languages. Fisherman's Wharf it will be, in sight of the Golden Gate Bridge arched against the sunset, and the prison of Alcatraz on its island fortress, and gutters crackling with prawn and crab shells, and everywhere the sharp smell of clam juice, which (they say) is the very essence of the Pacific itself.

Please write to me on reception of this, and give my regards to all the family. [. . .]

DON'T WORRY!

Love,
Oliver

* YMCA.

To Elsie and Samuel Sacks

SEPTEMBER 29, 1960

MOUNT ZION HOSPITAL, SAN FRANCISCO*

Dear Ma and Pa,

I trust you are keeping well, and in good fettle for the Fast† ahead. [. . .]

I went to Yosemite National Park over the weekend, which is about 200 miles from here. After a blazing summer, the waterfalls are dry and the vegetation pretty parched. It's a botanist's paradise in spring and summer (I enclose, especially for Ma's and Auntie Len's envious delectation, a booklet on the High Sierra flowers), a climbers' paradise in the summer, a skier's in the winter, a geologists' and pleasure lovers' all the year round. Last Sunday was a day of a clarity unknown in England, and one could see the whole length of the valley, 100 miles either way. To see distant objects so clearly is so out of my experience, that the whole scene assumed an unreality, combined strangely with its extreme precision. I went into the Mariposa Grove of Big Trees, and saw the "Grizzly Giant," 110 feet in circumference, and 4000 years old, so old you feel it must be aware in a sense, if only of light, and growth, and hurt. [. . .] No wind penetrates the big trees, and a total silence hangs inside the grove. It's easy to see why people worshipped objects so ancient, and huge, and beautiful. The sequoia cones are quite tiny by the way, though there are some pines in California with cones a yard long. The high Sierra receives its first snow towards the end of this month.

I moved into hospital yesterday and am slowly finding my feet. They are giving me board and lodging and laundry, though I cannot receive any payment yet. But Levin‡ will be giving me $60 a month, out of his own pocket, which should help. He's really been very sweet about things: he said, look if you need more, just ask me. Though I'll be pleased when I can receive a salary on a regular and normal basis.

Most of the house staff are Jewish, though their Jewishness does

* At the time, Mount Zion offered accommodations for interns.

† The annual fast for Yom Kippur, the Jewish day of atonement.

‡ Grant Levin, who, with Bert Feinstein, was in charge of Mount Zion's neurology department.

not extend beyond a love of chopped herring, and Jewish jokes, and passionate political arguments. It is the last of these, rising and falling in the wards, and the dining room, and the lounge, which is so different from the bland political apathy of my fellow residents in England. Everyone is very approachable, and the hospital more or less runs on first name terms. It seems to be a fairly paternalistic institution, since one can sign for free theatre and concert tickets etc. and has the run of U. Cal. and other institutions. There is a civic centre fifty yards away, with an immense floodlit swimming pool, which I shall probably be patronising frequently. The food is of high quality, attractively prepared, and unlimited in amount. This last is a potential danger, and I must rule myself with a rod of iron, otherwise I shall weigh 300 lb in three months or so. By the way, I enclose a picture of myself taken at Monterey, emerging like some hairy and overweight Venus from a pacific lagoon. Did I tell you about Monterey and Cannery Row in a previous letter or not? I spent about five days there, observing and eating marine biology.

Next week when I'm more into the swing of things, I'll start going to the countless other sessions run into the hospital, EKG conferences and proctological seminars and various other frightening things. They certainly have a very splendid programme of postgraduate teaching, and as I mentioned, one can by reciprocity attend all the University meetings also. [. . .]

Yesterday, a long fascinating session on a woman with an epileptogenic tumour.* Various parts of her brain were stimulated, (she was, of course, fully conscious): first the motor and sensory areas were mapped out, and later various parts of the temporal lobe where stimulation gave rise to elaborate hallucinations, resembling those which usually preceded her fits. It was a fantastic experience, seeing her sit there, fat and happy in the neurosurgical chair, relate in a matter of fact voice her grotesque hallucinations, while Feinstein was scratching her exposed brain with his beaded electrode.

I must break off soon, because I want to go onto the wards and get to know some of the patients and their ailments. [. . .]

* One treatment for such a tumor, which can cause seizures, is to use electrodes to probe the brain (while the patient is awake) to locate the epileptic focus, and then to ablate it. Since the brain itself has no pain receptors, this is mostly painless for the patient.

The first rain of fall came today, and the climate is now that of London in September. Though a few miles inland, in Sacramento for example, it is still in the high eighties. I shall also go into a supermarket when some funds come, and send you the most gigantic and interesting food parcel you have ever seen. Believe you me, John Barnes* is nothing compared even to a small-town supermart here, and the larger ones are staggering cornucopias of everything which can be eaten, swallowed, chewed, smoked, sniffed or drunk in the whole world.

I will write again soon, and in the meantime look forward to hearing from you again, and to receiving also the various forms I made mention of, without which I cannot start immigration proceedings. [. . .]

Regards all round.
Love
Oliver.

To Elsie and Samuel Sacks

OCTOBER 3, 1960
MOUNT ZION HOSPITAL, SAN FRANCISCO

Dear Ma and Pa,

I hope you have lost weight and acquired virtue over the Fast; there should be a clear run now until Chanukah, unless I am forgetting something.

I looked in briefly at one of the orthodox shools yesterday, and found it fairly empty (this was midafternoon), although all of them were packed out on Friday night, which is accounted much the solemnest portion of the Fast. There are, I should explain, three denominations here: orthodox, conservative and liberal (reform). In the conservative, there is a cantor of sorts, admixture of men and women, retention of some Hebrew and yammelkas: it forms therefore a useful intermediary group for those who are too illiterate to

* A London department store with a gourmet food section.

enjoy the orthodox, but too coloured by custom and timidity to go right over to the liberal, i.e. the majority of people here.

One of my bosses, Bert Feinstein, invited me over to his family to break fast with them.* They are all originally from Winnipeg, and his parents now run a hotel here, while his brother is a consultant radiologist. Of course they are not actually from Winnipeg; they left Rumania for the prairies in about 1920. His people have a splendid Spanish style house, with a glorious patio fronting the harbour by the Golden Gate Bridge. We broke the fast on Bourbon (a change from tea), which made all of us rather tipsy on empty stomachs, and the conversation was of the usual kind—namely, how offkey the Chazzan† was, how irrational the rabbi was, how hot the shool‡ was etc. They prepared a very splendid meal, with knishes, and cholent, and paprika stuffed with meatballs, and chicken stewed in wine, along with some more americanised items, like chicken fried in cornflakes (unpleasant) and corn pones (to humour a cousin from the deep south). [. . .] The whole tribe is going to Las Vegas next week, where the prodigal Bert is reading a paper on his work. I find it difficult to imagine neurophysiology in that exotic setting of gambling, divorce and generally glittering vice: L.V. is a fantastic place, so I'm told, being an entirely artificial oasis, and rising like some hashish vision of Samarkand, fountains and gilded towers in the heart of the desert. And a very favourite place for scientific conventions.

I was invited last night to a party on Treasure Island, an entirely artificial island in the middle of the bay, created for the 1939 exposition, and this morning went on a fishing trawler for a couple of hours. I forbade them to throw the "refuse" away, and spent some happy time sorting over the sea stars and whatnot entangled in the nets.

A couple of my fellow interns took me with them to a Football match this afternoon, the Frisco Forty-Niners against the Los Angeles Beefeaters or something. It was a marvellous setting for a match, the enormous stadium surrounded by the trees of the Golden Gate Park, a brilliant blue sky, with thin clouds chasing across it from

* The end of Yom Kippur is marked by "breaking the fast" with a feast of traditional foods.

† The chazzan, or cantor, leads the congregation in prayers, which are often sung.

‡ OS often used this spelling for "shul."

the ocean (which fronts the Park). There were about 60,000 people watching, mostly in the red caps of the SF supporters. I suppose you have seen pictures of American football. The players are enormously padded about the shoulders, which gives them a very top-heavy appearance, and at the start of the game face each other, pair to pair, in a low crouch, like the stance of fighters in Siamese wrestling. Despite their unwieldy appearance, they move as fast as a speeded-up film, and contract broken limbs with quite extraordinary frequency. Three players had to be carried off during the hour or so I watched. Between touchdowns, a brass band played, to the strutting of a group of drum majorettes. I was unable to repress a hoot of laughter at this, and earned some furious glances from the very serious people all around me. It is certainly rather scaring to hear the deep bay of a gigantic crowd, all bursting with the same emotion. It reminded me of an unpleasant science fiction story, in which people had discovered how to cause material changes by the exercise of their willpower, very substantial changes if they all willed together: criminals would be conducted into the centre of a vast arena, and there destroyed by the cumulative hatred of the people, all exerting their willpower to make them burn alive.

I left after an hour or so and retired to a secluded part of the GG park, with a cigar and a quart of icy Schlitz beer ("the beer that made Milwaukee famous"), and a translation of *Il Principe*, and decided that I preferred solitude to being one of a crowd. [. . .]

Yesterday I went on my first rounds with Levin and Feinstein. Far more informal than those at the Middlesex.* They have a very large range of involuntary movements under treatment (Parkinsonisms, dystonias, torticollis, tremors) etc. And a fair number of interesting emergencies come in. Being purely an observer at the moment, and unable to accept clinical responsibility, is quite a pleasant mixture. I saw one woman with a splendid parietal lobe syndrome, who draws extraordinary reversed clockfaces, always on the right side of the paper.†

I got a letter from Auntie Len two days ago which I shall reply

* The Middlesex Hospital in London, where OS had studied medicine.

† A classic neurological test is to draw a circle and ask the patient to write the appropriate numbers in the "clock."

to shortly. She seems to be having very bad luck with her back and legs.

I have also just got a long and extremely interesting letter from Jonathan, who enclosed a review of his show, in Edinburgh.* He is curtailing his Cambridge job in neuropathology to six months, and then spending a year in Showbusiness, so he can subsequently return to Medicine affluent, with it as a delightful hobby, not a mean bread-winning soulkiller.

Thank you, Pop, for your letter which came today, along with a list of policies. Surrounded entirely by Jews in this place you can hardly feel I am quite out of it. It would be hypocritical and pointless for me to profess any Jewish beliefs, theologically: but I am by temperament and training emptied of all religious type of beliefs. But I am conscious of our culture and uniqueness and will never lose contact with it.

yrs,
oliver

OS FIRST MET JONATHAN MILLER when they were both students at St. Paul's School in London. The two of them, along with their friend Eric Korn, shared similar backgrounds: Jewish, sons of physicians, scientifically minded, brilliant. Miller later went to the University of Cambridge, where he read natural sciences and medicine and became a member of the Cambridge Apostles, an intellectual society at the college. He then qualified as a doctor and worked as an intern for two years before going on the road with *Beyond the Fringe*.

* *Beyond the Fringe*, a satirical comedy revue written and performed by Jonathan Miller, Alan Bennett, Peter Cook, and Dudley Moore, debuted at the Edinburgh Festival to great acclaim, and would soon move to the West End in London, followed by a Broadway run in 1962.

To Jonathan Miller

SCHOOLMATE AND CLOSE FRIEND

OCTOBER 11, 1960

MOUNT ZION HOSPITAL, SAN FRANCISCO

My dear Jonathan,

I was delighted and excited to receive your letter, and also your cutting.* So much had been happening to me that I imagined everybody else's life was static in comparison, but I perceive that you too have been faced with important decisions, which may prove of far-reaching effect. I am glad in a sense you did not write before, because we would then both have been entangled in the essentially introspective intricacies of each other's "problems," and perhaps subsequently have found this a source of resentment whenever we regretted our decisions (and there are times when the best decisions seem disastrous).

First can I congratulate you on the Cambridge job, although this must have been almost a foregone conclusion. Even if there were no other considerations on the horizon, I think you would have been wrong and possibly wretchedly immersed in pathology for three years, and quite liable to find yourself deadended among the pots for good. As it is, it will be a delightful interlude, and you will also be a senior apostle who has made doubly good, and a bulwark of solid virtue and endeavour, conceivably. I like the way the papers are already speaking of neuropathologist Miller. It sounds a great deal more impressive, and funnier, than medical student Miller, schoolboy Miller, psychiatrist's son Miller. Miller. I wholeheartedly agree with you that Medicine is better as a delightful hobby. [. . .] All this, of course, is upon the assumption that you will return to Medicine, and of this I am not sure. Are you? [. . .]

I honestly forget how much I have written to you in previous letters, and so risk repeating myself somewhat. First about my present position. I started here about a week ago, as a research assistant and "guest" at the hospital. This nomenclature is important, because I have not got my immigration status yet, and so am not permitted to accept employment. My duties, so far as I have any, are to acquaint myself with what goes on in the neurosurgery and physiology dept.

* Miller had enclosed a press clipping, a review of his show.

and get involved in things as much as I can or desire: and otherwise, to examine what patients, attend what rounds and seminars I wish at the hospital. I cannot start internship even in California until after I have taken the State Board exam, and this damnably enough is not being held until next Spring.

The unit is essentially a stereotactic one concerned with dyskinesias, both from the experimental and the therapeutic side, with all the valuable interplay of the two aspects. Levin, the senior man, [. . .] is cool and tortuous in all his ways, excepting his sudden generous impulses, and his driving (he drives a V-12 Ferrari, which is possibly the fastest production car in the world). Feinstein, the younger man, [. . .] is brash and opinionated and never wears socks (unlike Levin, who not only wears socks, but has suits costing $300), but is volcanic with ideas, some of them penetrating, some of them amazingly obtuse, which he throws out with a fine lack of discrimination. Then there are some voluptuous secretaries whose legs you would slobber over, who live on peanut butter and saccharine tablets, and long to marry a doctor. And finally there is emeritus von Bonin, a wonderful old man [. . .] with an enormous red face full of intelligent wrinkles and German humour, who is the final genial court of appeal regarding any uncertainty about the nervous system, animal or human. As for the interns here, they all look like Groucho Marx, wander round the hospital wearing T shirts and smoking cigars insolently into their patients' faces, full of Jewish jokes and political vehemence, yet astoundingly open, and capable of unbelievable industry, except when the world series baseball is on, when everyone crowds before the TV sets and the patients have their coronaries and hematemesis unattended.

The World Series and the Kennedy-Nixon debates occupy everyone who has concerns outside his social group. After two months of haploid life* I have emerged with pleasure and I think finality from its claustrophobic intensity, and find it an extreme pleasure to be working and living with people who talk and think and act apart from their own sexuality.

* In medical terms, "haploid" refers to cells that have only one set of chromosomes, namely eggs and sperm. OS seems to use it here as a euphemism for his immersion in the relative freedom of gay life in San Francisco after his more secretive encounters in England.

Money can buy so much here, and is the subject of such respect, that one falls insensibly into growing more conscious and envious of it. I am sure that (for the right sort of person, you and me) its freedoms far outweigh its responsibilities and limitations. The money rolls in, abruptly, almost as soon as one finishes residency, though whether it will do so as easily as it does now in five years' time I can't predict. The US deliberately underproduces doctors, which gives them an immense power, as well as great opportunities to be corrupted. There are only too many doctors round here who will shark a patient for $1000 worth of nonsensical investigations. But even the honest ones do very well. I spent last weekend with a young neurologist who has been in practice three years or so, who already has a superb redwood "cabin" on Mt. Tamalpais half an hour out of the city, with fifteen rooms and a swimming pool, and a prospect of the whole bay through one immense window, and endless mountains through the other, and went yachting* with him later. [. . .]

I have probably seen and thought and felt more than in the preceding ten years, and damnably none of it is on paper, or will be now, for that matter. I have had the material for a hundred essays, a hundred New Yorker articles, as well as a batch of novels. Perhaps it has all been so closely woven with my own emotional convolutions that I shan't be able to disentangle it, I don't know. If only I had had the sense to buy a camera! I hope you have bought one, by the way. I even paid a visit to Cannery Row, which effectively discharged any lingering nonsense about it.† It no longer exists as such, because the sardines just stopped coming to the West Coast some years ago, and show no sign of returning. However there is a ridiculous Steinbeck theatre among the rusted desolate canneries, and the place has become a big tourist centre and is overrun with "intimate" bars and flashy seafood restaurants on the waterfront. However there exists an incomparable coastline for

* More likely sailing. As he commented in a letter around this time to his parents, "One of the lovely things about professional life here, is that one not only makes money, but the leisure and the facilities to use it gracefully. Half the staff here have sailing boats, which strikes me as much healthier than the stolid English consultants, who just toddle round a golf course."

† As he would write in *Uncle Tungsten,* OS had been beguiled, as an adolescent, by John Steinbeck's *Cannery Row,* and entertained dreams of becoming a marine biologist like Doc.

hundreds of miles south of this, with the richest tidal pools you could want. I found an absolutely monstrous chiton* washed up on the shore at one place. I enclose, I don't know why, a photo of myself at an idyllic cove near Monterey, eating, with a friend. [. . .]

I have yet to meet Thom Gunn.† He left SF for London the day I arrived, but is due back soon. But I am, by and large, disappointed by the werewolf crowd‡ here. Yesterday, I encountered on the bay's edge a familiar stooping figure, dressed in a very fine English suit with striped trousers and short, stiff collar, who was attracting a good deal of amused and surreptitious attention. I was, myself, in a pair of jeans (it was very hot) and nothing much else. Anyhow, it was Leslie Le Quesne from the Middlesex, over here for a surgical conference. When I plucked at him, he turned round with an arrogant air, and then amazement and pleasure overcame him and he yelled, "My God, Sacks!" followed by "how extraordinary" and "it's a small world" many times. We had a few beers at a very posh waterfront bar, making a very incongruous pair, and it was all very unreal, Leslie's intense uncompromising Englishness of face and dress and mind against such an exotic backdrop.

I am petering out in inanities. I look forward to hearing again from you when you start at Cambridge. Give my love to (? Dr.) Rachel.§

To Elsie and Samuel Sacks

OCTOBER 15, 1960
[MOUNT ZION HOSPITAL, SAN FRANCISCO]

Dear Ma and Pa,

Thank you for your long and welcome letter: more than a week had elapsed since I had heard from you, or for that matter from anyone, and I was getting mildly alarmed.

* A type of mollusk, OS's favorite group of marine creatures.

† A few years before, Jonathan Miller had given OS a copy of one of Gunn's books of poetry, *The Sense of Movement*, and OS shared Gunn's enthusiasm for motorcycles and the leather scene.

‡ A reference to rough trade, or perhaps just cruising.

§ Jonathan's wife, Rachel Miller, was pursuing her medical degree.

[. . .] I have as yet received no parcels from England, nor the small packet from NY. I imagine the Customs have a habit of holding things up unnecessarily. You have evidently been put to a lot of trouble packing up my numerous and slightly awkward belongings, and I am deeply indebted to you for doing so. Indeed, I don't know where I'd be without your help. However, I am surviving OK at the moment, for this benevolent hospital provides its dependents with white trousers (hence pants), shirts and coats: adding to these white socks and plimsolls we look much more like cricketers than doctors. [. . .]

Last weekend I spent idly by myself, lying in the sun and reading in the immense Golden Gate Park all Saturday (I finished Henry James' short stories, and devoured the autobiography of Charles Darwin). On Sunday morning, I went in for an outdoor weightlifting contest at China Beach and got placed third among the heavyweights. China Beach fronts a desolate surf beaten corner of the bay, and is in full view of Seal Rocks, so that one can watch the seals leaping around and basking while recovering breath between lifts. It is certainly a very romantic place for a weightlifting meet. [. . .]

I have indulged in one small extravagance this week, a prism of thallium bromide-iodide from an optical co. for $5, of very high density and perfect translucency. You remember how I used to have a strange mystique about such things, and would spend enchanted hours poring over the lists of densities and refractive indices and crystal lattices in the International Tables, during my eccentric adolescence.

I am going tonight as my bosses' guest to the quarterly hospital dinner, which will introduce me to those on the staff here whom I have not already met. It'll probably be a posh affair and give ample scope for my satirical pen if I get to writing about it. Tomorrow I shall probably be going to a quarterly meeting of the American Neurological Assn. [. . .]

There is a lot else to say, but I shall defer this to another letter. I hope you are both well at home. I am glad to hear Michael is doing some work at the hospital, and is on town parole: this seems to me a distinct improvement.* I hope that sooner or later I shall hear from

* Michael, who had become overtly schizophrenic as a teenager, lived with their parents except when he was hospitalized for treatment.

my other two brothers, to both of whom I wrote at Rosh Hashannah. I have written to Auntie Len last week, and am expecting her to reply soon.*

Love,
Oliver

To Elsie Sacks, Samuel Sacks, and Helena Landau

NOVEMBER 28, 1960
MOUNT ZION HOSPITAL, SAN FRANCISCO

Dear Ma and Pa and Auntie Len,

I have been remiss in writing, due to trying to do too many things in the same lifetime. [. . .] Things are quieter with Grant† away, though busier also, since I am starting to assist Bert Feinstein at the stereotaxic operations. I have now taken over the neurological work ups and follow ups on all patients and am designing various new rating techniques, both clinically and cinematographically, for evaluation of the operative results. The old anecdotal days of medicine are passing—"I once had a patient"—and one strives more and more to get results which can be "processed" statistically. In many hands this results in a nonsensical pseudo-quantification of evidence, and against this one must be on one's guard.

I know quite a number of the house staff and attending staff now, and have even been known to play cribbage in the evenings. However I am still a dunce over card games, and intend to remain one. I have enough time-wasting activities to my credit already. Last Thursday was Thanksgiving (for the safe landing of the Pilgrim Fathers and the founding of White America), which is a sort of secular Xmas,

* Almost none of OS's letters sent individually to his aunt survive, and those included in this volume are likely drafts, whether eventually sent or not. In the 1960s, most of his letters to England were on airmail "flimsies," lightweight sheets of paper that would fold to form their own envelope, to reduce postage costs. While occasionally he would type out letters using carbon sets, copies of most airmails to his parents survive only because they were returned to OS after his father's death.

† Grant Levin.

with turkeys and ham and decorations etc. I went to no less than four massive dinners, one at Allinek's (whose wife you delivered, Ma; they had some cousins there who were your patients in the East End, Pa),* and at the parents Feinstein (where there was no ham in evidence, but stuffed cabbage balls and smoked salmon). Xmas has been largely taken over by Jews here, and I have heard Xmas trees referred to as Chanukah bushes in all seriousness. Which reminds me: you will be receiving a package of assorted American wonders around the start of the New Year.

The bike has done 800 miles since I bought it two weeks ago, and so is nearly run in.† There is a speed limit all over the US of 65 mph, which is very frustrating: however one cannot really go faster if one tries, on account of the vast volume of traffic on the freeways—often almost bumper to bumper for hundreds of miles. Off the main roads however, there are enchanting country lanes, and winding mountain roads, and one can almost fancy oneself back in England, save for crimson succulent vegetation, which sometimes seems almost Martian. The weather has so far remained pretty good, save for a violent storm last week when, with the collusion of the house "mother" (who looks after us all in the Residency) I took my baby up to my bedroom, where it stood for a couple of days, leaking oil and fuming in the middle of the room. However this was received quite well at MZ, since it tied in with the notion of English students and house doctors they derive from our export films such as *Carry On Nurse* or *Doctor in the House*. I have taken to carrying a spare clutch cable, links for the chain, spare bulbs etc. for all long journeys, so that I never find myself helpless 200 miles from the nearest city. I will probably be going down to L.A. via the desert road in two or three weeks, and staying there for two or three working days, negotiating the future. I will come back by the superb coastal road. Both of these are comparatively slow compared with the freeway, but this would be boring and nightmarish, in turns.

The sun had a curious chromatic halo round it today, which probably foretells another storm. There was virtually no winter last year, or at least no rain, so this one is going to be hard, people prophesy.

---

* Elsie Sacks was a surgeon and gynecologist; Sam Sacks was a general practitioner.

† Or, as Americans might say, "broken in."

I have a pile of discharge summaries to compose, which I shall do in solitude now, among the deserted electronic apparatus, its purity sullied by the cigar and bottle of beer I have brought in to solace my work with.

Then a quick work out, and to bed. And up at six thirty, if the weather holds, for a morning ride.

Do you remember a Somerset Maugham story about a man who had a spell cast on him by some jilted Island girl, and developed a fatal hiccough? One of our patients, a coffee baron, has had a hiccup for six days following operation, intractable to all the usual and some very unusual measures, and I fear may go the same way unless we block his phrenic* or something. I suggested bringing in a good hypnotist: I wonder if this will work? Have you any experience of this as a major problem?

All my best, and write soon.
Love,
OLIVER

PS Could you send my large black pullover (the one which we had dyed), my blue and grey squared suit ("speckled") and my new dark suit with waistcoat.

Thank you

To Elsie and Samuel Sacks

FEBRUARY 2, 1961
[SAN FRANCISCO]

Dear Ma and Pa,

I am glad you are both in good form. [. . .] I was saddened, but not surprised, by your news of Michael's relapse. The most impossible cross to bear is the awareness that someone near and dear to you has

* The phrenic nerve controls the diaphragm's movement; blocking or cutting it would cure the hiccups but also make breathing more difficult. OS recounted this story of the hiccup patient in *On the Move*, prompted by rereading this letter to his parents many years later.

undergone a complete change of personality, of a destructive type: in a sense, almost any other sickness would be preferable. But I do not think it realistic to hope, at this stage, that Michael will ever be able to live more than a protected social life, with perhaps periods of having to be institutionalized. New and better drugs for controlling symptoms will undoubtedly appear on the market, but it is near certain that he has already undergone a good deal of irreparable change. Perhaps I am wrong to put this so crudely, but I think you will both be less torn if you cease to entertain extravagant hopes of his "getting well."

Thank you, too, for news of other members of the family. I am, myself, a family man at one remove. That is, I like the awareness that I am part of a huge ramifying family, and to hear news of it: but I am happy to live out of the sphere of family feelings and pressures. [. . .]

Let me tell you something about a conference at the University here last weekend, entitled "The Control of Mind," to which Levin and Feinstein had invited us all* ($25 times twelve: presumably tax deductible!), full of eminent people—neurologists, psychiatrists, historians, Jesuits, journalists, writers, philosophers etc., and destined to start an epoch, or at least start people thinking.

Wilder Penfield of Montreal (Nobel Laureate, etc.) started the ball rolling. Unfortunately, he has become mildly senile, so was rather disappointing. He was followed by Holger Hyden, a marvellous Swedish biochemist, who produced some very exciting results on the chemical bases of memory and learning, as well as intimating that he has some astonishing drug which can affect this directly. (This was taken up by an Irresponsible Press in a big way!) Hebb,† also from Montreal, summarized his work on sensory deprivation and, by showing that we are almost entirely a product of our environments, demolished the Rousseau Natural Man very effectively. Aldous Huxley gave a tremendous after dinner speech on Education. I had never seen him before and was amazed by his height and cadaverous emaciation. [. . .] Leaning forward, in intense concentration, he somewhat resembled Vesalius' skeleton in meditation. However, his marvellous mind is as good as

* Presumably, all the neurology interns and residents.

† Donald Hebb, a Canadian psychologist, studied the neurological underpinnings of learning; his work in the mid-twentieth century was hugely influential.

ever, served by a wit, a warmth, a memory and an eloquence that brought everyone to their feet more than once. On Sunday morning, the pharmacologists described some of the new psychotropic drugs and their social implications.* [. . .] On Monday, the journalists and politicians spoke, the computer men promised us moral intelligence in their machines, and the conference ended with a series of informal smaller groups. A very exciting weekend, I am very grateful to L and F for taking us all. Humdrum life seems rather tame now. [. . .]

Write again soon, and remind Auntie Len that I am hoping to hear from her also.

Love,
Oliver

PS I have not smoked now for six weeks. I can't say I miss it much.

IN THE SPRING OF 1961, having taken the exams that would allow him to begin an official internship at Mount Zion on July 1, OS traded in his Norton Atlas motorbike for a second-hand BMW R69 and set out to travel across the United States.

To Elsie Sacks, Samuel Sacks, and Helena Landau

APRIL 22, 1961
LAUSANNE SCHOOL FOR GIRLS, MEMPHIS, TN

Dear Ma, Pa, Auntie Len and All,

Finding myself, for the first time in two weeks, with a typewriter and some leisure, I thought I would let you know something of my trip. [. . .]

I am typing from the unlikeliest of places, the Lausanne School for Girls! A friend of mine from Oxford days, who bears the splendid name of Walter Raleigh Coppedge, is headmaster here, a very

* For a longer description of Huxley, as well as Arthur Koestler, see *On the Move*.

astonishing evolution. He is, I find, a very good one too, which needs a good deal of courage in a place like Tennessee, where it is still illegal to teach the "false doctrine of evolution," and where one is savagely slated by the press for any progressive idea. At present I am sitting outside, barefoot, in a pleasant Cowper-Dryden like woodland, with a lake to one side and the graceful school buildings to the other. As I write this sentence, Walter is conducting two grey-haired parents of a prospective student around the grounds, no doubt deflecting their gaze from the strange barefoot man who sits typing under a tree. The ground is covered with catkins and acorns, and if I can ignore the monstrously large ants and butterflies, I can easily imagine myself back in England for the nonce.

My journey, in outline, has been as follows: first down to LA for the weekend. Then through California to Death Valley, which is tolerable this time of year (the ground temp. rises towards 200 in the summer).* It reminded me very much of the Negev in certain ways, though is a great deal more spectacular: one can, for example, look up from Badwater (the lowest part of the world after the Dead Sea) to the highest mountains in America, less than a hundred miles away, and clearly visible in the still air. The air was anything but still when I was there—a huge hot shrivelling wind blew up from the valley, something like a monstrous hairdryer. From Death Valley I crossed into evil Nevada (no speed limit, no limit on anything: a bubble economy subsisting entirely on sale of liquor and gambling). My one evening in Las Vegas, mercifully perhaps, was truncated by a dust storm of grotesque violence, which twice blew me off the bike, filled my ears, eyes, nose, mouth, clothes with dust, and made me marvel that such a place of glittering twenty-first century skyscrapers as L.V. (mad and arbitrary in the middle of the desert) cannot protect itself against the elements when they really decide to "have a go." I passed rapidly across bleak horrible Nevada (no wonder they set off the first A bomb in its deserts) to Arizona, where I spent about four days. For colours, and for geology, Arizona can have no equal on the earth. [. . .] The Petrified Forest and the Painted Desert are amongst the most incredible natural phenomena I have seen—in fact, they

* This might seem an exaggeration, but it was quite possible even in 1961; the ground temperature at the aptly named Furnace Creek reached 201 degrees on July 15, 1972.

reminded me of the cover pictures of some wild science-fiction magazine. From Arizona I passed into New Mexico, mostly a "mesa" (a high plateau desert) around 7000 ft. I mustn't forget to mention the Grand Canyon, which inspired all the usual feelings. What engaged me more than the scenery of New Mexico were the Navajo Indians, who have a sixteen million acre reservation occupying a good part of N.M., Arizona, Utah and Colorado—they are the only tribe of Indians who are on the increase. [. . .]

From New Mexico to Texas, which is ridiculous—for you can drive through a hundred miles of solid wheatland (no fields, hedges or any of our petty European partitions), and hardly see another human being. Fearfully, in New Mexico, I spent thirty valuable dollars (it broke my heart, but I am sure it was wisest) in buying a new back tyre and tube (I am carrying the old one as a spare—the poor bike is loaded with close to a hundred pounds of impediments!): I would hate to get a ripped tyre in the middle of one of these deserts. From Texas through Arkansas—sleepy Sunday state, where the men and animals resemble each other, and watched my passing with sleepy incurious eyes (and where, indeed, the sleepiest and Sundayest place of all was a village called London—population: 353, produce: Ozark Mountain peaches)—to Tennessee. [. . .] *Probably*, I shall now go further South, to New Orleans* [. . .] and then wind my way along the coast, across Florida, and up the Eastern seaboard to Washington and New York, arriving there the first week in May. From there to Montreal and Toronto, and then back through the Northern States.

I have not yet heard about the State Boards etc. but will probably find information awaiting me at New York. I hope to be starting at Mount Zion in June or July (perhaps July, in preference, for this would give me a clear month to arrange, edit and type my new journal). Please let me know how things are with you back in London. If you write soon (i.e., if the answer will be back within a week), write directly to the YMCA, W. 63rd. St. N.Y. 23 and mark: TO AWAIT COLLECTION.

Keep well, and love to you all,
OLIVER

* Outside New Orleans, his BMW gave out, and he hitched a ride with some truckers. He later wrote about this adventure in a piece called "Travel Happy (1961)."

To Elsie and Samuel Sacks

[RECEIVED IN LONDON ON JUNE 24, 1961]*
[SAN FRANCISCO]

Dear Ma and Pa,

Thank you very much for your letter, which came today.

First—before I forget—the best returns for your wedding anniversary on the 22nd: may you have many more before you.

Second, before I forget again, here are some photos of me (one with the new bike)† taken in New York; Auntie Len might like one of the close-ups. Tell me if you like them; as you see (genes, hormones, race, age, rage, depravity) my hair is getting very scanty, but my smile is as charming as ever. [. . .]

The internship becomes, if anything, less tolerable daily. Not because of the "work," but because one is treated like a shit or a menial or both, by the majority of the attending men. Having fifty masters (a consequence of being in a private hospital) is difficult at the best of times: many of them are second, third and fourth rate (this is not extenuated by there being a few very good men on), and one sees the grossest examples of medical incompetence and immorality daily. How MZ has achieved so high and desirable a status I can't imagine: or, at least, I can—the others must be *even* worse. Surgical internship is particularly awful, for it consists of being "used" all day: sudden phone calls: "Doctor, you've been *assigned* to such and such a case: be up here in ten minutes." The case to which you have been "assigned" is one you have probably never seen before, and will have no chance of seeing again, i.e. you are deprived of responsibility: as like as not, you will be a fifth assistant on a mastectomy, so you lack even the consolation of imagining yourself useful. Appalling mutual hostility obtains between the attending and house staff—which was epitomized in the savage lampoons of the annual "show" last week.

However, something can perhaps be "done" about all this. I hope so, for otherwise this will be the most miserable and useless period of

* OS's parents saved all his letters and often noted the date when they were received or acknowledged.

† He had replaced the broken-down BMW with a Brough Superior, an English bike known as the "Rolls-Royce of motorcycles."

life. I can write nothing at the moment, which adds to my fury, and have been eating pathologically—gained 24 lbs. in fifteen days. Incidentally, Pop, delighted to hear that you are becoming slim—don't let this chance slip like all the others. [. . .]

yours,
Oliver

To Elsie and Samuel Sacks

JULY 6, 1961
MT. ZION HOSPITAL, SAN FRANCISCO

Dear Ma and Pa,

Thank you very much for your letters: your mammoth ones are rare, Pa, but well worth waiting for. This will be a slender reply, and a rather tardy one because I am in a mood of great vexation and uncertainty. [. . .] I am not a good correspondent, because I speak and write *at* people rather than to them. [. . .]

A new infuriating enactment of the State Boards now makes foreign doctors do *yet an extra year* of internship here; i.e., it really doesn't want them in California. I am not sure that there is any way of evading this if I am to procure a Californian licence. Indeed, I am wondering: do I in fact want such a licence. I see this loathsome internship prolonged indefinitely, and to no purpose. I am discovering, in a far more intense degree than ever before (these were mere intimations) an extreme aversion to patients, sickness, hospitals and particularly doctors. I cannot imagine that private practice would be any pleasanter: first, I do not enjoy clinical responsibility and would probably betray it sporadically; second, I want leisure, and a lot of it; third, I am not very competitive. I am thinking of returning, here and now, without further waste of time and temper, to academic work, i.e. physiology. [. . .] However, with Oxford in mind,* I am also none too sanguine about this: I am probably too temperamental, too indolent,

* He is referring here to a rather spectacular failure at a physiology project on Jamaican ginger ("jake") paralysis, described in *On the Move*.

too clumsy and even too dishonest to make a good research worker. The only things I really enjoy are talking, lecturing (when I have the chance), reading and writing. It occurs to me (with the precedent of Charles Singer* and many others in mind) to give thought to the subject of the history of medicine and science, and I shall be contacting the department at Cal in this subject. This way I might harness my long long education in science and biology and medicine with my other interests, and my ability to write, and make myself a reasonably pleasant and profitable life. The truth is, you know, I should never have become a doctor. However, I regret nothing: I have a great deal I can still capitalize on. By a curious telepathic coincidence, I was mentioning some of these thoughts to Grant Levin only yesterday, when I received a letter from Jonathan,† beautifully written and full of realization of similar issues. Let me quote a sentence here and there: "I am, like Wells,‡ enchanted by the prospect and paralysed by the reality of scientific research. The only place where any of us move nimbly or with grace is with ideas and words. Our love of science is utterly literary." Looking towards the future, he says, "I now feel I am ready, equipped with an extraordinary biological education, to turn a powerful instrument into a region where I never dared imagine myself at home." He thinks, utopianly, of some unit where psychology, sociology and neurology all meet. And I am thinking, in analogous terms, of scientific history and journalism. I don't know what will finally become of either of us, but I feel that this turmoil is a healthy sign, even if it disturbs you a lot. It would really be a great shame and rather a waste to turn into a mediocre doctor with no love for medicine. However, do not get too alarmed: I shall do nothing rash, and nothing without careful advice and help from others. In the meantime, this internship will give me bread, bed and money (100 dollars a month, affluence!) while I brood and plan.

It will be my birthday on Sunday, exactly one year since I flew from my homeland. It has, on the whole, been the best and most exciting

* Charles Singer (1876–1960), author of many books on the history of medicine and biology.

† Miller.

‡ Miller and his two best school friends, OS and Eric Korn, were all great admirers of H. G. Wells's. (Korn, who later became an antiquarian book dealer, specialized in the works of Wells and Darwin.)

year of my life, and one of a healthy bewilderment and discovery. But I feel old! But perhaps not too old to change.

My best love to you both—and to Auntie Len, who owes me a letter.

OLIVER

To Elsie and Samuel Sacks

OCTOBER 16, 1961
MT. ZION HOSPITAL, SAN FRANCISCO

Dear Ma and Pa,

[. . .] The chief thing I have done this week has been to go to the Brain Research Institute in LA over the weekend. This enormous building was officially inaugurated—the largest neurological centre in the West, and on its way to being the best in the States—amid symposia, lectures, banquets. Grant Levin had procured me one of the coveted guest tickets.

I rode down on Friday night, along highway 33, through a lunar landscape of coccidiomyces spores piled in vast desolate dustheaps. When I got to Coalinga, in the middle of nowhere, at three in the morning, I found all the gas stations closed. There was no light anywhere. I honked my horn up and down the main street and banged on doors; no answer. The town was bewitched, or in a coma. I managed to shake a pint of gas out of petrol pump hoses, but could not force the lock on any of the pumps themselves. Then, had any of the inhabitants been awake, they might have seen a strange spectacle: an enormously bulky and muffled motorcyclist, one of Mother [*sic*] Zion's favorite interns, dismantle his stethoscope and try to siphon some petrol out of the cars standing in the town square. However, I was unsuccessful, and just got mouthfuls of petrol for my pains. I tried to flag down cars, one or two passed in an hour, without success. So I went on slowly, hoping to make Blackwell's Corner, a sort of oasis. But I ran out seven miles from it, it could have been worse, and had to push the bike for this distance. [. . .] At last I got to L.A. and found that I had dropped one of my shoes en route. So I had to buy another pair. The conference itself was magnificent. Speakers had been invited from all over the world. [. . .]

In the evening we had a splendid banquet at the Bel Air Country Club. Everything in LA is excessive in size and monstrousness, and Bel Air was about the limit in outrageous country clubs. I got a lot of pleasure from riding through its gilded gates on my bike still covered with dust from the desert. The meeting had a great sense of the past: the ghosts of Ramon y Cajal and Charcot and Hughlings Jackson* strolling through the corridors.

Unfortunately it was so hellishly hot in LA over the weekend, the more so as I had on my best suit, complete with waistcoat and stiff collar, preposterous in a temperature of 110. For a week or so in October a burning wind comes up from the desert, properly called the Sirocco or the Santa Ana, but more expressively just THE BIG HEAT. It is a wind which drives men crazy. Crimes of rage and violence sweep over Southern California when the Sirocco blows. The traffic in LA was absolutely murderous. And it got no cooler at night: 100 all through Saturday night. On Sunday I decided that, Nobel laureates or not, I could take no more of the conference room, which was not airconditioned, and spent a lazy day down at the beach. Last night I rode back along the coast road, 101, and made LA back to SF in seven hours, which was fair going, considering how often I stopped to eat en route. The bike is well-called the Rolls Royce of motorcycles: at eighty it still has a weird silence and is completely cool and stable, so that one has supreme confidence in cruising at this speed. Whereas I always felt the Norton would either explode or just jangle itself to pieces through incessant nerve-wracking vibration. I saw a tremendous number of shooting stars, and also a small meteor, which I first took to be a rocket sent up by a ship in distress. A cosmic gala night. I suppose we must be passing through a shower of the things. I arrived back here profoundly exhausted, after three nights without sleep (for one cannot sleep when the sirocco blows) and eleven hundred miles of travelling. I hope there are not too many night calls tonight. Perhaps I could design a little cervical dilatometer which registers automatically, and so save myself the dreary business of having to

* Santiago Ramón y Cajal, John Hughlings Jackson, and Jean-Martin Charcot were foundational figures of nineteenth- and early-twentieth-century neurology.

check at all hours. For this is all I am in Obstetrics service: a greased finger, nothing more.* They allow the interns to do nothing, nothing whatever, except circumcision and this is ridiculous because circumcision requires a good deal of judgement and experience. I refuse to do them: I'm scared stiff of cutting the whole caboodle off.

I'm at the end of the paper now, and someone is 7 cm. (what you used to call a crown) dilated [. . .] so I must run. Please write and let me know what is happening back in London.

love,
OLIVER

P.S. I wonder if you could send me on two string vests, the ones *with* little arms. They don't seem to have heard of them in America.†

OS'S FRUSTRATION WITH REPEATING an internship he had already completed in England was leavened by the prospect of moving on to a residency the following year. He had applied for a position at UCLA.

To Elsie and Samuel Sacks

NOVEMBER 13, 1961

MT. ZION HOSPITAL, SAN FRANCISCO

Dear Ma and Pa,

[. . .] First: you will be pleased to hear that my months of miserable anxiety and maneuvering are over—the UCLA appointment has been confirmed, and I will be starting there on July 1. [. . .]

* OS is describing a rotating internship, in which young doctors are exposed to a variety of specialties: obstetrics, surgery, cardiology, etc.

† In England, string vests, usually sleeveless, were singlets made of netted material, which would keep one cooler in summer and warmer in winter.

I spent an interesting weekend boar hunting in the Santa Lucia mountains.* A great sport—only we saw no boars, only an ancient spoor leading to a dried-up waterhole. However, I shot a jackrabbit. I discovered a marked relish, and an even more surprising natural ability, in the use of knife, bow-and-arrow and rifle. Knowing how timid I generally am, and astigmatic, I should be the last person in the world to enjoy such things. It must be an ancient bloodlust coming out in me. We stayed (there were five of us) with an old German couple in their ranch at the top of a mountain. It was a Paradisal place, perpetually windless and bathed in brilliant sun, with storm and fog rageing all around. Like the garden of Eden it was filled with exotic fruit trees, and, idiot that I was, I gorged myself on unripe guavas, which are supremely delicious. [. . .]

I also got in some Scuba diving: they have great corals and underwater fishes near Monterey. I don't know now whether to save my pennies for a gun, or a diving suit, or a flight to New York, all of which I want equally badly, and unreasonably!

On the way back, we stopped off at Santa Cruz to see an uncle of Ted's (Ted, the weightlifting neurologist at Cal., organized the whole thing): this was a doctor, totally paralyzed by polio and living in a tank respirator, and engaged in writing a novel about the early Norse invasions of America. He was a marvellous talker, a sort of paralysed Homer. He suddenly crystallized a feeling I've had, incoherently, for some time: that America has everything except—a mythology. This gives things an almost moonlike desolation, an absence of human and magical intent. Yet there is so much which is the very stuff of legend, just waiting. Waiting, perhaps, for someone to compose an enormous spurious mythology of the continent. I am haunted by the idea. This seems an odd note to end on.

Love,
Oliver

* Likely he was in Los Padres National Forest, home to a number of feral pigs.

To Elsie and Samuel Sacks

FEBRUARY 4, 1962
MT. ZION HOSPITAL, SAN FRANCISCO

Dear Ma and Pa,

I'm afraid it is about a fortnight since I last wrote: I have very little to write, and have been in no mood for writing in any case. One of my periodic depressions: I see myself clearly as manic-depressive now; jewel-like spasms of sharp delight and sensitivity in everything, good humour, wild hopes and ceaseless writing, alternating with long horrible periods of sloth and misery. I dearly hope that the stimulating environment of UCLA next year will diminish the depressions, and lengthen the productive and scintillating manic phases. As Orwell says (in a very depressing little book, *A Clergyman's Daughter*), the devil's subtlest weapon is the sense of futility. Almost all of us at Mount Zion, in varying degrees, are afflicted with this; it comes of having nothing of any importance to do, and no responsibility.

Travelling is my chief balm and stimulant, as music was to Saul, and I look forward with enormous pleasure to taking off for Mexico, in four days' time (I will send you postcards en route). I wish I had longer: one week's vacation in the fifty-two is pretty minimal. The weather has been remarkable here: it was bitterly cold two weekends ago, and actually snowed a bit; the streets were full of children who had obviously never seen snow before, and were making their first tentative snowballs, by a sort of instinct. Now the whole state is gripped by a dense fog, and there have been hideous road-accidents everywhere; I have hardly ventured out myself. The weather should be lovely at this time in Mexico—never hotter than 70 by day, or colder than 40 by night. The roads are treacherous, and I will do all my travelling by day. [. . .]

Love to all at home,
Oliver

To Elsie and Samuel Sacks

FEBRUARY 23, 1962

MT. ZION HOSPITAL, SAN FRANCISCO

Dear Ma and Pa:

I hope you are both well, out of the cold weather and the smallpox scare. [. . .]

To go through various points you raise [. . .]. Sorry you found "too many four-letter words" in my trucking interlude!* But I could not, as you imagine, give verisimilitude to my picture without using their dialogue: and of course the four-letter words *mean* nothing—they are merely verbal condiments. [. . .]

I must tell you something more of Mexico: I penetrated about five hundred miles below the Arizona border and then turned back, and came back along dirt roads, through tiny squalid villages where the inhabitants would stand and stare and gabble as I went past on my extraordinary mechanical contrivance—the like of which they had probably never seen before! What with my red shirt, and my beard (have I told you—all the interns are growing beards for our centennial this year!), many of them would shout: see, Castro!! as I went past. Their poverty is inconceivable by European standards: you see men and pigs scavenging on the same garbage heaps. The whole population is pitifully thin and emaciated: I felt ashamed of my own bloated repletion while I was there. And in the streets of Culiacan I saw a corpse lying in the gutter, probably dead of starvation, with nobody paying any attention to it. When it starts smelling too high, they will collect it and throw it on the rubbish dump. This should give you some idea of the squalor which exists in Mexico. Much of this could be alleviated by major schemes of physical help (housing, roads, irrigation, mechanization etc) and, above all, education. However, all the government's money is spent on a handful of show-places, like Mexico City, Guadalajara, Mazatlan etc. There exist a small number of enormously wealthy Mexicans, whose wealth is flaunted in the middle of all this gruelling poverty. You would think that here are the ingredients of revolution: but no! So far as I could

* "Travel Happy."

comprehend, in my poor halting Spanish, there is a universal sense of passiveness, of laissez-faire: it is warm, the sun shines, the children play, there is always manana. There is an aspect of this attitude which is very pleasant: you see the men lying in the fields doing nothing, sleeping with their sombreros over their faces; it is all so picturesque, so romantic, so relaxing. California seems full of gibbering maniacs in comparison! They are an amazingly kindly and hospitable people: I was many times invited into their little shacks as I passed through the villages. Naturally I could not refuse the little food they gave me—though I suspect I may have become a carrier of *Entamoeba* in consequence. The children loved the bike: they would clamber onto it, sometimes two or three at a time, and I would take them along to the next village, eight or ten miles away, and then they would run back home!

Luckily my tires stood up to all this punishment, although this set have lasted barely five thousand miles in all. I had a lot of trouble getting back into the States: I had forgotten to re-register as a resident alien, and had to swear on the bible that this was a bona-fide mistake: they rifled through all my luggage, and even all my pockets (while I stood naked), impounded some tetracycline, codeine comp. and benzedrine I had in my emergency kit, and then fined me for not declaring them. Oh yes, a good time was had by all! I emerged however with my Mexican purchases, a gallon of fiery tequila (for my friends: I can't stand the stuff), a Mexican blanket, and a guitar (which cost thirty bob new!).

And now I'm back in harness again.

Love,
OLIVER

To Elsie and Samuel Sacks

MARCH 25, 1962

MT. ZION HOSPITAL, SAN FRANCISCO

Dear Ma and Pa,

[. . .] I took a train down to LA last Friday (to save my legs for the contest); a beautiful journey, through much wilder country than the road goes. I had forgotten what a very pleasant way of travelling train is; so few people use them here. [. . .]

The [weightlifting] contest went well, though not quite as I wished. I and my chief contender took first place jointly, and we both squatted with 575 lb. which was substantially above the old record (515 lb.). So I shouldn't complain, because I am now one of the (joint) Californian champs. However I just missed 600, by the skin of my teeth, which breaks my heart, for I have been obsessed with this figure for years, and consider that it more or less separates the men from the boys. I feel that if there hadn't been so much messing around at the contest (in which one "cools off," physically and mentally) I might have got it. I am determined to reach, or surpass, this figure in the near future, so I must (regrettably) keep my weight up for a little while yet. If I could do 650, I would be the 10th man in the world to reach this figure, but I think this would be too much to hope for in the few weeks of eating and training I have left, before the Pacific Coast contests on May 5th.

I hope to send on to you (and indeed delayed this letter in the hope that I would have received them) some photos etc. of the contest, and also of Los Angeles, which I took myself. In black and white this time—colour is much too expensive for general use, at least on an intern's income.

After our excessively rainy winter (three times the usual amount), I think Spring is getting the upper hand here.

However, to counteract this, I am now back on surgery for my remaining eight weeks, and this (as I have said many times before) is as hellish as one can imagine. I am happy to say that I have given the department a great deal of trouble first by my size, for they have had to get scrub-suits made to measure for me, and second by my beard, which in its exuberance cannot be covered by a regular mask. Since they suspect it of harbouring staphylococci, I am wearing for the moment a sort of reversed snood, until they can figure out a new sort of mask. Anyhow, this cuts down on my attendances in the operating theatre, which is all to the good. [. . .]

Best regards to Michael, David and Lillie, Auntie Len, and the kids.*

Love
OLIVER

* OS's brother David and his wife, Lillie, had three children.

2

Los Angeles

1962–1965

To Elsie and Samuel Sacks

JULY 6, 1962
UCLA MEDICAL CENTER, LOS ANGELES

Dear Ma and Pa,

[. . .] I am thoroughly delighted with the University and the neurology dept. so far. The campus itself is immense, ultra-modern, ultra-efficient, but by no means as unfriendly as these qualities might suggest. I am working in the NPI (neuropsychiatric institute), which was built only a few months ago. I am one of four residents, all of whom are intelligent and charming, and a pleasure to know and work with. The Professor himself, Augustus Rose, is a gigantic man (almost seven feet high, and weighing, I suspect, a good deal more than I do). A magnificent teacher who obviously enjoys teaching immensely, accessible, and generally very stimulating. It is really marvellous being in an environment like this, after that unspeakably horrible year at Mt. Zion.

I will not pretend that things are slack. I find I have to be on the wards soon after seven, collect all bloods etc. myself before rounds with the senior resident at eight. Then clinic till noon. Usually time for a quick swim in the dazzlingly blue NPI pool, and a sandwich lunch in the great tree-shaded patio enclosed by the four main hospital blocks. Rounds with the Professor every afternoon till about three. Then admissions. As things stand, one is on duty every 5th evening and night. It is hard work, but very stimulating, with none of the scamping which is done at a private hospital.

The rest of the campus is fairly deserted, this being the summer vacation. I have been very immobile myself, not having the bike; Los Angeles is inconceivably immense and diffuse, and its public transport is practically non-existent. It is in general a good deal more spacious than SF: the streets wider. There is a moderate amount of smog, but since the hospital is air-conditioned, one is not too conscious of the outside atmosphere. I think this should be a very profitable and enjoyable year.

How terribly quickly that month in England passed!* This is the way, what one enjoys passes in a flash, what one hates lasts forever. Next time I must plan things better, so that I am not committed to an incessant social round, and end up with no time to myself. I hope it is not too early for you to consider, and plan for, coming to California next summer. [. . .]

Love,
Oliver.

To Elsie and Samuel Sacks

JULY 19, 1962
[LOS ANGELES]

Dear Ma and Pa,

[. . .] I strolled down to the beach last Sunday. Or at least, I thought it would be a stroll. It turned out to be seven miles away, two hours walk. And after a pleasant swim (the water is about 75), and a reconnaissance of possible future places to live, I turned round to take the bus back—and found that I had forgotten to bring any money out with me. So I had another seven miles walk back! It was a blazingly hot day, and I was dressed far too heavily, and sweated like a pig. [. . .] In future, especially with my build and heat-intolerance, I must be very careful to walk slowly and dress lightly. My weight, you will be pleased to hear, has dropped fifteen pounds since I came here, and is now 248 (stripped). Another 20–25 lb. to go, and then I think I shall level out, i.e. around sixteen stone.

* OS had spent most of June in London. He would continue to spend a month or two there each summer for a decade or more.

I have a very interesting, and I fear tragic, patient: a young man with coccidiomyces meningitis. Cocc. is endemic in the San Joaquin valley, and to a lesser extent in the San Fernando valley, where he comes from. The systemic form is very rare, not more than one in a thousand or so of those who get the disease [and] the meningitis even rarer. The poor bugger wasn't diagnosed six months ago, as he should have been, and I think now that he's probably had it. He has very severe hydrocephalus, for which we will do a ventricular-jugular shunt, and frequent terrible seizures, which are due to his brainstem jamming in his foramen magnum. It is amazing that he continues to survive, without ill consequence, such a horrible business, repeated many times daily. Today I saw the spores (microspherules) grown out of his ventricular fluid, which has confirmed the diagnosis absolutely. We are starting him on Amphotericin B—a thoroughly toxic and nasty antibiotic, and the only substance known which has any useful effect against the fungus. In future, I shall avoid Highway 33, and the whole of the San Joaquin Valley I think. I shudder when I remember the nonchalance with which I slept one hot and dusty summer night near Blackwell's Corner, with the dreadful spores blowing all around me. Of course, the majority of the inhabitants in these areas *do* have the disease, asymptomatically: it is only a tiny fraction of unfortunates who get it in its lethal systemic or meningitic form. [. . .]

Look forward to hearing from you again soon,

Love,
Oliver.

OS SPENT HIS FIRST YEAR in Los Angeles living near Muscle Beach, sharing an apartment with Mel Erpelding, a young man he had met in San Francisco. They worked out together and traveled frequently on weekends, exploring California. OS had fallen in love with Mel, but their relationship, as intimate friends, remained platonic.

To Elsie and Samuel Sacks

SEPTEMBER 13, 1962
UCLA MEDICAL CENTER, LOS ANGELES

Dear Ma and Pa,

[. . .] I spent a very happy and healthy week camping out in the forests of Oregon and North California, becoming a considerable fisherman, and an adept at lighting a fire with a few damp twigs. I still covered quite a great distance—3500 miles in my round trip, in nine days—and saw a great deal of country I had never seen before. I came back through the Mother Lode, and had the excitement of recovering some minute flecks of gold on stones in some of their auriferous rivers. There are a few people there who still pan for gold: the yield is scanty, but they can just about eke out a precarious subsistence. In Amador City, one of the "ghost towns" on the way back, an old prospector showed me a 13 oz. nugget he once dredged up: it was a funny feeling handling it, and he kept his hand on his gun all the time I held the precious thing, so I wasn't too tempted to run off with it! Pop—you would go crazy about swimming in some of the lakes and rivers here: late summer is a perfect time, because the temperatures are around 65, and the water has an extraordinary transparency. In Summit Lake, in Lassen Volcanic park, you would see hundreds of brown and speckled trout swimming round a boat, if you took one out. I caught three, and made an excellent supper of them. Sooner or later, I may invest in a line of my own.

[. . .] I have now moved into the new apartment (address: 3021, Washington Boulevard, Santa Monica, California: telephone not yet connected), which I am sharing with a friend, Mel Erpelding, whom I knew for a couple of years in San Francisco: he's doing sociology. This way we only pay $45 each, and I think live much better than one person could do alone. We are just a quarter of a mile from the beach and ocean and have resolved to get up half an hour earlier each morning and get a swim: it's a good way to start the day. Better in the doing than the anticipation! I've borrowed a few pots and pans etc. from Joan,* which I will return to her as we accumulate some of our own. There's quite a capital expense in getting these things, and

* A cousin.

sheets and blankets etc, which we shall keep as small as possible. [. . .] The apartment is about fifteen minutes from UCLA by bike, which is really quite reasonable.

I wonder if you could now send on to me that parcel of books, and also send—if they have not been included in it already—my two vols of the shorter O.E.D.* As for the remainder of the books, we will have to think of these another time, when I am more permanently settled. [. . .]

That's my lot for now. Look forward to receiving your next letter.

Regards to everyone,
From your slender sixteen stone son,
Ol.

To Elsie and Samuel Sacks

OCTOBER 2, 1962
UCLA MEDICAL CENTER, LOS ANGELES

Dear Ma and Pa,

[. . .] I went to the local shool on Rosh Hashanah: I could hardly plead that it was too far away, for by a strange chance I live only three doors away from the shool! Being used to the Orthodox ritual, I found the reading of English, etc. rather distasteful. Not of course that I could *understand* the Hebrew anyhow, but still, I like the sound of it, and it's familiar, and one is a creature of habit. By chance, the Rabbi there is called Samuel Sacks! We also have a patient called Nellie Sacks. I never ran into so many goddam Sackses before! Went to a Rosh Hashannah dinner with one of the gang on Sunday, and that too made a pleasant change. My first chopped liver in three months: better than peanut butter which seems to be my staple these days (I am now a svelte 222 lb, and feeling almost emaciated).

The apartment is getting more organized. I am adjusting to the horrid realities of scrubbing floors, beating carpets, degreasing cooking pans etc. and even beginning to take a little pride in the place. I

* *Oxford English Dictionary.*

shall send you a photo of it some time. It's right down by the ocean, so I go swimming almost every night. The water is brightly luminous now, in these late summer days, the waves breaking into blue and orange as they topple over: wonderful sight. We have filled the apartment with jars of luminous seawater, and they glow eerily at night.* In fact the place is rapidly coming to resemble a small marine biology museum, and strangers can't bear the smell of decaying seaweed and formalin: but I like it.

A snippet for you, Pop. I went into an Army Surplus store a few days ago to buy something or other. Got talking to the fellow there who had an obviously London accent. He told me he came from the East End, near Commercial Road. Oh yes, I was born there myself, I said (a pardonable lie). Anyhow, after a while I said, Let me guess who your family doctor was? Well? Dr. Samuel Sacks. He stared at me as if I was a thought reader. Anyhow, we reminisced for a long time about you: he said there wasn't another in the East End to touch you. His father was called Solly Oder, and he had a barber-shop at 8, Grove St. Does this ring a bell with you? It's a small world.

What else? I've been fussing round, not much real travelling. A lot to do at the hospital. Swimming. Training. Buying household things—how many there are! [. . .]

This is a rather patchy letter: it'll be more solid next time. In the meantime, keep well (you didn't say how you were getting on now, Ma), make tentative plans for coming to California next summer, and give my regards to everybody,

Love,
Oliver

* Mel still remembers OS teaching him the name "dinoflagellates" for these tiny luminous organisms.

To Elsie and Samuel Sacks

OCTOBER 24, 1962

UCLA MEDICAL CENTER, LOS ANGELES

Dear Ma and Pa,

I was going to write last night, but thought I'd wait and see what happened off Cuba first.* So I drank a pint of rum toddy instead, and decided the first nuclear missile would find me cheerful. When I got to hospital this morning, I saw that all the clocks had stopped at six o'clock, a portent if ever there was one. However, the ships have retreated, so I suppose the immediate risk is over now. An agonising choice, with a fearful gamble at one end, and—I think—a great moral victory as the outcome. It even got through to the great American public, and briefly vied with the World Series Baseball in the attention it drew: the streets full of anxious (and expectant, and—who knows?—*eager*) citizens, hanging outside the plate-glass windows of television shops, hypnotized, and uncannily silent. [. . .]

I had a phone call from Jessie Fox† last Thursday. I had half been thinking of riding up to San Francisco for the weekend, and her call decided me. So we went up (Mel, who shares the flat with me, has now got a BMW of his own), and I met her on Sunday afternoon, and we talked our heads off about everything. She has a tremendous fondness and admiration for you both: "such *luvly* people!" she keeps on saying. [. . .]

Mel and I [had] made a late start on Saturday, and arrived in SF very hungry and perished with cold. But we called on Anne, a nurse I knew pretty well from Mt. Zion, and she made us an enormous kedgeree of rice and fish (this was around one in the morning), and we slept like logs on her floor. Sunday was one of those perfect October days which only SF can concoct. Breathtakingly clear and sunny, though cold, yet windless, and with great woolly banks of fog wreathing the Bridges and the Bay, and the sailing-boats dodging in and out of it, playing blind man's buff. In the afternoon, we all went over to

* October 24 was day nine of the Cuban Missile Crisis. President Kennedy had announced two days earlier that the United States would blockade Cuba in response to the presence of Soviet missiles there.

† Identity unknown; perhaps a fellow expatriate who knew OS's parents.

the von Bonins[*] in their little house high above the redwoods of Mill Valley. They seem to live in a Shangri-La of perpetual sunlight up there, high above the fogs and discontents of the rest of us. An idyllic scene: she playing Bach on the piano, as we came in, and he bent over a microscope. They are both nearer eighty than seventy, I imagine, but irrepressibly active, and curious, and young in spirit. Mrs. Von B tells me she had a letter from you. It was very nice of you to write.

We left SF around midnight on Sunday, and drove back through the night, each wearing four pullovers under our leathers. Though it wasn't so cold in fact. I love driving back through the night in California, when the stars are so clear, and the Milky Way lies like a great arc above one. We were back by seven: moderate time for over 400 miles, just a steady average around 65.

I have received the first parcel of books from E. Joseph.[†] The flat is beginning to look civilized now. I think too that I may buy a piano, a rather huge and ugly whitewood upright, which nobody will buy for this reason, but which has a lovely tone and perfect action: for $150. It would cost that to hire a piano for a year in any case. So I can get back to my beloved Bach: and perhaps a little Brahms too. Otherwise I am pretty busy. I have finished the abstract which has been sent in now: touch wood, there might be a chance of going to NY to read the paper at the Academy.[‡] I have been elected a member of the LA Society of Neurology. And I have to present a patient to the venerable Houston Merritt[§] on Friday, which scares the hell out of me. So the days and the weeks pass, and one lives one's life without knowing it.

Regards to all,
Oliver

---

* Gerhardt von Bonin (1890–1979) was a neuropathologist and neuroanatomist.

† A bookseller in London's Charing Cross Road.

‡ The abstract of his first paper, on hereditary photomyoclonus (a seizure disorder triggered by light), which he hoped to present at the American Academy of Neurology meeting.

§ Merritt was chair of the Neurological Institute at Columbia University from 1948 to 1967.

WITH SOME TIME OFF FROM his job at UCLA for the Christmas holidays, OS made a brief visit to Jonathan and Rachel Miller in New York. In the meantime, his parents had written to tell him that his cousin Abba Eban,* known to the family by his given name, Aubrey, was in Los Angeles.

To Elsie and Samuel Sacks

JANUARY 7, 1963
[LOS ANGELES]

Dear Ma and Pa,

I am very late in writing. A variety of things: going to NY [to visit Jonathan Miller], starting my new job here, not having a typewriter etc. etc.

First let me thank you for your long combined letter which came a couple of days ago; let me thank Michael for the H.G. Wells short stories—I shall write to him very soon.

I'll deal first with your letter and the points you raise, and then with the other news in general. It was of course marvellous speaking to you all over the phone—although, paradoxically, the line was not nearly as loud and clear as when I phoned you from San Francisco. Jonathan is developing an outrageous parody of transatlantic phone calls (everyone shouting: hallo! how are you? yes, I'm fine. I say, FINE! And how are you? eh? YOU. Oh. I'm fine. What? I said I'm FINE TOO. etc.) [. . .]

No, I did not know Aubrey was here. By the way, a little episode which will amuse you. As I was driving to the airport in New York, the cabby said, out of the blue: I don't know whether you've ever heard of him, but you have a voice just like Abba Eban! I don't know whether he caught the common Oxford-Cambridge brogue, or whether his ear had detected some subtle genetic identity between us.

My piano is coming this week. I looked up an old friend in New York (a piano teacher) and he has given me his series of patent exercises to help get back the feel of the keyboard, after my long lay-off.

SCUBA means self-contained underwater diving apparatus (i.e.

* Eban, the Israeli statesman, was OS's first cousin on his father's side.

compressed air). Here is a newspaper clipping from last Saturday's paper, showing me and three of my colleagues being instructed at the pool, in the Y. Fame at last! I will be going down with the group to Baja Mexico this weekend, below Ensenada, to scuba dive. I hope to get, first chance to use a spear gun there. The visibility is fantastic—you can see more than a hundred feet down.

As for my trips: I stayed in Chicago a couple of days: enough to be best man for Pete Weinberg and his wife (he was at mt zion) and renew acquaintance with them. They're a sweet pair. It was icily cold and blizzardy there (they call Chicago the Windy City). Then I went on to NY.

Rachel met me at the airport, because Jonathan had a bug and a fever and was in bed. I spent a marvellous week with them. They have a tremendously full and happy life. I wish I could be near them for more of the time.

Most days I just walked about the city. After London, New York is incomparable. Wild electric beauty, crowds rushing in and out of taxis, theatres, delicatessens (I put back 12 lb.!), skyscrapers, ghettos, everything packed into the tiny extent of Manhattan island. Eight swarming millions. It makes Los Angeles seem like a vacuum, a huge inane senseless mess of people, with none of the culture and vitality of New York. But then, New York is packed into the crowded eastern seaboard: no hills, oceans, deserts, anything like California. I could not give up California to live in New York. The only solution would be to have New York in California. I visited lots of museums with them, went to parties, and meals, and talked and talked (Jonathan has a mind like an atom bomb), and even baby sat.* I hardly had time to sleep. [. . .]

I've got to run now. I'll write again very soon. [. . .]

Oliver

---

* Having grown up in wartime England, OS felt overheated once the temperature rose above 60 degrees. As one Miller family joke had it, while he was babysitting infant Tom, even though it was January in New York, OS flung open the windows of the apartment, nearly freezing the child. (Tom remained unscathed.)

OS'S BROTHER MARCUS, A DOCTOR (three of the four Sacks boys had become physicians, like their parents), had immigrated to Australia after the war. He was ten years older than OS, and thus had been away at school during much of Oliver's youth, so they had barely spent time together as adults. In early 1963, on his way from Sydney to visit their parents in London, Marcus stopped in Los Angeles for a few days to see his youngest brother.

To Elsie and Samuel Sacks

MARCH 25, 1963

UCLA MEDICAL CENTER

Dear Ma and Pa,

This is a considerably overdue letter, but by now Marcus will be with you, and relaying all the news firsthand. I'm glad the stuff arrived. I was delighted to get Michael's letter, and will be replying to him in due course. He certainly seems to have devoured the books!

Last week was one of happy pandemonium, trying to be a resident at the VA,[*] to continue my paper,[†] and to be with Marcus as much as possible: three lives in one. Extraordinary how akin Mark[‡] and I are in many ways: the more so as we have had so little contact with one another during our lives. My attempts to show him Los Angeles were partly foiled by incessantly being lost: we must have travelled hundreds of miles, in the aggregate, in the wrong direction.[§] One of the residents lent me his little Volkswagen to drive around for the week, which worked out splendidly, since Marcus (inexplicably) would have nothing to do with the bike.

I realize that I am (genetically) almost exclusively paternal. Marcus has a good dash of your blood also, Ma. He is, for example, a great deal tidier and more systematic than I: he immediately took me and

* OS's duties as a resident included working at a nearby Veterans Administration hospital.

† On hereditary photomyoclonus.

‡ OS refers to his brother variously as Marcus, Marc, or Mark.

§ OS and Marcus shared a complete inability to recognize even the most familiar surroundings, and frequently got lost even in their own neighborhoods.

my flat in hand, and made a pretty good attempt at metamorphosing the pair of us. Perhaps he is right in a way: my grossness etc. is not a charming attribute at this age. And I can't expect to have the privileges of a large toy-bear in perpetuity. I took Mark to hospital, and he met Herrmann, and Markham,* and a few of the residents. He was very popular with them all: and our iron-hearted blasphemous Irish staff-nurse Miss Fogarty, conceived, I suspect, a passion for him. "Tis a pity he's so young!" she has been muttering today!

We went up to Mt. Wilson, or would have gone up had I not lost the way. We went to Marineland, after going forty miles off course, and Mark discovered a deep empathic feeling for the rollicking cetaceans there. We ate very well at a variety of restaurants, and both increased a few lbs. [. . .]

The week went in a fever: over almost as soon as it started. Perhaps our next meetings will be over the Barrier Reef.

I must hurry back to my flaming paper now. I am beginning to *hate* the word myoclonus! I gave a rehearsal or preliminary version of the talk at our Clinical meeting here last Saturday, and it went down well, and controversially. Some of my most crucial lab. results will come in only a week or so before the actual meeting, so I must have two versions ready: one with argument X, and one with its contrary. Like the obituaries one prepares in advance!

That's me lot for now—

Love—
Oliver

To Elsie and Samuel Sacks

APRIL 25, 1963
HOTEL LEAMINGTON, MINNEAPOLIS, MN

Dear Ma and Pa,

I know you have the utmost difficulty deciphering my writing, so I will write briefly now, and at length when I am back in LA.

* Chris Herrmann and Charles Markham, two of the neurology faculty at UCLA.

Read the [myoclonus] paper today, or, rather, put it in my pocket and just *talked.** Went very well (though perhaps my "British" accent disarmed them) and was followed by a lot of vigorous discussion. The *Press* was even there—both medical and lay—so may have a little write up. I enclose a page from the programme, with the preliminary abstract written several months ago.

This is really a marvellous meeting—the whole of American Neurology is here (bronzed from Miami, teeth chattering from Alaska); I have met dozens of interesting people. Things are *laid on* with a shovel: here I am ensconced on the ninth floor of Minneapolis' plushiest hotel, complete with bath plus bidet (*so* useful), television, escritoire, ankle deep carpet, and able, with an idle movement of my hand, to command an opulent breakfast in bed, or have my suit pressed. A lovely change! For a little while.

We all went this evening (300 of us, in five monstrous buses) to "Ye olde oak inne" (built 1957) on the shores of Lake Minnetonka (where Hiawatha wept for his beloved), were fed on roast ox and venison roasted whole on giant barbecues, and then herded in to see a charming old comedy, rather on the lines of *Arsenic and Old Lace*—thoroughly amateurish, and uproarious. Splendid evening!

The journey here is a story in itself. Came on the "California Zephyr," which is the last and perhaps the most magnificent of a dying breed, the leisurely cross-country train. (Everyone flies now.) Sitting up in the "Vistadome" you can see 360° all round you, and are so insulated from the usual noise and vibration of a train, that you seem to be floating along in silence, about fifteen feet above the fields. You *must* travel the US like this one day. We climbed right up to 10,000 ft. in the Colorado Rockies, and needed *four* locomotives to lug us along. Air incredibly sharp and exhilarating—I've never been so high.

And the foods on the train! Nobody buys train food—far too expensive. There were Italians with enormous pizzas, and Americans with hamburgers and thermoses of coffee, and Chinese with weird concoctions, and me with pumpernickel and garlic würst, all guzzling in the same compartment. A great cultural education!

---

* This would be OS's lifelong pattern: he was never able to simply read a paper in public. He would painstakingly sketch out an annotated outline of his themes, but as soon as he was onstage, he improvised.

That's me lot for now. I'll probably find a letter from you in LA, and will write from there. Is Marc still with you, or where now? My love to Mike and Auntie Len,

yrs,
OLIVER

DURING THE SUMMER OF 1963, Mel moved out of the Venice Beach apartment he shared with OS. The two would remain in touch until the 1980s, frequently corresponding and occasionally visiting each other. But Mel's departure was a blow for OS: as he described in *On the Move,* he had romantic and erotic feelings for Mel, and they had spent much of their free time together, talking about books and marine biology, and exploring the West Coast. OS resolved never to live with someone again, and he decided to move to a house in Topanga Canyon, whose winding scenic roads he would have known well from roaming around LA on his motorcycle.

To Jonathan and Rachel Miller

AUGUST 21, 1963
UCLA MEDICAL CENTER, LOS ANGELES

Dear Jonathan and Rachel,

[. . .] It was marvellous speaking to you again on the phone a couple of weeks back. It solves so many problems, just dialling. I didn't have time to collect my thoughts—I thought it would take a few minutes to get through, but I was talking to you within 30 seconds of dialling.*

I am constantly reading you or about you, and am overjoyed at the incredible opportunities opening out in every direction. I hope you will soon find the leisure and incentive to write a book, innumerable books—for incessant flittering in magazines, though it's a great way

* Long-distance telephone calls (to say nothing of transatlantic calls), like telegrams, were quite expensive then, and thus usually as brief as possible.

to start (didn't Trilling[*] and Wilson[†] etc. all start this way) may finally be rather destructive, and also tend to make you static in a way. I hope that when the Show is over you will be able to resist some of the commitments and temptations which will come your way, and get down to something "big." You are wonderfully intelligent, but you have come to a time for major formulations, above the level of sentence or review. (Rereading this paragraph, it sounds odiously condescending. I should be one to speak?)

As I mentioned, I shall shortly be moving into a small house in Topanga; it is the middle of wild mountainous country and a mile from the nearest neighbour. A few minutes drive from the City. The Ocean ten minutes away. Plenty of room. I hope I can persuade you to spend a few days or weeks up here: time to collect your thoughts, dash off a monograph on Hughlings Jackson, and see something of Monsterville, Horroropolis or Faecal City[‡] (variously named). It would be criminal of you to go back to England without seeing something of the country: the South, the mid-West and the West Coast, at the very least. If you have time, travel by train. They are superb, uncrowded, leisurely, and almost extinct.

Myself: I'm in Neuropathology for a few months: slicing brains, peering at slides. Satisfied with fossil cells, ignoring our colleagues in the electron microscopy lab. I shrug my shoulders, like Gimpel:[§] who is to say how many removes we are from reality? I find myself increasingly fascinated with degenerative brain disease (lipidoses, Alz., Jacob-Creutzfeldt, Canavan's, etc. etc.). I see so many, and almost none of them are diagnosed in life. People don't seem to be aware of how common these things are, how characteristic their presentations, how easily diagnosed in life. I have the material for half a dozen isolated papers, but am now wondering, ambitiously, whether I might not bring in my interest in the myoclonic syndromes etc. and attempt a monograph on degenerative brain disease. There has not really been such a thing, with the range and scope I am considering, for over fifty

* Lionel Trilling, author of *The Liberal Imagination.*

† Edmund Wilson, author of *To the Finland Station.*

‡ OS's pet names for Los Angeles.

§ Gimpel, the protagonist in a short story by Isaac Bashevis Singer, is widely tricked and mocked in his village but remains good-natured.

years. I realize increasingly, as you have of yourself, that I am purely an armchair scientist. I hate "research." On the other hand, I can rummage through a library, assimilate a huge variety of stuff, classify it, tabulate it, analyse it, talmudically, and spew it out in Jamesian prose.*

By the way my paper has been accepted by *Neurology*, and should be out in a couple of months or so. The original was 20,000 words too long and I had to revamp it completely. Herrmann cut out all my semicolons, commas and dashes, Aguilar (these are my co-authors) my wild images and metaphors. So it's sort of dull now. I will send you a copy, my first-born, when it comes out.

I have become huge and oedematous once again. Psychologically, it is difficult for me to stay at a normal size. I love to shake the pavement as I walk, to part crowds like the prow of a ship. I shall make a final try this year for a world's record in the Squat (want to do 750), and then stop messing around. I am becoming a mass of chronic injuries. Los Angeles is full of decayed weightlifters. Sheppard, world mid-heavy, a bloated alcoholic; Ashman, embittered, tubercular; Berger, imprisoned for rape; Ahrens, a grotesque psychopath.† My ex-idols. Moral corpses.

Sexually, I have retired. My motivation, never great, has diminished to vanishing. I am fat, balding, elderly.‡ There is no point staying around town. I am therefore ready to become a hermit in Topanga, and devote myself to Nature, my iguana and my milch-goat, and the typewriter. Incidentally, what do you think of James Baldwin? [. . .]

I have taken up photography in the past three months. It is becoming an insane preoccupation. I fancy myself the new photographic Hogarth of Santa Monica, perpetually recording its squalor, its ugliness, and relentless facetiousness. The construction of an album is wonderfully satisfying, both a logical and creative exercise, and almost a means of psychoanalysis, of retracing your own subterranean longings and associations. Grotesque amorous syllogisms. I enclose, for your delight, a photo of a gigantic advertisement in the desert, fifty

* Difficult to know whether he is thinking here of Henry James (whom he admired but less often referred to) or William James, one of his greatest heroes.

† Denizens of Muscle Beach in this era included world-class weight lifters Dave Ashman, Chuck Ahrens, Dave Sheppard, and Isaac "Ike" Berger. (Although Berger had been accused of statutory rape, OS was likely exaggerating his "imprisonment.")

‡ He had just turned thirty.

miles from the nearest habitation. It can be an introduction to the "wilde enormities"* of the West. I am also enclosing a particularly horrifying handbill which was thrust under my door. Southern California is stiff with racists, fascists, righters, Birchers, etc.

Readings: *Doctor Faustus*: tedious but monumental. Marvellous descriptions of neurosyphilis and imaginary compositions. *Catch 22*: the best and wittiest and only mature war-book I've ever read. Malamud: the Jewish monologue affects me deeply, then I forget him. Baldwin: wow! in grain, in literary and intellectual texture, second only to Faulkner. At best, magnificent, but patchy. I haven't read his latest.† Kazin and Hardwick, on your promotion, remain my general moral/cultural references.

Write me, and tell me your plans.

My best love to you both,
Oliver

To William Tunberg

ARTIST, WEIGHT LIFTER

[UNDATED]
UCLA MEDICAL CENTER, LOS ANGELES

Dear Bill,

[. . .]Went to the Wrestling Matches last Friday with Big Steve, Jim H. and Peanuts.‡ The four of us weighed about 1000 lb. Steve is becoming grotesque in size and strength, and approaching Ahrens on both counts, as well as in personality. There was a huge line outside the building. Steve loped up and down it, like a caged animal, and then he butted through, grabbed the huge iron Exit doors of the building, and tore them open. The three of us rushed after him through the opening, followed by two hundred of the crowd. They were impotent

* OS discovered the work of Sir Thomas Browne in his university days and often quoted this phrase, among others.

† James Baldwin's novel *Another Country* was published in 1962.

‡ Steve Merjanian, Jim Hamilton, and Bill "Peanuts" West.

to stop us, although dozens of police were called in and milling in all directions. The wrestling was a lot of fun, entirely ham, of course: Blassey [*sic*]* the favourite, an old veteran, was (let himself be) beaten by an immensely tall Negro fighter, and after his fall beat his breast and sobbed and yelled, a pugilistic Lear, a great tragic actor. I didn't realize at first that these fights are the lineal descendants of the old time melodramas, with their heroes and villains, homey morality, and clowning. There was a little old man next to me, very mild and watery in appearance, but absolutely bloodthirsty, who kept muttering: Go on, get him, pull his eyes out! Grab his balls! etc. in an undertone, an awful sadistic litany. After the fight, Steve was raging, vicariously: he suddenly let out a bloodcurdling Armenian war cry in the streets (surely an ancestral memory) and hurled his huge bulk into a STOP sign, which fell and toppled, like a felled redwood. [. . .]

But things are quiet once again.

I'll see when you arrive.

alles gut,†
Oliver

To Marcus Sacks

BROTHER

SEPTEMBER 12, 1963
UCLA MEDICAL CENTER, LOS ANGELES

My dear Marcus,

[. . .] Your visit revealed several things to me. First that we are very similar in an astonishing number of ways, though ludicrously different in others. Second that we are, I suppose, both adults, or as much so as we (I) will ever become. Thirdly, that as possibly the only responsible or conscience-ridden members of our own family we have certain

* Freddie Blassie, heavyweight wrestler.

† In German, "all the best."

problems in common. It is evident from the intimacy and completeness of your letter that you have come to the same conclusions. For one reason and another I tended to feel like an "only child" when I was growing up: partly age differences between you and David vs. myself, partly the war—and the separations this involved, and partly Michael's withdrawal and psychosis. I hardly ever had the sense of having any brothers, and so this is a very pleasant discovery to make so late. When you left England I was still an undergraduate, practically a schoolboy, in mentality, and you were a remote, and even slightly forbidding, figure (don't take this amiss! but such were my feelings ten years ago). And now I am, like yourself, a crusty bachelor of sorts, ageing, and obscurely discontented.

This brings me to some of the problems you mention. You have known Gay for some time now, two years (?), long enough for you to have a fairly shrewd idea of your own and each other's feelings. It sounds to me, from the tone of your letter, that if you do *not* marry her, you will regret it for the rest of your life. You are faced with two problems: first, considering differences of age, background etc. can you "make a go" of it? It sounds to me as if you are both fairly confident of this (I say *fairly* confident, because one leaves supreme confidence in anything behind with one's teens). Characteristically she, as the younger, is the more confident, whereas you, with twenty years' experience of being tossed around by life, and full of morbid Jewish timidities, are less certain. The second problem is Ma and Pa's reaction. Pa is the more flexible, Ma is a chronic brooder. It is a problem: what *will* he tell Rabbi Landy? And what will Rabbi Landy think? And what will everyone in NW2 think?

The answer is—nothing. There will be kefuffling and whispering and indrawn breaths for a bit, and then everyone will "accept" it and the world will go on as before. Pa will recover from the "blow" fairly quickly, Ma will take a year to do so. I have no real doubt that the passion to see you "settled," a wife, a home, and (best of all) the voices of little grandchildren, will bring them great and enduring happiness. And before you know where you are, Ma will be teaching your big blonde shikse how to make gefullte fish.

There is, of course, one obvious step which might minimize or eliminate all these vexations: that is, if Gay became a Jew. You naturally are reluctant to suggest such a step—it seems an appalling imposition. If Gay is a red-hot Roman Catholic, for example, it's naturally out of the question. But if she is as lackadaisical as you, then it is a move she might not take exception to. But then there are *her* people? What would they say? What would the priest say? etc. There must not be a contest of wills over a thing like this: this would ruin your future together. But *if* Gay is complaisant about "conversion," then this might be a very happy solution all round.

Emotional blockage. We all have it, one way or another: sometimes the more promiscuous, the more blockage, of any *real* feeling. I've no doubt our good parents have played an unwitting role in turning all their children into emotional simpletons, part passive, part predatory, and dreadfully timid ("shy," "well-bred," "self-contained," "diffident"). I think you are doing an excellent thing going to a psychiatrist: if he's any good he should help you a lot. The fact that your letters to Gay are becoming, *pari passu,* warmer and more intimate, is hardly a coincidence. It takes so much to give oneself to anyone wholeheartedly. I have a feeling that this is something Ma never did, conjugally, although her feelings for her children have been overpowerful, possessive, dependent, helpless. All of us, in consequence, have been "mother's boys" to varying extents, rendered emotionally ischemic by apron-strings.

How right you are about 37.* The house has a terrible force of its own, like those decayed Southern mansions in Tennessee Williams' plays. It's full of ghosts, and the ghosts are dated—roughly—1938.

1938! Last golden year before the War: Marcus and David, rowdy and boisterous, in stained cricket shorts, wrestling on the back lawn. Michael, already perhaps a little reserved, reading a book in the library. Oliver, a tiny pudgy boy, running out of the front door and leaping into his mother's arms. [Ma's] brothers and sisters—Annie, Isaac, Abe, Dora, Birdie, JoeVic, always visiting, and writing, and talking, buzzing with vitality. Vitality! Michael running into their room in the summer holidays at Felpham (this is *his* story) and pulling Pa out of bed for an early morning swim. Pa huge, grumbling,

* The house at 37 Mapesbury Road in Northwest London where OS and his brothers grew up.

happy: "What do you mean, pulling an old man of forty-three out of bed for!"[*] The huge number of patients, carelessly encompassed by a huge vitality: Ma's absorption in the EGA.[†] Erev Shabbat: Father Duck and his four ducklings going to shool. "Shalom, Sammy! Ooiiy, vot beautiful children!" The old man with rhinophyma[‡] who pinched our hairless cheeks. 1938! Dreams, ambitions. Brilliant boys, great careers, nice Jewish girls one day (but a long way off), the daughters-in-law becoming daughters, the grandchildren, all living very close, all in NW2, in a perpetual almost incestuous intimacy.[§] [. . .]

IN A 1992 HISTORY OF the UCLA Department of Neurology,[¶] a former colleague, Charles Markham, described how OS tested the patience of their boss, Augustus Rose. Markham wrote that OS "was a weight lifter with great muscular mass which frequently led to the emergence of his shirt tail from the back of his pants when on duty. [. . .] At one point he grew a beard much to the displeasure of Rose, who made him remove it."[**] And as OS continued to compete as a weight lifter, he also continued to consume prodigious amounts of food, often wolfing down numerous double cheeseburgers and milkshakes at the hospital cafeteria, where food was free to residents and interns. He was sometimes known to help himself to food from patients' meal trays in the middle of rounds.

---

* Later, in his essay "Water Babies," OS related this story as an episode that happened to him, not to Michael.

† Elizabeth Garrett Anderson Hospital.

‡ A bulbous, bumpy red nose.

§ Letter incomplete, possibly unsent or an early draft.

¶ *History of Neurology at UCLA*, privately published.

** And yet, Markham added, "Despite his strength and bulk he was always gentle and tender with and liked by his patients." Rose himself later characterized OS as "a mix of non-conformity, [a] touch of irresponsibility and brilliance coupled with intense drive and great capability."

To Augustus S. Rose

CHAIR, UCLA DEPARTMENT OF NEUROLOGY

OCTOBER 21, 1963
[LOS ANGELES]

Dear Dr. Rose,

I acknowledge receipt of your letter regarding the consumption of hospital supplies. I will in future regard them as sacrosanct. I feel compelled to make you a reply.

First: I have no clear recollection of there being any hospital regulation forbidding the consumption of hospital stores. But perhaps my memory is at fault here. Second: there has existed a cordial tradition, in all hospitals that I know of, that a Resident on duty may take a cup of milk or coffee etc. if he is thirsty. I have always regarded this as a courtesy, or unspoken privilege, rather than as an "infraction of regulations." And this too is the view of many of my fellow-residents here, who have habitually availed themselves, as I have, of hospital milk from time to time. None of us have had any thought of crime or censure being involved. Third: it seems grossly reprehensible to me that the kitchen should report me in this fashion, without *first* intimating to me the state of affairs. To find oneself spied and informed upon in this fashion, while in complete ignorance of the situation, smacks more of a police-state than a University. Fourth: as my Professor, it is your duty to reprimand me, or deal otherwise with me, for infractions of hospital regulations or derelictions of moral standard. But as a man who has sponsored and supported me in the past, I might have hoped that you would at least have discussed the matter personally with me, before having your secretary type so accusing a letter. Fifth: I take strong exception to the moral aspersions implicit in your letter—if you regard me as a thief, I will make full restitution for the milk "stolen" (it must total at least $3), and take my leave at the end of the academic year.

I will no longer delude myself regarding the auspiciousness of my career at UCLA, having spoiled my academic record by the heinous sins of untidiness, unpunctuality, and slaking my thirst at the Ward Refrigerator. But you must know, as well as I, that a man is not the sum of his minor misdemeanours, but of his best endeavours. You are well aware that I am not devoid of intelligence or of serious interest in the neurological sciences, and I shall still nourish the hope that you will not entirely abandon me if I seek your support in finding myself a position elsewhere next year.

Yours truly,
Oliver Sacks

To Mr. Hobson

LOS ANGELES DEPARTMENT OF MOTOR VEHICLES

DECEMBER 3, 1963
[LOS ANGELES]

Dear Mr. Hobson,

I have tried without success to contact you in the past couple of days, and so will now take the liberty of writing to you.

Despite scrupulous efforts to avoid the slightest infraction, I am afraid that I have compromised myself twice since my interview with you: once, by making a U-turn in a business district (a momentary inadvertence, on an empty road, for which I have been fined $10), and a speeding violation (75 mph) returning from San Francisco last weekend. In both cases the carelessness of an instant has blotted out the otherwise continuous caution I have exercised, and in both cases—unhappily—a patrol car materialized with the promptness of an avenging Fury! I am profoundly apprehensive of the effects of these two further violations, and am writing to you first to emphasize that both were due to momentary lapses for which I am exceedingly sorry, and secondly to enquire about my probable situation at this juncture.

There is little I can say in this letter which has not already been discussed by us. You are aware that I have accumulated a large number of violations in the past year, all of which are trifling in themselves (minor excesses of speed, for the most part, with no suggestion of reckless or dangerous driving), but which constitute a lethal indictment in their cumulative effect. You are aware that I have driven for fifteen years, more than 600,000 miles, under all possible conditions, without incident or accident, thus indicating my essential care and competence. You are aware, thirdly, that my life and career in California are critically dependent upon my having the use of my own transport: I have to commute between a number of hospitals here (UCLA, USC, Children's, Pacific State, etc.), and the distances are vast. If my driving licence is suspended, I will have to give up my residency, tear up all my roots in California, and settle elsewhere. All of which would be tragic, brutal and unnecessary. Finally, I must hope that your intuition tells you that I am a serious and responsible member of the community, and have shown myself over the years an equally serious and responsible driver.

I shall strengthen my resolve to stay within the strictest letter of the law, and "make assurance doubly sure."* I shall force myself to concentrate and watch the speedometer etc. sixty seconds in the minute, rather than fifty-nine, and so do all I humanly can to avoid any further violations, and further embarrassment for you, me, or the Department of Motor Vehicles. I hope that I may receive charitable treatment, even though I have doubly trespassed on the patience of the Department.

It is, I fear, conceivable at this point that I have nevertheless passed beyond the tolerated limits, and that the suspension of my licence is now probable or certain. *If* this is so, I would like to be apprised of the situation as soon as possible, in order to start plans and preparations for leaving this state. I sincerely hope, however, that you will reassure me that the sands of time haven't *quite* run out on me, although they have almost done so. I look forward to your early reply.

Yours truly,
Oliver Sacks, M.D.

* He is quoting *Macbeth* (as one does).

To Elsie and Samuel Sacks

MAY 16, 1964
UCLA MEDICAL CENTER, LOS ANGELES

Dear Ma and Pa:

[. . .] I had perhaps concealed from you some of the vagaries of life at UCLA. There is, in general, less formality, but more rigidity, in American universities, compared to English ones. In consequence, some of my (comparatively harmless) idiosyncrasies—viz. untidiness, unpunctuality, huge size, waddling gait, mode of transport etc. etc.—had attracted amused but inimical attention, and even placed the tenure of my job in jeopardy. I was told that I had better reform, or wd find myself out of a job. With this sword of Damocles above my head, the more painful because I knew myself to be basically a good and responsible doctor, I naturally found myself in something of an anxiety state. An anxiety state which sometimes seems to be, let me add, a common and chronic symptom with many people in competitive academic circles.

On the other hand, there is one thing which they go for here even more than a bland conformity, and this is advertising. If a man shows himself an accomplished and prolific writer, if his papers are readily accepted for publication, then he is a most useful asset wherever he is, and some of his foibles may be passed over in consequence.

So far as I am concerned, Denver* marked something of a turning point. I had been in very bad odour prior to the Academy meeting. But I scored (if I may be immodest) a triumph at Denver: I was widely complimented on my paper and received a number of tentative job offers the same day, from scattered centers here and there. I rather suddenly ceased to be Sacks the embarrassment, and became Sacks the ornament. Rose, my professor, who had been thoroughly upset about me, became overwhelmingly genial, discussed various things pretty freely with me.

The upshot of all this is roughly as follows. I find in myself a mounting interest in neuropathology. Cells being easier to deal with

* OS attended the annual meeting of the American Academy of Neurology in Denver to present a paper on hereditary myoclonus with his co-authors.

than people. I think I probably want to become a neuropathologist, or at least take a minimum of two years training in this subject. Good neuropathologists are excessively rare, and can command handsome posts and salaries wherever they go. People enter neuropathology from different areas—some from general pathology, some from neurology, some even from psychiatry. [. . .] On the other hand, it would be utter folly for me not to finish my neurology residency, of which I have my last, and senior, year to go. Rose, in effect, has said to me: stay on at UCLA on your own terms. You will be, in name, Senior Resident in Neurology, and thus complete your formal residency requirements with us, but you may spend the year doing your beloved neuropathology. You are good at this, and your untidiness, unpunctuality etc. etc. really doesn't matter too much over there. [. . .]

In addition, it has been arranged that I should do a couple of Neurology Clinics a week (thus keeping my hand at clinical aspects, without being burdened by patient care on the wards), and also a number of complete postmortems, thus giving me at least a modest background in General Pathology. [. . .]

You ask whether I wish to stay in America. A difficult question. I cannot conceal that there are an enormous number of things I detest here. And yet, there are enormous opportunities, not merely professionally, but for leading a full and fascinating life in many other ways. It would hardly be fair to judge my environment by the experience I have had thus far *as a resident,* i.e. as someone on a shoestring salary, with very little leisure, and a large number of acute and chronic anxieties, as well as the overall feeling of being unfulfilled. A more responsible job, higher financial rewards, and the still greater rewards of creative scientific work, will all come, I hope, in not too long; and with these, I may find that life is really very pleasant here. I will confess, however, that I have nostalgic thoughts of London. It is a magnificent city, whatever people say of it. The only comparable city here is New York, where I have a feeling I may come to roost.

I left England under peculiar circumstances, hurried, furtive, deceitful etc. A London which ceased to please an irritable superannuated student in his mid-twenties could become the ideal home for the prodigal son, when he returns as a professor. I phantasy! Seriously, I have had thoughts of returning. It would not, however, be sensible unless I returned to something good: in effect, to a consultant or

Professorial position secured for me by any reputation I shall have made in the States. If I were to go back now, I would merely find myself in the interminable middle of the registrar rat-race: probably behind all my contemporaries, because of my absence, and any dubiety which might exist about American residency training. [. . .]

I will look forward to your next letter, and endeavour to be a better correspondent. My best love to Auntie Len—it's *her* turn to write now!

To Augusta Bonnard

PSYCHIATRIST*

MAY 31, 1964
UCLA MEDICAL CENTER, LOS ANGELES

Dear Dr. Bonnard,

Thank you very much for your kind letter. I too enjoyed your company immensely; I came to discharge a duty, and found a friend. It's been a very long time since this happened! [. . .]

I think you may have hit on an important and unrealized motive for my having left London, viz. the profound discomfort at having a psychotic brother around the house. In more general terms, I have retreated from a position which drove him right out of reality, and was driving me into a state of perpetual anxiety and conflict. The dilemmas posed (it sounds hideous to "blame" my good parents for this) stem largely from my parents' rigid Jewish ideas on propriety. Thus it seemed eminently proper to them that I should become a doctor: for myself I wanted to be, in turn, a mathematician, an inorganic chemist, a zoologist, a neurophysiologist—anything but a doctor. I never had the courage or conviction to oppose them. Only now, at the age of thirty, after a decade of idiotic vacillation, am I finally turning away from the clinical life, into something more congenial. I am really quite terrified of clinical responsibility. My parents envisage (what Jews do not?) my marrying a "nice Jewish girl," taking over

* Bonnard, who had long been friends with both OS's parents and Jonathan Miller's parents, had come to Los Angeles for a yearlong study leave.

part of my father's nice Jewish practice, and settling in Brondesbury for ever and ever, with my uncles and aunts and cousins and children and grandchildren, in a sort of cosy incestuous intimacy. Unbearable idea: the negation of self. Yet if I stayed around, it might have happened, representing as it does the line of least resistance. Although I wax eloquent on America's great virtues—her geographical splendour, her energy, her opportunities, her exciting newness—this is mostly pretence. America could have been almost anywhere else, so long as it was well out of range of England. Many of these very virtues have become utter dross: the physical magnificence is vanishing with each passing day (and perhaps will soon cease to exist outside the National Parks); her energy no more than a robotic busyness; the loose social organization emphasizes human apartness and solipsism; and the novelty evaporates in stale gimcracks. There remain only the professional opportunities: and perhaps I should know by now that a good man finds opportunities everywhere. Only I don't know if I'm a good man.

To Elsie and Samuel Sacks

JULY 1, 1964
UCLA MEDICAL CENTER, LOS ANGELES

Dear Ma and Pa:

I am exceedingly sorry to have been so dilatory in writing for the last few months. The reason, no doubt, is that these have been months of considerable anxiety and uncertainty about the future, as well as being hectically busy. Today, July 1, is the first day of the new academic year, the first day of my Senior Residency, and the first day of what I hope will be a tranquil and productive career in Neuropathology. My clinical responsibilities will be minimal (one afternoon a week in the Neurology Clinic, to keep my hand in, and to satisfy the requirement of the Neurology Boards); I do not regret the years of clinical work behind me—I am sure they will always stand me in good stead. I am no fool at diagnosis, and I have earned the affection of many of my patients. But I have always felt, in a sense, anxious and diffident with patients, and can only give of my best in something like neuropathology, where I am my own master, can do things in my own time and way, and have no instruments but a microscope and a typewriter. This, at least, is my present feeling. If I am mistaken, nothing has been lost; for in any case I will be eligible for any Neurology consultants jobs at the end of the coming twelve months. Actually I will be doing a double residency, since I shall also be doing all the autopsies at Brentwood Hospital, a large State mental hospital affiliated to UCLA: this will be accredited to me as a year of General Pathology, so that when I am through with my training, I can expect to have formal qualifications both as a neurologist and as a neuropathologist. With these I should find the highest positions open to me. [. . .]

I have seen a good deal of Augusta Bonnard, whom I admire more and more. We have had many talks together about everything under the sun, and particularly about life in England compared with that in the States. She herself cannot tolerate professional life here, and as you know is returning to England this September. I too share many of her doubts and dissatisfactions about life here, and am also feeling a considerable urge to return to my own country. On the other hand, there is no point returning to England when I am only half qualified. [. . .]

Otherwise life is fairly quiet. I entertain a fair amount; am having a small dinner up in Topanga for Augusta Bonnard tomorrow. I go up to San Francisco every few weeks, where I always see the von Bonins, that marvellous couple whom I am so fond of. They will soon be going to Europe for a year or so and I hope you will be able to see something of them when they are in London. I shall be taking a few days off next week, for some skin diving down in the clear waters of Mexico (the sea of Cortez). [. . .]

Keep well, and write soon.

Much love,
Oliver

IN SEPTEMBER 1964, OS began seeing a psychiatrist, Seymour Bird—presumably at the urging of Augusta Bonnard (though psychoanalysis was in its heyday, and OS would have known a great many analysts and analysands). He had long been plagued by mood swings and especially by a fear that he, like his brother Michael, might be schizophrenic or perhaps manic-depressive. (He frequently used the term "psychotic" of himself and others in this period but usually applied it more loosely than we would today.) By December 1964, when he wrote the following letter to his brother Marcus, he was also facing the decision of whether or not to return to England once his residency ended in the summer of 1965.

To Marcus Sacks

DECEMBER 22, [1964]
1840 IOWA TRAIL, TOPANGA, CA

Dear Marcus:

[. . .] You and Gay sound enormously happy together, or so my intuition tells me. And God knows you're overdue for this sort of happiness. Although nobody, or no factor, can be held "culpable" in any direct sense, I think all of us—but perhaps especially you and I—have remained infantile Jewish dumplings for far too long. David got drawn into a strong woman's ambit quite early; you late, after years of housekeepers and superficial consolations; I'm still pending, as you might say; and poor Mike is wandering round somewhere in the psychotic recesses of outer space. Second: you sound, if I am not mistaken, as sick of Australia as I am of the States, and both very anxious yet very reluctant to "return." One half accepts the conventional view that return is a sort of failure. One had imagined coming back in a blaze of glory, like a prodigal son; but fears that one is really coming back like a beaten dog, with its tail between its legs. I think the crux of the matter is a psychological one. Neither of us emigrated in any sort of heroic spirit; we weren't going to seek our fortune in foreign places, and see the world, and all that. We might have thought so. And certainly, the professional scramble being what it is in England, could rationalize our departure this way, to ourselves and others. But I am coming to entertain a different, and rather awful, thought: namely, that we (or certainly I, and I sort of assume you) left in a negative way, left because we couldn't stomach the home situation, left, in short, because there is something intolerable and destructive about having a psychotic around the house. And though one hesitates to use the word "tragedy" about one's own home (it sounds so theatrical), I think there has been a real tragedy in our parental lives consequent upon keeping Michael at home. Sure, one can entertain, in a rather limited way; one can go out occasionally; one can leave for short, slightly fearful, holidays—but I think Ma and Pa have been forced to lead painfully constricted lives *because* they kept Michael at home; have been in a state of chronic, helpless embarrassment; and, not infrequently, in actual physical fear of him. I was eleven when Mike had his first psychotic break, and I remember I was profoundly terrified, both physically and mentally. Physically, because he was larger and older and *irrational,* and therefore—I had to presume—potentially violent; mentally, because I sensed his appalling isolation and vacuity, and because I feared for my own sanity, by analogy, or identity. Perhaps I am talking out of the top of my head. One is in deep waters here, and I know no way of determining the truth, even though the truth is, simply, a matter of *feeling,* but I am persuaded that this is one of the reasons we left, and consequently one of the reasons we would hesitate to return.

I had no intention of launching on this immense tirade. It sounds shockingly callous. What are *your* feelings on the subject?

Myself. I am spending an externally tranquil, internally rather violent, year, my last, in this Neurology residency. I am spending it in Neuropathology, with a little soupcon of clinical work. I veered away from clinical work, because the grinding anxiety and responsibility of looking after patients drove me nuts. But now, although I love neuropath, I find I rather miss patients, and stand in something of a dilemma: which shall I stay in?

Part of the trouble is that I have been a resident for far too long, denied real responsibility, so that I am now almost terrified of it. [. . .]

To Elsie and Samuel Sacks

MAY 26, 1965
[LOS ANGELES]

Dear Ma and Pa,

I am so full of good resolutions about writing, and so incompetent about realizing them. But here I am, actually started: so here goes!

First, JOB: letters have been sent, recommendations solicited, sent ("he arrives late, he rides a motorcycle, he dresses like a slob, but he has a good mind tucked away somewhere, and maybe you'll have better luck with him than we have had. I like him, but he has given me a lot of grey hairs"), and everything ratified. I will be starting at the Einstein as a fellow in Neuropathology on September 1.* Or—if I can get a little extra time to go to a conference in Vienna in the first week of September—a bit later. My first year there will be divided, half and half, between classical neuropathology and neurochemistry. My stipend will be an initial $7000, which can be supplemented by one or two thousand more by doing one or two Neurology clinics a week. Not princely, but by no means miserly. Undoubtedly I could earn twice this by starting in a staff position in Neurology, or three times this by practicing privately, but so far as I am concerned I shall be doing what I want to do at Einstein, and working with the best and most creative team in the country in Neuropathology. And if and if and if (and how frightened I am of the reverse side of it all!), if things work out, I should find a degree of absorption and satisfaction in work which I have not known for many years. Terry, my professor, is a comparatively young man, fiery, ebullient, sardonic, Jewish, volatile, creative, neurotic, and probably the best electron microscopist in the world.† Unfortunately he himself will be in Paris on sabbatical for the academic year 1964–5, and this is one of the reasons why

* The Albert Einstein College of Medicine, in the Bronx, had been founded only a decade earlier, as a part of Yeshiva University, and attracted many American Jewish doctors, as well as many European refugees. In its early days, it was set apart from more established institutions by its interdisciplinary approach and its esprit de corps.

† The charismatic Robert Terry was a pioneer in using electron microscopy to study Alzheimer's, which was seen then as an exceedingly rare form of early-onset dementia. Terry and his colleague Robert Katzman realized that the more common condition that had thus far been considered "age-related" or "senile" dementia was not a vascular condition, as once thought, but in fact a form of Alzheimer's disease.

I am electing to spend this year [. . .] doing Neurochemistry, in his absence.

I was also offered a position by Abner Wolf and David Cowen at Columbia. Abner Wolf is really the founder of neuropathology in this country, and ex-president of the Neuropathological Association, while David Cowen is the editor of the *Journal of Neuropath[ology].* I was, needless to say, immensely flattered and flustered by their offer (coming the day after Terry's) and could hardly believe that after dragging out a fifth-rate, depressed, submarginal, barely tolerated existence at UCLA, I should find myself flying to New York impulsively, and immediately being offered the two most coveted trainee jobs in Neuropathology in the country. [. . .] Despite the great eminence of Wolf and Cowen, the classical tradition they supremely represent, and the great prestige which would accrue to working with them, I finally decided that the atmosphere at Einstein, the ebullience and creativity and informality of the school, and in particular the personality of Terry whom I immediately took to, would offer not only the best excitement and incitement to work, but also (what with my chronic history of unreliability and depressions and ups and downs etc.) the best chance of survival. Einstein has accommodated greater odd-balls than myself, and somehow helped to realize their potentialities despite the trouble they give: and I was very fearful that the cool, monastic, crystalline atmosphere of Columbia would be both intolerable to me, and intolerant of me. Bird* has been of immense assistance to me in settling some of my "problems," but I am well aware that I constantly threaten my own existence by impulsiveness and rages and this and that, and I must take this factor into consideration in the choice of the most prudent position open to me. Einstein offers a superb opportunity, and I hope to God I am ready to make full use of it. My ignorance of neuroanatomy is still unspeakable, and I never will be a good anatomical pathologist. This, of course, is one of the reasons why I wish to venture somewhat into the very new field of Neurochemistry.

[. . . As] a result of paying Bird his $300+ a month since September, and pouring about $700 into major repairs on the motorcycle (I should of course have sold it when it was at the 50,000 mark, and

* Seymour Bird, the psychiatrist OS had begun seeing in Los Angeles.

hadn't started to crack up: but, as is inevitable when one is on a pretty marginal budget, I have been constrained to patch and repair, which is never economical), and my general sloppiness and disorganization, I am now in a shaky state financially. Indeed, had not Augusta Bonnard insisted on leaving me some money before she left, I could not have continued the analysis, as a start. I have now taken some $800 from the nest-egg which she left me, and hope that I will not have to deplete it further. I think that by selling the piano and what remains of the motorcycle, I should just about have enough to pay my fare to England, though nothing left over. I am arranging to get a new BMW (the de-tuned, rather slower version, instead of the R69S, the fast but ultimately expensive and troublesome model which I bought in 1962). [. . .] I will, of course, have neither the climate nor the necessity of using it for daily travel in New York, where the subways represent by far the most convenient method of getting anywhere. I can therefore reserve the bike for occasional and weekend use, which should be a lot less hard on it. I confess that after eight years' motorcycling, I have little wish to change to anything else: I have survived over 250,000 miles driving without incident, and feel that I am as safe on a motorcycle as anyone is, or as a motorcycle ever permits. [. . .]

The Einstein is in the Bronx, which is emphatically no place to live, being rather like the drearier areas of South London. I will probably get an apartment on the West Side (near Central Park) or in the Village: both of these are within 30 minutes train ride of the Einstein. Apartments are relatively expensive in Manhattan, probably 50% dearer than in Los Angeles, and I shall obviously have to get something far less commodious than my house in Topanga. I think I have had my fill of being a recluse, and am returning to urbanity in all senses of the word. [. . .]

The Exhibit,* I am delighted to say, seemed to be a great success, and attracted a lot of favourable attention, and also I suppose, indirectly, the job offers from Columbia and Einstein. We are putting it up at UCLA now, where it will be my swansong and farewell to the University. [. . .] I don't think I have the capacities, and I am sure I don't have the temperament, to make a first class research worker,

* OS presented a poster exhibit on Canavan's sclerosis, a degenerative disease of the brain, at the 1965 meeting of the American Academy of Neurology, in Cleveland.

but talking and writing, lectures, and exhibits, and I hope one day, books and monographs, seem to be my forte. Perhaps my only forte. I see myself as a sort of middleman, with few ideas of my own, but capable of expressing other people's ideas rather better than they do it themselves. Scientists tend to be a rather inarticulate lot, and I hope that I can somehow survive as a teacher, a sort of neuropathological Talmudist, despite indifferent practical abilities. This, at least, is how I see things at present. I shall be surrounded by millions of people, which will be a change, and I think a very congenial one, after my two years' communion with the rodents and rattlers of Topanga.

Love,
OLIVER

3

Jenö

1965

IN AUGUST 1965, OS WENT, AS usual, to London for a few weeks. He was due to take up his new position at the Albert Einstein College of Medicine (AECOM) in September—but notified the college that he would be arriving a few weeks late, so he could attend a neurology conference in Vienna. In the meantime, he met Jenö Vincze,* a charismatic Hungarian theater director living in Berlin, with whom he spent a short-lived but passionate and amphetamine-fueled few days in Paris and Amsterdam (missing the Vienna conference entirely). He returned to New York City at the end of September, taking up temporary residence at a hotel in Greenwich Village.† Over the following few months, OS and Vincze exchanged numerous letters. At first these were exalted, exultant—they were madly in love and planning a life together.

To Jenö Vincze

[OCTOBER 4, 1965]
[ALBERT EINSTEIN COLLEGE OF MEDICINE, BRONX, N.Y.]

My dearest Jenö:

I have clutched your letter in my pocket all day, and now I have time to write to you. It is seven o'clock, the ending of a perfect day. The

* In *On the Move,* OS refers to Jenö by a pseudonym, "Karl."

† The Hotel Earle on Waverly Place—now the Washington Square Hotel—was then a rather seedy apartment hotel in the heart of Greenwich Village that attracted musicians and writers. During the mid-1960s, regulars included Bob Dylan, Joan Baez, and Bo Diddley.

sun is mauve and crimson on the New York skyline. Reflected from the cubes and prisms of an Aztec city. Black clouds, like wolves, are racing through the sky. A jet is climbing on a long white tail. Howling wind. I love its howling, I want to howl for joy myself. The trees are thrashing to and fro. An old man runs after his hat. Darker now. The sun has set, City. A black diagram on the sombre skyline. And soon there'll be a billion lights.

We're alone in the lab, just two of us. The coffee's boiling on a Bunsen Burner: terrible coffee, but that's how you make it when you're here at night. The cryostat makes a friendly bubbling. There's a smell of hot wax, and formalin, and the thousand concoctions we need for our stains. Beautiful smell, full of memories: back to my first dogfish, long ago. And slides all around me: the jigsaw of a human brain.

You're so right, Jenö: I don't feel the distance either, only the nearness. We're together all the while. I feel your breath on the side of my neck. Your hair is tickling my nostrils. I have that tingling in my penis.

Your letter! What are you doing to me? I tore it open, and found myself trembling between tears and laughter. I pretended I had something in my eye, and rushed out of the room.* You have said what I had intended to say. I wrote in my diary (my "diary" is really an endless letter to you, a conversation) after we'd been on the phone: I stride along the road too fast, driven by the rush of thoughts. My blood is champagne. I fizz with happiness. I smile like a lighthouse in all directions. Everyone catches and reflects my smile. They nod, and grin, and shout "great day."

The mountains skip like rams, the hills like lambs.†

I read Psalms in profanity, for the joy they contain, and the trust and the love, and the pure morning language.

And suddenly, Chaucer again. Did Chaucer know he was the first? Grabbed Aquinas and Venerable Bede: dust, theories and dead history. Beowulf. Heroes. There are no heroes. Great Chaucer who looked and felt and shat and fucked, the beginning of real literature. I feel beginnings in my blood, a vast field has opened beneath my feet. The air, real air. Voices. Bodies. Intensely real. And you the realest of the real.

* OS was receiving his mail at the college.

† From Psalm 114.

My letter, my diary, is a sort of madness. I write so much. I want to catch everything and share it with you. You will be deprived of all your social life, your sleep, your food, condemned to read interminable letters. Poor Jenö, committed to a lover who's never silent, who talks all day, and talks all night, and talks in company, and talks to himself. Words are the medium into which I must translate reality. I live in words, in images, metaphors, syllables, rhymes. I can't help it. Occasionally, just occasionally, I can catch a moment, a feeling, with the camera, and short-circuit the enormous flood of words.

I enclose such a photo. Boy, about to dive. Moment of incredible grace. Intimation, epiphany. I don't want a movie sequence of the dive. Fuck the movement. I want the moment which precedes the movement, yet contains it. About to become. [. . .]

Friends. Jonathan is the only one that matters. No, two fine men in Los Angeles. One of them, Tom,* wrote me today from the seashore, and said he wished we were together. He writes: "Tidepools, which are my laboratory, are the best therapy for the constricted soul." He's right. I long for tidepools, for the intense happiness of those long long days. There'll always be a feeling of regret, somehow attenuated into a film of happiness. We'll go to the tidepools together, Jenö. Tom was a good friend and a married man, and square, but fine. But you're everything. I want to share my joys with you. To see the green crab scuttling for the shadow, translucent egg cases hung from seaweed. A little octopus, just hatched, jetting for joy in the salty water. Sea anemones. The soft sweet pressure if you touch their center. The chalky hands of barnacles. And polychaetes in their splendid liveries (they remind me of Versailles), moving with insensate grace. And dive with me under the ocean, Jenö. Through fish, like birds, which accept your presence. And scarlet sponges in a hidden cave. And the freedom, the complete and utter freedom of motion, second only to that of space itself.

I'm lost; I keep dreaming all the while. Friends, I was saying. I have none in New York. They seem thin as wafers. Their actual presence is less than your imagined presence. [. . .]

No, I don't have an apartment yet. I have looked a little, rather

* Tom Dahl, whom Oliver met at UCLA, was interested in photography and marine biology, two of OS's longstanding passions.

indolently. I don't mind staying in a cheap hotel. I have the freedom of the streets and cafes, and the freedom of my skull. The Village is fun. Phoney, commercialized, not what it was: these are the comments you always hear. But it is intensely alive. Lights, cafes, bookshops, people; lovers, children, bums, and queers. A sort of gimcrack St. Germain des Prés. The weather has been of supernal beauty. The day steeps everything in golden liquid. I saw a van Gogh reproduction last night. I don't know its name. A sidewalk cafe in the evening, with a wonderful amber light flooding through the doors and windows: huge, mad stars in an indigo sky. For this, you have to be great, crazy, or wildly in love. I don't feel mad; I have thrown away those hateful drugs; I do feel great; and I'm gone for you.

I never saw that golden light before we met in Paris.

Whorls and diadems of light, the mirage of New York, the silence of the empty school, unreality. The only reality is the clack of the typewriter, my laboured attempt to speak to you.

You asked me questions [about my job], and I put you off with fancies. [. . .] An incredible warmth pervades this place.* So different from the aloofness, the meanness, the idiocy of Los Angeles. Norton,† my other chief, would speak to me—if the need arose—as ruthlessly as you did by the Seine, and as tenderly. There is a fantastic creative furor in the place. Not the ugly pressure, the competitiveness, of Los Angeles. But an impatience to find out the truth.

Terry's assistant, another coincidence (so many of them!), is a Magyar. I liked him at once. I want to learn Hungarian with him, and German on the train each morning, so I can talk to you in your native tongues. [. . .]

Do they treat me well? How shall I answer. They treat me as a man. They speak straight. They don't pretend. They're not out to make impressions. None of that awful childish ring (guilt: punishment; command: disobedience; guilt: punishment) which seemed to be the round of my life in Los Angeles, where they treated me as a child, and I reacted as a child. I think I shall love it here. I think it will be the making of me. I see a future rise above the horizon of hypothesis. [. . .]

* Terry's lab.

† William Norton, a neurochemist.

Go out and have a coffee, Jenö. You've got pins and needles, your mouth is dry, you've been reading this letter for much too long. But I want to tell you all. I have been greedy of every moment.

I didn't feel too much last week. I was busy, frightened, and confused. I hadn't heard from you. And little uncertainties—how shall I say it—fluttered like moths about the lamp of uncertainty. Your voice was anguish and beauty and strength. It made sense, it lent strength, it comforted me in the vastness of New York, and the vaster vastness of a new life.

I never admitted it in full, though perhaps you guessed. I was drugged the whole summer. In a close tight cell of amphetamine madness. A pharmacological Bastille of my own creation. And you led me out. Doctor Manette: "restored to life."* Two sides to every question. Paris was the positive radiant answer. [. . .] I doubt if I shall ever take drugs again. [. . .]

Nine o'clock! Two hours gone? Forgotten supper, forgotten gym, and the phone calls I might have made, and . . . Another time. This is the time for you. I feel closest when I read your letters, and when I write to you.

I went down to the Docks on Saturday after your call, and sat on the Quay and watched and watched. How the huge ships squeeze through a narrow neck. And the statue of Liberty (false bitch!), which you so want to see. I took a picture of a young man who was also sitting on the quay. Perhaps he too was dreaming of a far-off lover. [. . .]

Late afternoon when I rose to my feet. A dark Guinness in an Irish bar. The light is glinting in its foam. The long shadows of afternoon. Irish boy behind the bar, a callipygous† pig, a sexy moron in his dirty apron. Silent drinkers: faces in a tarnished mirror.

Another dockside bar, brown and smoke-filled, radiant in the evening sun. You remember that bar in Villejuif, with its pure translucent essence of France? (I have a photo of it. It came out.) Well, this bar was America, powerful, tarry. And I fumbled with my camera like

---

* OS often referred to Dickens's *A Tale of Two Cities,* a favorite book from his childhood. The first part of the book, about Dr. Manette's release from the Bastille, is titled "Recalled to Life."

† Having shapely buttocks.

a fool. A man came out, H. C. Earwicker,* Finnegan the hodmaker in all his glory, a gross primordial man with a wad of tobacco in his mouth—he came out, and spat at me (fatuous tourist with his silly Nikon). He was right, of course. I must be covert. Joyce knew. "The artist must refine himself out of existence (in drama), above, behind, beyond his creation."† Something like this, I forget exactly. [. . .]

Even the typewriter ribbon is grey and tired. But not I. I am filled with an indomitable ferocious energy. The title of a vulgar hit: RAGE FOR LIFE. That's it. Look at your handiwork, Jenö: this creature you have set in motion. Pygmalion: are you pleased with your Galatea? [. . .]

I must finish. I could sit up all night. I could write you a book. I want to write you a book. If ever I write a book, it will be for you. It will be two books in one: a book for the public, and a book for you. A book of amphibolies, double meanings and private references. [. . .]

To Jenö Vincze

[OCTOBER 6, 1965]

ALBERT EINSTEIN COLLEGE OF MEDICINE, BRONX, NY

Jenö!

I love you insanely, yet it is the sweetest sanity I have ever known. I read and reread your wonderful letter. I feel it in my pocket through ten layers of clothing. Its trust, its warmth, exceed anything I have ever known. [. . .]

How constantly I talk with you! I write to you, continuously, in my Diary. My last letter was overloaded, it had been building up for eight long days; it was congested, and in places false. Some of the things I wanted to describe for you had gone stale, or dead, as "set pieces." I have never learned to write a letter. I have a nasty tendency to write

* The protagonist of James Joyce's *Finnegans Wake*.

† Joyce wrote, "The artist, like the God of creation, remains within or behind or beyond or above his handiwork, invisible, refined out of existence, indifferent, paring his fingernails."

at people, instead of *to* them. It is histrionic, egocentric; one is consciously "erudite," "charming" or "fey." One quotes oneself, applauds discreetly. One keeps a carbon copy, for posterity! One pre-selects opinions, moods. Forgive me—I will try to change.

Your own letter was of the most profound and beautiful spontaneity. ("I am amazed, as I never had the intention to write what I am writing at present.") I am so glad you *did* write what you did! When you spoke in Paris (with such knowledge, such experience, such a man-of-the-world matter-of-factness) of a time, x weeks, or months, a year, whatever, a *period* after which we would have explored each other to the bottom, the *inevitability* of boredom, of parting company—I couldn't bear it, I was speechless with misery. I'm a simpleton, I know; I have never had [. . .] an "affaire" before. But I can't bear to think of weeks or years, of personality in pints or pounds. I believe we are both infinite, Jenö. I see the future as an endless expansion of the present, not the remorseless tearing-off of calendar leaves. [. . .]

Jenö—I love you, I can think of nothing else.

This is the first Yom Kippur I have ever enjoyed. I had a sinful breakfast, and a sinful work-out; I am wearing my sinful jeans, and I have no intention of visiting the Synagogue. I had two beers in Mooney's Bar—in company with three redfaced, blue-eyed, whisky-gobbling, baccy-spitting old blasphemers. And now I am writing to you, my Hungarian "shikse," whom I love more than God. [. . .]

Even tho' I left my home and country, and gave up even a token regard for Jewish custom, Yom Kippur was always an uncomfortable, restless day. I could never fix my mind on anything. I was anxious and depressed. Today I feel great. I will never spurn my Jewish roots, but Judaism has retreated in the past few weeks; no longer an obsession, or uncomfortable joke, it has become a language I learned in my childhood and still remember; funny, old-fashioned, and unimportant. I'm no longer a Jew, I'm a man. And if I have a God, it's you, my love. [. . .]

[This] brings to my mind (you must forgive the grandiose analogy—it's more an association than a parallel) the story of *Billy Budd.* You told me you were reading *Moby Dick;* you wd enjoy this strange other book of Melville's.

Billy (in a sense, yet never explicit) is the ideal of homosexual beauty, radiance and virtue; his tormentor represents the perversion of lust, the hideous jealousy and malevolence, of which only a thwarted homosexual is fully capable. Their fantastic confrontation in the cabin is the moral climax of the book; they become saint and devil at this moment.

Innocence, faced with an outrageous slander, the revelation of total evil, is speechless with horror; he is totally beyond the world of words; Billy's soul *becomes* an action; he clenches his fist, in imperative moral reflex, and fells and kills the other man. (Then God hurled Lucifer out of heaven.) But he has transgressed the Law of the land. In his exquisite, unassailable purity Billy dies, bewildered, among the weeping men who have to hang him.

What I am also saying, Jenö, is that you have made me feel differently about myself, and therefore about homosexuality as a whole. I am through with cringing, and apology, and "explanations," and pious wishes that I might have been "normal." Through with searching my past history for the aggressive mother-figure, the engulfed father-figure, the this, the that, behind my "malady." I am not interested in such cowardice any more.

You have taught me that one can be glad and proud of being homosexual. One is not imprisoned in the Social *matrix*. Intelligence and sensibility are much less fettered. There are special freedoms, yet special responsibilities. You have made me see—above all, by the example of your self—that there is utterly no need for isolation, shame, concealment. Indeed, one's potential for friendship, for living, is immensely widened.

I am so happy I want to shout from the housetops. So proud I want to describe you to all my friends, even the "normal" ones, from whom I always kept a certain (I suspect now, unnecessary) distance.

I must bring this giant letter to a close. Am I killing you with words?

I enclose the photo, and something I wrote—about a year ago—in a time of profound depression and isolation. With your help, Jenö, and your love—never again.

Your Oliver

To Jenö Vincze

OCTOBER 11, 1965
ALBERT EINSTEIN COLLEGE OF MEDICINE, BRONX, NY

My dearest, distant Jenö,

It is four in the morning, how perfectly called the "dead of night." I am obsessed by some awful words of Rimbaud: "The clock of life stopped a while ago. I am no longer in the world." I have risen from my sleepless bed: one nightmare after another. Nightmares of being drugged again, nightmares of exposure, failure, and disgrace. I thought I was so happy. Indeed, I *am.* But the old old fears come back in the night. Are these things ever exorcised, or are they part of being human?

Four things: nowhere to live yet, not *really* in my work, fear and weakness when alone, the momentary dissipation of that marvellous happy beauty—in the coldness, deadness, emptiness of the night.

For the first: I leave this hotel tomorrow I trust. I have taken (too quickly, too impulsively, but I need *a place*) an apartment in a modern block. It has the fine, antiseptic, impersonal quality of these flats you see outside Copenhagen or Tel Aviv. I don't give a damn. I am indifferent, at the moment, to *domestic* surroundings. I am not one of yr little fags who need to have the right *address,* and little *chic* Victorian rooms, and a colour scheme in mauve and *puce.* I am Oliver Sacks, whose life is in the mind, the outside world, and *you.*

The apt is solid and comfortable, without pretention. Close to the Village, the Station, the Road. Close enough to lights and cafés and all-night people. All New York is huddled close. But this isn't the real reason why I took it. My mind was resolved when I saw the *water.* I am only a furlong from the Docks and Wharves. I'll gaze in the evenings over the water, imagining Germany through the mist. I'll watch the great ships as they pass my window, and scan their decks with my huge Russian lens. My room is the room of a man who is *waiting.*

Second: fiddling enough, Jewish Holy days over, tomorrow I *plunge* into work. I may not be able to write the vast letters I wrote to you last week. They were written, in part, when I *should* have been working.

Third and 4th: The horrible spectres are already going. I feel warmer and better as I write to you. [. . .]

I enclose "99."[*] I had the sudden need, this weekend, to dig up the scribbled notes of my first bike-journey in the US, and crystallize it into sharpened images. More clearly now than ever before (and this is largely due to you), I am finding it easier to distil my sharp perceptions from the *rhetoric*, the *self-obsession*, the clumsy *irony*, the *clichés*, which have bogged me down. A good friend of mine, and wonderful poet,[†] said I must learn to write with a "fierce discipline." 99 is a serious try.

I hope you have received the Camus. I'd like to send you many more books, but suddenly fear you'll think me *patronizing*. And write, Jenö, write! A week is too long for me to wait. If only I could phone you every day!

I love you, and I want your love

Oliver

To Jenö Vincze

OCTOBER 12, 1965

ALBERT EINSTEIN COLLEGE OF MEDICINE, BRONX, NY

Jenö?

Still there? Still the same Jenö? Forgive me—it is eight days since I've heard from you. Eight days is nothing; one is busy; important business; the vagaries of the post; did I *myself* not keep *you* waiting for longer at a time of great uncertainty? True, true; completely true. I am too anxious, too demanding (like my mother who wants a letter every week: God forbid I should be as possessive as her!). But I feel like a fool writing to you—*yet again*—in the absence of response. And then I say again—a week, pfft! he hasn't had time to respond. And then darker layers of thought come through: You idiot, Oliver! How long will you delude yourself? Jenö's had his fill. Can you not hear him, among his *gifted* friends: "Liebling, the *last* one was a

* An essay about motorcycle life that OS was working on but never published.

† Thom Gunn.

motorcyclist, intriguing but bizarre!" And then, yet deeper layers of trust and faith. What have you done, or said, or written, that I should *ever* doubt you? Have you not shown me, by your own example, the very essence of trust and honesty? And then, deeper still the distrust I have of myself, the fear that I am the one who will betray. And, deepest of all, the fear that there is nothing certain, nothing solid, in the world: *Tout ça change*. And the cool voice of Dr. Bird, when I become *extravagant* of word or feeling, absurd in action or expectation. "No, Oliver, no! *This* is reality." He wd give me "reality" (the analyst's gift), and I wd see how far I had strayed from it . . .

Jenö: why have I written all this shit? It wasn't any intention. I *SUCK* at you for reassurance, like a baby at the breast. You are my lover, but not my confessor or my analyst. I talk too much (the same in writing), and demand too much. Perhaps I should *divert* my sick questions to an analyst. I've been reluctant, though, feeling I wanted to focus every thought, every feeling, on *you*, and you alone. You are bored, angry, bewildered, aghast, as well you might be, being the target of such a monstrous onslaught! Or is this just the delicious agonising dilemma of every lover, as he plucks off the daisy petals one by one . . . He loves me! He loves me not! He loves me? He loves me not? HE LOVES ME

To Jenö Vincze

WEDNESDAY [OCTOBER 13, 1965]
[ALBERT EINSTEIN COLLEGE OF MEDICINE, BRONX, N.Y.]

Jenö!

Your marvellous letter today (I was not at hospital Monday and Tuesday, but at a conference). Totally unexpected. [. . .]

I shall start an answer now, for the day and circumstances are propitious. I am sitting under a gnarled Chekhovian cherry-tree. A perfect garden, running down towards the copse: slender birches and aspens, great *he-man* oaks and willows, and the maples—a *glory;* there is nothing in the world, in our Europe or elsewhere, to compare with the New England Fall. No description, no photograph, can describe this *gift,* this utterly amazing beauty. Even the memory cannot hold it, so it comes as a wonder every year. The light so pure and clear: in Eliot's phrase, "a grace of light." And an old whitewood farmhouse, weathered, of great charm, where my host and hostess* of this afternoon are living. Seeing my happiness as I opened yr letter, and the mood of delight I was in, they offered me their garden, and a pad of paper, for the quiet of afternoon. (A blue jay darted through the undergrowth.) Excellent, most perceptive, people! The wife I knew ten, God! twenty, years ago: one of our grubby Hampstead set. I had forgotten her existence, till we met again in Dr. Terry's laboratory. I found her tranquil, creative, happy, glowing in the sun of Einstein. "Being here has *made* me," she whispered in the car.

A lovely lunch, an omelette and a dryish wine; their son, a lusty three year old, has decided I'm an "apple"—one of the better class of human being in his world. I have been playing nursery rhymes for him on the Bechstein grand, upsetting that fine instrument with outrageous variations, playing with my knuckles, and my nose; telling him stories about sea-elephants, Kangaroos and boojum-trees. Enjoying, not the taste of paternity as I used to think, but the easy love of a man for a child. (The love Mark Twain describes so well.) And your letter running through my mind.

Music? You ask. [. . .] My "relation" to it? We "live apart," as the Law Courts say. I grew up in a house of Bach—Francesco Ticciati, a wild Italian genius, taught my brothers and myself. I speak now of 1939. The War was no time for music. After the War, romantic phase! I filled the house with Liszt and Chopin, Schumann, Grieg. I loved Schumann best: his black moods, and his sudden childlike innocence of sound: esp. *Phantasiestücke* op. 12, and Carnaval. My childhood love of Bach (Francesco would teach us nothing else) was forgotten for a Time. This era (12–17) was a time for sitting in the "gods" (the gallery at Covent Garden), which only cost a shilling, and hearing *Aida, Boheme, Otello:* I would always close my eyes to hear. Many of my friends "dug" Wagner at this time, but I could never stand him. But I loved the voluptuousness of Richard Strauss.

Oxford. An element of snobbery entered in. People would say, "But, really, my dear, *Chopin*!" until I felt ashamed. I went back to Bach. I played the organ in the college chapel. How I used to love that organ,

* A fellow English émigré and her husband, identity unknown.

in the darkness of the empty chapel! My thoughts became music, power, without a conscious effort. I tortured myself with polyphony and atonalities—Hindemith made me gag! I admitted the mysterious seductions of Fauré and Debussy. And I went mad, utterly rapt and crazy, over *The Firebird.* I still do. It's a sort of musical hashish for me.

And then, Jenö (how this flows on, as tranquil as the afternoon), then the long barrenness, the boredom, turning inwards. My musical life came to an end 10 yrs ago. I ceased to go to concerts. I never possessed a gramophone. I couldn't bear to play.

As such, music has very little place in my existence now. Sometimes, improvising in the organ-loft, I would say "Music! this is my life." Now I feel like any other un-sensitive and un-gifted person, whose *fingers* received an expensive education, but are stiff now with disuse. And then again, a snatch of music drifts across the housetops, and I quiver like a setter at the hunt. I would like to *re-enter* the world of music; to listen from the "gods" once more; to wield the organ; with Jenö Vincze at my side.

To Jenö Vincze

OCTOBER 18, 1965
[ALBERT EINSTEIN COLLEGE OF MEDICINE, BRONX, N.Y.]

Your letter, Jenö, today: at last!

[. . .] I have been writing to you, in effect, for 48 hours, starting from the blackness before our phone-call. I am going to put my *bits* together: conflate them, that's the word.

SAT MORNING: I am so cold, Jenö, so horribly frightened. I am a thermometer in your hand. Infamous week! Each day I waited for your letter. Each day I sank into deeper despair. I can't eat, or sleep, or work—I wear the face of a murdered zombie. Your "presence" has vanished from my side. I feel so alone, so completely abandoned. [. . .]

You have brought me the most exorbitant happiness, Jenö, a happiness so intense its extinction is death, or the feeling of death. *I love you and I need your love.* [. . .] My strength was such a thin illusion.

I feel I have *betrayed* you day and night. Not with the mind, but the body, I pretend. I suspect that *this* feeling—projected—is part of my fear of *your* desertion, and the basis of my own self-fury. I too have been an "ugly duckling"; you tell me now that I'm virile and attractive. Must I "prove" it on every street-corner? I must catalogue my vileness. Do I do this to lacerate Jenö, or find forgiveness from him? Such trash. You would never believe! I despise them; I am utterly impotent; I wait for them to go. Item: ageing pedant who sucked me humbly in the darkness. Item: Gaunt Southerner, all bones and penis. Item: Spanish youth, mascara'd whoreish eyes. Item: a Jewish Porker (horrible! a parody of myself), his bandy legs in tight white jeans, yellow with Vaseline about the crotch: a "fine" collection of pictures, dildoes, whips. This crew? These clowns? Why do I force them down my throat? Forgive the enumeration of this pointless shit. I have to mention this, to vomit it up, like pricking an abscess in my conscience.

I do love you, Jenö. I am for you. You are for me. This is the centre of my life. All else is idiocy and neurosis, *without real meaning.*

SATURDAY AFTERNOON: You have made the sun come out again! If only I could phone you every day. [. . .]

Evening: walking and walking by the river. The New Jerusalem, its billion windows golden in the setting sun. The creation of the New York *night,* a monstrous ganglion coming to life. Whose lamp is this, and whose is that? Cameos of ten million lives.

The fantastic Cubist backdrop of the New York waterfront. So patched and piebald, these broken buildings. So full of life and ghosts. This building is 1888; *that* street is pure 1820; this area is surely Emersonian. This is the profound and beautiful difference between New York and London: that New York is a *mosaic* of the past, like a jigsaw puzzle, while London has risen layer by layer—the present is like a film, a transparency, *through* which you see the past. Stand with me, Jenö, on one of London's seven hills: Below you 1965. The cars and buses disappear; dust flies and horses neigh—it is 1765. The sound of trumpets, the panting deer. There's Elizabeth, in a great white cape, with Essex by her side: 1565. And back and back and back and back, to the ramparts of Londinium, and that winding Neolithic valley, where a bearskinned Vincze loved a shaggy Sacks.

This, essentially, is why you, and I, will return to London. London is a face thousands of years in age. She bears, in her wrinkles, her alleys, her sudden beauties, the total experience of this time. She has evolved, layer built on layer, like the crust of the earth itself. Whereas New York, though exciting, jazzy, is entirely different: a *synthetic* city, if you wish, with an oddly provisional, *unmerged* quality: Jews here, Italians there. Crime on 85th. Dignity on 86th. This area is very pricey; that is clearly running down. Westchester for my married colleagues, Christopher St. for my queer queer self.

If only you were with me, Jenö, walking by my side. I shall write out Eliot's "preludes" back at the flat, and send them to you with this letter. They are the most beautiful poems of London ever written. And equally well, they could be any city which smacks of greatness. Somehow, the IIIrd Prelude reminds me of our little room in Rue Cannette (I have a picture, from the window).

SUNDAY: Day of reflection, recollection.

You are you are you are you. Of course. But you have had *precursors,* Jenö. A bad word, but I know of no other. The last thing I mean is that *they* were somehow fragments of you, or that *you* are but an aggregate of them. Two of major importance in my life.* I never had any real relation with them. Not only no *physical* relation, but—I suppose I was too much of a baby to relate at all. They were perhaps abstractions, more than human beings, to me:

Richard Selig (his very name means "happy"), a dozen years ago. Gifted, handsome, American poet. The *first* to open life for me. I realize suddenly that *my* favourite poets (Rimbaud, Villon) were really *his:* he would read them by the fireside to me. Oxford. 1953. Poor Richard. He had so little time. He had tasted only a hundredth of life's sweetness, he was just coming into the manhood of his gifts when he was struck by a fast malignant cancer, and died a few months later—just 27, even younger than his favourite poets. (It was then that I decided to leave zoology, and become a doctor. I had such a hatred of disease and death! This is my only motive for being a good doctor, or good scientist. I shall fight against mortality. Unbearable that a Selig, Vincze, Sacks, should die!) [. . .]

The second was Thom Gunn. It must have been in '58 when Jonathan read me a poem of his, a marvellous poem about motorcycles. It

* It is curious that OS omits Mel from this list, unless it is because Selig and Gunn, like Vincze, were part of the intellectual scene; Mel seems to have occupied a different place in OS's emotional life.

was titled "On the Road,"[*] and below it, "California 1957." It must have been in the *Observer* or *Sunday Times*. I determined, at that moment, that—*sooner or later*—I too would go to California, and know and love Thom Gunn. He had formulated, with such power and precision, the babble of my own emotions. So fine an intellect, and yet so "with it." Phrases from his poems constantly recurred (recur) to me: the streetcorner kid who "presides in apartness"; the late-night city "here she is at her loveliest: extreme, material, and the work of man"; certain clothes and implements "emblems to recall identity." And above all the motorcycle poems. Judge them for yourself. I shall be mailing you a copy of *The Sense of Movement*,[†] probably later this week. (Along with *Prater Violet*. Fool that I was, I took the Camus' Notebooks to the hospital mailroom, ten days ago, for packing and airmail. No doubt they have made a balls-up of it. I hope, too, they didn't look too closely into it, for it was filled (in the margins, the flyleaves) with frantic endearments of a Subversive and Unamerican type!)

I found and gave a hundred other reasons for my flight from England, two years later: the family, the draft, professional bottlenecks. And why? Was I now going West to see the world, and make a fortune, and go dusky in a tropical sun? Etc. Etc. Nonsense, lies. These were the ostensible, *public*, reasons. My real reason was to meet Thom Gunn. I lay in wait for him, for weeks and months. I questioned everyone I met, in a casual, knife-sharp (but how "give-away"!) way. I finally met him at a party. I was drunk with happiness, and drunk with drink. I didn't dare to meet him sober. (I suddenly think of an analogy, as I write: how I dared not meet *you* sober, that first Sunday night in Paris.) I remember, somewhere in my half-forgotten orgy, quoting one of his poems to his face: and he laughed, and stroked my cheek, and said: You know, I'm *me*, and not my poem![‡]

---

* The actual title of this poem is "On the Move."

† Gunn's 1957 collection, which includes "On the Move."

‡ A letter from Gunn to his partner Mike Kitay circa February 1961 indicates that Gunn first met OS at a leather-friendly bar in the Embarcadero. Gunn wrote, "There is a queer, colossally big London Jew called Wolf, a medical student, and friend of Jonathan Miller, who says my poetry changed his life—it caused him to get a bike and wear leather, and he tears around like a whirlwind—and came out to be a doctor, here because I live here." (Quoted from *The Letters of Thom Gunn*, ed. Michael Nott, August Kleinzahler, and Clive Wilmer.)

I think Thom was flattered, amazed, amused, to have this leather baby in his hands. We had no further physical contact, ever. And like Selig, Thom was accessible only to a certain point. I had no claim on him beyond this. I loved him in a way—but felt uneasy in his presence, like a parasite, or child. And he, perhaps, found something endearing in me, but embryonic.

It was Thom who first brought home to me that I did have "literary" gifts of worthwhile calibre. And he would point out again and again (did ever a man find such a ruthless, tender, honest critic?) how I *spoiled* what I had written, with rhetoric, and extravagance, and self-pity, and cliches, and dishonesty, and phantasy, and laziness and, and . . . and . . . As *you,* thank God, have also said to me. But at this time—I am speaking now of 1960—I could cope neither with myself nor any criticism. I conceded the truth of what he said: I needed and appreciated what he said; but I *wilted* underneath it. I became embarrassed and ashamed, afraid to see him. I felt worthless, futile and dishonest. My last contact with him was almost two years ago, when I sent him the last piece I had written, one of the few things ever written in Los Angeles: an allegory called "Machine": about . . . no matter. Thom blackened the paper with his comments ("a *good* word . . . a *lousy* word . . . too much *regret* . . . yes! this is factual"). He wrote (and I feel the words as a sort of legacy, as if Thom had followed Richard to the grave): "Where it is good it is magnificent, the germ of the best thing you have written yet. But you need . . . a fierce *discipline.*"

Forgive me for quoting this business *in extenso* (I seem to think aloud with you, as if we shared the same room, or the same head). But it is of importance to the sort of demand I *may,* no I *am,* also making from you. (Or from Analyst Bird, or Dr. Terry.) I plead, in some fashion, for a "fierce discipline" from them all. Ultimately, of course, all discipline comes from within. [. . .]

To Jenö Vincze

OCTOBER 20, 1965
[ALBERT EINSTEIN COLLEGE OF MEDICINE, BRONX, N.Y.]

My dearest Jenö:

This is going to be a long and horrible letter. But I promise you it has a happy ending. I am writing it partly in fulfillment of some internal need (the need to record, tabulate, analyse, convey, all of the vagaries of my soul), partly in order to clarify for myself the events of the past two weeks, and partly because I *owe* you such a letter, for you are closer to me now than anyone else: indeed the events of the last two weeks have been a sort of nightmare, paradoxical proof of this, and with the realization of this has coincided the resolution of my crisis.

In the beginning (if I may use this classical beginning), in the beginning I was clothed in an armour of happiness. Everything conspired to bring me joy. Perfect September days, cool yet brilliant, in which everything *shone* with an exquisite beauty. Einstein too was delightful, exciting: I adored my bosses—Norton and Herzog: little gods even if the great God Terry was away. [. . .]

I *will* call it happiness, not euphoria or intoxication. For it was vital, full of content. I have never been in better shape. I was warm and smiling, infectiously gay ("walking on air," a colleague said). I delighted people by the *sun* of my being. How carefully, each morning, I shaved and polished my shoes! The delight of choosing a suit, a shirt, a tie—different and perfect for every day. An elegance now mine by right, for I felt and *was* the New Man in me. Too happy to look for a flat. I could live in a hovel or filthy slum, and stay as happy as a lord. Beautifully organized in my life and work. What I *had* to do—for I grudged every moment not thinking of you, not writing to you, not being with you—what I had to do, I did in a moment, with perfect instantaneous clarity. Like the review I wrote on the "Axonal Dystrophies"; it fell out, on the paper, like a jewel from my head.

And, above all, I had you, YOU—almost constantly at my side. The sense of nearness, almost presence. The sense of completeness, of togetherness, of sharing the world.

Saturday came. We spoke on the phone. I was mad with the beauty of your voice and words. Sunday came. I phoned my old friends, the few I really liked; across the country, in Los Angeles. Golden weekend, and your letter on Monday, such a LETTER—who was as lucky in the whole of the world? And did you not say you would write the *following* day? I too wrote to you on sunny Monday, a letter afire with the fever of living, the Nuova Vita and the Brave New World.

Now comes Exodus, the fall from Grace, the loss of the Garden, the descent into Hell. [. . .] Tuesday I ran to the Office and looked for your letter. No matter. Wednesday again. So Jenö was busy. Thursday no luck. [. . .] Friday, Saturday, Sunday, MONDAY! What could be wrong? No letter yet. I made a nuisance of myself throughout the hospital. Insisted they search in all sorts of places. Perhaps it was here, or here, or there: mistakenly sent to another "Sacks."

And then the days of the second week. All the simple, innocent explanations gradually vanished from my mind. It was borne in on me, in sudden spurts, and by greater degrees with each passing day, that you had no serious interest in me. I felt I had made a fool of myself. I imagined you, the wittiest in a witty cocktail party, talking of "an English motorcyclist, bizarre you know, but quite intriguing." The converse was true. I became less certain of my thoughts for you. You were no longer so present by my side. I got up later, I laughed much less. My clothes became sloppier. I did a minimum of work. I wrote no letters. I didn't trouble to have a phone installed. I met a young man, a congenial friend. But this too was a source of pain and indecision. For I feared some feeling might be *diverted* to him (as if feeling were finite, like a water-supply). I felt the sense of betrayal within me. If only, IF ONLY, I could have communicated with you! But why should I write? I was proud, I was angry. I had written two letters, and you made no reply. [. . .]

My mood was black. I had to phone you, I had to know. Through my fear I could hardly hear you on the phone. I heard depression in your voice too, though disguised and overlaid by your concern for me. I think you probably realized how frightened I was; for you suddenly spoke of coming for Xmas (a thought unmentioned in your letter). I put down the phone. Brief three minutes! Happy, much happier: but not entirely happy. I realize now I was already, by this time, too afraid, too depressed to be entirely reassured. [. . .]

In the evening I walked by the river with Dick, the young man whom I have mentioned to you. [. . .] A beautiful walk. A lovely quality of the light and air. Everything lucid, suspended in air. I had not felt so good since two weeks before. But (I question, I question, I fear, I'm not sure, I make dirt, I obscure) I said to myself: Oliver, you whore, why do you feel so good this evening? Is it because you phoned Jenö today? Is it just the perfect weather? Or are you feeling *too* happy with this young man? My fear, my depression, returned right then.

And when we returned from this marvellous walk, *he* seemed subdued and excited at once. I asked him to tell me what he was feeling, although I had a fear, a hope (one *must* not deny the ambivalence of feeling). He said it had been a wonderful walk: that he loved being with me all the time: that I was just his "cup of tea," and he feared he was wholly "infatuated" with me.

These are words which are very precious, if they speak the truth from anyone's heart. One feels humble and grateful, and flattered and proud. I had all these feelings at that moment. And I also had an appalling fear, a confusion, a sense of betrayal—I don't know how to put it. Dick, I said, I am very fond of you. . . . He interrupted me. Don't force idiot words out of your mouth, he said. I know you're fond of me. And I know you love Jenö. I am no threat. I expect nothing, I demand nothing. But I like being with you—relax! Or words to this effect.

We parted early Saturday night. He didn't stay. We had no sex. I slept abominably, worse than usual. I had been on a rising dose of chloral hydrate. Even with a hundred grains, I could get only two or three hours of stormy sleep. [. . .]

I rose at dawn on Sunday morning. No point further tossing in bed. [. . .] I had arranged to see Dick at my place at five [but] Dick wasn't there. I thought of the previous evening's talk. First Jenö, I said, now Dick: how they desert me, one by one. (You will perceive that by this time, projection and paranoia had become quite clear.) Clear? To whom? Not clear to me. [. . .]

A huger dose of chloral that night. Restless nightmares of shame and decay. I saw my mother *putrefy* before my eyes, become a mass of rottenness like cheese slimy with infected insects, like M. Valdemar, in that vile Poe story.

Monday, I dragged myself to the hospital. [. . .] I made no pretence of doing any work. I sat all day before the typewriter, writing my enormous letter to you. I strove hard to delete the pain and fear, but they kept coming back, in different forms. I sent it off by Monday evening. I felt none of the *elation* I had previously on receiving a letter from you, or sending one to you. I couldn't be bothered to go to the gym: it seemed pointless: I felt old and tired.

Dick came by on Monday evening, cooked a fine supper, brought some wine. I could taste nothing. I forced a smile. I forced a compliment to my lips, I forced a kiss, a caress to his body. A sort of woodenness was the only sensation I had. [. . .] Orgasm was mechanical, strictly impersonal: an insult to Dick, to anyone; I was hardly conscious of his presence, even though he lay beneath me.

That night, OF ALL NIGHTS, I decided not to take my chloral: although in a state of terrible agitation and depression, and habituated to a monstrous dose of the stuff. In Prufrock: "Have I the strength . . . to force the moment to its crisis"? Strength, weakness, I don't know. But deprivation of my chloral was the final thing: it forced the moment to its crisis. Caused an explosion of my overwrought brain. [. . .]*

I wrote in my journal:

"I'm OK now. I am humble, I am grateful, that I'm sane. I have *meddled* too much with my consciousness in the past. No more! It is too good and too precious to be risked. I have learned from today: the greatest privilege is REALITY—not drugs, or dreams, or hells, or raptures—but a *cool brain and a generous heart.*" [. . .]

I have finally, in my notes, and this letter to you, described my maelstrom journey into the abyss, and am now finding myself—miraculously, almost—on dry land again.

What can I learn? No doubt whatever of your supreme importance to me, for the absence of your letters, the feeling of *bereavement,* has been a main factor in the whole depression. Of course, other things: not being properly *involved* yet at Einstein, and ambiguous guilty feelings about nice Dick. [. . .]

NEVER have I entered so extreme a depression. And NEVER has it achieved, if only for a few hours, psychotic calibre.

---

* The following page of this letter is missing. OS later described, in *On the Move,* the florid hallucinations he experienced on a bus, and how his friend Carol Burnett, a fellow doctor, sat with him through four days of intermittent delirium—but this was likely a few weeks later as he continued to fight his dependence on chloral hydrate.

Clearly I'm not crazy, nor will be. And yet, equally clearly, there were real disturbances of thought and perception, especially in the train.

The following partial explanations strike me:

1) I have been in a state of the most intense emotional excitement in the past three weeks, chiefly but not exclusively associated with *you.* A state the old doctors called "overwrought."

2) I have further lashed my emotional self in my reading and my writing, living on an exclusive diet of Rimbaud and "99." Pushing all the while for absolute superlatives, epiphanies, ecstatic phrases—not at all a solid diet. Heady, like champagne: for a while, I'll stick to literary CORNFLAKES.

3) The suddenness of the psychotic phase, and its brief duration: and the relative preservation of insight: alike suggest *intoxication.* Such a state would result, for example, if a small dose of LSD were given to an *unwitting* person. I stress—unwitting—for the anxiety and paranoia are largely reflections of not knowing what is happening to one.

I haven't taken any amphetamine since that last Dunkirk night. But I have been taking enormous doses of chloral hydrate to sleep with. I am highly suspicious that *this* may have played some part in the whole business, or its *withdrawal* on Monday–Tuesday night, suddenly, when I was so habituated to it. This, at least, is one hypothesis. It is supported, I think, by the fact that I still notice some *mild* perceptual changes today, although I am *not* depressed. As with any drug which has been taken to a toxic level, I will withdraw it slowly, under the protection of a tranquillizer. This is standard practice for these things. And, in the future, *no more sleeping mixtures.* I have cut out everything else. The chloral too must go. For, I always take things to *excess.*

4) One other factor. This is myself. I am gifted, with a lively imagination of "ecstatic" type (Rimbaud-like, I prefer to think). Second I have had an enormous amount of actual experience with hashish, LSD etc. and so *know* what a lot of these changes are like. Thirdly, I have a schizophrenic brother, whom I have observed with fear and fascination for many years. A primary symptom in my own pathology is the *fear* of becoming myself psychotic. Armed with this fear: my imagination: my "material" for the construction of a psychotic

or pseudo-psychotic world: and an intoxication/withdrawal toxicity from chloral . . . given all this, I find the necessary ingredients of yesterday's state, and pronounce the verdict SANE.

I love you, my only Jenö. I wish to God you were here!

THURSDAY Last night I found the following sentence in an old Pharmacology book:

"Sudden withdrawal (of chloral) may occasionally produce a clinical picture similar to delirium tremens and known as *chloral delirium.*"

My answer! You may imagine my relief. And you may also now regard much of the huge preceding letter as a description, from the inside, of a classical and rarely-seen delirium.*

Why did I not make my own diagnosis sooner? When I found sleep *impossible,* and thoughts boiling in my brain, images *alien* to me: when I woke to a world inexplicably ugly and frightening; when my hand shook and my memory faltered in the braincutting session; when I saw everyone round me enclosed in depression or paranoia, and finally when I crossed New York in that nightmare journey of hallucination and unreality? Why was I so slow to realize?

The first answer is that *this* is the nature of a delirium. The mind is bewildered in its own phantasmagoria. The patient with alcoholic D.T., at least if he is having it for the first time, does not come to his physician and say: Doctor, I am delirious. He says, Doctor, I feel a horrible pricking on my skin, as if *ants* were crawling over it.

I might also say that in ten years of taking chloral, on and off, I have never had even a hint of such a reaction. But then *never* before did I suddenly stop taking a most massive dose. I cannot pretend that I did not *know* of the dangers of a sudden withdrawal. Indeed, I *did* have forebodings, which I immediately repressed. Let me quote physiologist Sherrington:† "To refrain from an act is as deliberate an

* In a letter to his parents around this time, OS wrote: "Had a very restless night, felt peculiarly tremulous and frightened in the morning, and started to have bizarre perceptual changes in the afternoon—buildings tilted and quadrangular, people looked insect-like with sudden nystagmoid jerkings of the eyes. Ghastly! I thought, not unnaturally, that I was going crazy. Then I thought of the chloral."

† Charles Sherrington, a physiologist and Nobel laureate also known for his writings. In his book *Man on His Nature* (1940), he writes that the brain is "an enchanted loom

act as to commit one." The whole business was undoubtedly *deliberated* at some level of my being.

Why? I cannot give a total answer. This is a matter for an accomplished analyst. One can never be conscious of *all* one's hidden motives, for this very reason: that one hides them *from oneself.* [. . .]

Anyhow . . . the huge explosion, and this huge letter, have acted as CATHARSIS, have cleared much of the fear and doubt which was clouding my mind.

I feel, now, cool, clear-headed: less dependent, yet more patient: certain of our love for one another: and of the happiness of the future. Write to me when you feel the wish. I shall no longer *suck* letters from you.

For the moment, the ecstasies and the agonies are *put away.* I will endeavour to live *calmly*—a cool neutral life, the colour of this unobtrusive, yet very pleasant, dove-grey suit I wear this morning. UNTIL YOU COME. AND THEN WE STRIKE FOR THE HORIZON TOGETHER.

Thank you for reading this monstrous letter!

DESPITE VINCZE'S ASSURANCES, OS had been thrust into what he characterized in one missive to his lover as a "violent oscillation between absolute certainty and absolute doubt."

To Jenö Vincze

DECEMBER 9, 1965

[ALBERT EINSTEIN COLLEGE OF MEDICINE, BRONX, N.Y.]

Dear Jenö:

This will be an impossible letter to write, but write I must. [. . .]

Jenö: what shall I say to you? I wrote you a letter about a week ago, hesitated to post it, discussed it both with Mark* and with a

where millions of flashing shuttles weave a dissolving pattern, always a meaningful pattern though never an abiding one; a shifting harmony of subpatterns."

* Mark Anstendig, a friend of Jenö's.

(female) intimate of mine, and then filed it away as testament to a reaction too brutal to transmit. Masquerading as a detached, clinical, analysis, it was in effect a devastating, a murderous, attack on you—denying your dignity, your manhood, your sanity, and finally your very identity as a human being. There were elements of truth, grossly travestied, in the substance of what I said: but the way I said it was inadmissible. No human being can turn on another like this, without a total moral sabotage. I hope I can be more reasoned in this letter, and more charitable. I must not be enslaved by my feelings, which are currently RAGE and PANIC.

I am up against something in myself which defies conscious control. I control it one way, it emerges another. You saw in Paris, with our poor old harmless Hoteliere, a minuscule attack of the same rage-panic reaction—and she, poor thing, had done little enough to precipitate it. In the past month I have been turning on people right and left. I shun them, I flee from them, but I carry a machine-gun in my flight. It is *this* of which you are frightened, this which your intuition warned you of.

I find the pleading, entreating, tone of your last letter intolerable, even less tolerable than the jealousy which preceded it. And yet you write with dignity and passion and warmth. I must tear you out of my system, *because I dare not be involved.* You are not the first, you will not be the last. I long for contact, but when it becomes too pressing, when it threatens my "independence" I am filled with rage and terror. I really cannot imagine life *with* anybody. We differ profoundly on this score, and the difference is basic and irrevocable. I have always lived in a one-man world; I fear I always will. Briefly, (oh, so briefly!), and to my own incredulous astonishment, I had a sense of sharing and togetherness. It has vanished, vanished so utterly that I can no longer conceive how it ever existed.

You do *not* need me. My strength is as illusory as yours. Our paths met for the briefest moment, and now they must diverge. I want no further contact with you, for like you, I am all-or-none: an attenuated relationship is worse than no relationship. So too would be one rent by distrust, jealousy and suspicion. I do not know what the word "love" means. Perhaps I loved you, but I no longer love you. I have only a sense of panic and revulsion, a categorical necessity to *get away.*

It is not exclusively to *you* that this reaction is directed. It is to anything and anybody with which I fear to be *involved.* The point of parting between us was that fatal little paragraph at the end of your letter—that irrelevant, insane, paragraph which started: "Does this mean divorce?" You fluctuate between grandiosity (I will be president of the US in fifteen years: I can walk through the traffic without danger) and paranoia (people are turning on me). Your mind is black and white. (Are you *for* me or *against* me?) I catered to you, to your phantasies: and you catered to mine. We were locked in a classical folie-a-deux. (We will revise the world. I will get a Nobel prize, you will be a world-famous director. I am a genius, you are a genius.) All this CRAP.

And now the cold light of reason seeps in. The play is over, the lights are up. [. . .]

I am torn in two, and must yield to my fear. Our menage really never had a chance. Perhaps I came so close to you knowing that reality would not permit it for very long; I yielded to a dream, knowing I would wake. I have woken. The dream was beautiful and terrible. It was everything while it lasted. I have unappeasable regret. But I am awake. The alarm clock is ringing. The world is waiting. I must shave, I must dress. The dream is forgotten.

Jenö: we no longer have any point of contact, and there is no point *in* contact. This is my final and farewell letter to you. I wish you what you wish yourself; I implore you to do nothing rash, and to make no reply: Mark, by the way, is a good and true friend to you: he did everything he could to hold us together, but I can no more be held than a mad rhinoceros.

adieu,
Oliver

4

Analysis

1966–1968

AUGUSTA BONNARD, OS'S PSYCHIATRIST FRIEND, had returned home to London but continued to see OS during his visits there. In a letter of December 13, 1965, she analyzed his continuing denial of drug use and proposed that his very fear of psychosis drove him to seek artificial, drug-induced experiences of it. He had not gotten very far in his analysis with Seymour Bird in Los Angeles, and now she suggested that he find an analyst in New York to help him, as she put it, "be the 'mensch' that lurks within you." He must have agreed with her, and realized that his out-of-control amphetamine use, at least, would soon kill him. On New Year's Eve, as he relates in *On the Move,* he resolved to get help.

Soon thereafter, he began seeing a new psychiatrist, Leonard Shengold. Luckily, perhaps for both of them, it turned out to be an ideal match: one of Shengold's special interests was early trauma and parental abandonment.* And OS, having been "exiled" from his parents at the age of six, sent to a boarding school in the countryside, had been deeply scarred by what he saw as his own abandonment.

* In later years, OS credited Dr. Shengold with saving his life many times over; in 1985 he would dedicate his fourth book, *The Man Who Mistook His Wife for a Hat,* to him. He continued his therapy with Shengold twice weekly for forty-nine years, until the final year of his life.

To Elsie and Samuel Sacks

FEBRUARY 11, 1966

ALBERT EINSTEIN COLLEGE OF MEDICINE, BRONX, NY

Dear Ma + Pa,

This will be a short note but better than nothing.

I am sorry to have caused you anxiety, and fear that my telegram—selfishly—may have upset you further. [. . .]

Perhaps the less said the better—after brief elation, I became rather morbidly depressed, felt estranged from everyone, hopeless, worthless, empty, at the end of the line. All the conventional melancholic symptoms.

Fortunately I think I am surfacing now, partly with some psychotherapy, and significantly from a realization (persistently denied) that poring down a microscope drives me nuts, and I need some return to *clinical* work. I talked this over in preliminary terms, with people at hospital today, and will probably move into a clinical setting in the summer (I can hardly imagine summer in these sub-zero days). I am not in a position to judge further ahead. I really have no sort of life in New York at present, but *as I am,* would have no sort of life anywhere. It is a question of changing or perishing.

Yr parcels are welcome, and form a slender tie across the ocean. I must send you my best + belated wishes for both yr birthdays and the New Year. And don't worry, my capacity for survival is remarkably strong.

PS Very happy to hear Michael has a job. This is excellent + encouraging news.

A WEEK LATER, OS DRAFTED the following letter to his Auntie Len, though it is not clear whether this letter was sent.

To Helena Landau

FEBRUARY 18, 1966

ALBERT EINSTEIN COLLEGE OF MEDICINE, BRONX, NY

My dear Auntie Len:

I hope this letter finds you in good health and spirits, and in your usual state of enormous activity.

I am sorry for not having written to you since leaving London.* This is not a specific inattention to you, but part of a general failure of communication with everybody. For one reason and another I nose-dived into a very intense, quasi-suicidal depression, from which I am now (I hope, definitively) emerging. I can't "blame" New York for this, because this is really a marvellous city, rich, exciting, unlimited in range and depth—as London is, although the two cities are profoundly different. New York is punctate, scintillating, the way all cities look from a plane at night: it is a *mosaic* of qualities and people and dates and styles, a sort of enormous urban jig-saw. Whereas London has much the quality of an *evolved* city, the present like a transparency overlying the wafers of the past, layer on layer, extended in time, like Schliemann's Troy, or the crust of the earth. But then again, for all its sparkling synthetic quality, New York is strangely old-fashioned, archaic. The huge girders of the "El" are a railway-phantasy of the 1880's, the crayfish tail of the Chrysler building a pure Edwardian vanity. I can't see the Empire State Building without the vast silhouette of King Kong scrambling up its sides. The East Bronx is like Whitechapel in the early Twenties (before the diaspora to Golders Green).

No, I can't blame New York, although the weather (which, amongst other things, makes use of my beloved bike virtually impossible), the strike,† the horrid paper-thin quadrangular cell of an apartment I got myself into etc. have added their quota. Nor can I blame Einstein, which for its combination of creative enthusiasm and human tolerance is as fine a place as I am likely to find anywhere.

I had, to put it bluntly, no life whatever in New York. That is to

* I.e., the previous September.

† Employees of the New York City Transit Authority walked out for two weeks in January, immobilizing the city.

say, no friends, no one I could go to, speak to, trust, enjoy; my "interests" became stale and empty to me: hospital a burden daily harder to bear; harder still to dissimulate my feelings of boredom and despair. I couldn't eat, or sleep, saw no conceivable future, saw my whole past as worthless and hateful; I compounded a deteriorating situation by drug-dependency, and—but why upset you with this horrible catalogue? I found myself, in short, finding existence intolerable, seeing no point whatever in further existence. More than this, I found myself struggling with what was, in effect, an insistent, a peremptory inner demand for acts of self-destruction, culminating in self-murder. I was in the grip of what one might term a *lethal* neurosis.

And though it was only in 1966 that I came finally to breaking point, and stared disaster in the face; yet I could see the same pattern of behaviour—bewildered, enraged, self-destructive—going back ten, fifteen years, back perhaps into my childhood: although this, strangely enough, was blotted out from my memory, by a dense sprawling amnesia covering the first twelve years of my life.

I went to seek some psychiatric help around the beginning of the year, for it was clear to me that I had become acutely disturbed, and would not survive (physically) unless some successful intervention could be made. [. . .]

Let me introduce this most painful idea which has arisen in the course of intensive psychotherapy, and at a time, as I have said, when I feel myself in extremis. Time and time again, I would demand of myself: What has gone wrong? Here I am, richly endowed with physical and intellectual gifts, with fine parents I must be proud of, and every opportunity etc. etc. What has happened to me en route so that I am slipping down the greased path of withdrawal, discontent, inability to make friends, inability to have sex, etc. etc. towards suicide in a New York apartment at the age of 32? What is the nature and reason for a neurosis of such dangerous proportions? And has it anything to do with Michael's schizophrenia?

There are many bizarre, schizophrenia-mimicking aspects to my own behavior: in particular, repeated association and macabre entanglements with schizophrenics, and second, repeated self-dosage with LSD and similar drugs which induce a temporary chemical psychosis. A million times I had to ask myself: WHAT IN GOD'S NAME IS REALLY GOING ON? And the final, desperate, question: WHO AM

I? What sort of person am I? Under my glibness and my postures and my facades, my imitations, what is the real Oliver like? And, is there a real Oliver? [. . .]

I have always felt transparent, without substance, a ghost, a transient, homeless, or outcast. [. . .]

A further hypothesis suggests itself to me, and it is in relation to this that I most seek your opinion. It seems to me that I do have a few, scattered, happy memories of early childhood, all before the age of 5. Of tea in the garden, and Kew Gardens, and the tops of buses, and maternal hugs, and Ma coming up when I said my evening prayer in bed and kissing me goodnight; a general vague memory of family warmth, and ease, and affection, somehow agglomerated into a sort of pure feeling. *Summer 1938, the last time we were all happy together.* At other times, I wonder if this is a sort of incorporated memory or feelings drawn from the family, rather than one provided by my own memory. Perhaps the distinction is not too important. Summer 1938. With Marcus and David wrestling on the back lawn, and Michael—already studious, quiet—reading a book in the drawing room and pudgy Ollie, the afterthought, tugging a hollyhock and laughing, and . . . And then. BANG CRASH. War. Raids. Blackouts. Lawn torn up for allotments. Big coloured window blown out of the hall. Family disrupted. Marcus to Leeds, David [to] Lancaster, Michael and Oliver [to] Braefield. And when it was all over—they couldn't put Humpty Dumpty together again. Things were never the same with the Sacks Family. Mark went off, David married, Mike went crazy. A sort of ugliness, a discomfort, fell upon the big house.

But I ask myself: Was it the War? Was this the natural, inevitable, evolution of the family? Or did some change come over Ma, the insidious onset of her paranoia, probably between 1938 and 1940, an emergence of symptoms doubtless exacerbated by the War and the stresses it imposed? I know there was general evacuation etc. but I must wonder whether it was really necessary, whether it was the right thing to do, to dispatch us all in different directions?* And I have, not even half-memories, but hints—I wonder whether her real affection

* Parents in London were exhorted to send their children away from the city to avoid German bombings. As physicians themselves, Elsie and Sam worked especially long hours, but they must have agonized nonetheless about sending their two youngest boys away. OS later wrote about his almost daily beatings at the hands of a sadistic

for us, and for me in particular, dwindled and died sometime in those two years, to be replaced by the morbid, demanding, yet frigid possessiveness of paranoia.*

To Marcus Sacks

MARCH 6, 1966

ALBERT EINSTEIN COLLEGE OF MEDICINE, BRONX, NY

My dear Marcus:

You are the only brother, the only member of the family, with whom I can talk with any freedom. We have never seen too much of one another, but I am conscious, as you must be, of our similarity in many ways. This is hardly surprising, for we have been moulded by the same forces, and our reactions have been similar, though very different in degree. This is going to be a "shocker" of a letter; I can't judge how well you will be able to stomach it; and I must leave it to your discretion as to whether you wish to show or discuss its contents with Gay; under no circumstance do I want you to divulge what I shall say to our darling, damnable, parents.†

I will try and give you some of the conventional "news" at the end, although, so far as I am concerned, this is subordinate, ghostly, compared to my inner "news."

Where shall I begin? I will begin with the first information available to me, chronologically: doubtless things go back further, but I am in no position to trace them. Marcus Landau, then, our maternal grandfather: a brilliant man, of overwhelming personality, who sired eighteen children from two passive cow-like women. Perhaps I am wrong about the women, but this is the impression I gain from

headmaster and the meager, unpalatable rations of wartime. He found solace, as much as he could, in reading and his passion for prime numbers.

* The existing copy of this letter ends here. It is unclear whether it was ever finished or sent, or if a page is missing.

† Indeed, this letter shows OS in an unusually condescending and even vicious mood. It is likely he never sent it, and possible that he was high on amphetamines when he wrote it.

Ma (a profoundly biased source, of course): of admiration amounting almost to reverence for her father, and little mention of her mother. [. . .]

And so, by a "commodious vicus of recirculation" (my mind is a sort of Finnegans Wake), to our own Mapesbury menage. Ma, of course, consists of two people, a normal and grossly pathological self. "Basically" simple, almost to ingenuousness, kindly, and very warm-hearted. This is what her patients see, and the rare strangers who ever come to Mapesbury: a charming hostess, though very shy. And they say: your mother is a marvellous woman—and so she is when she does not act out her appalling conflicts. Now there is the other side: the woman who is profoundly antisocial and desperately lonely, who alternates between a frigid-sadistic-obsessive hatred of her family and an overwhelming possessive *engulfing* love for them. A woman who is prone to explosive *rage,* outbursts of a profoundly irrational and destructive kind, and who is also prone to extravagant depressions, at least one of which, to my knowledge, was of plainly psychotic intensity. She is always bewildered, frightened and remorseful after her rages, when she is "herself" once again. She is a woman both overwhelmed and terrified by sexuality, so much so that she has made of it her profession: sexuality becoming a matter of rubber gloves, surgical instruments, and clinical detachment. Her medical stories (Jonathan Miller gives a beautiful imitation of how she always talks of food while she is operating, and "cases" at mealtimes: soup with the pus, pus with the soup) are a sort of repetition compulsion, a sort of exorcism or disguised return to a taboo but endlessly alluring subject. On the subject of homosexuality in particular, she has a fanatical horror, and I would surmise has much of this deeply repressed in her own make-up.*

But put her at home, in the garden, or reading a Somerset Maugham story: making gefilte fish, or presiding over the family on a Friday evening; give her what she needs and she becomes, once again, her simple and lovable self. This is our mother, or today's formulation of

* OS described in *On the Move* how, when he was eighteen and admitted his attraction for other men, his mother blew up, saying, "You are an abomination! I wish you had never been born."

her (my ideas are in a state of violent day-to-day, second-to-second, flux).

Now, our father. Look at that photo of him when he was seventeen: look at the demanding yet over-protective attitude of his sisters to him (Lina especially), and one realizes that he must always have been rather bewildered, rather helpless, rather passive. There are many reasons for supposing that he was rather a gifted young man: he qualified at an unusually early age, he was Henry Head's houseman,* he was deeply interested in Ibsen, at a time when Ibsen must have been fairly esoteric. Then what happened to him? He swum into the centripetal whirlpool of our mother, and got sucked right in. When I was feeling somewhat overcharged a few days ago, I wrote in my journal: he was engulfed, leucotomized, murdered. What I would suspect to have been a very promising professional career was subordinated to the need to run second place to our mother. She was to be the brilliant woman, the well-known surgeon, he to be the dim, bumbling, general practitioner. There is much evidence to show that he was most exceptional, even *as* a G.P. I constantly hear stories of his excellent diagnostic acumen, and his versatility: it was not for nothing that he rapidly became one of the most prominent G.P.'s in the East End: the "King of Whitechapel." But superimposed on this has been our mother's contempt for him, *incorporated* in ourselves as the notion that Pa is really dim, a nonentity, or appendage of his brilliant wife. There is reason to suppose, as I have mentioned, that Pa was literate and versatile as a young man, but he has not opened a book (non-medical, or non-Hebraic) for forty years. One way and another he has tried to break away: with the childlike pleasures of his motorboat and motorbike, at one time; with the inordinate time he spends with his patients, an endeavour to get away from Ma; and, in the last ten years, increasing immersion in his Hebrew books, often till four or five in the morning. He has tried to be a good father to us all, and many of my happiest memories are of going with him on his calls, when I was a kid; of swims together; of playing duets together; and occasional, terribly rare, talks we have had together. He has tried,

* Henry Head was a distinguished neurologist under whom Sam Sacks did his internship (that is, was his "houseman").

and he has been defeated. His life has been geared to failure, to being overwhelmed. His pathology is diametrically opposite to Ma's: where she is demanding-obsessive, he is masochistic-conciliatory. He has been decapitated, castrated, by her, cut down to a bumbling old fool. And yet, how they have stuck together, how they have needed one another. They have formed what is almost a folie-a-deux, a symbiotic yet destructive relationship.

Now what about us, the four Sacks Boys, those paragons of normality: what has happened, what is happening, what will happen, to us; and to our children, if we have any? What part, conscious and unconscious, will *we* play in the neurotic-psychotic transactions which have been transmitted and incorporated from one generation to the next?

This is primarily why I am writing to you, for I am no longer able to keep my agonized introspection to myself. And though I tend towards hyperbole and a variety of other distortions, you will recognize what I saw as being essentially true, and have no doubt formulated this, to a greater or lesser degree, yourself.

You then, the eldest of the four. You are of very superior intelligence, brighter than either of our parents. You were a Hebraic-Oriental-scholar-in-the-making during your schooldays, but "owing to the War," as it was rationalized, took to Medicine instead. You did this, no doubt, with mixed feelings, although the degree of conflict may not have been, may not be, evident to you, being covered by denial and rationalization. Certainly, however, you would not have responded to the external pressure to enter medicine (parental pressures etc., family tradition), had it not been matched by an *internalized* pressure, an order almost, to the same effect. I know little of your student days, days as a young doctor, in the Army etc. Around 1950 you felt you *had* to leave, and not only to leave, but to go as far away as possible. And here again, you rationalized (as I did, in *my* flight, ten years later) that professional prospects were better in Australia, that you wanted to see the world etc. etc. No doubt these factors were *also* present, but the primary motive was flight, flight from a situation you obscurely felt as intolerable, as dangerous, as engulfing. [. . .]

I have just re-read your pathetic letter of a few years back, when you confess your inability to sustain relationships, your compulsion to break them off when they become too intimate, too threatening: I recollect also your self-depreciatory images as a little Jewish gefullte fish etc. You were deeply neurotic, deeply torn by internal conflicts: but functioning pretty steadily, and on the whole, with the aid of your intelligence and ego-strength, tending towards some tolerable modus vivendi, although a significantly crippled one in relation to your potentials. You mentioned that psychotherapy was of considerable assistance to you, and it is my dearest hope that you have met in Gay someone who will help restore your self-esteem, give you some much clearer image of yourself, and enable you to combat the near-lethal "inheritance," you, we, have received from our family situation. I think it vital that you do not underestimate the degree of actual and potential disorder which is present in you, and I feel *it is of the utmost importance* that you enter, or continue, intensive analysis with an absolutely first-class man. [. . .]

I know this is an impudence for me to speak like this. But you may rest assured that marriage *alone* is not enough. It is vital for you to get at the roots of what is happening, otherwise you will be condemned to *recapitulate* certain situations over and over again; only by bringing the nature of these things to consciousness can one deflect the helpless and bewildered compulsion to act out, and repeat, the primal emotional confusion which has dominated our lives.

David now, our "charming" brother, who dresses like an ambassador, and plays the heavy father with some conviction; who specializes, like some Leacock character, in diseases of the rich; who is ambitious, vain, frigid, and superficial—although there is genuine warmth, genuine kindness, somewhere underneath all this. David [. . .] has adopted a radically different "solution" (pseudo-solution) to the conflicts which he too, undoubtedly, contains. [. . .] Deep emotional conflict has been repressed, and replaced by a frightening *blandness*, an ingratiating, superficial, condescending, somewhat contemptuous, attitude—a terrible poverty or thinness of emotion. Superficially he functions extremely well, to the envy of us all. [. . .]

Now Michael: totally friendless and absorbed in phantasy from his earliest childhood, Michael became floridly deluded and psychotic at sixteen or seventeen. What is more serious is that he has never got really better. In fact, this isn't in the least surprising, for shooting someone full of insulin or Thorazine, and then throwing them back into the unchanged environment which precipitated a psychotic reaction in the first place, is bound to have this effect. Michael continues psychotic because this is the only tolerable modus vivendi for him, and at some level, he is perfectly aware of this: he showed, last summer when I saw him, an amazing perspicacity, but one which was quite dissociated from his actual and continuing condition. Ma and Pa, but especially Ma, obscurely and unconsciously aware of their responsibility in his disease, display the most amazing denying obtuseness about it all. I remember vividly that ghastly Yom Kippur in 1959, when I came home from Birmingham, and was told that Michael was being "naughty" again. I went to see him, and the poor bugger was wildly deluded and hallucinating. Ma and Pa refused to accept what I said, and added that it was Yom Kippur in any case and no time to make any move. That very night, as you know, Michael ran out down Maida Vale, in his pajamas, full of Messianic delusions, was hauled in by the police—eugh, a terrifying business! Why do they (Ma, in particular) keep him at home, and why does he stay at home? By keeping him at home, by his staying psychotic, Ma has, in effect, a *captive son,* an appendage, a parasite, bound hand and foot, who has a snowflake in hell's chance of ever getting away. We, at least, tried physical flight: Australia and California. David has taken refuge in the haven of denial, psychopathy. Michael, in whom the conflict was most blatant, least tolerable, was driven to delusion-formation, replacement of intolerable reality by substitute constructions (hallucinations are secondary to delusions, although they reinforce them by a positive feedback). You remarked on the intense, the desperate way, our parents put aside money for Michael. The reason is not mere altruism or solicitude, it is the intense though unformulated *guilt* of their part, their complicity, in the creation of his disease: the money they put aside is blood-money, a sort of penance. It is true that when they die Michael will not survive: he will either become permanently and profoundly catatonic, or he will commit suicide, probably the former.* Let me add that the intense conviction of guilt is quite

* In the event, he did neither. After their parents died, Michael lived in a group house in the Mapesbury neighborhood, where OS continued to visit until Michael's death in 2007.

unrelated to any actual *blame* in the matter: it is really a delusion of guilt, which characteristically the parents suffer for the children's psychoses, and the children for the parents' psychoses. [. . .]

yrs,
OLIVER

To Seymour Bird

PSYCHIATRIST

MARCH 12, 1966

ALBERT EINSTEIN COLLEGE OF MEDICINE, BRONX, NY

Dear Dr. Bird:

I wonder if you would be kind enough to send me, for income-tax purposes, a receipt for money spent on psychotherapy between January and June of last year.

I have often wanted to write to you, but refrained, characteristically, because of my inability to steer a middle course between a "nothing" letter and an exorbitant one. I live, as always, in the chaos of my self-created crises, but am undoubtedly moving in the direction of greater insight, diminished self-destructiveness and knowledge and control of reality. After a series of quasi-suicidal crises on returning from London, massively regressive reactions precipitated by the disruption of an extravagantly symbiotic and destructive "love affair" contracted in Europe, I went to see Austin Silber, who in turn referred me to his colleague Lionel [*sic*] Shengold. In the past two months with him we have got far nearer the volcanic heart of my problems than I could ever have consciously conceived before. This has involved the re-animation of the most atrocious, and frankly murderous, childhood memories, mostly associated with my mother, who of course "incorporated" almost as lethal a neurosis/psychosis as mine. But I venture to think that I am getting over the hump, so to speak, and will avoid, though by the narrowest margin, the "demanded" psychosis/suicide equivalent so rampant in the family, and make it as a human being. At the moment I am relating to everyone round me, friends and colleagues, with the incontinent, remorseless, violence I have always inflicted on myself, but some assurance and control are beginning to show through.

I shall never forget the debt I owe you for puncturing the first crust of my denials and defences (intensely ambivalent, like all feelings of indebtedness!), and making possible the sort of Palaeozoic eruption which is going on at present.

Warmest regards,
Oliver W. Sacks

To Elsie and Samuel Sacks

MARCH 14, 1966
ALBERT EINSTEIN COLLEGE OF MEDICINE, BRONX, NY

Dear Ma and Pa,

Thank you for your letters. I am sorry to have generated so much anxiety, but I am now, one way and another, getting "over the hump," and will once more write fully and frequently. [. . .]

I had a call from an nth. cousin, a Mrs. Gardner,* and will be going to supper with them in a few days' time. Pesach maybe. I get nostalgic on the Seder nights. They embody a feeling of warmth, of oneness, I have almost forgotten. Jonathan Miller was here for a brief meteoric visit a month ago. He will be returning soon for a couple of months, casting and directing a play. I look forward immensely to his company. In some ways he has been my only friend. The rest seem like wraiths. I had a couple of well-meaning but utterly bizarre letters from Augusta Bonnard. I fear she is cortically hypoxic from emphysema. Incidentally, Ma: can you obtain and send me that enormous family tree drawn up by an industrious cousin in Jerusalem? I have become suddenly interested in genealogy and sociology and politics. Delayed social awakening: should have happened thirty years ago.

* Madeline Capp Gardner. OS would become close to the Gardner family, joining their holiday and family celebrations for decades.

Briefly, and sparing you gruesome details, my chronic state of crisis became acute for a while here, a sort of epilepsy of indecision. It was really quite impossible to communicate with anyone or do any work. I even lost forty pounds and became positively skeletal. Fortunately I have secured myself an excellent analyst, and we are really getting to grips with things. The last two weeks have witnessed an exhilarating restitution. I have become suddenly busy at hospital, and am simultaneously in the middle of four projects. In fact thoroughly manic, but this is better than depression. I seem to be writing and talking and arguing in an incontinently volcanic logorrhoea.

I have actually written you a great many letters. Ditto also to Auntie Len. To everyone I know. They litter my room. I never try and post them, because I feel I will regret it. This is what I mean by an epilepsy of indecision. But now I am writing, you may surmise that my activity is becoming less chaotic, and more controlled.

It is Spring, almost Spring. Tiny green leaves in Washington Square. Ice floes cracking on the Hudson, the river stretching like a man awakening. Warmth, sun, hope, germination. Finished the winter of discontent.

Work. Clearly I should never have chosen neuropathology. Sheer perverseness, chronic and characteristic. I have such a hatred of anatomy. Next year I will be a junior Consultant on the Neurology wards, and continuing research in neurochemistry which I enjoy. Takes me back to happy chemistry days, eighteen years ago. Eighteen years! It seems like a second. No sense of time. The quality of a dream. I feel my bald spot and mutter, thirty-two. How little it seems to register. Past and present flow together in my mind.

Ma, before I forget, I lost my driving licence in the washer. Could you conceivably get a duplicate for me from Middlesex County Council and forward it airmail. [. . .] It is inordinately complicated getting a New York licence. Simpler to stay with my English one, as permitted for a year. [. . .]

I am exceedingly happy to hear that Michael has a job. He was astonishingly lucid, painfully lucid, about a lot of things when we talked together in the summer. I think he would be capable of a degree of restitution far greater than you imagine. His illness is nothing to do with "chemistry"; rather any chemical changes are secondary to the psychic ones, and these are at least partially and potentially reversible. He needs intensive treatment with a first-class analyst familiar with his sort of problem.

As I do, for my analogous though milder problems. I may have to borrow some money from you, for I am spending most of my salary keeping sane. But I will be more detailed in a following letter. I am sufficiently glad to have written *this* one, which I shall undoubtedly post.

Love, happy Pesach to you all,
OLIVER

LATE IN 1966, OS BEGAN a number of new part-time jobs in the Bronx: working in a headache clinic at Montefiore Hospital and seeing patients at Beth Abraham Hospital as well as Bronx Psychiatric Center. He also taught a group of AECOM medical students, doing rounds with them at Beth Abraham, often followed by a discussion in the nearby New York Botanical Garden.

To Mel Erpelding

FRIEND

[CA. DECEMBER 1966]
165 CHRISTOPHER ST., NEW YORK

Dear Mel,

I started a reply the day I received your letter, and then another, and another, and got furious with myself, and confused, and never did get to the postbox. You know the syndrome. They call it: CONFLICT! [. . .] How I miss the mountains and deserts of the West now I've left them! And New York is strictly murderous for a two-wheeler: the roads cobbled, shattered, cratered, greasy, four months of squall and slush and ice in the winter, and the traffic unspeakably violent and deadly. I confess I have had my fill of it, and am getting a car—a Volvo probably—because I don't think I will survive another winter on the bike here. Not that I'll give it up; bikes will always remain my dearest symbol of some sort of freedom I sometimes feel but can never formulate—but I'll keep it for weekends, for decent weather, for the country, for friends, and give up the nightmarish, daily 25-mile ride to Hospital. [. . .]

I thought at first I could survive on a two-day work week, and hopefully write and write for the rest. It worked—for a month: during October I wrote over 200 pages of a Beach novel. It started as loose, easy, Lardnerish* sketches, and then became more involved, more introspective, more contrapuntal, until I got suffocated in the labyrinth of my own indecisions: emotional indecisions, that is, as to where *I*—the liver, the writer—stood in relation to my "characters," my friends. I had in mind something analogous to *Ulysses* in a way, the shifting web of the Beach, its dozens and hundreds moving in and out of each other's lives, its timelessness, its rootlessness, its restlessness, against the groundtheme of the Ocean. All the action to be kaleidoscoped into a single day, given extension through the memories and phantasies of the characters, but utterly superficial, like the froth on the Ocean. The whole to take place between two Tides, and the movement of the sea to be the movement of the book: starting with a sort of birth, the adolescents, as I would see them every morning, in the grey dawn, rocking and rocking on their surfboards, rocking in the lap of the Sea (*Thalatta, thalatta,* our great green mother!), and ending—but I never even approached my ending: in the phosphorescent waters of a summer night, swimming far, far out, two of my characters (you guess which), recovering at the end of the long, frantic, restless day the sense of reality, of the strength and the calm of the Ocean, and at the moment of their knowledge, being drowned, and sinking into the green and changeless abyss. Me and my bloody Death-ecstasy obsession. Anyhow, I've destroyed the whole fucking thing. And on destroying it, suddenly found my hands empty, and the sense of panic and emptiness all round me again, and the sense of loss, and of having destroyed what meant so much to me.

And now? I am, for the moment, less saturated in the Past, less committed to it. My clinic, my patients, give me a sort of satisfaction I haven't had for a long time, and I now have teaching appointments

* Ring Lardner (1885–1933) was known for his often satirical short stories and gift for dialogue.

both at Einstein and a smaller, satellite hospital.* With patients and with students I feel, at least to some consistent degree, at ease, which I could never do in "research"; for research, last year, and before that eleven years ago,† has always crystallized into a malignant love-hate struggle with my boss, alternately teasing and masochistic, and utterly destructive. I have some sort of life at the moment, and, starting Jan 1, will be earning about $20,000 a year, without any overwork. I will be moving into a fairly decent apartment, instead of the hateful tenement I have condemned myself to for the last year, and—maybe, maybe—there is the possibility of a future. I have a horror of "success," as defined in Manhattan professional terms, seeing it as a constriction, a mutilation, a sort of leucotomy. But these are *my* fears, not the fears of anyone who knows me.

And I must have a piano, which I have desperately missed in the last year. I have just been looking at a beautiful interview with the 90-year-old Casals, who has a little house by the sea in Puerto Rico. Every morning he walks on the beach, and then sits down at the piano, as he has done for eighty years, and plays from Bach's Preludes and Fugues. "Bach and Nature," he says, "are two approaches to the same thing: a wonderful way to greet a new day. The amazing thing about the Preludes and Fugues is that even though I have known every note of it all my life, listening and playing it *every single day,* there is always something left to discover in it."‡ Isn't that an incredible, an utterly beautiful thing for a man of ninety to say? I am reminded of the great old man in Kazantzakis.§

And a Hi-Fi, so I can listen every day to Casals, and Bach, and supplement my wretched inability to play decently. And a darkroom, because I have terribly missed the happiness of creating pictures, of saving thoughts and feelings, which I can sometimes do as a photographer. It is less easy to lie in pictures than in words, and New York—so ugly, so monstrous, so wicked, so cruel—is full of sudden

* Beth Abraham, a hospital for chronically ill patients.

† OS is referring here again to his postgraduate studies at Oxford's Laboratory of Human Nutrition and a failed experiment on jake paralysis.

‡ OS related this story about Pablo Casals (then the most famous cellist in the world) again nearly forty years later in *Musicophilia,* apparently from memory.

§ Nikos Kazantzakis, author of *Zorba the Greek* (1946) and *The Last Temptation of Christ* (1955).

and fantastic beauties, like jewels bedded in a dung-heap. One of the jewels—I hope we can spend some afternoons there together—is the Natural History Museum, where some unknown and anonymous genius has created marvellous dioramas of America and Africa, caverns, sunsets, ice cliffs, deserts: the mostest in vicarious travel. There is one diorama of that huge bony tumulus in Wyoming, which I saw and partly climbed in '61. It is pure reality; I think I viewed it from that very spot, and it brings back that strange, tortured, wonderful ride round the country I had.

How we have travelled, Mel! Always fleeing, always seeking, always deceiving ourselves, never arriving. Anchored to the Past, dreaming of the Future, and—in some fatal, blind sense—oblivious to the Present.

Lists of regrets, manufacture of dreams! I say to myself: Why did I leave Santa Monica? Beaches, white foam, Topanga, friends, the exhilaration which saturated every day. Bah! But what madness to leave San Francisco! Aerial, hilly, New Jerusalem. Or why did I not stay in the wilds, in the backwoods of Canada, a lumberman and poet? Or was not the real spitefulness ever to leave London—my only, wondrous London—my home, and the home of my people? Or was the real and ultimate sadness to *grow up*, to leave the Magic Region of childhood, the time of wish-fulfilment and infinite power, the feeling of love and an endless future?

IDIOCY! It is all idiocy and vain regrets. Fatally easy to transfigure the past, to see in it millennia of epic happiness followed by cruel unmerited expulsions. It is the myth of Genesis all over again. In anticipation every move is an Exodus, and in retrospect every move is a Genesis. Did I not cross the ocean to the New World, the Vita Nuova, on the 27th birthday, the symbol of Rebirth? And was not California prejudged as El Dorado or the Fleece of Gold? And New York, in turn, as the Ultimate Metropolis—rich, brilliant, ablaze with promise—that same false promise which drew Dick Whittington* to London Town six hundred years ago?

And in you, Mel, for reasons beyond articulation, that feeling of

* Whittington, a poor country boy who came to London and eventually became its Lord Mayor, was, according to legend, attracted by the rumor that the big city's streets were paved with gold.

spiritual twinship (companion in hope and misery, co-Hero in the world of Promise, co-Victim in the world of Disillusion); that feeling which it seems to me I have always had, and shall never lose.

Slowly, terribly slowly, with pain and elation and backsliding and progress, we learn. I have been finding my analysis invaluable. Much of the grandiosity has been punctured, but something solid, something substantial and indestructible, something Real, is beginning to emerge. There is no Promised Land, but there is promise in myself. All the power I need, and all the power I need to destroy. That awful, predestined turning against self, the violent smashing of self and others, the perpetual sabotage of all one does—I am beginning to realize its demonic force, its roots in the past, and to grapple with my unseen enemy. Will we allow ourselves to survive?

The snow is falling past my window. I have been typing two hours in a sort of trance. I can't help myself, there is such a feeling of happy-miserable turmoil.

Please write me a long letter about yourself, and life in Florida, and scuba diving. I feel sure you did the right thing going there, and I have confidence that things will work out—not suddenly, or miraculously, but by degrees, and learning, and starts and stops.

And when will you (both)* come to New York, and see your old friend? Christmas, New Year. I am always waiting, but I wait without impatience, because there is a feeling of nearness which makes distance irrelevant. How I hated that feeling of estrangement last year! Or I am in most evenings, so give me a ring.

Time to stop. Time to send you [. . .] all my love and best wishes for the coming year, and to hope that we see plenty of one another.

Yours (for good and bad),
Oliver

ALTHOUGH OS WAS MAKING PROGRESS in his analysis with Dr. Shengold, he had not yet fully given up amphetamines. One February night in 1967, he came across a rare volume in AECOM's

* Mel and his girlfriend, Blossom.

library: Edward Liveing's nineteenth-century book *On Megrim,* which he read through in a single sitting. Enthralled by its rich narrative style, and in what he later described as "an amphetamine induced epiphany,"[*] he resolved to write his own book on migraine, based on the patients he was seeing at the Montefiore clinic. This new role as a physician, listening to and sometimes helping patients, teaching students, and, especially, combining this with the creative engagement of writing, changed his life. In *Hallucinations,* he would write, "The joy I got from doing this was real—infinitely more substantial than the vapid mania of amphetamines—and I never took amphetamines again." Freud emphasized the importance of work and love; OS, though he had given up on romantic love, had fallen in love with his work.

To Elsie and Samuel Sacks

JUNE 19, 1967
10 SHERIDAN SQUARE, NEW YORK[†]

Dear Ma and Pa,

[. . .] Like all of you at home, I have been hanging on the news of the Middle East Crisis.[‡] A fantastic *muddle,* almost inextricable labyrinth of grievances fancied and just, lies, equivocations, bravery, cowardice, glory, dishonour, hysteria—perhaps Israel is the only country that thought clearly and acted honourably. I found, to my intense surprise, a tremendous feeling for Israel welling up in me. I may have entertained phantasies about rendering medical aid to North Vietnam, clinical aid to Bertrand Russell, but I practically got a ticket to Jerusalem. More than thirty of my students went, bless them. I felt

* In *On the Move.*

† OS had recently moved away from the riverfront and into a sublet in the heart of the West Village, just a few buildings away from the Stonewall Inn, where, two years later, riots would mark the beginning of the modern gay civil rights movement. (By then, however, OS had moved to the Upper East Side, to be closer to his work in the Bronx.)

‡ Earlier in June, the 1967 Arab-Israeli War, also known as the Six-Day War, resulted in Israel's annexation of the West Bank, the Golan Heights, the Gaza Strip, and the Sinai Peninsula.

very proud of you, Pa, when David told me of *your* almost going there too in the Crisis. It is so strange that *violence can clarify*, and ugliness point the way to sense and virtue. I was assailed by long-buried (but not forgotten; there is no such thing as "forgetting") memories, feelings, suddenly remembering the 137th Psalm in its entirety, suddenly weeping at a picture of soldiers at the Wailing Wall. Even fingering the Hierusalemic knickknacks which are part of the "furnishings" of my new apartment.

The new apartment is absolutely delightful: high, airy, spacious, as much as my former one was troglodytic, box-like, beastly, penitential. [. . .] It is fully, indeed excessively, furnished by its present owner [. . .] in solid, Jewish, middle-class, style (I dare not say "taste"!), and contains, in addition to superabundant furniture (so how many chairs could she sit in?), quantities of Jewish mnemonia—terracotta models of Rachel's tomb, beaten Druze brass, and Graetz' *History of the Jews* in 127 abnormally heavy volumes.* It commands a splendid view in several directions, from the Statue of Liberty to the West (when I stand on the stove) to the Empire State Building from the bath. I am really extraordinarily happy, indeed hypomanic, since I moved in. So nice to have a place one isn't ashamed of; to entertain one's friends and colleagues; to sit at my broad desk composing the definitive monograph on "Migraine"; and to lie on my bed with the windows open, bathed in the sulphurous vapors of New York. [. . .] The sense of *home*—albeit a temporary one—is very congenial, very necessary, and something I forewent entirely in my disastrous first year in New York.

The mention of "home" brings me to the subject of Michael. Indeed it has been partly sympathy if not identification with his predicament, as well as not knowing what to say, and knowing that whatever I might say would be attended with guilty and angry reactions all round, that has delayed my writing to you.

In a matter as confused and as charged as this, there are few certainties and fewer "right things to do."

I think I would like you to show this letter to Michael, partly

* Here OS exaggerates: Graetz's history, eleven volumes in the original German, was published in six volumes in English. But the number 127 would have pleased him, as it is a prime number.

because I have difficulty in writing to him directly, as he would have difficulty in replying to such a letter. Indeed, this failure of communication has been central to his problem. It *is* very difficult for Michael to speak to you, fully or meaningfully, and almost equally difficult for you to speak to him. There is, partially, and tragically, a *mutual incomprehension.*

His inability to express himself in a fashion (or for you to listen to him in a modulated fashion) made it necessary, I would guess, to "explode" the way he did. It was a sudden violent eruption of discontent and accusation which must have been boiling up in him (unknown to you, and quite probably unknown to himself) for a long time. Following his passionate accusations and threats, he felt panicked, annihilated, and fled the house. He walked and walked, guilt-ridden, like Orestes, till he collapsed. He then maintained he would *prefer* to be institutionalized rather than remain at home.

This was violently, unfairly and cruelly expressed; he was enraged, not fully in control; but you cannot call it "delirious," nor should you call it "chemical." He is, in a sense, entirely right. He apprehends the mutilation, the humiliation, of remaining "at home," and the fact that continuance of an intolerable situation can only fortify his "illness." He is caught in a web of tragic ambiguities; one part of him wants to remain at home, to be looked after, to be treated as an appendage, the other part is crying for liberation. Yet the "outside" is felt to be very dangerous; he fears he lacks the "savoir faire" etc. etc.

I don't think Michael is in any sense "burnt out," altho' he has coped with violent and conflicting feelings for many years, by obsessional, constrictive, ritualistic practices etc. His "calm" over the past few years has not represented a definitive equilibrium.

I am inclined to think that he *should* live away from home, *should* have a job, and *should* have first-class psychotherapy on a regular and moderately intensive basis. That he *is* (potentially) capable of supporting himself, of living independently, and of resolving at least some of his very major problems, and of achieving at least a measure of emotional clarity and fulfilment.

He has important assets which have not been in the least eroded by his "illness": he is intelligent, he is imaginative, he is painfully honest, and he has obvious capacities for giving and accepting affection.

I do not gainsay the use of tranquilizers etc.—they may be quite useful in cutting down acute agitation, insomnia etc—but I think it a major error to *label* him as "organic," "schizophrenic" etc. and leave it like this. It is an error, let me add (for I aim to clarify not to accuse) which has been fortified by 150 years of medical jargon and incomprehension and is enormously widespread. Certainly one can be "crazy"—but this is a way of reacting; Michael became violent because he had to make his point and I think he must be listened to.

Maybe I've said too much. Perhaps you will be in no mood for the rest of my "news."

The last 3 months, bye and large, have been busy, happy and fairly productive. I shall stay on at the Headache Unit, though with less work and more pay. [. . .] Migraine, seemingly a dull, trivial subject at first, has opened a door for me into an immensely complex and fascinating area inhabited by all the "functional" illnesses, and I believe that I have some ideas of first-class worth in this area.

I have been asked to stay on, and moreover "take over" Beth Abraham,* which I will do in conjunction with a considerable increase of student teaching. My fondness and flair for teaching has been appreciated, and expanded in the warm light of recognition, and the College will be placing a major portion of undergraduate Neurology tuition in my hands.

I have also enjoyed consultancy at the Bronx State Hospital† (an immense psychiatric (and epileptic) "back ward") and hope I can consolidate my position there.

I can look back at the last few months with thankfulness, for I was really a *beggar* back in October, timidly asking for part-time work, and I am now, to some extent, being recognized and even in a position to make demands. I must ascribe this largely, tho' not exclusively, to psychotherapy. My potential is exceptional, but so also have been a variety of neurotic and (transiently) psychotic predicaments. I don't feel, in any final sense, "out of the woods," but I have a measure of clarity and hope. [. . .]

I saw a good deal of David and Elizabeth‡ last week, and enjoyed their charms and company immensely. I was also struck by the

* That is, take over as the hospital's neurologist.

† Also known, more formally, as Bronx Psychiatric Center.

‡ OS's brother David and his daughter Elizabeth.

quantity of odd, violent, if not downright crazy feelings they both contained, as indeed we all contain, both of you, as well as myself, as Michael does. It takes 5 or 10 generations to make a "character," neurotic or otherwise, so there should be less blaming and naming, and more understanding.

Write soon,
Love,
OLIVER

PS Buy *The Divided Self* by R. D. Laing (Pelican), a remarkable and illuminating book.

IN LONDON THAT SUMMER, OS proceeded, as he recounts in *On the Move,* "to write a book on migraine over the course of a couple of weeks. It spilled out suddenly, without conscious planning." He took the manuscript to Faber & Faber, an English publisher, and they liked it. He returned to New York in September and hoped to complete his revisions swiftly. But as the months went by, those plans went awry.

To Elsie and Samuel Sacks

MARCH 24, 1968
234 EAST 78TH ST., NEW YORK*

Dear Ma and Pa,

It is, as you say, as I regret, more than four months since I have written to you, or, indeed, since I have written to anyone. It is not that I am lazy or have nothing to say, but that I have felt myself in an unpleasing situation and frame of mind (more or less, with ups and downs, since my return in September), and thought it best to keep my feelings to myself. [. . .]

It is very difficult for me to do any writing, any creative work,

* Early in 1968, OS moved to an apartment on Manhattan's Upper East Side. It was only a few blocks from Madeline Gardner's family. Her daughter remembers how OS would routinely drop in for an ample breakfast on his way to work.

coolly and consistently. I do it in sudden excited bursts, when conditions are right. They are rarely right, probably not more than 1 pc. of the time. They were right in the summer, in London, at home, in the garden, in what you (Pop) called an "exalted" frame of mind. It wasn't exalted, it was merely healthy, and urgent, a brief liberation from tensions which interfere with creative work. I made a major error in returning when I did; another two weeks and I would have had the revised MS complete, and in Faber & Faber's hands. As it was, my sense of dread and premonition regarding return to New York was entirely borne out by the course of events. Jonathan Miller warned me about this, and he was entirely correct in his ominous prognostications. My momentum carried me through one more chapter when I returned, and that was it. I ran into several sorts of impasse, in particular my boss Friedman,* who saw in the manuscript an originality he lacked and hated, and a promise of success and quality which threatened to throw his own countless (174) trashy, trivial publications on the subject into obscurity. Not that this would have been the case, but this was how it was formulated in his suspicious, envious, spiteful and ungenerous mind. He adopted an attitude of frank paranoia—"Sacks" (he used to call me Oliver before), "you've jumped the gun. You can't do this. It is umm, presumptuous. I refuse to let you use any clinical material from the Headache Unit," etc. etc. [. . . I] chose to ignore everything I had heard about him from everyone who had ever worked with him. I had been told, over and over again, that he demanded a passive, mindless, servility from anyone who worked in the HU, and immediately turned against them. [I tried to] "bring him round," to argue with him, to revise the MS extensively, to compromise, to accommodate etc. It was entirely futile. I even went so far as to write a further paper (on "Migraine Aura"), at his specific request, for the yearly handbook on Headache. This provoked a literally insane personal attack on me: how dare I quote so many "ancient obsolete sources" (I provided an extensive historical introduction in order to place the entire subject in a meaningful context); how dare I criticize the "accepted" theories

* Arnold P. Friedman was director of the Montefiore headache clinic, as well as chairman of the headache sections of both the American Medical Association and the American Neurological Association.

of a local vascular origin; [. . .] how dare I use so many semi-colons and dashes etc. etc. Regarding the *substance* of what I had said, the arguments and evidence I had marshalled, he eschewed comment. [. . .]

It should be evident to you, as it has finally become evident to me, that he is an impossible man, and that I can only proceed with the book when I resign from the HU and escape his orbit. What was my fault, and deeply neurotic, was to put up with all this shit, to comply with his demands for silence and inactivity, to turn myself off, and, in effect, to destroy my own creation. [. . .]

At Beth Abraham I have sole (neurological) charge of over 400 patients, and am beginning to get immersed in a number of research and therapeutic projects. The latter include the setting-up of special model wards for three classes of patients in whom I am specially interested, and who respond, to an astonishing degree, to loving and patient attention: namely, (senile and pre-senile) dements, aphasics, and Parkinsonian patients (here, I am choosing especially the younger and post-encephalitic patients). Some of the research work which is gaining momentum has regard to some chromosome studies in a number of unusual hereditary diseases, immunological work in disseminated sclerosis, and some enzyme-studies in ALS. I have become neurological tutor to the entire senior year of medical students at the College, and they come over to me twice weekly at Beth Abraham. I love teaching: it is perhaps the only area where I consistently function at my best and most un-neurotic level, and from which I always derive pleasure. This seems to be increasingly recognized at the College, and I am gradually being given more teaching responsibilities. [. . .]

Incidentally, I had a flattering letter yesterday, out of the blue, from the Chairman of the Neurology Dept. at Chicago, saying he had heard about me from "someone who knows me personally," and would I be interested in a full time (assistant professorial) teaching position, one of the few such in the States, where teaching is generally relegated a poor second behind "research," and the gifted teacher/poor researcher like myself doesn't have a chance. I shall thank him for his kind letter, but explain that I am not considering any move at present. Indeed the only move which makes sense to me now is to return to England.

If I returned to England I would probably like the sort of work I am doing now, namely working in chronic disease institutions. The majority of doctors can't stand this, and indeed such positions carry a lowered "status" compared with fully academic positions. However, it is what I like, what I am good at, and what I can derive much information from. [. . .]

Of course everything would have been much happier and greatly facilitated if I had been able to proceed with my *Migraine* book. Ideally, I would finish this, abstract it to a [doctoral] thesis, follow it up with the Parkinsonian book, achieve a modest critical success with both, and return to England esteemed and desired. Ideally. However, what is possible theoretically is often impossible in practice. I may have to come back with my tail between my legs, as it were, i.e. without any special publications to prove my worth. I have to admit that the last few months have been wretched and unproductive, and that I am far from confident that this turn of affairs can be altered. Further, my tolerance of New York and this country has worn about as thin as it can go. For much of the time, I have the feeling of being in some awful cross between a kindergarten and a concentration-camp. The country is seething with violence, and moral ugliness. The continuous barrages of lying and double-talk come daily closer to the conditions of *1984*. [. . .]

I will carry my neurotic conflicts and pains wherever I go. *These* will not be resolved by migrations (as they were not resolved by the last eight years of flying here and there). However, above and in addition to these, I have an unappeasable nostalgia, a wish to return to family and friends, to a city and a country and a civilization which means something to me, which formed the backdrop of my memories, development, feelings, and which, if bankrupt, is at least gentle, and given to reminiscence and compromise, rather than this country, which, first, is essentially foreign, alien, and without depth or meaning to my eyes, and which, secondly, seems embarked on an irreversible and hideous cultural psychosis, currently corresponding to the state of Germany in, say, 1937.

I must wind up this letter. I shall certainly come to England in the summer, and I shall take off at least six weeks, if not eight. This would cover July and August. I need a massive break from things here. I need time to write, because it occurs to me that only conditions approximating to last summer will allow me to start writing again, and I must realistically look round for future positions. [. . .]

Enough. Please write soon.

Love to everybody, and tell Auntie Len I will write her soon.

OS CONTINUED SEEING PATIENTS AT the migraine clinic, but Friedman forbade him to take home copies of his notes about patients. In June 1968, OS managed a way around this; he arranged with a janitor to get into the clinic at night, so that he could make copies of his own consultation reports, by hand. Shortly thereafter, he left for London.

To Mike Warvarovsky

FRIEND FROM MUSCLE BEACH

JULY 24, 1968
[37 MAPESBURY RD., LONDON]

Hello Mike,

This is just a "hello" letter and minor chat. Not another thesis, I promise you.

I cut some things out of today's *Times* which I thought might interest you. A notice of a huge Art exhibit in Munich, an obituary of Henry Dale,* and a letter by Bertrand Russell and a galaxy of English Eminences. Dale, who just died at 93, was going strong to the moment of his death. Russell, who is 96 or 97, is absolutely lucid and crystalline in his thought, as you can judge from the letter. Marvellous Russell, the godson of John Stuart Mill—I sometimes regard him as the last living embodiment of reason and social passion. It excites me to think of these extraordinary, chimerical old men, intensely creative in their tenth decades. One has to think of Picasso also, and of Stravinsky. And of Margaret Murray, the archaeologist, who wrote a charming autobiography entitled *My First Hundred Years.*

I can't say any nice things about myself. I am terribly restless and

* A Nobel laureate in medicine.

irritable. I can't work. I can't stand my parents—my mother brooding, paranoiac and depressed, the walls oozing with accusation, everything suspended between blame and guilt. My father, clownish, scholarly, doing his best to avoid the house.

And my brother Michael, both deeply schizophrenic and completely logical and truthful, crucified in attitudes of spite and vengeance, his humanity monstrously extended at both ends of this spectrum, into animality and divinity. *He* makes sense, tragic sense. [. . .]

So, I can't stand THEM—my family, although they are blood and flesh, and my past, and most of myself; yet I long for them so desperately. I hate and love them with a vengeance. Moreover, jobs are terribly tight in England. It occurs to me that I may have to stay in America, even though I can't stand *it,* either. I could oversimplify the situation, and put it diagrammatically, and say: I have a good living but no (possibility of) life in the States, whereas here I could have a (possibility of) life, but no living. I think this is what they call a Double Bind. The only way out, I can imagine, is to write and write, and become sufficiently recognized to make my own terms wherever I am. But then, I am too angry and too spiteful at the moment to let myself write. I don't know the answer. But I must somehow hold on, and not regress into rage, depression, drugs, etc.

Write soon,
Yrs,

To Mike Warvarovsky

AUGUST 31, 1968
37 MAPESBURY RD., LONDON

Dear Mike,

A hurried note to say, first, that a friend of mine—Orlan Fox (I believe you met him in New York) will be in Munich* for a couple of days this

* Warvarovsky had become a military doctor and was stationed in Munich.

week. [. . .] I have taken the liberty of giving him your address—he's a pleasant, open chap, I think you might enjoy meeting him, and could perhaps show him round a bit.

The rest of my own news I can also put briefly. My first month here (July) I spent, as you know, in neurotic quarreling with my parents, and also awaiting some move by my chief, Friedman, who had warned me not to proceed with my book. (He is insanely paranoid, jealous, petty.) Around the beginning of August, I had made some sort of internal accommodation, and was able to start a little, desultory work. Mid-August, a bombshell arrived: Friedman fired me. There was no warning, no reason given, no compensation; just a curt letter. I was abruptly deprived of my patients, whom I enjoyed, and also of half my salary (10,000 a year). I flew into a violent rage, but was of course impotent to do anything. I wrote a ferocious letter, a little masterpiece, but it was too actionable to send. Then, not unnaturally, I turned all the rage inwards and got very depressed. So much so, in fact, that I took to my bed for two weeks, and refused to get up, and felt constantly nauseated with a horrible sense of foreboding and impending suicide.

Now, start of September, I have got up and out, and have decided to stay here an extra two weeks, and am working hard on the book. Not in a state of creative fury like last year—I cannot command this sort of inspiration, and my interest in the subject has been killed, so to speak. But I have already done all the thinking and organizing, and its redrafting is pretty much a routine job which I must force myself to do—for I have been hagridden and guilty about it for more than a year.

I return to New York about September 15. It will be my last year, because I can't stand the American Scene any more. (Did you read about the clubbings, etc. in Chicago? When Nixon comes in, as he must, I see that the lights will go out all over America, and a brutally repressive police-state will become a daily reality. There is nothing to choose, these days, between Russia and America—or the Pope, for that matter.) I think Western Europe, poor, dilapidated, pessimistic as it is, is now the sole repository of the possibilities of civilization. The rest of the world will be resolved into monstrous, lying, warring Superpowers, and their pitiful squashed satellites.

I will also be changing from neurology to psychiatry. I plan to make my double transition next summer.

Do write and tell me what's doing with you. And whether I can expect to see you in New York soon.

Yrs,
Oliver

ON SEPTEMBER 1, STILL IN London, OS challenged himself to rewrite the manuscript completely and submit it to Faber & Faber within ten days. In *On the Move* he describes how this—which he accomplished without drugs—produced a manic, exalted state that lasted for about six weeks, well into October.

To Jonathan Miller

SEPTEMBER 23, 1968
234 EAST 78TH ST., NEW YORK

Dear Jonathan,

[. . .] It has been an incredible three weeks, a concentrated, maniacal, *annus mirabile.* I have had a superb eruption of creative powers—I will never know more, generically, than I know now—and in so doing, as perhaps I always feared (and good reason to refrain from the danger!) I have utterly blown my ego to pieces. I suppose it might be called an acute schizophrenic psychosis, if labels are useful. Certainly I have been fantastically, beautifully hallucinated in the past week; the entire world has been no more than a *tabula rasa* on which I would project my metaphors as hallucinations. I have been literally ecstatic (ecstatic paresthesiae here and there, wherever I want, with the flow of my thoughts). I had an incredible, Adamic feeling of rebirth, at 11.20 last Monday, and seemed to emerge into the full shadowless sunlight of utter reality. I saw all the faces about me darkened, partial, anxious, ignorant, superstitious, credulous, stubborn and blind. I wanted to shout "Eureka" or "Hallelujah," and run amok, praising and saving: but I realized it would have been awfully bad *taste,* so I refrained. I have kept my marvellous psychosis a secret, between my analyst, a few friends, and myself: for it is important to eat and sleep, to observe bounds, and to "function" externally.

I have had similar, if milder, "episodes" before. I would call them *manic,* because they did no work: their insights were denied; they resembled moral and emotional and existential *loans* with no capital in hand—and they were paid back by awful penitential depression, stupors, sense of meaninglessness and accidie. This one has done triumphant work, both in the book, and in me. And I will never be the same. Some, I must, and have already, re-repressed—one cannot live in an acute psychosis/ecstasy. I retreat from the Icarus position, to that of father Daedalus. I am in full possession of my powers. I feel like a troubadour, jester, or Shakespearian "fool": I feel utterly transparent, consumed by the truth, and with the need to spread it in witty, beautiful and valid metaphors.

As for my ego???? I have no sense of anxiety, shame, cringing, diffidence, distrust, paranoia, or those beastly sadomasochistic cravings which made my life hell. I feel utterly liberated from the cage of neurosis, from the four dimensional meshwork of paranoia (family: Jewishness: Medicine: America). Although my state is libidinously driven, I feel no explicit sex-cravings at the moment—neither homo- nor hetero- nor fetishistic, nor anything. But the whole universe is my Kunt. My readers will be my Kunt. My words and intelligence are my Penis, etc. etc. Everything crystal-clear, in shadowless consciousness.

I can't predict the future, and am making no grandiose plans. But this is where I stand, on Sept 23. And it is not all narcissism: for I love my friends, and among them, most especially you.

Yours,

WHILE OS CONTINUED TO CORRESPOND frequently with Augusta Bonnard, they had entered into mutual recrimination centering on a debt incurred in 1964, when Bonnard had lent OS $2,000 from her American bank account before leaving the States.

To Augusta Bonnard

OCTOBER 10, 1968
234 EAST 78TH ST., NEW YORK

Dear Augusta,

Thank you for your letter. It was not entirely unexpected, and I am not wholly sure what to reply.

It is unfortunate, for you, that you made it so difficult for me to repay you when I was in a position to make a partial settlement, in July.* For the following month, I was sacked from Montefiore Hospital—without reason, warning, or compensation—and at a single stroke deprived of half my livelihood. This being so, I am now in a state, not of penury, but in which my expenditure considerably exceeds my income. Being in a position of professional and emotional equilibrium, in other respects, I am in no hurry to secure myself an indifferent job simply to make up the balance. I am therefore, as you will understand, in need of the $2000 which I managed to save last year to see me through these relatively lean months. [. . .]

I see no immediate prospect of my being able to repay you in dollars. You want the money, which is reasonable enough, and I, indeed, cannot tolerate the existence of a debt which has nagged me since 1964, and which was not of my soliciting in the first place. If you are agreeable to the prospect, I will ask my parents to pay you the money in sterling, and you will have to add whatever increment seems to you reasonable to buffer you against possible debasements of the pound. They are in a position to repay you, and I am not. I will have to repay them in time—but this is as it should be: debts are best kept within the family, and you are the only creditor I have ever had outside my family.

Perhaps you will write again and state your feelings in the matter. You would be foolish to refuse repayment from my parents, for further demands from me will necessarily be fruitless, and will only aggravate the enmity between us.†

Yours sincerely,
Oliver Sacks

* Bonnard had insisted that OS repay her in American currency rather than British.

† Bonnard continued to refuse repayment from OS's parents, but eventually, in April 1971, he was able to repay the debt himself.

To Mike Warvarovsky

OCTOBER 14, 1968
234 EAST 78TH ST., NEW YORK

Dear Mike,

I received your letter with enclosures (it was curious seeing my attitudes of six weeks past—oh, what an eternity ago!—reflected back at me from my own letter), and a postcard from an American Military Base (God Bless America!!!) which arrived this morning. I will ask you to forgive the faintness of the typewriter ribbon, which is just about typed out after several hours of my two-finger pounding every day, and any incoherencies which you may detect, which are due to delirium, for I may have a highish fever at the moment, and get a little mad when my cerebral cortex becomes overheated.

I have no photos of myself and new bike, nor am I likely to get any, because I have given my cameras away (the Kodak as a gift, the Nikon as a sort of permanent loan to a friend in London): for I am through with photography in this continent. I know all about it, its beauties and uglinesses: I have sucked it dry. I am full up and overflowing, and need no more records: perhaps, most pertinently, I no longer feel I am really here at all. I am in the enchanted timeless world of my own thinking (which has shifted now the *Migraine* book is over) to problems of willed movement and the manifold paradoxes of Parkinsonism, my current study and next book; beyond this, to problems of representation and the spoken word (my third book will be on Aphasia, or rather all the Asymbolias), and beyond this, to the metaphoric and dramaturgic representations of madness (my fourth book: "The logical structure of Madness"). But it is not a lonely world—far from it! The voices of the past are always with me, Wittgenstein, Liveing, Wiener, Gowers, Hughlings Jackson; in the eighteenth century, my grandfathers Robert Whytt and Cheyne; in the seventeenth century, my great grandfather, the most admirable Thomas Willis (of the circle of Willis, which you will be studying in your 48 hour week, and all sorts of other things), and the labyrinthine mind of Sir Thomas Browne. Behind them: Ptolemy, Pliny, Heraclitus ("One cannot enter the same river twice"), Aristotle, Plato (I shall *think* like Aristotle, but *dream* like Plato) back to divine Pythagoras. And in bed, every night, I read *The Odyssey,* sometimes weeping, sometimes laughing, and knowing that time has no real existence: that all life is a circularity of metaphors in an endless moment. No, it is not a lonely life, I have never felt less lonely in all my life. In my new life—and I feel so new, yet so old!—the stupid distractions and confusions and blindnesses vanished, I hope for good. I have forged some friendships deeper (those based on truth), and I have brought others to a close (those based on craving, claiming, sucking, and envy); I have travelled more than 2000 miles since my return, hither and thither, in what country has not yet been polluted by its vile owners: and two or three times a week, I go to the Opera, or to plays, or concerts. I have never enjoyed life so much, nor felt that I deserved to enjoy it before. [. . .]

Memories crowd upon me, all memories have returned to me: memories I thought I had shed as a tree sheds its leaves. Nothing is lost, nothing forgotten: only concealed from awareness, when there are motives for concealment. And especially, your letter reminds me, especially memories of California, of the wild beauties of Big Sur. (I would invest it with a mythical past, with all the scenery from Homer: rocky Ithaca: those rocks near Santa Lucia, those were the Symplegades whose clashings Odysseus was forced to skirt: and a little octagonal café, fifty miles north of San Luis Obispo—that is Eumaeus' hut; and the old dog there that licked my hand, was that not Argus, the old hound of Odysseus who knew his master after twenty years? And that other café, gilded, garish, surely King Antinous' court, where royal myths and memories unfurled like music!) Jade Beach, where I dived with, scrambled with Mel (beautiful, lost Mel! what has the wide world done with you?). The timeless beauties and absurdities of Muscle Beach, the voices I still hear, the monstrous bodies and postures I can summon to my mind. (Shall I see them again, shall I hear them, shall I touch them? Shall I go to the West Coast this Christmas, or let them retain their permanence in the amber of my memory?) The myrtle-trees of Oregon, globe-like, solitary, silhouetted on the enormous plains . . . sweet Lois* with her violet eyes (she died of an overdose, some damn psychedelic drug). San Francisco, rising dreamlike from its misted ramparts—how vividly, how agonizingly I bring back its image to my mind, San

* Identity unknown.

Francisco on a Sunday morning, September 3, 1960, when I first saw it on my first day in America! Memories, memories! The weights of the old Dungeon,* clanging like cymbals on warm winter evenings, Dave Sheppard, bloated, beautiful, stooped in some shuttered mindless meditation by the battered valveless radio in the now-defunct Dungeon. And that great wave which smashed me . . . I see the slim surfers in the grey waters of Malibu, I see them as I would see them when I spiralled down the loops of Topanga Canyon, sitting on their boards, rocking, rocking, to and fro, endlessly, happily, on the vast bosom of Mother Ocean, and their fearless male figures riding her waves, the first and oldest riding in the world; and Mother Ocean, huge, impatient, tossing them off her mighty body, somersaulting her tiny lovers, smashing her children—oh, with such power! with such an ineffable lack of human malice!—on the timeless yellow Beach. [. . .] Voices, voices, like the fluttering of leaves. Sunlight bubbling in a thousand valleys. Memories of faces which I would not see as they were, but as I wanted to see them: memories of a thousand wishes and fears. And, perhaps because I am writing to you, and a particular chord of memory vibrates in resonance, I am haunted now by the image of Mel, but it is a haunting without guilt, without remorse, only the sense of sadness and loss because we were both so blind, so ignorant, so innocent, and so suspicious.

But I am indulging myself in this letter and this mood. I have put away my photos and writings of the first thirty-five years, for they have about them a dreadful circularity, as if I were whirling round and round, passively, on a phonograph record. I am finished with this merciless gyration, and I welcome the limitations no less than the infinities of my conscious future.

The days are endless, tranquil, full of excitement.

yrs,
Oliver

* Joe Gold's gym, fondly nicknamed "the Dungeon," had been founded in Venice Beach in 1965.

To Mike Warvarovsky

OCTOBER 22, 1968
234 EAST 78TH ST., NEW YORK

Dear Mike,

Here are the two photos we took in Germany, on that strange Saturday morning—oh! incredibly long ago.

I've been rather low since my last letter to you, largely (at least I hope largely) due to having had the 'flu, which is rampaging in New York at the moment. I should have gone to bed, read science-fiction, and been tenderly looked after, which is the sort of regression one *needs* when one is sick. But, as we exiles find out, one misses home most when one is ill in some lonely foreign comfortless country, where nobody gives a fuck etc. So I didn't go to bed, and went to work every day, with a highish temperature etc. and have generally, one way and another, lost about ten days. I think it is partly in place of a depression: for I needed some sort of come-down after the elated hyperactive month I had had. I suppose there is some wisdom in the body and mind, which decrees that a creative or elated bout will go on for so long (as long as is needed), and then be followed by a period of rest, sleep, depression, hibernation, whatever, as the systole of the heartbeat is succeeded by diastole.

I'm not actually depressed in the old way i.e., feelings of hate, fear, guilt, accusation, worthlessness, etc. None of that, or very little at all events. But a sense of *loss,* of grief, of waning power, of loneliness, of sadness. More a feeling of *mourning* for a lost loved object: the object here being not a human lover or mistress, but the book, the state of mind which created it, and that part of myself which, quite literally, I have shed with the writing of it. I *have* lost part of myself, although the loss will be recouped, a thousand times over, in the gain which I receive from the publication of the book: the knowledge that my ideas will be scattered like seeds (I almost said inseminated) into the minds of others, and the sort of wide half-sexual intimacy with my unknown audience that this will give me. But all this has not come about yet. There are months of waiting, of fiddling, of arranging this and that, and perhaps, inevitably, I will have feelings of sadness and loss during this time.

Otherwise, I ignore my environment as much as possible. I no longer have the overwhelming paranoid feelings for it I had before—hatred, terror etc. Rather, an unchanging sense of disgust and pity, and a hunger for beauty and order and open country. I escape, physically, at weekends, by leaving the city, and—in spirit—at concerts, theatres etc. in which, for the first time in my life, I am fully indulging myself. I almost try to make-belief that I am not here: to seclude myself with my interests, my thoughts, my books, my correspondence, my friends here and there, art, nature, exercise, anything which will blot out the sad and ugly reality of Amerika awaiting its Nixon.

Do write,
yrs,

5

Coming to Life

1969–1971

SOON AFTER HE STARTED WORKING at Beth Abraham Hospital in 1966, OS began to notice dozens of patients, scattered among the wards, who were virtually immobile and unable to communicate. Going back through their records, he realized that they were all survivors of the pandemic of encephalitis lethargica that had swept the world for several years after World War I. This still-mysterious disease was often fatal, and those who survived were often left with strange syndromes that could seem like an extreme form of Parkinson's. Huge care facilities had been built to house many of these people, but there was no clear understanding of what had caused the epidemic, and no cure (either for this postencephalitic syndrome or for "ordinary" Parkinson's disease, which was far more common).

By the 1960s, the epidemic itself had been forgotten and so, for the most part, were these postencephalitic patients, some of whom had been hospitalized for forty years. In 1968, though, the medical community was galvanized—as was the press—by the news that people with Parkinson's could be helped by a new miracle drug called levodopa, or L-dopa. Wondering whether this drug could also help his own immobilized patients, OS applied to the FDA to use it as an experimental drug.

To Elsie and Samuel Sacks

[CA. MARCH 1969]
[NEW YORK]

Dear Ma and Pa,

[. . .] You ask what "FDA" stands for: it stands for Food and Drug Administration. I solicited FDA for a special "investigator's licence" so that I could inaugurate a research project on a drug called "DOPA"* in Parkinsonism (there is some evidence, questionable at present, that DOPA may be as useful and as specific for Parkinsonism as insulin is for diabetes; my prime concern would be to substantiate or refute this belief). I had intended to incorporate the DOPA trial in a broad-based study of that form of Parkinsonism which has hitherto been most refractory to treatment and least understood in mechanism: so-called "akinetic" Parkinsonism. As soon as the FDA gave me a number (a privilege: for they turn down *300* requests for the experimental use of DOPA *every day!*), and as soon as I had drawn up the skeleton of a research project, I was immediately approached by Labe Scheinberg (Professor and Chairman of the Neurology Dept. at Einstein) and by Irving Cooper (the neurosurgeon who has operated on more Parkinsonian patients than anyone living): both wished to join forces with me, and run a combined study. The suggestion is a flattering one and could potentially be of the greatest mutual benefit: on the other hand, I must beware of any pressures to dislodge me from my central position in the investigation. You will understand that when two powerful men, who have some claim to be considered as the first neurologist and first neurosurgeon in New York, appear on the scene in this fashion, there is a risk that someone of negligible formal status such as myself may get squashed between the two Leviathans. Therefore I have to make it clear, without ruffling tempers or jeopardizing my own position, that the project was *my* suggestion, must be controlled, finally, *by me:* and that when credit is apportioned, it is *I* who must be first name on any or every paper which appears. This is not arrogance: it is survival. [. . .]

* OS uses "DOPA" as a shorthand here for L-dopa.

Please tell Auntie Len, also, that I have sent her a birthday present by *surface* mail.

GENUG.* Love to all,

To Elsie and Samuel Sacks

MARCH 26, 1969
234 EAST 78TH ST., NEW YORK

Dear Ma and Pa,

I hope you are keeping well and in good spirits. [. . .]

Things at Beth Abraham continue on their blundering course. The DOPA still never came (my chief, a schnorrer† at heart, is trying to beg for some, instead of buying it, despite the fact that he is rolling in funds specially donated for its purchase), and I can hardly bear to face my poor Parkinsonian patients who have been promised and promised, and let-down and lied to, a dozen times in the past six months. You will not be surprised that this stupid situation fills me with rage and guilt. Fortunately, there are a few other patients whom I can study quietly, in my own time, in my own way. It is obvious, however, that I can depend on Beth Abraham for nothing, or rather nothing more, and I shall be seeing [. . .] this week about getting myself another (additional) part-time job elsewhere—preferably in a State Hospital (i.e. the equivalent of Shenley).‡ [. . .]

Yes, before I forget, thank you, Ma, for the motorcycle leathers which arrived a few days ago.

I imagine Spring is arriving in England, and things are bursting from the ground.

* "Enough," in Yiddish or German.

† A schnorrer, in Yiddish, is a moocher, someone who is always happy to take rather than give.

‡ Shenley Mental Hospital, outside London, had a reputation as a progressive treatment center and was a leading training hospital as well.

Keep well, write soon, say a special hello to Michael, Auntie Len, David and family, etc.

Love,

To Elsie and Samuel Sacks

[CA. APRIL 1969]
234 EAST 78TH ST., NEW YORK

Dear Ma and Pa,

[. . .] My three patients are doing extraordinarily well on the DOPA: one of them, who was virtually unable to talk or move, a very severe post-encephalitic, is now chatting and toddling down the corridors. I am inclined to think that DOPA may indeed turn out to be as useful for Parkinsonians as insulin for diabetics, or nearly so. In the meantime, I have been studying the three patients closely, and am gathering some novel and fascinating insights into their state. I will be putting another five patients on the drug shortly. Although a great deal has been written on Parkinsonism, as it has on migraine, I think that by the summer I shall be in a position to write a short monograph which may be of some use. Anyhow, whatever I say, I have my own distinctive *style* of saying, and this should help.

As I had reiterated many times before, I need *patients* at all times, and conditions which allow me to work quietly and gather together the information for monographs. I do not demand a glittering, "stimulating," highly competitive, academic environment—indeed, I can hardly survive in such conditions, even were they within my reach. However, sooner or later, if I make my mark in my books, the most desirable posts will be offered to me, and I will finally be released from the vicious circle of abjection and failure.

For *time is on my side:* in the civil war between creativity and destructivity, between talent and neurosis, time is on my side. And each day and week and month and year that I can hold off my devils (depression, self-hate, self-destruction, etc.), I am further from perdition, and closer to some fulfilment.

E. M. Forster, in a brilliant essay on Cardan,* remarks that to him "the greatest things in life were work, self-examination, and the hope of immortality. Human intercourse was unimportant beside these." I do not compare myself too closely to Cardan, but I do have to concede that the chances of satisfactory human relationships *of any kind* are most difficult for me to believe in. I am, by turns, credulous and distrustful; I fear proximity and intimacy, seeing it as a dangerous invasion, a potential annihilation of my own identity; and then again, in the tally of contradictory feelings, I expect constantly to be abandoned or deprived of what I have—these things may be softened by analysis, but they are too fundamental in my character to permit radical change. But, I have a great compensation: I can think, I can sublimate, I can channel my thwarted ardours into work. And it is in terms of work that most of my hopes lie.

As usual, I have said more than I should; but you will gather, if nothing else, that the emotional Barometer (to which all other things are secondary) has moved to FAIR: occasional showers, but sunny intervals.

Do write soon,
Love to everyone (as the Hippies say),

To Elsie and Samuel Sacks

MAY 17, 1969
234 EAST 78TH ST., NEW YORK

Dear Ma and Pa,

[. . .] I now have 15 patients on DOPA, and am staggered and gratified at its ability (in many though not all cases) to re-animate patients who had been virtually petrified for years: they are literally (like Dr. Manette after his eighteen years in the Bastille) "restored to life." This, of course, in turn leads to a very complex state of affairs: one cannot restore the potential of movement and independence to someone

* Girolamo Cardano, a sixteenth-century Italian polymath and physician.

who has been helpless and dependent for decades without creating a most complex, unprecedented situation for them and everyone associated with them; I am fascinated by *this* aspect, amongst others.

My hours and salary have been raised to ¾ time at Beth Abraham, which means that I can now break even financially: more would be desirable, but is not necessary at the moment. In point of fact, I have been working like a madman for the last fortnight, staying at hospital till 9 or 10 in the evenings. The usual working hours are full of fuss and hysteria, and I can work tranquilly and alone in the evenings. I am gathering almost more information than I can deal with and starting to nibble on the huge Parkinson literature, which is even larger than the Migraine literature. There are times when I feel overwhelmed by how much has been said and thought on the subject, and think: Who am I? What can I contribute? Everything which can be said has already been said! etc. Other times when I feel the greater part of this literature is garbage, and that there is always room for saying something new on an old subject. I begin to have the feeling that ten weeks of intense work will put me in possession of the facts and thoughts I need to write a first draft on a Parkinsonism monograph—and I begin to feel that I will be in a position to do this (in London) in August.

It makes me furious to think that, in effect, six months were wasted before I could get to grips with the subject, and this largely through the folly and stubbornness and indecision of a fond[*] old man who happens, alas, to be my Director: and as largely due to my own inability to handle same old fool. However, things *are* going OK now. And if I complain of the six months I have wasted, I could as well bewail the thirty-five years I have wasted. Moreover, my methods of work at all times tend to be sudden and peculiar—and perhaps I can do more in ten weeks of concentrated effort than in ten months of regular work.

Messeloff[†] still wants to get everyone involved with these patients—Dr. X and Dr. Y and Mr. A and Miss B—in his stupid American

* OS uses "fond" here in its archaic sense of helpless, weak, or silly, as Lear calls himself "a very foolish fond old man."

† Charles Messeloff, Beth Abraham's medical director.

concept of a Big Deal and a Multidisciplinary Approach, etc. I try to ignore all this and arrogate the patients to myself. I am a superb clinical observer, and I neither want nor need anyone's help. I find more and more that if I am to accomplish anything it is necessary to regard virtually everyone about me as either stupid or hostile obstructions to my own activities. I am not impolite, but I have to go my own way. Dependence on other people is lethal for me—for my work, for my self-image, for even the possibilities of independent existence.

I have very little to say otherwise, for I have indeed had very little life outside the hospital lately. [. . .]

I hope you are all keeping well.

Love,

To Mike Warvarovsky

MAY 25, 1969
234 EAST 78TH ST., NEW YORK

My dear Mike,

I should have replied earlier to your charming letter written in Venice, all bubbling with the freshness and joys of travel. I adore the wide-openness, the *innocence,* with which you see the world—it is an innocence which exposes you to all sorts of dangers and fears, and wilde phantasies, but an innocence, also, which nineteen people out of twenty have never known, save in their lost and forgotten childhoods. How right you are about catacombs and secret thoughts!

Everything we build is an allegory of ourselves: the whole human world a metaphor of the human state. My own love is for city walls (oh, they were so beautiful in San Juan where I went for a weekend), those high stone walls which enclose, defend, and unify a city, as we wall off the citadel of ourselves: but then again, the opposite is true, and they symbolize the prison in which we spend our lives ("The world is a prison from which the prisoners are daily led forth to die"—Pascal). [. . .] I love privacy and walled gardens: yet how much more, perhaps, would one love a world without walls, an infinite garden. Music is an infinite timeless garden: all art, and love, and God.

We long to be as solid and substantial as stones, and, then again, to diffuse into the universe like the finest aether. I find my whole life torn between these opposite desires, and I am sure that the same is true, in even greater degree, of yourself.

I last wrote to you, I believe, in early March, when I was in an ugly frame of mind, bereft of work, poisoned with rage and hate, and longing to be—not dead—but asleep, away, for a while. I am in a much happier mood now, for I have started the intensive work on my Parkinson patients [. . .] and I have had the intoxication of seeing l-DOPA, a ravishing drug, restore to an incredible life those pitiable patients who had been almost literally turned to stone, speechless, motionless, and even thoughtless, for twenty years or more by the horror of their disease. My first patient, a man of fifty, who had had the severest Parkinsonism since the age of fifteen and had not spoken or moved for more than ten years, said (his first words): "I am reborn. I have been in prison for thirty-three years. You have released me from the custody of my symptoms." And I could not stop myself from weeping. And in using the drug, I have learned what I suspected before, the nature of the WILL. There is no such thing as "pure will," a sort of switch which one turns on by choice. Will is the ability to create stimuli to which one must react: to create them, and then to subordinate them. One cannot conceive of action, pure action, active action: all action is passive (and I include thought as action) and follows from certain stimuli. The Parkinsonian patient may complain (like a psychotic patient) that he is deprived of will power, that his muscles will not obey him, etc.; with DOPA, and with methods which I am learning, we can restore to him the ability to create stimuli to which he will then react. Too many stimuli, and he will be *driven* to incessant motion (akathisia) as before he was held in immobility (akinesia).

At their highest levels, these forced reactions to stimuli are felt as *fascination* and *compulsion:* the Parkinsonian who cannot move his eyes from an object says, "My eyes are magnetized by the object. They are transfixed and fascinated by it. I *cannot* move them. But the chance passage of someone walking in front of me will break the spell." Similarly, he may be compelled to make certain movements, although knowing, and saying, that the movements have no intention, no content, no goal. Thus the nature of drive in the nervous system unlocks the secret of will. And here I am enchanted by my work, for neurophysiology shows what poetry and philosophy have always known: that we are—in a very fundamental sense—automata, reflex-machines; and also that we are *composite*. Knowing that we are necessarily passive and composite, we can then make ourselves, for daily purposes, active and unified. But there is no "soul" and no "will"—these are fictions, universal fictions, like the Garden of Eden.

Oooh! I am sorry, I didn't mean to ramble ahead like this, and monologue at you. But I am suddenly so caught up in work—working at the hospital till late at night, every night, and then dreaming of my work when I go to sleep, that, in turn, I feel myself fascinated and compelled, and almost devoid of other life. But this is a good compulsion. And only good compulsions can save one from evil and tragic compulsions, to which we are driven for so much of our lives. I have been inexpressibly shocked and saddened (but perhaps, also, excited) by the fate of a friend and former colleague—I knew him in San Francisco, many years ago, and once wrote a little story about him entitled "Dr. Kindly."* He was deeply sadistic in his sexual life, but exquisitely controlled and tender: he understood, better than anyone I have ever known, that pain and humiliation can redeem from guilt and confusion. Three years ago (and this is *my* guilt) he was introduced to Methedrine, and *trapped in its excitement.* He started to give his lovers and patients—victims all—huge shots of Methedrine, and to tie them up and beat them in sadomasochistic orgies. They, perhaps, survived: but he became the final victim of himself, lost the razor edge of control, and finally disintegrated into a demoniacal madness: he saw his cat as the devil, murdered it, cut it up into little pieces, and asked his friends round. He is in a state hospital now, their most dangerous patient, consumed with a murderous frenzy. And my shock etc. is due to identification: the feeling that there, but for the love of God, go I. Except it is not the love of God, but creative work and insight which can save one from such a drama exciting one, and drawing one, to destruction. [. . .]

To get more chatty: No, I didn't finally go to California. First it was shortage of money (now remedied), and now it is involvement in work. I will have to leave it to the summer, I think. [. . .] My bike

* This piece does not seem to have survived.

is fine—I try to find some tranquil country to drive to every Sunday, away from this shit-filled, hate-filled monster of a city: but mostly I am so wrapped up in work and thinking at the moment, that I hardly notice my environment. So I have very little other news. [. . .]

IN APRIL 1969, OS HAD cautiously begun giving small amounts of L-dopa to a handful of patients, and when they did well, he offered it to several more. At first the drug did seem miraculous, enabling nearly motionless patients to walk by themselves again, to talk, even to dance and sing. But over the next few months, unpredictable side effects began to occur, and it became more and more difficult to modulate the effects of the drug and find the right dose for each patient. These reactions put OS in a profound predicament on a human and therapeutic level, but he knew that they also presented him with a rare chance to document and understand the workings of the brain. As he writes in the next letter, to his parents, "what I started, reluctantly, as a trial of just another drug, has proved an almost incredible tool for the dissection of a vast range of human behaviour, from the most primitive postural reflexes to the most complex psychotic reactions."

To Elsie and Samuel Sacks

JULY 30, 1969

234 EAST 78TH ST., NEW YORK

Dear Ma and Pa,

A very brief note, whose arrival I may myself precede.

You sound like you enjoyed Malta very much—islands have something lovely about them, and churches and villages give one a sort of feeling which Cannes and Parnes* can't. Do read (if you haven't done so) one of the most beautiful books ever written about an island: D. H. Lawrence's *Sea and Sardinia*—it's Lawrence at his lyrical best

* A town in France not far from Giverny.

(as opposed to his ranting worst), and full of exquisite descriptions of little hill-towns and people. I would so much love to go to Malta, or Sardinia, or the Greek Islands, or a Scottish (otter) island, for a week or two this summer. I nearly blue [*sic*] everything I had to go for a voyage to the Galapagos Islands with the Darwin Society, but alas! have had to content myself with reading *The Voyage of the Beagle* as my bedside book—but I do not feel *entitled* to take any holiday. Indeed, I have not felt entitled, since about May 1, to go out for an evening, to take off a weekend, to go to a show, to go out to a dinner, or anything. I used to be the idlest person who ever lived: but now I seem to live for work. The last three months will turn out to have been, I suspect, the most interesting and productive in my entire life: what I started, reluctantly, as a trial of just another drug, has proved an almost incredible tool for the dissection of a vast range of human behaviour, from the most primitive postural reflexes to the most complex psychotic reactions. I have been too busy with the daily discovery of new things, and their immediate meanings, to be able to do much reading or writing. This is one of the essential reasons why I must come back[*] and write for a month. I have the material for three monographs and about twenty papers: the three monographs I see my way to combining into a single book—it will be a whopper, I could not condense my material and thoughts into less than about 500 pages (its title and substance will be something like: "Coercion, Compulsion, and Compunction: A clinical study of primitive and forced behaviour in post-encephalitic Parkinsonism and psychosis, and its modification by l-DOPA").[†] My case-histories alone, which will constitute the first third of the book, will occupy about 150 pages—say, fifteen minutely-detailed histories of ten pages each. [. . .]

I hate to leave my patients for a month: partly because they will not be adequately looked after in my absence, partly because I am still, as it were, in the acute phase of observation and discovery. But I *must* leave now, because various stresses, exacerbated by the absence of Shengold,[‡] are tending to push me towards breakdown (breakdown

* I.e., to London.

† This very academic-sounding book eventually became *Awakenings*.

‡ Shengold, like most New York analysts, did not see patients during the month of August.

in me, as in Pavlov's dogs, takes the form of alternating exhaustions and excitements, with unstable and rapidly-shifting attitudes), and because I am almost literally bursting with the need to sit down and organize and present what I already have. It is most unfortunate that all this didn't get started right back in October, as it should have done, but there is no point bewailing this now. Messeloff nags me to try and get out a couple of snappy papers quickly, so as to get priority in the literature etc: his urgings arouse my defiant refusals, for I am realizing, more and more clearly, that my own metier lies in *pondering*, and expressing myself monographically when I feel ready to. Whenever I am nagged to produce something quickly, I remind myself of the attributes which Bacon recommended: "patience to doubt, fondness to meditate, slowness to assert, readiness to reconsider, carefulness to dispose and set in order." [. . .]

Tomorrow, Israel Shenker (whom you met at Ray and Fred's, and who works with the NY Times) is coming to see me and my patients and Beth Abraham. He may write an article on it. For myself I deplore and fear publicity, but the hospital—which is terribly bankrupt and unknown—is in need of some: in effect I am being forced to schnorrer for the sake of my patients.

I will be flying to London on Monday night, arriving at home probably around 7 or 8 on the morning of the 5th. Will look forward greatly to seeing you all, and to a calm month of (alas! it has to be almost entirely) work.

Love,

To Tom Dahl

FRIEND FROM UCLA

AUGUST 8, 1969

37 MAPESBURY RD., LONDON

Dear Tom,

It was a great surprise, and a greater pleasure, finding your letter on my arrival in London a few days ago. Just before I left I was making myself out a new address book, transferred the familiar "Lourdes" address, and wondered what was up with you and when I should write.

First, your new work and post sounds beautiful, very exciting, very overdue, and earned over and over again. You sound very happy, and I feel very happy for you—the long years of dragging around, subordinate, and enslaved, and smouldering inside, are clearly over for you, and your new sense of freedom sings out from your letter. And Cetacea*—what could be more entrancing to work with! (I have a whole secret library on porpoises, and a pin-up of one above my head, along with a long quote from *Moby Dick:* I automatically think of them when I get misanthropic.) So I *do* share your feelings, some of them at least, and am happy that you remembered me, and are sharing your excellent news. [. . .]

For myself, things are also going well, and promise better, and for the first time in my life I can count myself fortunate, and have ceased to be tormented with endless regrets, recriminations, and absurd illusions. Three years ago, I returned to Clinical Neurology, and sought myself an analyst—and these two acts were life-saving for me in almost every possible sense. I know myself fairly well these days, and have ceased to hate myself: I know my powers, and I know what I love to do—and this is the most vital knowledge, and if I have obstructed the usual advancement academically by my eccentricity and seclusiveness and paranoia and lack of technical know-how, I think I am making some sort of parabolic incursion into a decent life and position by way of a series of monographs. I wrote a massive one on Migraine and some allied disorders last year [. . .] and am waiting impatiently for its publication. It's not a great book, but I suspect (and a few critics have been kind enough to suggest to me) that it is the best on the subject for some time—it is a very old-fashioned book: my strength is not in gathering new data, but in tenacity and pondering on the implications of well-known and well-established data.

This book seems trivial to me now compared with my present endeavour, which is to marshal an immense array of observations on Parkinsonism and psychosis (especially as these occur together in post-encephalitic patients and may be modified by l-DOPA and other drugs), and to fit these into a very general Pavlovian and Luria-like

* The order of marine mammals including whales, dolphins, and porpoises.

framework: these post-encephalitics are incredible to work with, and I have had the privilege of seeing what an immense range of innate behavioral mechanisms may be dissected by diencephalic disease and stimulation, and of the acute constructions and reconstructions of attitude and behaviour which these infinitely-conditionable patients show from day to day. I will probably entitle this book "Compulsion and Compunction," and will dissect out a hierarchy of coercive systems from the mid-brain to the super-ego. Or at least, I hope to! [. . .]

Do write again soon, and regards to your wife and family,

OLIVER

To John Z.

FORMER LOVER

AUGUST 14, 1969
[37 MAPESBURY RD., LONDON]

My dear John,

For others a postcard, but to you I cannot send less than a letter. But what there is to write, or what I should write, is another matter.

[. . .] It was a lovely flight (once we were off the ground), and I had a beautiful soaring sense of freedom as soon as we were on our way, which became sort of ecstatic (and made me weep to myself) as we floated down, at the end, through great soft fleecy clouds (I kept thinking of them as pastures or meadows, suspended in the sky, Edenic fields of eternal sunshine) down to funny little Gatwick Airport, which is so rural that we almost landed in a flock of sheep (they scattered, bleating, as we neared the ground).

London is dilapidated, disreputable, and wholly delightful, although its indolence would elicit censure from all right-minded energetic Americans. England has lost all her possessions and all her fierce ambitions, and has become, in her decline, almost infinitely gentle and civilized. Nobody seems to be shouting or rushing or pushing or blaming—the Imperial blazing summer is over, and there is a lazy timeless feeling, like a long September evening. And along with the indolence, people seem so cheerful, so *frivolous*—there is none of that harsh, deadly seriousness, that tone of urgent importance, which seems to invest American life these days.

I have spent ten days—how? I am not quite sure. A lot of time with my old friends, afternoons in Kew Gardens (which have not changed in 200 years—and I love gardens), swimming in the lakes on Hampstead Heath, idling in Soho and admiring the beauty of today's youth and their freedoms, smoking an infinite number of delicious Havanas (which one cannot, of course, buy in the States, because Cuba is the "foe"), and doing a certain amount of thinking, reading and writing for my book. I think I will have to start on its serious composition almost at once, but the whole subject has extended itself so much in my mind, and excites me so much, that I am afraid to take the plunge: it is exactly the feeling one has before leaping from the top diving-board.

And again, I have been enjoying my idle freedom, and the sensible orderly routines all round me: the bustling of old family servants (to whom I am "Master Oliver"), and the rituals of the week (the fish fried on Thursday, the lawn mown on Sunday, the family gatherings on Friday, etc.).

And all the local tradesmen and shops which have hardly altered from my childhood—although I am intellectually radical, I am socially conservative, and love the sense of a calm social order in which everyone has (and enjoys) their place and role. I miss all this so much in America, where I feel strictly nobody surrounded by anybodies, and everything is shifting and changing all the time (the nightmare called "Progress"). I long for the ease of an indolent, stable, liberal tradition.

I have thought of you a certain amount, and have wondered what was a-doing with you. I hope we can construct some meaningful but not-too-intense relationship when I return in September, something calm and civilized, and not at all devouring-torrid-enslaving, etc. We will have to see how things work out. I hope you are keeping well, cheerful, etc. and will appreciate a note from you if you have the time or inclination to write,

Yours,
OLIVER

To Charles Messeloff

MEDICAL DIRECTOR, BETH ABRAHAM HOSPITAL

SEPTEMBER 12, 1969
[NEW YORK]

Memo to Dr. Messeloff:

I think it essential that the administration ruling confining our patients on medication to the Hospital be modified with regard to our DOPA patients. Several of our patients have been going home regularly [. . .], others have gone on summer camps [. . .] and others on day-trips. In all cases the patients have been given l-DOPA and other medications to cover their needs while out of the hospital. There have been no problems whatever. *Is this all to stop?*

Going out is the very breath of life for these unfortunate patients who have been not only institutionalized, but imprisoned in their symptoms, for up to fifty years. Miss [P.]* has been a different woman since Mrs. Noden took her out on a little shopping expedition a few days ago: Miss [P.] had not been out of the hospital for three years, and wept with pleasure at being able to do so. Two of my patients, Lillian [G.] and Tillie [A.], have been institutionalized for more than 35 years, and have not seen the outside world for more than ten years. They were both looking forward eagerly to going to the beach on September 10, and were both heart-broken when this was suddenly vetoed at the last moment.

It is a great cruelty to activate these patients and then deny them activity, to animate them and then deny them the amenities of life. *I consider excursions an absolute necessity for their well-being*—and had, indeed, hoped that the number of these might increase. There have never been any technical difficulties about providing medication for them to take with them, for its administration *en route* or at their destination.

L-DOPA is an unprecedented drug with unprecedented effects, and unprecedented measures must be adopted for those patients who

* Names abbreviated for privacy.

thrive on it. [. . .] This should be discussed at today's (Sept 12) Staff meeting.

O. W. Sacks

To Jonathan Miller

SEPTEMBER 13, 1969
234 EAST 78TH ST., NEW YORK

Dear Jonathan,

I imagine that you are once more surrounded by wife and family (you look terribly *lost* even in a few days of enforced bachelordom), and working up for your opening of *Twelfth Night.* I hope the student group turned out more satisfying to work with than you felt at first.

For myself I am working about as hard as I could hope to do for a sustained period without break-down. It involves, amongst other things, getting up at 5 each morning to write for three hours which I hate to do (it's against my *cycle* as a start), but is the only certain way I can think of for getting good undistracted writing time every day. I've been diverted onto some papers, but I have written three of these, and have another to go, and then will be able to settle back to my book with a good conscience. [. . .]

When I returned [from London], it was to find myself the focus of a little minor fame, the consequence of Shenker's *NY Times* article.* I had about two hundred letters from patients (and their doctors) as far afield as Australia, a request from the BBC to borrow some of my films and tape-recordings, and from a radio-station here to tape an interview, and from *Der Spiegel* to reproduce a photograph, etc. This has brought me a little pleasure (I confess), but more distress. Especially as it has all gone to the head of Charlie Messeloff, the old fool who runs Beth Abraham. A magical promise, a magical triumph, has proved irresistible to his impotent senility, and he has flipped into an acute grandiose senile paranoia, with the Parkinson

* Israel Shenker's article, headlined "Drug Brings Parkinson Victims Back into Life," appeared on page 43 of *The New York Times* on August 26, 1969.

patients and DOPA and *me* as the nucleus of his megalomania. In my absence he assumed dictatorial powers, summarily fired the Chief of Medicine (an excellent man, and friend to me, who was looking after the floor in my absence), capriciously altered the medication, and did endless harm. On my return I found that *five* patients of mine had broken hips: Charlie, in his senile logic, had thought that if they were doing well on X Gm. daily, they would do twice as well on 2X Gm: they were wildly activated, hurled themselves to and fro on their weak and contractured legs, and of course went down like ninepins. Nobody knows better than you the dangers of a broken hip in a frail, elderly, chronically-disabled patient. The whole hospital is in danger from someone who has in effect become a brain-damaged Dictator whose merest whim can break or make. I have endeavoured to calm him, to reason with him, to compromise with him, and above all to keep him at a distance from myself and my patients: it is all useless. One sees the horrible, inadmissible sexual fascination at the centre of paranoia: my ward, and my patients, and above all me, myself, have become the emotional centres of Charlie's life. He keeps touching me, damn him, and saying, "You don't realize, Oliver, but I'm your best friend, the very best friend you ever had!" or "I want to be with you all the time, we must have an intimate working relationship, just like a marriage!" etc., etc. It is ironic and even tragic that my position at Beth Abraham should be rendered intolerable by him, at a time when everything else is going beautifully. I think that either he or I must go, and I may just have enough power now (with the strong backing of [. . .] many other colleagues who have suffered at Charlie's hands) to get him to go, by bringing the matter to the attention of our Board of Trustees. I have a little but not much patricidal guilt at the thought. I think that Charlie, not through malice, but through poor judgement and caprice, has already threatened the lives of many patients, and is a mortal danger to the hospital. And that he *must* go. And yet it is a hateful business to be involved in. [. . .]

I do hope that your poor father is out of his nightmare-complications, and being restored to his former state.

Drop me one of your rare lines—

Yours,

To Elsie and Samuel Sacks

SEPTEMBER 26, 1969
234 EAST 78TH ST., NEW YORK

Dear Ma and Pa,

Thank you very much for the books & papers which arrived [by] air-freight yesterday, in perfect condition. [. . .]

Fortunately, I think that broadcasts, newspaper articles etc. are now over. I share all your uneasiness about these. Medical articles and books are fine, but the other sort of crude publicity frightens me, although I have to say that Cotzias, Melvin Yahr* and everyone else in the field, over here, apparently have no hesitations or compunctions about it. But even though the AMA† may be lax, my own conscience is ruthless, and I cannot take any liberties with it. One ethical dilemma I have yet to resolve is whether I should see any of the many Parkinsonians who have written to me in consequence of the *NY Times* article. On the one hand I don't want to profit from publicity, but on the other hand I think I can probably assess and treat a case of Parkinsonism as well, if not better, than most of the people around here. I will probably compromise, and do a very *little* private practice, e.g. see three or four patients a week, and see if I can get a couple of hospital beds assigned to me so that, if necessary, I can bring in a patient and stabilize him on l-DOPA. [. . .]

Meanwhile I tolerate Messeloff as best I can. It is a great misfortune that he is behaving like this, but I see little chance of ousting him. I think I will have to accept the melancholy fact that my relation with, and management of, my patients will be partially undermined, and try to transfer most of my zeal to my writing. I manage to control my emotions, which is something, but I feel quite exhausted and ill after dealing with his meshugas all day, every day. This week I have had tonsillitis, and yesterday my back went out, though not too badly. I attribute both of these minor ills to the wearing influence of a tense daily situation. It's a dreadful thing to say, but I do wish Messeloff

* George Cotzias and Melvin Yahr were pioneers in the use of L-dopa to treat Parkinson's.

† American Medical Association.

would get ill—not seriously, but just enough to keep him out of circulation for a few weeks or months. [. . .]

So? I wish you all a very good New Year. I should drop a note to David and family, who sent me a New Year card.

To Mike Warvarovsky

OCTOBER 3, 1969
234 EAST 78TH ST., NEW YORK

Dear Mike,

Thank you for your postcard (postmarked Sept 2) which reached me on my arrival here from London, where I had spent the previous month. You too, I gather from your card, had also just returned from a trip to your home. [. . .]

I had hoped that the proofs of my "Migraine" book would be ready for correction when I went to London in mid-August, but they were not; my Publishers have been very dilatory about the whole thing, and I fear that my brainchild won't see the light until May 1970. This means, one way and another, that there will have been a total delay of almost three years since it was first written: some of this due to external obstructions, some to inner resistances of my own. If, however, I have learned to combat both, to some extent, the experience will have been a valuable one.

In the last six months or so I have been almost continuously involved and buoyed up by my current work (on Parkinsonism and psychosis, and the use of l-DOPA; my work and name was mentioned, very briefly, in *der (die, das?) Spiegel* recently), and now feel like a treasure-ship staggering into port laden with almost more than it can carry. A unique group of patients (those with the residues of the encephalitis lethargica, which ceased in 1927), a unique drug, and a unique observer (myself) have combined to give me a remarkable insight into human behaviour, from the most animal and automatic, to the ultimate structure of conscious mechanisms, and obsessions, etc. I have planned out a 3-volume work [. . .], of which I have written part of the first volume: mostly very elaborate case-histories, some as long and intricate as novels. I also have the material for about fifty papers, of which I have written a few. I have no doubt about the importance of all this work, and (if I am allowed to complete it) it will establish my reputation firmly and finally. [. . .]

My self-respect is steady (but not exalted) at the moment, my state of mind calm, and my hopes for the future fairly firm. I missed my September "mania" for the first time in five years, but on the other hand I have been able to count on almost continuous calm-of-mind and creativity for several months now, and hope that I can continue to do so. The manias are reactions to despairs, and after some years of both, I find myself less in need of, and less attracted to, the extravagance of either state.

On the other hand, I have to pay a price for this exemption from conflict and turmoil. I hardly exist, now, apart from my work. From getting up at 4:30 each morning to write, to my exhausted sleep—which is usually filled with dreams of Parkinsonism and the teeming theories of the moment—I have little time to myself, and almost none for human relationships or ordinary pleasures. I have become almost a monk in my devotion to work, and in my sense of a huge, privileged *task* which I must work at and finish. In my sense of being, as it were, dedicated to a supra-personal task (or mission), I can escape from the suffocating confines of other desires and guilts, which (as you know, and as you experience in yourself) are almost insane in intensity. In a paradoxical way, my existence has become tolerable and pleasant by having become, in a sense, impersonal or suprapersonal. There is much less fear, less craving, less blaming, less claiming, and all that. Monk-like, the Spinoza of 78th St. (Spinoza has become my idol, my ideal of calm and noble self-sufficiency), I write and think day and night, and when I am finished, will feel that my life has not been wasted, and that my thoughts and personality (such as they are) will not die with my own physical death. And this is a very comforting feeling, which makes my current back-breaking labours tolerable.

I confess I am lonely, and have no shoulders to weep on, no breasts to fall asleep on, and all that; and that I am neglecting my poor body—overeating, depriving it of sleep and exercise—but all this, perhaps, is the price one pays for too complete a sublimation. Perhaps I can achieve a more reasonable balance with time. I confess, too, that I still nourish nostalgic dreams and visions of California, of trips—physical and psychical—and doubtless I shall do to the end of my days. But I feel that my present journey, although it is accomplished mostly at the typewriter, can adequately compensate me for all others.

So, you see what a strange fate has befallen your old friend! The passion for truth has driven all other passions out of my life. But I promise myself, next summer, if my labours are over, to go away somewhere for six months—feeling, for the first time in my life, entitled to go wherever and do whatever I want.

To Mike Warvarovsky

JANUARY 19, 1970
234 EAST 78TH ST., NEW YORK

Dear Mike,

The arrival of your card this evening, and your New Year's greetings, bring me a sense of welcome from afar, and reminds me of my dereliction in our correspondence. [. . .]

The Californian Urge I empathize with, and submitted to a short time back, when I impulsively flew there for the Christmas–New Year week. I had a most beautiful ten days, molten with nostalgia, with the sense of Return and Re-Union, etc. Loved people, loved places—a tremendous sense of closeness, of immediacy, the sudden cracking-open of the hard, protective, estranging schizoid husk which I have built for myself in New York: I found myself embarrassingly close to tears for much of the time, but they were tears of joy, sorrow, memory, closeness, affection, not the hateful hard tears of rage and spite which I sometimes shed in New York. The joy of driving slowly up the Coast: of seeing a vast sun sinking into the ocean at Piedras Blancas, the golden rushes of Lucia, the cypresses at Carmel tormented into shapes of anguish and beseechment: everything seemed steeped in gold-and-greenness, outside time, eternal, almost allegorical. And lovely, at a human level, to meet again with Dave Sheppard, Frenchie (who hugged me when I came into the bar where he was drinking), and Peanuts, and Bill Tunberg.* I have felt utterly lonely in New York,

* Fellow weight lifters from Muscle Beach.

and the sudden feeling of friendship almost overwhelmed me when I went out again to the West Coast. And not least, an entirely unsolicited and very exciting job offer, which I must think very seriously about. What do I have to think about? I loathe New York, I always have. But, but—my analyst is here, and he has saved my life many times, and I suppose is felt as the only possible saviour of my soul. But I hate my dependence upon him, upon anyone or anything: and yet my feelings of dependence can become limitless: I depend on some people and things to give me my very sense of existence. And I hate to leave my patients—some of them, at least, whom I love, from whom I draw nourishment, to whom I give nourishment, and whom I would like to put on a big Flying Fortress, a Parkinsonian jumbo-jet, and whisk with me to California. I am very uncertain of my plans for the moment.

I am afraid that my creativity, like most of the good things in life, and the good loving mood, has been *occluded* for the last four months, roughly: I say occluded, for as soon as the prevailing mood of hate and spite and guilt etc. is drawn aside, it shines as brightly as ever. I so hate myself, I have to ration the good things to an absolute minimum. And I need it so much, to combat the dangerous feelings of worthlessness and folly which take hold of me, and command me to destroy myself from time to time. Never have I felt such stubborn resistance to what is best in me, and this reflects the felt danger and impermissibility of the triumph which is almost in my hands. Every rush of creative thinking is followed by a demoniacal rebound, a monstrous horde of punitive injunctions and (so cunningly is fatality disguised) masochistic temptations. I thought I was going to lose my hands over Christmas: I took a biggish overdose of ergot,[*] and almost vasoconstricted my fingers (to say nothing of my toes, nose, ears, chin, and penis) into gangrene. Eugh! I can hardly bear to think about it, and the very thought makes me nurse my fingers (still pale, cool, swollen, analgesic) and think how near they came to mortification. Whether or not Freud was right in postulating a basic Mortido[†]

* Ergotamine is an alkaloid derived from the ergot fungus. An overdose of this can cause not only gangrene but hallucinations and even convulsions. At therapeutic doses, it has been used to manage severe migraines.

† Death instinct.

in everyone, I assuredly have my fill of it (and so, probably much more than I, have you).

Having ruined my peripheral circulation, I think I must now make a virtue of necessity, and give up all the dear poisons I have been hooked on these past six years—nicotine, amphetamines, mescaline, cocaine, etc. —all of them, the whole bloody lot, because they're all vasoconstrictors. I cannot take 1 mg. of speed[*] without inducing an intense spasm in my finger arteries. So I have to be *good* from now on. I wonder what will fill the moral vacuum left by drugs? And also the moral and emotional vacuum left by a sort of perverse love-affair, never consummated, which has chained me in masochistic bondage for ten years?[†] I think it'll have to be God. ("The best cure for dipsomania is religiomania.")[‡] Which reminds me that I met Howard Heidtmann[§] (now Bhrendan Christenson!), who has undergone a profound, and probably permanent, conversion: he is quite insane I suppose (I take all Faith to be delusional wish-fulfillment) and also obviously redeemed: loving and gentle where he used to be hateful and violent, innocent and forgiven where he was consumed with guilt and reproach, etc. The Sin against the Holy Ghost (which is the inadmissible desire to be fucked) at once gratified and forgiven in its ultimate sublimation: becoming the beloved, the "vessel" of the Lord, the soul, passive and feminine, swooning in the more-than-carnal, redemptive embrace of its Heavenly Father. I was greatly reminded of Schreber's *Memoirs,*[¶] which you must read. So let's all get God, man!—the ultimate trip for all of us wicked sinful paranoids.

And on this note, dear Michael, I conclude, and wish you all you wish yourself for the New Year.

yrs,
Oliver

* So much for OS's vow to give up amphetamines, unless he is speaking historically here.

† He seems to be referring here to his love for Mel.

‡ He is paraphrasing William James.

§ A bodybuilder and a model.

¶ Daniel Paul Schreber, a German judge whose 1903 description of his own schizophrenic psychosis, *Memoirs of My Nervous Illness,* influenced psychiatrists from Freud onward.

To Michael Sacks[*]

BROTHER

FEBRUARY 7, 1970
234 EAST 78TH ST., NEW YORK

Dear Michael,

I was very happy to receive your short note of January 4, and have been guilty and angry at myself for not having made any reply to it in the month since its arrival. For reply it certainly requested, despite its appearance of a bald statement of fact. [. . .] My reactions (like your own, though less so) are contradictory, and this has made the writing of a return-letter difficult.

One general thing is quite clear, and a requisite for your survival (as a person). You must have some third mode of existence, some middle-ground, between living at Hospital (which, with the exhaustion of its initial promise, is barely tolerable for you) and living at home (which is less tolerable): you must have some middle-ground between feelings of doom (yourself the victim, the persecuted, the cross-bearer) and phantasies of redemption (yourself the messenger, the chosen, the evangel and bearer of redemption). Above all, you must secure some release from your conscience, which is the most merciless I have ever known: like Kafka, you labour constantly under a sense of infinite guilt. You can find nothing right or loveable about yourself, and with this nothing forgiveable in those around you. One's moral space, so to speak, can be filled with accusation, and leave no room for reality.

I know, I spend a great deal of my life in just this paranoid posture, filled with feelings of hate, vengeance, abjection, hopelessness. And I too (though obviously much less than yourself) regard every moment as a moment of crisis, full of threat and promise, full of destructive and redemptive potential.

You pretend, constantly, that you have no "savvy": untrue, absolutely untrue. You have lots of savvy, lots of common sense and just judgement of people and situations: but you will hardly permit

[*] In a way that might seem odd to a twenty-first-century reader, OS rarely wrote to Michael about his schizophrenia, though they sometimes spoke to each other about it in person. For the most part, though, they seem to have regarded it as a delicate subject, and OS often feared saying something that might upset his brother.

yourself to believe this, or use it. You imagine, again, that you produce only feelings of revulsion and contempt in others: untrue, utterly untrue again, you have a very real capacity to inspire affection and esteem in others, but this too you decline to believe.

I have just heard (in a letter from Ma and Pa) that you have secured a job in the West End, partly inside and partly outside. I am very glad to hear this. Please let me know more of it.

Besides a job you need (not economically, but for your very sense of existence, and for the possibility of release from the knots of hate and spite which bind you, and our parents, at Mapesbury) a place of your own—a hostel, a little furnished room, *something you can get away to,* someplace where you can put a few belongings of your own, and establish some autonomy. There is no human being, there is no animal, which does not need some territory it can feel as its own: and you are no exception. At Mapesbury you are engulfed, swallowed whole, nullified, reduced to a sort of enraged slavery and dependence, and this is death for you. [. . .]

You must move out of the world of obligation, as much as you can, for obligation is bondage, and splices you to attitudes of compliance and defiance. And all obligation is hateful.

As opposed to devotion, which is an expression of love. I won't nag you to start psychotherapy or anything: it may well be pointless at this stage, and God knows you have had enough negative experiences of [psychiatric treatment]. What you *do* need (and what I find myself moving towards) is a sort of religion—and here I mean religion not merely as a set of ceremonials but as an impregnable basis of loving and devoted feeling. For there is nothing in the world which makes one feel really good other than the feeling of fulfilling some task (mission, or whatever) towards others.

I had meant to send you, and you should read, the most beautiful book in the world (to my mind): Spinoza's *Ethics,* which after a fantastic (and I think, absolutely true) analysis of the nature of all human feelings, moves on towards a sort of transcendent psychology, theology, the nature of true acquiescence and freedom, and what he calls the intellectual love of God. It is a marvellous book, the nearest I have to a bible.

Anyhow, I have said much too much, and sounded pompous and patronizing, which I was afraid I would.

Do drop me a line, when you feel like it, and tell me how things are going with you.

Love,
Oliver

To Jonathan Miller

[CA. APRIL 1970]
2512 BRONX PARK EAST, BRONX, NY*

Dear Jonathan,

Lovely to get your letter yesterday. Your letters are indeed rare, but most prized. [. . .]

There is obviously *so much* to do here, with my present patients. In March I wrote four long letters and four articles which I sent to the *Lancet:* they accepted my letters (I enclose copies of two) but felt that the articles would overload them. Now when I tot things up, I realize that first I have another dozen articles to write, and [. . .] *then* I must get on with my next subject (I do one subject at a time, and do it definitively), which is "dementia" (which I will demonstrate and construct in terms of hierarchical cortical functions and intellectual structures, Luria† and Piaget in reverse), and *then* "schizophrenia." So I see years of work stretching ahead, which gives me a good privileged feeling (the "task," the "mission," etc.) but also a feeling of being walled-in, and having to shelve the wild wanderings and wanderlusts and extravagances of youth, neurosis, drugs, etc. But it is not a bad bargain. Thank God I have a remarkable ability to sublimate and to become dedicated to things, otherwise I would end up catatonic, like my brother in Bedlam.

Physically life has become a good deal better, now that I have a car (a 1967 Rover) and a pleasant apartment. The hospital is renting me

* OS had just moved into an apartment next door to Beth Abraham, owned by the hospital and reserved for physicians on call. He had agreed, essentially, to be on call continuously, the better to observe his patients at all hours.

† A. R. Luria, the Russian neuropsychologist, was a mentor to OS (he is introduced as a correspondent in the editorial note on page 261).

this for a fairly token sum. I have six rooms (which is large for NY) which occupy the ground-floor of an oldish house, with a garage and basement, and also a little garden fore-and-aft thrown in. [. . .] I hope you will consider staying over for a day or two when you next come to NY.

Naturally (or naturally enough with my degree of ambivalence) I also feel aghast at the thought of such comfort and security etc. and feel that a) I will be punished for it, b) that it will suddenly be taken away, and c) that it will cause an immediate stultification, petrification, mummification, (? lunefaction), for I have adopted the working hypothesis that any productivity I have is proportional to, and dependent on, my insecurity and misery, and is, so to speak, *permitted* by this. I think, however, that the opposite will turn out to be the case, and that a reasonable degree of inner and outer comfort and security will make a steady productivity easier, and release me from the former necessities of drugs, manias, self-indulgent depression, sudden flights *to* and *from* (the ideal, the intolerable) etc. Essentially because it becomes easier for me to detach the products of my imagination from my own wishes and fears, which liberates the imagination on the one hand, and also liberates one from one's personal phantoms and unconscious imaginings. You, obviously, have had a liberated imagination for many years, but with me it is fairly recent.

Though I do ask myself: liberated for what? Almost every night, if not several times a night, I dream about excitation, and inhibition, and ultraparadoxical states, and points of "breakthrough" (you will see I have gone very Pavlovian lately). One can get so *consumed* by one's work, that there is precious little left over. And I begin to wonder whether I am becoming inattentive to the phenomenal world, its beauties as well as its horrors, through inner concentration: so that at times I feel (I will not tell you whom I am paraphrasing)* that I am becoming neither more nor less than a very good machine for deriving general laws from large bodies of data. I think you are more kaleidoscopic than I am: my metier seems to be a sort of concentration, a cohabitation with my data.

And the resistance one must fight to find new ideas! So much emotion, so much attachment, goes with the old. And one feels like

* Miller would have known this famous quotation from Darwin.

a murderer when one must smash down one set of ideas to make room for a new set. The real work is this smashing down. Once one does this, the very questions evaporate: you see what is true facing you, no longer obscured, and now it is obvious, really tautologous. And so there is a sense of let-down. And an intense sense of surprise that other people *bridle* at what one has said, because one forgets all of *their* resistances. [. . .] I used to think it either a false modesty, or a concealed grandiosity, when Locke wrote: "it is ambition enough to be employed as an under-labourer in clearing the ground a little, and removing some of the rubbish that lies in the way to knowledge," but now I think it was an exact and just estimate of his role. And I want no other role myself.

This letter has blathered on much too much.

Write again soon.

My love to the family—I will hope to see you all in August.

Yrs,
Oliver

To Elsie and Samuel Sacks

MAY 6, 1970

2512 BRONX PARK EAST, BRONX, NY

Dear Ma and Pa,

This letter will arrive, I imagine, about the time you return from Israel. I hope you had a good, relaxed time there: what is the general atmosphere like? From the outside, of course, the whole Middle East seems to be boiling in a particularly dangerous way, or about to blow up. Indeed the whole world seems that way at the moment—tensions are extremely high here, following the invasion of Cambodia and the brutal shooting of those students in Ohio. Groups of people, complete strangers, stop and talk to each other in the street; most of the colleges throughout the States are putting on strikes or boycotts; there is a very fearful, ominous atmosphere, something like that in the first days of 1939, I imagine.

It seems almost improper to pursue one's personal life, when everything around one is so agog. If March was a lovely calm productive month for me (and I realize, looking through my papers, that I wrote more than a thousand pages during March, although only a fraction was submitted for publication: the *Lancet,* by the way, have sent me the proofs of the four long letters I sent them, and these will be appearing in the next few issues), April was one of flux and turmoil, associated especially with moving house, or rather with setting up a household of my own. The bustle has been quite a pleasant one in many ways, although the consumption of time, labour, money etc. has been far greater than I would have imagined. In a sense I have had to start from scratch, for my life in New York has really been a sort of living-out-of-a-suitcase existence in furnished rooms hitherto, whereas now I have a "home" of sorts, or so I must regard it. And so I do regard it, with all the concomitant feelings of security and alarm which anything permanent or permanent-seeming always gives me. [. . .]

There are two quite massive (or by New York standards quite massive), airy rooms, in the front of the house. One of these I am turning into a dining room. [. . .] In the front room I have, as yet, no furniture, but a lovely grass-green shaggy carpet going from wall to wall. I may just put cushions, or a very low sofa here. It'll be a sort of spacious lying-around room. Its furnishings I want to be all "natural," i.e., I have some nice lamps with wood and cork bases, and shells and plants and so forth. I shall make a sort of plywood platform running crescentically round the window side of the room, which [. . .] will be filled with pots and earth and mosses and ferns and other greenery, and resemble a woodland bank. At least this is the sort of motif I want. I have felt absolutely starved of nature and living objects and greenery for the last five years, and will try to create a sort of conservatory-atmosphere in the front room. There is also a square of ground in front of the house, and another behind it. These I must divest of the broken glass and other rubbish they have at the moment: I will then try to hedge, wire off or otherwise enclose them, and make a miniature or two miniature gardens. Fortunately I seem to have "green fingers," and the loads of ferns and other primitive plants I have been digging up and transporting from nearby woods seem to be doing well, and can probably be transplanted out fairly soon. So, a new "domestic" side of me seems to be emerging. I'll never become very domestic, if only because it wastes so much time: but it's a good feeling pottering around one's place, and imprinting oneself and one's tastes on it this way and that. I feel that I have lived an anonymous, squalid, uprooted etc. sort of existence far too long. [. . .]

One way and another, and even a year ago I would never have suspected it, my beloved post-encephalitic patients, and their infinitely varied reactions to l-DOPA, have led into a vast landscape of clinical study, really into a life's work. And by the time I am "through" with them, I will have derived an experience and knowledge of primitive (sub-cortical) behavior matched by nobody else around, to say nothing of the "bonuses" like the study of tics, obsessions, mannerisms, perceptual distortions, memory upsurges, etc. as these occur in the patients. I seem to be getting a bit verbose, and not a bit grandiose. Actually I don't feel grandiose, I just have a sort of feeling of being fortunate, in that such a treasure of knowledge and understanding has come my way, and is now beginning to offer itself up for harvest.

The business of my car, however, has been much less felicitous, and has caused me a great deal of rage and anxiety. I mentioned, I think, that I was beguiled from my original intention to buy a new BMW into getting a 1967 Rover, ostensibly in "showroom" condition. This car, it is true, has excellent bodywork and a good engine, but I have had a packet of infuriating troubles with it in other ways. [. . .] I am exceedingly sorry I didn't pay the extra $1500 and get a new BMW to start with. One has to be very car-smart and lucky in buying secondhand cars, and I am afraid I am neither, and so got cheated, etc. [. . .] Part of the trouble, of course, is that I am now so *vexed* about all this. The actual trouble is with the car becoming a nucleus of neurotic feelings, which tend to become more and more dramatic: e.g. feelings that the car will fall to pieces at any moment, that the people who sold it to me were criminal psychopaths, that I want to take physical or legal action against them, phantasies of revenge, etc. Once feelings of this sort get inaugurated, I find it quite difficult to maintain my calm (or my "cool," as they say now), and to deal with the whole matter in the reasonable unemotional business-like way it demands. [. . .] And it's such a flaming waste of time to get insomniac and paranoid about a mere lump of metal. I think, however, that the matter will be resolved in a day or two, and will then cease to fester in my mind.

I feel myself almost diagrammatically divided into a "good" self and a "bad" self. The good self is full of love and wonder and praise etc. This is the part which makes me see my patients as miraculous concatenations of disordered brain-systems, and to feel endless sympathy, understanding, compassion, etc. for them. The good part is the creative part, and must be in the ascendant if I am to write or even to work in any really-involved way. The bad part is full of hate and fear and blame, and is as selfish and destructive as the other is altruistic and constructive. I suppose everyone has good and bad "halves" in this sort of way, and it is just that they are *both* rather hypertrophied in me, and that analysis etc. has made me acutely conscious of them. [. . .] Messeloff tends to bring out my bad half, which is why I try to steer clear of him. On the other hand, teaching, patients, poetry, philosophy, Nature, etc. bring out the good, lyrical part of me, which is much pleasanter for me, and for everyone concerned. You will be pleased (amused/amazed?) to hear that I was given quite an honour by the student body last night. There is a sort of "elite" group—the "AOA"—Alpha Omega Alpha, consisting of the top ten students in each medical school throughout the country. Anyhow the Einstein branch of the AOA asked me to be their guest at their annual dinner last night, and made me an honorary member of the AOA, and gave me a certificate and a key and other little tokens. I think, however, that one of the best parts of the evening was when my old enemy Labe Scheinberg, now Dean of the College, praised me fulsomely, and said that I was the best Neurology teacher the college had had since the late lamented Saul Korey. All of which made me blush, and feel that they were all crazy etc. But I also felt very good, and feel that this is, in a sense, perfectly true. I think that I am a good (and, very rarely, at magic moments, a great) teacher: not because I communicate facts, but because I somehow convey a sort of passion for the patient and the subject, and a feeling of the *texture* of patients, the way their symptoms dovetail into their total being, and how this, in turn, dovetails into their total environment: in short, a sort of wonder and delight at the way everything *fits* (and it does all fit, so beautifully, like a wonderful jigsaw puzzle). So yesterday evening was a good evening, and brought out the good part of me again, after this had been eclipsed by ten days of anxiety and paranoia about the car.

This letter has become much too long and garrulous. It was lovely to hear you all over Pesach, and I will look forward to your next letter.

To Augusta Bonnard

JUNE 8, 1970
2512 BRONX PARK EAST, BRONX, NY

Dear Augusta,

Putting together a short paper on the peculiar oral behaviour activated in some post-encephalitics by l-DOPA, I suddenly bethought myself of your paper on "The primal significance of the tongue," and want to say how full of interesting observations and insights I have found it. Several of my patients have developed (usually in the context of other tics) "tongue-sucking," "tongue-swallowing" and complex hand-mouth mannerisms (thumb-sucking, etc.). These acquired, gratifying practices seem to evolve from innate tongue-movements and oral behaviour (tongue-pulsions, licking and lapping, retching, etc.) identical with those elicited by Hess* with diencephalic stimulations. [. . .] I have no doubt that you have run across a good many post-encephalitic patients: they seem to me the most interesting (and in many ways the most tragic) neuropsychological patients there could be, and studying them (especially with something like l-DOPA) has been my chief pleasure and task for the last two years.

After four years of analysis (I should say nearly five years—I am "forgetting" Los Angeles) I find myself possessed of certain strengths and stabilities I never knew I possessed, and released, for at least a good part of my time, from the dreadful neurotic bondage which had me in thrall, really, between the ages of 16 and 33. I can't say I'm too successful with personal relationships (and doubt now that I will ever be), but I find a profound absorption and pleasure in my work, and stay fairly well-sublimated for most of the time. I look back with horror at the days of drugging and narcissistic absorption, which would surely have destroyed me had it not been for analytic intervention. I am even slowly and finally becoming solvent, and think it at least possible, and even probable, that I will be able to repay my debt this year. Hitherto, with an inadequate income (45% of which was absorbed by analytic costs) and high costs-of-living in Manhattan (I have now moved into the Bronx, a stone's throw from the Hospital where I work), I could barely keep my own head above water, let alone repay debts. [. . .]

* Walter Rudolf Hess, a Swiss physiologist and author of *The Biology of Mind.*

My parents both seem very well and active for their ages, my father especially. I must say this: that for the first two years of analysis, I was consumed with feelings of reproach and even hatred against my parents, seeing them as all-culpable vehicles and inducers of neurotic misery. These feelings I have worked through, and now I find myself loving and admiring them for a quality (indeed, the most valuable human quality) which has been the centre of their lives, and is now the centre of mine: namely, a feeling of *devotion* and dedication to their patients and their problems. This feeling of devotion (which gives me a sense of self-worth, an absolution from self-hate, and an intellectual and moral task or "mission" which will last me as long as I live) is something I have only really experienced in the last three years or so: And it is obvious, from my knowledge of you and from re-reading your paper, that something very similar and intense must be a mainspring in your own life, and (while in action) a means of transcending personal conflicts and loneliness.

I will be coming to England this August, which will coincide, I hope, with the final (intolerably slow and obstructed) publication of my "Migraine" book, and will hope to see you then.

Best wishes,
Yours sincerely,
Oliver Sacks

*To Paul Turner**

PHYSICIAN

SEPTEMBER 12, 1970
2512 BRONX PARK EAST, BRONX, NY

My dear Paul,

[. . .] It was an enormous pleasure seeing you in London again, and—I realize in retrospect—a meeting which certainly coincided with, and possibly prompted, a change of mood. Because I had been feeling rather depressed, and sullen, and unwanted etc. during my first

* Turner and OS had been young doctors at Middlesex Hospital together.

ten days in London, and you gave me such a feeling of *welcome*. I had a funny feeling that you spoke for all England when you said: "Why don't you come back to us one of these days? Join in, be part of us," etc. Because one of the most constant (bad) feelings I have is the feeling of exile, an alien in an alien country, etc. and your welcome (to say nothing of your kind efforts to arrange a talk, a radio-interview, and what-not) appealed in the directest way to my desire to come back. I spent most of my holiday walking round London, round my old haunts rich in association, and reading Dickens, and reminded myself once and for all that I am a Londoner born and bred, that I am steeped in London, and that I will inevitably feel uprooted and deprived if I live anywhere else. And perhaps the old buildings of Barts,* under the shadow of St. Paul's,† reinforced the feeling, as did the splendid solid *dignified* meal at Simpson's, and the brown-raftered pub, and the walk over the bridge, and the prospect of the Thames, and reminiscing together.

I went to see Calne‡ at Hammersmith, the day after meeting you, whom I found very bright, eager, but essentially pedestrian. That's a mean word to use after his receiving me so pleasantly, and giving me an afternoon of his time, and letting me sit in on his Parkinson clinic, and arranging a visit to Highlands§ for me, and showing me an advance copy of his book (which came out a few days ago). Seeing Calne, sitting through his clinic, and reading his book, were also highly significant experiences for me, for they brought into the sharpest focus the antithesis of two styles. He sat in his clinic, alert, neat, quick, with a pile of ruled test-sheets in front of him, rapidly reducing his patients to test-scores, and paying (unless I do him an injustice) almost no attention to them as people: tremor 3, rigidity 2½, akinesia 4, next patient please, etc. I was irresistibly reminded of Henry James' impression of Ellis Island, of the immigrants "appealing and waiting, marshalled, herded, divided, subdivided, sorted, sifted . . . an intendedly 'scientific' feeding of the mill," and of the folly of this

* St. Bartholomew's Hospital in London, familiarly known as Barts.

† The cathedral.

‡ Donald Calne, a neurological colleague and distant cousin.

§ Highlands Hospital, on the outskirts of London, was home to about two hundred postencephalitic patients, the largest group of such patients OS knew of beyond his own patients at Beth Abraham.

sort of "objective" examination which replaced human judgement by stereotyped questionnaires. [. . .] In some very fundamental sense, when all the gradings and rating and testings and measurements have been done, and all the test-sheets and schedules filled up, the very flesh of the subject has escaped, like a jellyfish through a tea-strainer. It is so senseless to describe symptoms and signs as if they were fixed emanations from the patient, when they are no more than reactions at a particular moment to a particular stimulus ("Symptoms are *answers,*" writes Goldstein,* "given by the modified organism, to definite demands . . . (their appearance) depends on the method of examination").

I am forgetting that this is a letter, and getting all hot-under-the-collar and polemical. Actually, I think I am rehearsing the preface to *my* book on Parkinsonism, which I have just started writing, and which perhaps Calne's book has, in part, stimulated. I am sure that a wholly different mode of description can be used, must be used, in place of these horrid test-figures with their derivative graphs and statistics. The classical mode of description is *narrative.* I see no way of picturing the innumerable influences and factors involved except in a narrative. Can you imagine Dombey or Jarndyce† reduced to a few figures and diagrams? I think there is something almost inhuman about the *mechanical* intelligence which has taken over so much scientific and medical writing: it is reasonable enough to describe physical and chemical and simple physiological processes, but cannot possibly indicate the repertoire of higher behaviour. I was reading *Hard Times* a couple of weeks ago, and find myself thinking of Calne's approach (which is the accredited and almost universal approach to patients, and mode of presenting data) as pure Gradgrind. You were good enough to use the word Oslerian‡ of one of my *Lancet* letters: I assure you that I was not indulging in deliberate archaism, but using the only mode of presenting the data which could begin to do justice to it.

To get off the subject (I'm sorry for *riding* it, quite inappropriately,

* Kurt Goldstein, author of *The Organism.*

† Dickens characters, from *Dombey and Son* and *Bleak House.*

‡ OS admired the work of William Osler (1849–1919), a Canadian physician and renowned diagnostician who believed in teaching medical students through patient histories rather than textbook descriptions.

here: and I think you probably agree with me anyhow), I visited Highlands last week, and had the most intense, the most uncanny, sense of déja vu I have ever had, seeing patients so similar to my own, sometimes down to the smallest mannerisms. I think they are a miraculous group of patients (and obviously much better cared for and healthier than ours), and I am quite clear in my mind that it is with *them* that I wish to work when I return to England. But how, when, in what capacity, by whom funded, etc. I can't possibly say. My flaming publishers have again postponed the publication-date of my *Migraine* book (Faber's, though very old and highly thought-of, are apparently notorious for this endlessly protracted muddling and bumbling along), to January 25, and I will make sure that I am in England on that day, and for a few days before and after, to look after my child, to boost it, if I can, and to see reactions. So I will surely see you then. But let's keep in touch, by letter, in the interim.

Thank you again for everything, and looking forward to hearing from you.

Best regards to your wife,

To Barbara Waymouth

FORMER STAFFER AT OS'S SCHOOL

SEPTEMBER 30, 1970
[2512 BRONX PARK EAST, BRONX, N.Y.]

Dear Barbara,

[. . .] Your phone-call, suddenly, out of the remotest past, set me to thinking of Brayfield* and those days generally. My memories are very patchy and biassed, because so much of that era has been blotted out of my memory. A traumatic amnesia, I suppose, because like so many

* OS referred to the school itself as "Braefield" (spelled variously) in *Uncle Tungsten* and elsewhere. Officially called St. Lawrence College, this noted prep school had evacuated during the war to Courteenhall in Northamptonshire, a few miles from the village of Brafield-on-the-Green. OS and his brother Michael were sent there at the outset of the war, when OS was only six. It seems that the Sacks family may have referred to it by its location.

evacuees I had the feeling of having been deserted or marooned by my parents, and the recollection of this is too painful to be allowed into consciousness. And yet, it was a very pleasant time in all sorts of ways. I loved the village—and all sorts of non-human memories: of the little shop, and the walks, and the churchyard, and the nest I built in a tree with Lawrence Greenbaum, and the mushrooms, and the mild cow-y meadows, and the smell of horses etc. etc. came flooding back to me when I recently read a book about a village (*Akenfield*—beautiful book). And in *some* ways I liked Vernon House; it has ever since stood in my mind both as the image of a benevolent institution, where one could be loved and protected, and also as the image of a terrible institution, associated with memories of starvation and repeated beatings. I am sure it is no coincidence that for the last ten years I have opted to work in institutions; where my patients, in some ways, are *like* pupils, or perhaps children. I have always liked *you*, and you provided a warmth, and a fairness, which I felt extremely in need of (and which was *not* provided by the Buckeridges,* whom I have since considered, perhaps most unjustly, as Dickensian monsters). And so, twenty-seven years later, if you will allow me, I will thank you for what you gave me, and what you stood for, at a dark time.

My very best wishes to you,
Yours,

IN SEPTEMBER 1970, *The Journal of the American Medical Association* published a letter from OS in which he reported on the effects of L-dopa in sixty of his patients and pointed out the complex and unpredictable side effects they had confronted. He expressed his reservations about FDA approval of the drug for general use, in light of these adverse effects and uncertainty about L-dopa's long-term safety. His colleagues, stung by this criticism of the much-heralded miracle drug, responded with skepticism and, largely, dismissal.

* Anthony Buckeridge, the headmaster of Braefield, later became a well-known author of children's books.

To Leon J. Salter

CORRESPONDENT

NOVEMBER 6, 1970

2512 BRONX PARK EAST, BRONX, NY

Dear Mr. Salter,

Thank you very much for your kind and encouraging letter of October 20, and for the fascinating (and terrifying) bundle of data you enclose. [. . .] As you note, the so-called "side-effects" of l-DOPA are endless in number (and the tally is by no means completed yet).

Of course, physicians have no right to use the term "side-effects" in this context. The very use of this term is based on assertions and assumptions which are quite without backing. All the so-called "side-effects" of l-DOPA are essential and inevitable effects, and this would be fully realized if there were adequate comprehension of the nature of Parkinsonism and the effects of continued stimulation of a damaged nervous system, on the one hand, and no intrusion of redemptive phantasies, on the other hand. Not the least part of the trouble is that everybody (and here patients, as it were, collude with physicians) has phantasies that some "miracle" drug, discovery, invention etc. will come along and make everything wonderful, will restore primal health, strength, virility, stability, etc. It is a phantasy which goes back to the Elixir of Life, and the Golden Apples of the Hesperides, and Nectar and Ambrosia, etc. And the greatest people (Freud, for example, in his early eulogistic essay on "Cocaine") have fallen prey to such delusions, and lent the weight of their authority to them. I will be publishing my own data on the use of l-DOPA (which I believe to be the most complete available), and my thoughts on the whole matter, in book-form. I have completed the book, but I can't think of a title. Or, rather, I can think of too many titles! What is finally so tantalizing is that l-DOPA *does* have the most extraordinary therapeutic effects to begin with (even in patients who have had Parkinsonism for 50 years), but that these are succeeded—at best—by relapse, and at worst by an endless and incalculable nexus of "side-effects" (like antibiotics, insecticides, etc. etc.).

I suspect we think and feel alike on the matter.

I am not a "campaigner" myself, by temperament, and I think that *for* and/or *against* positions are both very tricky. My own chief concern is to tell the truth *about* the matter—which I have endeavoured to do in my book. But there is no doubt that campaigners with good heads and hearts—like yourself—are immensely valuable, and do a heroic and often thankless job trying to undo the dreadful muddles which we (as physicians, as human beings, as creatures of phantasy and greed) are constantly creating.

Thank you again, and the best of luck to you in your endeavours,

Yours sincerely,

To Elsie and Samuel Sacks

NOVEMBER 7, 1970
[NEW YORK]

Dear Ma and Pa,

It seems a long time since I wrote or received a letter, but I don't suppose it can really be so long—it must be an illusion created by intense pre-occupation and busyness writing for the past five weeks. [. . .]

In the meantime, I have decided to split my originally-planned book into two. The [first part . . .] will deal specifically with l-DOPA, laying stress on its dangers, on the reasons why these have not been adequately reported, on the dangers of being seduced by phantasies of "miracle" drugs which make everything come alright, etc. I think it will be cool, but white-hot (if you know what I mean), deadly, very topical, and very important. And I want it published in a hurry—if only to save the minds and lives of patients who would otherwise lose these in consequence of getting l-DOPA. I have been in touch with Faber's, and hope to send them off at least a rough (i.e. not retyped manuscript) this week or early next. The immense detail of the data, its flawless organization, etc. should greatly add to the book's effect, and make it impervious to attack. And I have no doubt that it (and I) will be attacked, for I have not concealed my opinion that most of the published work on the subject is drivel, if not hum-bug, and has amounted to an enormous act of deceit and self-deceit, and a blot on the medical conscience, etc. And one can hardly state such opinions and be loved for them. But state them I must. I am borne up by a feeling of moral duty in the matter, and by the knowledge that I am not myself in any sense a campaigner or combatant. [. . .] I think (I hope, and doubtless I fear) that this DOPA-book will be something of a bombshell. [. . .]

I have held my hand for nearly two years, refraining from any premature publication of large studies (as [. . .] my own commonsense advised, although Messeloff, of course, was always pestering and nagging me to publish "something positive," in less-than-no-time, etc.). But now I have seen the whole melancholy round of l-DOPA (and it is *round*—like the Earth: a man cannot escape from his Parkinsonism any more than he can walk off the surface of the earth: at the point when l-DOPA seems most effective, when the patient shows anti-Parkinsonism, or is in the Antipodes of Parkinsonism, at that moment his course is already rounding itself to return to where he came from)—now I have seen the whole round, I am ready to pour out a spate of publications. [. . .]

As you gather from my content and tone, I have really had no other activities or interests of any kind for some weeks. [. . .] I have scarcely seen the autumn change of colour—and it has been an unusually lovely Fall here. And I've been too busy for any exercise, and over-eating and smoking, which is bad—as soon as I get the packet off to Faber's, I think I will take a few days off, and devote some attention to the other things in the world. Things seem appreciably cheaper now, and I have got enough to pay off Augusta [. . .] and withdraw that particular thorn from my flesh. [. . .] I should be grateful, Pa, if you could refrain from mentioning my DOPA-book to anyone at the moment, especially cousin Calne. Do write and tell me how you all are, and what is happening in London.

OS CONTINUED TO WORK ON his L-dopa book. Meanwhile, in January 1971, after various production delays, Faber & Faber finally published his first book, *Migraine*.

To Rita Henryk-Gutt

REVIEWER

[CA. APRIL 1971]

37 MAPESBURY RD., LONDON

Dear Dr. Henryk-Gutt,

You have recently reviewed my book on *Migraine* in *Nature* and the *British Journal of Hospital Medicine.* [. . .]

I need scarcely say that I have found your reviews both distressing and disturbing. I am grateful that you give me credit for a "considerable reading" in philosophy, biology, history of medicine, and psychology, and that you allow that I have had "extensive experience" in the field. But you make it sound as if I have used whatever knowledge I have as a *weapon,* as a means of discouraging research. And you must feel this very strongly, or you would not have written *two* hostile reviews.

And yet, perhaps, you are partly right. A most eminent neurologist recently said to me: "Your book is fascinating, and awfully well-written and all that, but of course it's *irrelevant.*" I was somewhat taken aback, and enquired what he meant. He replied: "There's no point *talking to patients,* asking about their symptoms, when they have attacks, and so forth. All that matters is to find the final common path of migraine in the brain, and *block it.* And that's that!" It was partly *in reaction* to this sort of de-humanizing outlook that I laid so much emphasis on the migrainous *person,* and his symptoms, sufferings, feelings, etc. and so little on his *chemistry.* [. . .] I think that you have also been unfair in concentrating your attention on the Chapter which most offended you, and failing to give me due recognition for the earlier parts of the book. You should, if only in passing, allow that I have provided a very careful and detailed account of the phenomenology of migraine, and that this has real value. There are the biological phenomena of migraine, and there is the physicochemical substrate, and one *cannot* be reduced to the other. [. . .] By training and temperament I am a biologist and clinician, that is to say, I deal directly with biological phenomena and not with physicochemical phenomena. If I were a laboratory doctor, it would perhaps be the other way round.

So, I'm sorry if you received the impression from my book that I wanted to do down neurochemistry, etc. Of course I don't. I know perfectly well that an event like a migraine *must* be associated with profound neurophysiological and neurochemical changes. If I pleaded for a wider approach to research on migraine, I meant only that its very important biological aspects should not be ignored. Indeed, my original title was "The Biology of Migraine," and perhaps I should have steered clear of the dangerous neurochemistry! It is certainly improper for a biologist or a clinician to under-rate the profound importance of physicochemical processes; but it is equally improper for a neurochemist, for example, to under-rate the biological processes.

I do hope you won't take this letter amiss, and I hope that you will give me (and my book) its due.

With kindest regards,
Yours sincerely,
Oliver Sacks

6

In the Company of Writers

1971–1973

OS HAD BEEN INTRODUCED TO the celebrated English poet W. H. Auden (1907–1973) by a mutual friend, Orlan Fox, at a cocktail party in 1967, after which Auden occasionally invited him for afternoon tea at his apartment on St. Mark's Place in the East Village. Auden had a keen interest in medicine; his father had served as a medical officer during the initial outbreak of the encephalitis lethargica epidemic. In 1969, OS invited Auden to visit Beth Abraham and meet his patients.

In 1971, Auden asked Fox to send OS a new poem, "Talking to Myself (for Oliver Sacks)."*

To W. H. Auden

POET

MAY 15, 1971
[NEW YORK]

Dear Mr. Auden,

I have been greatly moved by the magnificent poem which Orlan has transmitted to me, and your wish to dedicate it to me. The poem itself has resonated in all sorts of ways in my mind—indeed, it has

* Though the poem seems to be addressed to his partner, Chester Kallman, Auden had told Fox that he felt OS was "partly responsible for this poem." It makes references to many subjects OS and Auden certainly would have chatted about, including migraines, randomness, mood swings, and neural activity. See Edward Mendelson, ed., *The Complete Works of W. H. Auden*, vol. 2, *Poems: 1940–1973* (Princeton, N.J.: Princeton University Press, 2022).

evoked so many images and feelings that I find it quite impossible to write a proper letter (I have written about twenty and torn them up!). It is marvellously accurate, and evocative, and as densely-woven as Nature itself.

I would be most deeply flattered and honored to have such a poem dedicated to me, and it would afford me a pleasure which would last as long as I do.

Thank you for the poem, and for everything. I shall try to write you a longer and more coherent letter soon.

With best wishes,
Yours sincerely,

To Thom Gunn

POET

MAY 25, 1971
[NEW YORK]

Dear Thom,

I don't know which gave me more delight—your letter or your book.

I had seen some of the poems in *Sunshine,** and had read "Moly" and a few of the others in MS when you were in New York (you were just about to send them off to Faber's, I seem to remember). You fuse style and content so beautifully—I was going to say, so easily, but I know that every poem and thought and image is the fruit of an enormous and invisible tree. And you have become so *warm* (so Sunshiny) in your poems without sacrificing anything of the exactness and sobriety and accuracy you always had—but it's a warm accuracy now, a loving sobriety, where before (I think, and hope I don't offend) there was something cold and constrained and almost ferocious which used to look through some of your poems.† It is lovely to see someone and their work grow and ripen, as you and yours have

* The book's title was actually *Sunlight.*

† One wonders whether OS was consciously aware of how exactly this criticism mirrored Gunn's criticism of OS's early travel journals.

been doing in the last decade—beyond belief. I congratulate you on them, and love you for, in, and through them—and without any of the somewhat frightened, somewhat repelled, somewhat abject feeling for you and your poetry which I used to have.

A phrase in your letter puzzled me: that my book* terrified (?) you. I think it is partly because there is something in it (and me) which relishes my patients' symptoms, and which admiring their structure forgets the suffering they represent. Some of my patients would come to share this feeling, and would present me with *bouquets* of symptoms. And, perhaps, it is partly after years of analysis and coming to feel myself an explorer rather than a victim of human underworlds. I don't *think* it's a sadistic feeling—I think it is a sort of reprieve from the old sadism, and a relish for human nature in every form. Or, at least, I hope it is. You put it perfectly about the necessity for one's judgement being part of one's style—I suppose words like "appreciation" and "relish" conjoin these. I used to have delusions about a perfect scientific language, a calculus, with the meaning of every term fixed and finite, for itemizing and inventorying reality. I now realize that such a calculus would be like a net which would let most of one's perceptions and feelings slip through. This terrible Gradgrind passion for defining and itemizing and reducing to commodities—so much of my scientific education, so-called, consisted of this, and it is deadly, as well as being nonsensical. I think I started to wake up when I started to see things in my patients which completely passed the bounds of any such "objective" description, which were almost ineffable—yet perfectly distinct and recognizable (like Chartres, or a symphony, or a complex ramifying joke or metaphor).†

* *Migraine.*

† Subsequent pages of this letter are missing.

To W. H. Auden

JULY 25, 1971
[NEW YORK]

Dear Mr. Auden,

You have been almost incredibly kind to me, and I have been almost incredibly boorish. I can only say that I have been in rather an ugly, depressed mood for several weeks, and virtually unable to write letters; I am going to London today, for the summer, and am feeling a sudden lightening of mood, and a sense of release; and the first thing to do, obviously, is to write to you.

Your Review* was magnificent—I think you have that rarest of capacities, the capacity to be interested in absolutely everything, and to approach everything in constantly new ways. Your review even made *me* look into my book with renewed interest, which I hadn't done in months! I already have intimations that your review has introduced the book to a set of readers who would never, in the normal course of events, have heard of it, or taken any interest even if they had heard of it. What I most want to do—in anything I write—is to try and restore something of the communication which seems to have been lost between scientific writing and the common reader. In migraine, and (perhaps even more) in the subjects I am currently about—Parkinsonism, tics, trances, etc.—one sees phenomena which at first seem utterly strange, and irrelevant to common experience; and then one sees that they are highly relevant to familiar things, and can illuminate these in all sorts of ways, and be illuminated by them; and, above all, one needs common language to connect them. And everywhere one sees that beautiful concord (in which every molecule and cell and organ and function has its place and role, and is perfectly fitted for it, and *knows* it)—so that, as you say in your poem, the body becomes a metaphor for the Cosmos.†

It was particularly thoughtful of you to send me Dunbar's poem on his Heid-Ake,‡ which is very fine, and witty. I am gradually building

* Auden reviewed *Migraine* for *The New York Review of Books*.

† A reference to a line from "Talking to Myself": "You can serve me as my emblem for the Cosmos."

‡ The Scottish poet William Dunbar (ca. 1459–ca. 1530) wrote about his own migraines, describing his pain and sensitivity to light.

up a remarkable collection of allusions to Migraine—they are almost incredibly numerous and varied, from Sumerian inscriptions to Mrs. Gradgrind. If I can ever make an anthology of them, they would represent every conceivable sort of commentary people have made about the Megrims. The Dunbar poem will be a particularly prized specimen.

[. . .]

To W. H. Auden

AUGUST 18, 1971

[37 MAPESBURY RD., LONDON]

Dear Wystan,*

Your letter was forwarded to me a few days ago, and it (or your poem, or you) was the best of palliatives. Does there come a point (if one is very lucky, or has the right gifts, or grace, or works at it) when style, feeling, content, judgement all flow together and assume the *right* form? Your "Anthem" seems instinctively and effortlessly lyrical, and absolutely natural, like an organic growth; and yet obviously has the most careful and sophisticated and exquisite choice of words—and no feeling of any "joins" anywhere, of artifice, of manipulation. Marvellous. I will treasure it.

Yes, I thought the Coleridge quote was a real find, and so to the point. And I agree (I *feel*) absolutely with the Novalis one.† In some sense, I think, my medical sense is a musical one. I diagnose by the feeling of discordancy, or of some peculiarity of harmony. And it's immediate, total, and gestalt. My sleeping-sickness‡ patients have innumerable types of strange "crises," immensely complex, absolutely specific, yet completely indescribable. I recognize them all now as I recognize a bar of Brahms or Mahler. And so do the patients.

* Auden signed his letter of August 2, 1971, to OS "Yours ever, Wystan," so OS now felt allowed to address him in kind.

† The German Romantic poet Novalis wrote, "Every disease is a musical problem; its cure a musical solution," an aphorism OS quoted frequently. Unfortunately, there is no record of the Coleridge quotation in the Auden correspondence.

‡ Another name for encephalitis lethargica.

Such strange physiological harmonies—I hope I can find some way to describe these, because they are unique states, at the edges of being, beyond imaginable being, beneath comprehension, and when the last of the sleeping-sickness patients die (they are very old now) no memory will be left of their extraordinary states. Writing seems more of a struggle now—maybe I'm trying something harder—I find meanings go out of focus, or there is some sort of "slippage" between word and meaning, and the phrase which seemed right, yesterday, is *dead* today. [. . .] And medical jargon is so awful. It conveys no real *picture*, no impression whatever, of what—say—it feels like to be Parkinsonian. And yet it's an absolutely specific, and intolerable feeling. A feeling of *confinement*, but of an inner constraint and confinement and cramp and crushedness, which is closely analogous to depression (although it is not emotional as such), and, of course, is very depressing. And a painful inner *conflict*—one patient called it the push-and-pull, another the goad-and-halter. It's a most hateful condition, although it has a sort of elegant formal structure. But no book that I know of brings home that Parkinsonism *feels* like this—they just reduce it to an unevocative listing of symptoms. I hope Osbert Sitwell* didn't have it too badly.

I've been reading some Goethe (for the first time, really) in the last week or two. Starting with his *Italian Journey*—thank God I *did* start with that, or I might not have got any further. And then the Pelican *Faust*—maybe it's the same with any translation. I *must* learn German. And Mann's fabulous essay on Goethe and Tolstoy. And *Elective Affinities*. And that great, meandering, affectionate Lewes biography.† There is one point (I think in his chapter on Goethe's philanthropy) where Lewes says that he could "eat Goethe for love"—and I think these are beginning to be my sentiments too.

I *hope* I can join Orlan‡ on a lightning visit to Vienna.§ There is nothing I would like more, but I am awfully fretted with my current book, and may not be at liberty (or *feel* myself at liberty) until I have

* Osbert Sitwell (1892–1969) was a writer and a brother of the poet Edith Sitwell.

† George Henry Lewes was the author of *The Life of Goethe* (1855).

‡ Fox.

§ Auden spent his summers at a farmhouse in Kirchstetten, outside Vienna.

finished it. I would love to see you in your own Kirchstetten, but if I cannot come I will surely see you in New York a few weeks later.

Yours ever,

To George L. Fite

SENIOR EDITOR, *THE JOURNAL OF THE AMERICAN MEDICAL ASSOCIATION*

AUGUST 28, 1971 [UNSENT]
37 MAPESBURY RD., LONDON

Dear Sir,

I must acknowledge, somewhat belatedly, your letter, and your return of my manuscript on l-DOPA.

And, of course, the list of "comments" provided by your consultant(s). The latter is a quite remarkable production, not for its intrinsic value (for it seems almost entirely valueless as a rational critique), but for the light it casts on the state-of-mind and emotional attitudes of your consultant(s), who—if I do not misunderstand you—must be regarded as representing, after a fashion, the "official" attitude to l-DOPA. I need scarcely say that I was at first distressed, and even shocked, that innumerable observations based on years of daily contact with patients, and the probable implications of these, could be so ignorantly and wantonly dismissed, or "wished away": the attitude immortalized in Dickens' Mr. Podsnap* ("I don't want to know about it; I don't choose to discuss it; I don't admit it! . . . I have also said that if such things do occur (not that I admit it), the fault lies with the sufferers themselves"). Your consultant(s) do not make, and are obviously unable to make, any substantive criticisms of my work, and have therefore descended (very much like Dr. Ryan in his letter which you published last December) into petty jibing, pomposity, vapid rhetoric—in a word, Podsnappery. Indeed, if I did not know it directly, I could infer the importance of my own work from the

* In Dickens's *Our Mutual Friend.*

very intensity of this threatened, denying, and defensive reaction. Your consultant(s) seem to hold their own dignity and integrity very cheaply, in that they are prepared to abuse their own positions of authority to suppress the results of an independent investigator, and to try and smother the very spirit of free enquiry. I do not, Mr. Editor, apply these considerations to *you,* for I realize that you are *ex officio* detached from opinions expressed in the JAMA, and opinions expressed by your consultants. And I also realize that you could hardly print something in defiance of the strongly-expressed opinion of your own consultant(s).

To get on to a pleasanter topic: I was most gratified to see a very fair-minded and generous review of my book on *Migraine,* in JAMA of July 5, 1971. It was a delightful surprise after I had got so worked up about the DOPA business, and I would be most grateful if you could transmit the enclosed letter of appreciation to your reviewer, Dr. Griffith. [. . .]

Please forgive me for an over-long letter—but I know you will appreciate that feelings must be discharged and expressed once in a while. With this letter, therefore, I shall regard the JAMA "episode" as closed.

Yours sincerely,
Oliver Sacks

To James Hughes

FELLOW TRAIN PASSENGER

SEPTEMBER 16, 1971
[37 MAPESBURY RD., LONDON]

Dear Mr. Hughes (or James—I'm not sure which to use after meeting you once),

I enjoyed talking to you immensely on our journey down from Harwich together. You were so open and sensible and direct—like a breath of fresh air, after all the confused, devious, contradictory, mixed-up people one usually sees (at least in Big Cities). I was very moved by your telling me about your nautical Grandad, and *his* father, and your father, and yourself—all deeply attached to the sea, one way and another, which perhaps we should all be. I know, for myself, that looking at the sea, or swimming in it, or crossing it, gives me some very deep, elemental sort of satisfaction, an intense feeling of realness, and a sense of return—perhaps everyone has this; perhaps it is connected with the fact that our lives start in a sort of sea, and that we necessarily see the sea as our origin or mother. I don't know.

I do hope you passed that silly exam. And it *is* a silly exam, like most exams. You are so obviously competent to manage your work, in practice, and you have shown this so clearly, that it seems ridiculous to make you take a silly exam which in itself means nothing, and provides no real qualification for anything; but that's the topsy-turvy way it is. Once it is out of the way, you'll be your own man and past the obstructions. I must say—just from speaking to you—that it is plain *to me* that on a ship you would keep an eye on everything, and miss nothing, and keep calm, and handle things in the best way; you radiate reliability; and it annoyed me to think of a pack of fools or self-styled "experts" in London forcing you to put on a suit and take a train and worry yourself sick with cramming, all for the sake of pleasing *them*. The only real test would be for one of them to come to Harwich and go on a real journey and see you in action.

After having lived in New York for seven years, I sometimes think that everyone is mad. And perhaps they are, in New York. And it is a rare pleasure to meet someone like you, someone, that is, who is eminently sober and sensible and down-to-earth, on the one hand, and full of adventurousness and love-of-life on the other: in short, a *real* person. I wish you the best of luck—and you must drop me a line one of these days.

Yours sincerely,

P.S. I enclose, with best wishes, a copy of my "baby"—*Migraine.*

To [first name unknown] McEwen

REVIEWER

SEPTEMBER 24, 1971
BETH ABRAHAM HOSPITAL, BRONX, NY

Dear Dr. McEwen,

I have just received a copy of your review of my *Migraine* book, and want to express my appreciation of your generous, acute, and always just comments upon it. I hope you will not think it an impertinence of me to write to you in this way—one of the joys of (a first) publication is that it opens the way to all sorts of unexpected contacts and correspondences.

I am, as you say, an "enthusiast," although I have never been sure whether this is a good thing or not. I am not an enthusiast for any particular theory or formulation of migraine—and, if anything, I over-simplified my formulations when I wrote the book (which was in 1967). What *does* arouse my enthusiasm is the wonderful range and variety of the phenomena, and the endless ways in which these are arranged (and used). Migraine fascinates me in the same way as, say, London fascinates me. I see it as a sort of field or landscape, with innumerable aspects. My book is really a series of sketches, an album, which tries to picture some of these aspects. Having seen and thought a good deal more in the four years since it was written, I realize how inadequate it is; but I also realize that it could be a hundred times fuller, and would still fail to do justice to the subject. This sort of approach, in a quite fundamental sense, precludes simple definitions. Obviously one can "reduce" migraine and select certain aspects, and make a rough-and-ready definition: one *has* to do this for diagnostic purposes. But finally, one can't "define" it any more than one can "define" London. One can only say—here are certain features, certain patterns, certain possibilities, a region of behaviour which overlaps with, and abuts upon, innumerable other regions. I didn't fully realize or admit this in the book. I still had a notion, in 1967, that one could, as it were, peel off various accidental and adventitious features, and expose the "essence" or core of migraine. I stopped thinking this way when I read Wittgenstein's beautiful and funny description of a man pulling the leaves of an artichoke in order to find the "essential" artichoke—and, of course, ending up with nothing at all. I do appreciate that one has to make a compromise between the actual and irreduceable complexity of biological phenomena, and practical, everyday needs. I may have got carried away somewhat, and probably should have put in a section on differential diagnosis, etc.

Obviously, the book could have been shorter and clearer and more succinct; but, equally obviously, it could have been longer and fuller. Like all books, it ended up as a sort of compromise. I selected 84 case-histories out of a much larger total. I *could* have put in 1200 case-histories, but then the book would have been about a million words in length! Again, I *could* have dwelt in much greater detail upon the analysis of a relatively small number of case-histories; but I feared that if I made the histories too detailed and circumstantial they might expose too much of the patient, and cause embarrassment to them (or their friends and relatives) by exposing intimate information. My preference is increasingly towards intensive rather than extensive studies, and in another book which I have in press (on post-encephalitic patients, and their reactions to l-DOPA, etc.) I do in fact confine myself to a handful of case-histories each several thousand words in length.

There are so many *possible* ways of writing a book, and every way has its advantages and disadvantages. But I would certainly alter my emphasis and discussion in various ways if ever the book runs into a second edition, and in this I will be helped very greatly by the comments of critics as thoughtful and perspicacious as yourself. I do thank you, once again, for a most valuable and constructive review.

With kindest regards,
Yours sincerely,
Oliver Sacks

OS HAD INDEED WRITTEN THE first nine case histories of *Awakenings* in London two years before, during the summer of 1969, but then misplaced his "only" copy, as he was wont to do. At some point, however, he had given Jonathan Miller a carbon copy for safekeeping, and had then forgotten about it. By 1971, unbeknownst to OS, Miller had shown the copy to his neighbor Colin Haycraft, a renowned editor, who felt it was worth publishing.

To Colin Haycraft

PUBLISHER AND EDITOR IN CHIEF, GERALD DUCKWORTH & CO.

OCTOBER 25, 1971

[NEW YORK]

Dear Mr. Haycroft [*sic*],

It was a very great pleasure meeting you last month, and an immense encouragement (and surprise) that you were interested by the case-histories I wrote in 1969. I am exceedingly sorry to be so tardy in replying to you—it has been most rude of me, and I don't really know why I have been so slow. [. . .]

My chief reason for holding back, though, is connected with the circumstantiality of the histories, and my fears that publishing such detailed information about living people might be distressing to them. A while ago I published a short letter in the *Lancet* (copy enclosed), which got released to Reuter's and then to the daily press in England and the States, and I got very upset that the patient I described would see the thing and get very upset. Actually, I have found that my own scruples and fears, in this respect, greatly exceeded those of most of my patients—*they* are often quite happy, so it seems, to be the subject of a medical communication or memoir. I have spoken to most of the patients whose case-histories I gave you, and none of them have raised any special objections. I have wondered whether I should actually show them what I have written, but feel this would be unnecessary and maybe cruel. As you will gather, I have been struggling with my own ambivalence for the last month or so. Of all the things I have written, the case-histories are my favourite—the most vivid, and concrete, and real; and I think they combine scientific and human interest. My misgivings have been only with regard to the propriety or otherwise of publishing such intimate matter: it was due to such misgivings that I put the histories away in a drawer after writing them.

I have just been reading *The Three Christs of Ypsilanti**—I don't know if you know the book—which may have some analogies to *Awakenings,* and which also caused the author considerable hesitancy about publishing. And Freud's "Dora" analysis, which *he* put in a drawer for five years after writing!

I *hope* therefore that some of my conflicts being resolved, I can get on with the necessary revision and emendations which the histories require. My general feeling is that they should stay substantially as they are—despite some manifest errors—because they represent authentic first impressions undistorted by subsequent formulations. I will write relatively short summaries or epilogues bringing each case up to date—this is bound to add a rather tragic element to many of them, because in many cases the superb effects of the drug were not maintained, or otherwise compromised. [. . .]

To Elsie and Samuel Sacks

FEBRUARY 2, 1972

2512 BRONX PARK EAST, BRONX, NY

Dear Ma and Pa,

Sorry—it is a while since I've written [. . .] Things are going quite smoothly these days, without (alas!) being wildly productive. (This doesn't mean that they aren't being incubated inside.) The chronically-inflamed situation at Beth Abraham has cooled down very pleasantly, and I have been able to maintain an adequate (and necessary) distance from Messeloff. I find my children's clinic,† which I am doing twice a week, fascinating, and a vital counterbalance to my work at Beth Abraham. I am keeping private practice at a low ebb, not through lack of requests to see me, but because I have to conserve my energies (and evenings) for writing. I am seeing three or four patients on a regular basis (twice a month), and also do perhaps three or four consultations a month. I confine my private practice to Wednesday

* A 1964 book by Milton Rokeach about three schizophrenic men in a state hospital.

† At Bronx Developmental Services.

afternoon and early evening. It is not particularly profitable (partly because I cannot bring myself to charge more than a quarter of what most of my colleagues do), but I find it satisfying and interesting, and (as you have always said, Pop) a needed complement to hospital and clinic work. I earn enough, without being over-worked or spread too thin, and have a feeling of financial (and other sorts of) security which I have not experienced before. [. . .]

I am spending a lot of time and energy swimming—I have joined the YMHA (the Jewish equivalent of the YMCA), where they have a large pool, the largest in New York. I try to swim a hundred lengths (1³⁄₇ miles) every day, which takes me about an hour. I weigh exactly 17 stone stripped and am in better shape than I have been for two years.

I go out occasionally—went to a Bach concert on Saturday, and to *Ghosts* on Sunday. How incredibly topical, or timeless, Ibsen is. I can see very well why he must have appealed to both of you so much in your student days. There is a marvellous biography of Ibsen by Michael Meyer (published in 3 volumes, I think, in England) which you might enjoy very much. I think Ibsen can appeal to everybody. I had quite a long chat with Al Capp* about *Peer Gynt* when we met at dinner some weeks ago. And Auden, whom I dined with last week, is a great Ibsen admirer (he wrote the introduction to Michael Meyer's translation of *Brand* several years ago).

Which reminds me, it was fascinating meeting Aubrey† at lunch (when David‡ was here). The room was filled with security men (completely invisibly, of course), and every so often Aubrey would throw a strange quick look at the door, (which he always faces when he is in public). His posture is curious—almost motionless, very few of the fidgets and movements people usually make, but extraordinarily vigilant. And such a strange mixture of diffidence and fluency. He is obviously an exceedingly shy man—there were long, painful silences from time to time—but a marvellous, witty talker once he gets going (as impressive in anecdote as in generalization and

* The noted cartoonist was Madeline Gardner's brother and thus a remote cousin of OS's.

† Abba Eban.

‡ OS's brother David.

analysis). Everything he said could have been broadcast or printed without change of a word. He is so at ease and accomplished in his public role, and so ill at ease and awkward (or so it seemed to me) as a private person. The combination didn't surprise me—indeed, it was very much what I expected. His son was there, obviously highly gifted, but also painfully diffident. And so was Isaac Stern,* bird-eyed, exophthalmic, with swift bird-like movements in total contrast with the grave, attentive immobility of Aubrey. He talks almost as well as he plays. Which reminds me, I enjoyed a very nice Xmas dinner with Charles Rosen, a gifted pianist of whom you may have heard. He has recently published a book on *The Classical Style,* a study of Mozart, Haydn, Beethoven, brilliant.

I think (at least hope) I will get writing soon. It's not that I haven't been thinking. I am continuously revolving thoughts and images in my mind. When I finally put pen to paper, the result may startle some of my colleagues, because I shall do a good deal of the writing in the form of short paragraphs full of images and metaphors. It will look more like poetry than physiology. But I am persuaded that that is the only way I can begin to express my thoughts without doing violence to them (or to the subject).

Meanwhile, the *Migraine* book seems to have a steady if unspectacular sale—I expect my next royalty cheque fairly soon. [. . .]

My love to Michael, and Auntie Len (I always *mean* to write to them), write again soon,

love,
Oliver

OS SPENT SEVERAL MONTHS IN London during the summer of 1972. Rather than staying at his parents' house, he rented a flat at the edge of Hampstead Heath, so that he could easily walk to the swimming ponds nearby. In June, he joined his brothers and their families at a celebration for their parents' fiftieth wedding anniversary. He got to know his new sister-in-law, Gay—she

* The celebrated violinist, who had deep ties with Israel.

remembers how they would go to the local cinema to watch Hollywood westerns together. After Marcus and his family returned to Australia, OS took up a routine: following a morning swim and a walk, he would write up the stories of his *Awakenings* patients and spend the evening reading his work to his mother. Her reactions, he later wrote, were crucial to him, and in *On the Move* he recalled, "She would listen intently, always with emotion, but equally with a sharp critical judgment, one honed by her sense of what was clinically real. She tolerated, with mixed feelings, my meanderings and ponderings, but 'ringing true' was her ultimate value."

To Orlan Fox

WRITER

AUGUST 5, 1972

37 MAPESBURY RD., LONDON

Dear Orlan,

I have answered all my correspondents *except you*—but this is not a singular rudeness so much as the knowledge that I need to think, that I want to think, when I write to you. Perhaps now, especially, since we are both wrestling with the *demons* of creation.

Thank you, first, for enclosing Auden's poem. I enjoyed it immensely, like the last one. Indeed, they (and perhaps also Auden's absence)* moved me, for the first time, to read a good deal of his earlier work. You're so right—he does bowl one over, especially in the mass, when his amazing gifts of intelligence and style seem to multiply themselves without limit. And (almost always) he has such perfect taste, so that he is always friendly in his approaches to the reader, but never invasive, and never aloof. I have tried (ah, but it is so difficult! And the difficulties are so treacherous!!) to keep bearing in mind his self-comments in his foreword to the *Collected Shorter Poems*—those which he had to throw out because they were dishonest, or bad-mannered, or boring. I find, particularly with my present book of case-histories, that these are the constant dangers—falsification and

* Auden had gone to his house near Vienna for the summer.

hyperbole, on the one hand, the drivelling on-and-on-and-on, on the other (the bad manners bit comes later, when I have to cross swords with a few colleagues and institutions in the Introduction).

My first five weeks here were both guilty and heavenly. They were, especially, weeks of *health* and constant activity—walking all day, every day, over and around Hampstead Heath (which is so close by, and so dear to me), swimming in the ponds there whenever I had the impulse, and having a revelation (Thought, Epiphany) every seven minutes or so, which I would excitedly enter in my notebook. In the last two weeks, however, I have *buckled down* to "real" work, with the help of a secretary (which—as you have so often said—is quite invaluable). I *now* have "something to show" (in a mass, a *weight,* of well-typed pages) for my activities, but also have a stale-dreary-heavy feeling, a deprivation of the flying feeling which was so glorious before I shut myself in with my typewriter and notes and secretary and laborious, tedious paraphernalia of composition. Do you have the same alternations? Do you follow a routine, and—if necessary—*grind things out* however you feel, or is your book only written when the daemon comes, and images *swoop* to the page before you?

Auden (like his favourite Goethe) can *summon spontaneity,* I think, which I can't do, in the least. And yet, absolutely my best thoughts are spontaneous—those which come in a flash, unexpectedly (yet so appositely), and already clothed in the best-possible words. They come to me when I'm walking around, or swimming, or shopping, or sleeping; they often come when I'm talking or teaching; but they *don't* come when I sit in front of this f——g typewriter!

Tell me—fellow-writer—how do *you* work this out?

I envy you (but will have the impertinence also to warn you) regarding your *involvement* with your Hero, Protagonist, Faustus, or whatever. One *must* be swept up in the currents of feeling which one's *characters* feel, or those characters will never become real. They are, in a sense, informed by one's feelings, and they derive their force from *enacting* one's feelings—and yet, they are not one-and-the-same as oneself. They are different, they are *themselves*—imaginative, sympathetic creations, not uncontrolled, phantastic projections. There's such a world of difference between imagination and phantasy, or between sympathy and identification. You're no more Faust than Goethe was Faust: but I'm preaching. Maybe, partly, out of a sense of envy, because—at the moment—I can't get even properly involved with *my* "characters," my flesh-and-blood patients whom in a sense I must fictionalize. I find too much tendency to *lay on the drama,* on the one hand, and to lose myself, on the other, in the sequence of small-but-dreadful-pitifulnesses and insignificances which *do* constitute so much of their lives. My mind keeps going back to a piece I wrote (but never showed you)—so help me!—just about twelve years ago, in my first summer in America—a study-at-close-quarters of a trucker and his mate,* which Thom Gunn said was the *truest* thing of mine he had ever seen, as well as the most interesting. Which brings us right back to Dishonesty and Boredom, or Truth and Interest. Oh well, forgive my blathering. Do keep going—*defy* the summer, your neurosis, and the rest of it—write again soon (to my Mapesbury address). I miss you, and am especially sorry you won't be able to come this summer, but—there will be other times and other summers. I wish I could write to Auden, but for some reason I daren't, perhaps you can convey my regards to him, and I will see you in September.

To Orlan Fox

AUGUST 24, 1972

37 MAPESBURY RD., LONDON

Dear Orlan,

How pleasant (as always) to get your letter of the 15th—and the enclosed Auden poem about the electric brougham, need of which (sentiments, after having been gassed, blinded, deafened, and almost killed thirty or forty times by traffic in Hampstead this summer) I thoroughly endorse. [. . .]

I have been pushed/consumed by my own demon in the last few weeks. Part of the trouble was a nice letter from the editor of *The Listener*—you may know it, a superior *Statesman*-like weekly with a circulation of 70,000, and a particularly high quality lately—asking for an article on some of my "Work in Progress." This f . . . . . g

* An abridged version of this essay, "Travel Happy," appeared in *On the Move.*

article—which I thought I would write in an afternoon and be done with—took me *nine* drafts, 60,000 words of typing, and 5 weeks of tortured fiddling-around i.e. it was more work and trouble than the whole of my *Migraine* book. I think I went rather overboard with it. The *first* version would have done, but I became a sort of mad Loom, creating worlds with each up-stroke, and destroying them with each down-stroke. Anyhow, it's done now—about 4–5000 words—and accepted: should be out in 3–6 weeks. I have really come out in the open in it, and placed myself fairly and squarely in opposition to the profession of specialists and specialization to which I belong, and have come out with a beautiful Leibnizian tri-une view of Nature (emerging from l-DOPA, and spreading to everything else). It's a bit megalomaniac etc, but it's *good,* it will make people think, and it will be the first publication of any fraction of my "General Theory." Strange that the Relativity of Medicine should make its appearance in a literary weekly, of all places, but such is the case.

I have just about finished the 19 case-histories (totalling about 55,000) for *Awakenings,* and am now working on the Introduction and Prologue. I think I will *just about* finish this, and dump it in Duckworth's hands, before my return. *IF* I finish any earlier than I expect (e.g. around Sept 8–10), I might perhaps fly to Vienna, and take a few days off, and—of course—pay my respects to Wystan, and present him with a finished typescript copy of *Awakenings* (the only one besides my publisher's and my own). [. . .]

Keep well and writing,

OS ARRANGED TO STAY ON in London for most of September 1972, hoping to produce a final draft of *Awakenings* as well as his article for *The Listener.* In October he returned to New York, only to find that his situation at Beth Abraham had changed. First, the hospital's director informed him that he would have to vacate the house he had been living in next to the hospital—because the director wanted to move his own mother there. When OS complained, he was summarily fired, depriving him of his position, his salary, his home, and, most devastatingly, his patients.

He began grappling with these new circumstances, but a couple of weeks later, on November 13, he received a phone call from his brother David saying that their mother had died of a heart attack while travelling in Israel. She was seventy-seven.

OS went back to London to be with his family. One bright spot, though, was seeing an admiring letter to the editor of *The Listener* in response to the article OS had published there about his postencephalitic patients. The letter was from an eminent literary critic, Frank Kermode.

To Frank Kermode

LITERARY CRITIC

NOVEMBER 24, 1972
37 MAPESBURY RD., LONDON

Dear Professor Kermode,

I was overcome with pleasure and gratitude when I saw your letter last week. [. . .]

I have been struggling for years towards a means of expression which does no violence, or not too much violence, to clinical observations; I felt strangled by the sort of "objective," dead language usually used by neurologists, which is the only language most of us approve or allow; I felt a great need to express myself and my "data" in a wholly different way and to a wholly different audience; I was delighted—and scared—when the *Listener* commissioned my article; and breathed the biggest sigh of relief I have ever breathed when I saw that someone like yourself approved. [. . .]

In particular, your favourite paragraph was *mine* also. It was the only one which "wrote itself," which didn't get over-written and congealed by efforts at condensation and abstraction. If one gives them a chance, patients always describe their Parkinsonism or whatever in terms of worlds and landscapes. They *depict*. It has taken me years and years—far longer than I like to think—to learn to listen to them, and to try and feel my way into the nature of their experiences. There is, these days, such a veto on this sort of empathic-metaphysical approach, such a Gradgrindian insistence on definitions and enumerations and "facts." And, absolutely, one must gain impressions, and stick to impressions, and not try to reduce them to formulae and facts. All patients are naturally poets: they try to express their experiences in images and metaphors, and we "experts" try to squash these back into "symptoms" and "signs."

I have found the problem of language the most necessary and difficult in trying to understand or convey what goes on with these post-encephalitic patients, who are uncannily strange, and uncannily familiar. [. . .] Back in 1966, I saw my first post-encephalitic patient—he had once been a librarian—with a look on his face at once infinitely clenched and infinitely remote, but I couldn't begin to imagine his state until he whispered "Panther" and suddenly put Rilke's Panther* into my mind.

I have tried to convey the sort of lives some of these people have in my forthcoming book *Awakenings* (Duckworth). It is bound to be a partial failure because one really needs to be a novelist as well as a neurologist, but I hope it will convey something of what these extraordinary, tormented, involuntary explorers of the depths experience.

I do hope we can perhaps correspond occasionally. I would love to meet you, but have to return to New York almost immediately.

A FEW WEEKS LATER, back in New York, OS wrote to his brother Marcus, who had been unable to make the long journey from Australia for his mother's shiva.

* "The Panther," by Rainer Maria Rilke, is about a caged panther; OS might have been thinking about the second stanza:

> As he paces in cramped circles, over and over,
> *the movement of his powerful soft strides*
> *is like a ritual dance around a center*
> in which a mighty will stands paralyzed.

(From *The Selected Poetry of Rainer Maria Rilke*, trans. Stephen Mitchell [New York: Vintage International, 1989].)

To Marcus Sacks

DECEMBER 30, 1972
11 CENTRAL PARKWAY, MT. VERNON, NY*

Dear Marcus,

I was going to phone you, and then to write to you, immediately after Ma's death, to exchange sympathies and fellow-feelings, but found it difficult. I was much moved by your deeply-feeling letter to Pa and to Auntie Len, which they let me read.

Like you, I had, in a sense, been "expecting" a parental death for years, and always feared what any long-distance phone-call might bring. I had felt unusually close to Ma, this last summer, and I think it was *her* happiest and fullest summer for several years; I dallied in London till almost October, partly because I had some sort of premonition or intimation that I would not see her again; and I think you had a similar feeling, when you left in June. The suddenness and completeness of the abruption was very shocking—there seemed to me a period when it was not to be borne or believed. And I think that what I went through was nothing to what poor Pa, and—perhaps even more—Auntie Len went through: I thought Lennie herself was going to die of grief in those first few days. In a sort of way, perhaps, one takes people for granted when they are around, and one feels the intensity and the detail of their presence only when they are gone or taken away: certainly the sense of the absence of Ma's presence was overwhelming at first—I constantly dreamt of her (particularly a tormenting dream that she had been taken ill in Israel, but fortunately made a complete recovery), and thought I heard her voice, and constantly expected to see her in the lounge or the breakfast-room; I am sure one's deepest and most elemental perceptions are perceptions of *presence* (and of its changes, or of its vanishing). An astonishing, and in a way terrible example of this was shown by Butch,† who returning to Mapesbury from the kennels, immediately burst into loud howling, and then after some hours lay quite still and lethargic, and then became ill—one could not doubt for a moment that he instantly

* OS had found a new apartment in Mount Vernon, New York, not far from AECOM.

† The family dog, a boxer.

perceived the (permanent) loss of Ma's presence, and then became sick with grief.

The Shiva was marvellous (if one can use such a word), and I am sorry for your sake (as well as ours) that you could not come to it. There was an amazing, beautiful outpouring of feeling and sympathy, all day long, a most vital, genuine, real, necessary, *warm* exchange of feelings and memories; there was nothing (I think) strained or straining about it, and Pa and Auntie Len were visibly supported and vivified by it, as I was as well, and taken out of that terrible solitude of grief which is perhaps its least tolerable aspect. It is lovely to talk and weep together with complete freedom, with complete absence of any shame or neurotic constraint. I will never again doubt the profound emotional (and, as it were, epistemological) wisdom that is embodied in sitting Shiva. And it is an epistemological accommodation one must be helped to, for there has been a sudden amputation in one's world-picture, the foundation and progenitor of one's world has gone. One has become an epistemological amputee, and one desperately *needs* the presence of kin and friends, of nearness and dearness, and to see how they too have been bereft and affected in their different ways. [. . .]

Michael stood up astonishingly well during the Shiva and in the disorganized fortnight thereafter, showing a common sense and a gentleness and a *real-ness* which I had never seen in him before; but he was obviously tortured by an anguish to which he could give no expression. (At one time he burst out to Auntie Len, "It's OK for you, you can weep. I can't."), and in my own last week in London was visibly entering an intricate, labyrinthine psychosis (he is in Friern Barnet* at the moment; I think, probably, he *has* to break down, like this, before a deeper adjustment can start). I spent a great deal of time talking to Uncle Dave,† who stayed with us through the week of the Shiva. He is a wonderful old man, with a most amazing, *alive* memory of the past; we talked for hours and hours about the early years and days of the Landaus, and the Landau grandparents, and especially about Ma's childhood and youth. I had a sense of Uncle

* Friern Hospital, a psychiatric institution in London.

† OS wrote about his uncle Dave, one of Elsie Sacks's brothers, in *Uncle Tungsten.*

Dave as a unique treasure, the only remaining repository of these memories, and of the entire span of Ma's life. Also his strength and independence shone out: at a time when I felt so anguished myself, I was supported by seeing how he had endured so many bereavements and losses, and his strength, like an old weathered rock. He showed me (to put it crudely) that life can go on, even when one thinks it can't. [. . .]

The week after the Shiva ended, Pa took up most (if not all) his previous work, and visibly came together again as he did this. I am sure work is vital to him provided he does it only insofar as it is enjoyable, and not to the point of strain to which he (to which all of us) sometimes goes. [. . .]

Ma loved you very deeply and wept on and off for several days after you had gone (she felt, I think, as you did, that she would never see you again). She also had a deep and affectionate feeling for Gay and the children, and this too she expressed freely after you had all left. It is sad that she found some difficulty in expressing it when you were all actually present, and that her real and positive feelings were sometimes overshadowed by fretfulness and irritability, etc. (At least, I gather that this was the case, especially during the earlier part of your visit, partly from Auntie Len, and partly from what seems to me a touch of aggrievement in your own letters to Pa and Auntie Len.) I may be quite wrong here (and perhaps it is impertinent of me to "read through the lines"), but I think you may have had, or still have, some feeling that Ma never "accepted" Gay, resented your marriage, felt no warmth to your children, etc. This is emphatically not the case, as you would know for yourself had you heard her speak of you all in the later part of the summer. But, during May and June, Ma was exhausted and over-strained, partly from the excitement of and preparations for the Golden Wedding, and partly from her hernia which was continually prolapsing and paining her; things easily became "too much" for her at this time, and she became irritable and over-sensitive; but I hope you will not let any unpleasant memories of this over-cloud your earlier and happier memories, or imagine that there was any deep-seated rancour or lack of affections towards you and yours.

For myself, in a sense, things have been looking up. I failed to finish my book in the summer partly because I got side-tracked onto doing an article for *The Listener.* But, maybe, this was a blessing in disguise, for the article obviously hit the right spot in an appreciative and responsive audience; there has been a lovely public dialogue and exchange of letters, following it; and I have been enjoying a mild but definite celebrity, of a sort which my *Migraine* book never achieved for me. In the last 2–3 years, I have mostly experienced unwelcome, and misunderstanding, and antagonism from other neurologists, which (as you know) made me more than a little paranoid. The *Listener* episode, on the other hand, gave me a lovely feeling of welcome and warmth and understanding and appreciation, and has shown me that in future I can most hope to appeal to a cultivated general audience, and as for my fellow-neurologists, they can go to blazes (neurology has been dead, imaginatively, and fundamentally, for half a century anyhow). My publisher, perceiving the interest that my article had aroused, took the extraordinary step of saying that he would have my book set in print within a *week* of my finishing it. (A week, as opposed to Faber's two years!) I have already corrected the galley-proofs of the case-histories (which I wrote in the summer), and I finished the rest of the book, in the ten days following the Shiva. It was curious—as if grief had replaced my usual guilt, I had a sense of intense sobriety and clear-headedness, a sort of intoxication of sobriety, as if all the nonsense and neurosis had been knocked out of me, as if I was nakedly and boldly face-to-face with the things which really mattered (health, sickness; life, death). And in this suddenly-procured metaphysical clarity I found no difficulty whatever in writing at full speed, and finishing in a few days what I failed to do in the entire summer. Duckworth's have put the book (*Awakenings*) first on their Spring-list, and it will be out in March or thereabouts. I have a great sense of relief at having got it out, for so much of what was in it was pent-up inside me, under terrible pressure, for years, precluded from expression by a hateful sense of veto or interdict. Now—and it is, I think, not merely coincidence that this should follow Ma's death—that sense of veto is gone, or at least much attenuated. Perhaps this is a way of saying that I no longer feel like my parents' child, or anyone's child, but like a person in my own right, suddenly free to do or say what I want.

After a month of being cosseted and cherished in London (partly as a mourner, partly as a writer), I have a certain feeling of let-down and vacuity here—I felt so nourished by conversation, by social intercourse in London, and so starved of it here—but I am settling down to a more satisfactory and stabler *modus vivendi* than has been possible for me in the past. I have a splendid new apartment,* and for the first time in New York am getting some sense of "home" from where I live. My work is not wildly exciting, nor is it meant to be; essentially I do just as much as I need to make ends meet, and no more, for my real life and satisfactions, increasingly, come from thinking and writing, and not from clinical practice. (Although this has its own joys, as real as its ordeals, and I would never give this up, even if I could afford to.)

I did enjoy seeing you (seeing you *all*) a lot in June, and I hope we can correspond and see more of each other than we have done in the past. There used to be so much difference between us, but now—when I am in my fortieth winter, and you in your fiftieth winter—we are closer together; and I also feel a temperamental kinship with you, a sense of a certain likeness between us, which I don't really have with David (who is so outward, so denying of inner life) or Michael (who is so imprisoned in his own inner life). I do hope, especially now that our family has been visited by death, that we can draw closer, and appreciate each other's presence more. I hope you will write back soon, and (more remotely) I also entertain the hope that I can visit you all in Australia this summer. (I have been reading *Kangaroo*† in anticipation).

Love,

* The new place was in a suburb just north of the Bronx and much more spacious than the Manhattan apartments OS had lived in before.

† A novel by D. H. Lawrence.

To Richard Lindenbaum

SCHOOL FRIEND, GENETICIST

JANUARY 12, 1973

11 CENTRAL PARKWAY, MT. VERNON, NY

Dear Dick,

Thank you very much for your charming New Year card [. . .].

My mother died suddenly on November 13—an instantly-fatal coronary, without premonitory symptoms—and I flew at once to London. [. . .] The terror and solitude of the abruption was appalling, infantile, between hearing the news on Monday evening and arriving in London on Wednesday morning; absolutely I felt as if the foundation of my own world had utterly vanished. [. . .] I don't know that I had any specific premonitions, but then I have had (one has had) presentiments of parental death since . . . I can't remember a time when I did not have them. But in the days preceding her death (she died in Israel, a cause for odious congratulation by some 'frum' relatives, but an additional grief, and perhaps anger, to me, because she was a Londoner through and through, as London as a London plane-tree, and I know she wished to die, quietly, sitting in the garden, instead of being roasted to death under an alien sun, on a foolish medical-Zionist conference of a sort she always dreaded, which she'd been pestered into joining) . . . In the days preceding her death, then, I experienced an absolute compulsion to read about the death of Lawrence's mother, and the letters and poems he wrote at the time, the intensity of his grief, and of his relief.

My father was shattered (David went to Israel to collect him, and the body), and himself looked pre-terminal in those first awful days; and my poor, darling Auntie Len—who had been so close to Ma, closer than anyone else, for an unimaginable three-quarters of a century—I thought she would die of grief. Michael became instantly catatonic, for a few hours, and then, in the most remarkable and unprecedented way, emerged into a tenderness, a melting, a solicitude, a sense, a realness, such as I had never seen in him before. For myself, I was infinitely comforted by the Shiva. Never again will I doubt the profound emotional and epistemological sense of that marvellous communal grief, a continuous pouring-forth and streaming of mind and feeling, the realest sympathy and relation I have ever known. [. . .] Our dog, my mother's dog, Butch, who had been away (as usual) at the kennels, came back the weekend after the Shiva, seemed to throw an instant all-seeing glance around him, and then (unprecedently) burst into a loud, terrible canine howling, and then after some hours, sank down motionless, and then ceased to eat. [. . .] Absolutely, without a doubt, that dumb loving beast perceived in a moment the taking-away, the permanent gone-ness of Ma's loved presence, and at once became sick with loss and grief.

Thank God (so far, and for the most part), I have felt mostly grief, with very little anger or reproach, although I feel and fear this coiled up inside me. Grief is so unlike depression: it is so filling and real and expanding and uniting and—(it sounds an almost blasphemous word)—nourishing, as opposed to depression, which is constricted and constricting, sharp, pointed, alienating, disuniting, hateful, tormenting, starving, stuffing.

Mostly, so far, I think I have felt grief, with only twinges (but terrible, even as mere intimations) of depression and reproach. I grieve for what was real in Ma. I can't summon up a single image of her kindness, or tenderness, her absurd-dear (perhaps Victorian) simplicity, her fondness for telling stories, her delight in babies and children, her oddly-mixed taste for Schubert-lieder and turn-of-the-century music-hall melodies (which came to her, again and again, with amazing frequency AND accuracy, quite involuntarily, in the last two years of her life, and always impressed me as a sort of unconscious preparation for the end: "In my end is my beginning").* I can't think of any of these real images, not for a moment, without breaking into tears; and yet they are good tears; lamentations without a touch of shame. But for Ma's other side—her dark, censorious, suspicious, cruel, punishing, side—I don't grieve for this in the least, I am glad it is gone (and wish only it were not so deeply incorporated in myself). Oddly, in my thoughts, and especially my so-vivid, many-times-nightly-dreams, I never see my mother as an old woman, but in her prime, with a streak of black in her hair, in her late thirties or very early forties (*my* age!), as she must have been when I was an infant; or still stranger, I dream repeatedly of *her* girlhood of the vast Landau family, somehow suspended in an eternal impossible summer afternoon in the old garden

* He is quoting from "East Coker" in T. S. Eliot's *Four Quartets.*

at Highbury (but how absurd! for the old Highbury House was pulled down in 1915), somehow, miraculously, her parents, my grandparents, whom I never knew, *and* all her seventeen siblings, and all my first cousins, and second cousins, and third cousins, all somehow, impossibly, contemporaneous, until the sense of impossibility grows upon me, and when it becomes sufficiently intense, awakes me! How strange! In my mother's handbag (the only one of her "effects" I went through), I found a picture which *she* must have carried with her throughout her life, of *her* mother, as a 16 year old bride, taken in the Spring of 1872.

What a sudden sobriety, a sudden jerking into social and historical sense, it all gives one. There was the sheer spread of generations*—from a telegram from Ma's last remaining uncle (101, or so, in Paris), and her great-great-great grand-nieces [. . .], twins whom she had brought-into-the-world three years ago, her last deliveries; so many relatives; and so many students, ex-students, former patients, those she had delivered ten or thirty or fifty years ago. I think that I had never had any sense of how far-reaching a life can be, genetically, socially, every way, and there was nothing especially far-reaching about Ma's. (It occurs to me, if I die now, how few people really know or like me, how few would attend my funeral.) The milling of the generations gave me an odd, comforting [sense] of Ma "passing to history" (though what that means, since one is *in* history, one *is* history, all the time, I don't know).

I poignantly regretted (and regret) that I didn't, don't, won't have a family of my own, that biological continuity—at least genetic continuity—must come to a stop with me. (Ma was marvellous with grandchildren, and they were with her: there can be such a lovely, un-neurotic, un-pressured, un-competing, and almost biologically-crucial relation between grandparents and grandchildren, the real continuity, in a way, across the generational fighting. I always regretted, being the youngest of the youngest, that I never knew any of my grandparents. Aubrey—Abba Eban—the only one of my first cousins to have escaped the otherwise universal severe neurosis/psychosis, was in fact brought up by his "dear and learned grandfather"—the Oliver Wolf Sacks I am named after, but never knew.)

* At the shiva.

But perhaps I have some compensations in my own sort of generation—generations of students, readers, books, thoughts, images—I hope so, at least. [. . .]

To Colin Haycraft

FEBRUARY 10, 1973
11 CENTRAL PARKWAY, MT. VERNON, NY

Dear Colin,

I have returned to find a distressing and potentially calamitous situation at Beth Abraham—and one which is reflected in thousands of Beth Abrahams all over the country. Nixon's latest budget has upped defence (I mean "defence") expenditure by twenty billion or whatever, and has made a merciless cut-back in monies available to schools and hospitals, especially hospitals catering for chronically ill "welfare" patients. I found on my return that Margie (our devoted speech-therapist), all the physiotherapists, occupational therapists, etc. had been summarily sacked. There has been an abrupt (and probably permanent) closing-down of our Workshop, our Rehabilitation Department, etc., as well as the dismissal of three-quarters of the nursing staff and social workers. I went up to the ward and found almost every patient speechless-motionless with fear. And they have good reason to fear, because the facilities which have been "cut back" were, by and large, the only hopeful and reconstructive ones in their lives. I saw mirrored in face after face that stunned, shocked "death-look" which augured the deaths of Sam H., Julien S.,* etc. Indeed, this *is* a death-sentence for many of them, and for other patients in the hospital, and for many of the 3 million patients in the US in so-called "extended care" institutions, which will now become, in effect, death-camps for them. Horrible. And the fact that it has come from the highest level gives one an absolute sense of impotence.

Again, I feel like [adding] a footnote†—but perhaps it would not be right at this time. Anyhow, I shall (if I can bear it) continue to visit

* Postencephalitic patients.

† To *Awakenings,* which was already in press.

the patients (as it were socially, as a human presence myself), and will observe the effects of *this* deprivation. I fear it will undo the work of the last four years, if not the last forty years. The bad things which have happened to them in the past will be as nothing to this. I saw that the patients looked absolutely shattered *before* I heard what had happened. And when I did hear, the words (I think from a Sanskrit poem) which came into Oppenheimer's mind when he saw the first atomic bomb at Alamagordo came to me: "I am become Death, shatterer of worlds." This will also gravely affect my work with the children I have been seeing, for most of the school guidance-counsellors, etc. have also been cut back, and medical services likewise. As in the double-helix, welfare and real-ness in this country will dwindle to less-and-lessness, and terribleness will surge to a level which will make the McCarthy era seem like child's play: this, at least, is what is being said by many people. And although I am generally blind and deaf to current affairs, my moral antenna cannot miss *this* affair.

Enough said. I will cease to shoot letters and footnotes in your direction for a while. I will try and cool off, and will await further news from you by and by (or is it bye and bye).

Best regards,

To Jean Cunningham

EDITOR, FABER & FABER

MARCH 6, 1973

11 CENTRAL PARKWAY, MT. VERNON, NY

Dear Miss Cunningham,

Unexpectedly, and simultaneously, I have just received your letter of March 1 and Mr. du Sautoy's* of March 2. Let me say, first, that I feel most complimented, delighted, at the suggestion that I might try my hand at a book on *Pain* for Faber's.

But what a subject! How huge, how deep, how various, how many-sided—how on earth would one tackle such a subject? It would, I

* Peter du Sautoy, chair of Faber & Faber.

admit, be a fascinating challenge. And the subject fascinates me, as it must fascinate everybody. It is so basic, so elemental, and—so utterly mysterious. Whenever I wade through some of the existing tomes on the subject (like the quarter-million words in White & Sweet's book), which try to reduce everything to pathways and mechanisms, and to tabulate the mystery of pain out of existence, I think of that marvellous bit of dialogue in *Hard Times:*

"Are you in pain, dear Mother?"

"I think there's a pain somewhere in the room," said Mrs. Gradgrind, "but I couldn't positively say that I have got it."

I have been thinking of pain, so it happens, a good deal lately, partly with the anguish of bereavement, and—at a more humdrum level—with the minor neuralgic torments of a disc which is playing up at the moment. Indeed, just yesterday, when I had burning paresthesiae in one hand, I jotted down in my notebook some of its qualities, its basic characteristics (I wrote: Pins and Needles. Quanta of Pain. Prickling, flickering, sparkling, tingling, ringing—a tinnitus of pain (cp discrete and continuous pain to clonus/tonus, jerks & spasms). Needling, acicular, nagging, goading. Pungent pain, pungent and titillating, like snuffs and curries. Pain as elementally, irreduceable, *irritating,* a basic aspect of biological irritability.) This is the sort of thing I jot down at odd moments, interspersed with odd quotes, aphorisms, diagrams.

But jottings are one thing, and a book is another. Assuredly not a day goes by without my patients presenting me with pains—from banal prickles to strange unlocalizable sensations, or states-of-mind, like those of the dying Mrs. Gradgrind.

I have always been interested in Pain. I suppose every neurotic is: it is his passionate interest in pain and suffering, his experience of it, his penetration into it, which gives Schopenhauer his edge over Kant, and gives (at least some chapters of) his masterpiece* its almost-unbearable eloquence and intensity.

Pain. Twelve years ago another author of yours (Thom Gunn) suggested I try to write a book on Pain. I forget what got me started, but I *did* get started on one of those huge monologues, or, more accurately, soliloquies I am prone to, a soliloquy on Pain, which was called forth by something or other Thom had just said. And he listened to me

* *The World as Will and Representation* (1818).

for two hours and said: "That's your subject. I wish we'd had a tape-recorder around—one day you'll write a book about Pain."

The first paper I ever wrote was on Pain—also, I think, in 1961. But, alas!—like so many things I have written—I lost it, and didn't have a copy. [. . .]

I had meant to write you a two-line letter—and I find myself taking-off in all directions. I must think about the subject—that is, as a possible subject for a book. [. . .]

To W. H. Auden

MARCH 31, 1973

11 CENTRAL PARKWAY, MT. VERNON, NY

Dear Wystan,

When I got your lovely letter of February 21,* I was filled with a rush of affection and gratitude, immediately wrote an answer, put it in an envelope and stamped it, placed it under a volume of the OED to flatten it, took it the next day meaning to post it—and I *cannot* be sure whether I actually did so or not. I have been tormenting myself for more than a month by trying to visualize the exact moment of posting (or non-posting). These things happen. I find they happen especially frequently with me if my feelings are aroused. Since obsessional indecision *cannot* be resolved in its own terms, I am (what perhaps I should have done a month ago) taking the liberty of writing *another* letter. [. . .]

First then (and if I am repeating myself, it is better than not saying it at all), thank you immensely for your magnanimous reaction to my book. You are the only person, other than my publisher, to whom I have shown a copy; and there is *nobody* whose favourable response could make me happier than your own. I *did* do a Glossary when I was in London in January, and I certainly hope that the book secures some readers outside the profession. Perhaps my experience has been an atypical one; perhaps I have been provocative, or sought

* In this letter, Auden wrote, "Have read *Awakenings* and think it a masterpiece. I do congratulate."

alienation; but the positive experience with the *Listener* article, like the five years of negative experience which preceded it, persuade me that it is all-but-impossible to have any real and fruitful dialogue in Medicine and medical circles (especially the barren neurological ones to which I belong), whereas there is obviously a mass of real, *alive* people outside Medicine who *will* listen to me, and with whom I can enjoy the delight (the *necessity*) of real converse, what (if I remember correctly) Dr. Johnson called "a streaming of mind."

Essentially, at the moment, I am mourning, and I am awaiting the publication of *Awakenings* (originally scheduled for April 26, delayed by rail-strike, *now* re-scheduled for May 17—but who knows?) I do not feel fully alive. I felt diminished almost to zero by my mother's death; the greater part of me seemed to have died with her (I was, no doubt, too closely attached to her); and it is only in the last few days, perhaps, with the signs of Spring all round me, that I have started to feel a re-stirring of my own sap, and a re-realization that I exist in my own right, that—unexpectedly, amazingly—I am still here, and the world is here, and—it scarcely seems possible, but more than four months have passed since her death. I am beginning, I think, to recapture some sense of real-time and history, and of seeing the past as past, never to be forgotten, but no longer forming a spurious present.

For some reason or other, I have found it almost impossible to concentrate on anything except D. H. Lawrence, and especially some of his lovely, earlier stories which are so full of life and love, and where his bitter-ranting-mad part is not in evidence. I do a great deal of pondering, and jotting in my notebooks—those notebooks which are *so* much better, fresher, finer, than anything I can achieve in my "finished" writings—but have not put pen (or typewriter) to paper on anything in particular. In addition to the book on Parkinsonism which I will be doing for Raymond Greene of Heinemann—a book to which I shall give the most orthological form I can, and which will (I hope) have some resemblance of the "Investigations"* in style and intent—I have been excited by a suggestion from Faber's that I do a book on Pain for them.† It seems to me that I have been

* Wittgenstein's *Philosophical Investigations.*

† Neither of these books ever came to fruition.

thinking about pain (that *one* thinks about pain) all my life—and it is exactly the sort of deep, elemental (and generally misunderstood) subject I should like to deal with: indeed the night after hearing from them, I dreamt I was an amoeba, feeling my way through the world, feeling regions of possible hurt—too hot here, too acid here, etc.—experiencing some indistinguishable, primal mixture of fear-pain-shrinking through which I palpated the world, and judged what I could do, where I could go, etc.*

I am delighted that *Love's Labour Lost*† went off well in Brussels, and full of admiration for your energy in undertaking so much, in so many different ways and places. I wish I had your resoluteness—you seem to go and *do* things at once, while I continually hover and waver, and inspissate myself in coils of indecision. [. . .]

I do not know whether you will still be in England when I come there in May. [. . .] If (as I fear) I will miss you then, I will at least hope to hear from you, and to see you here, there or somewhere soon. I remember, at the airport, on that painful Saturday,‡ when a stranger came up and greeted you, I asked you whether you thought of the world as being a large or a small place; and you replied, "Neither. A *cosy* place." So—the world being a cosy place—I am sure I will see you in England, Austria or New York within the next few months.

P.S. I will now go and post *this* letter instantly, lest it get caught up in indecision like the last one!

* This extraordinary image reveals OS's persistent interest in feeling and in "pre-human" animals and plants, which he returned to in a 2014 essay titled "Sentience: The Mental Life of Plants and Worms."

† Auden and Chester Kallman had written the libretto for an operatic treatment of Shakespeare's work by Nicolas Nabokov (a cousin of Vladimir Nabokov).

‡ A reference to Auden's 1972 departure from America, where he had lived for decades, to return to England and Austria. Orlan Fox and OS had helped Auden pack his things and taken him to the airport, and Auden had given some of his belongings to each of them. To OS, he gave his stereo and record collection, as well as a volume of Goethe's letters.

To Richard Lindenbaum

MARCH 31, 1973
BETH ABRAHAM HOSPITAL, BRONX, NY

Dear Dick,

I *think* I wrote to you a few weeks back—but [. . .] I do want to thank you, most particularly, for the *Treasury of Mathematics*[*] you gave me, which has more or less become my bedside reading (strangely companioned by Sir Thomas Browne and D. H. Lawrence), and has led me into some strange seas of thought. How wrong most of our education was—on the whole! The masters are so much easier to understand than the rotten, lifeless, second-hand, ground-up stuff we were crammed with at school. In particular I enjoy Sylvester (I was so struck, on the train back from Didcot, by a passage from him, that I quoted it at length in my Epilogue).[†] I seem, increasingly, to be able to *see* Invariants and Covariants floating in the air, wonderfully pure and distinct, like Platonic Forms. Gauss[‡] is tremendously difficult, partly I suppose because of the intrinsic difficulty, partly of his unrelenting rigour and terseness—but so rewarding! I found the section on curved surfaces—"surfaces not as boundaries of bodies, but as bodies of which one dimension vanishes . . . flexible" has gradually allowed me to grasp, to *visualize* (and I can't properly hold anything in my mind until it becomes an *image* of some sort) increasingly everything—"inner" and outer space—*as* curved, closed, continuous, variously warped surfaces. I can understand (a bit more of) Einstein much more clearly now I have read Gauss; and in particular the cartoon-like deformations of structure-space which seem (to me, anyhow) so characteristic of neurosis, Parkinsonism, and other such violently warped and polarized states. There is a wonderful passage in Proust where he speaks of his own two states (the aesthetic-epiphanic vs. the neurotic-egocentric): "I learned to distinguish between these two states which reigned alternately in my mind . . . contiguous, and yet so foreign to one another, so devoid of means of communication, that I could no longer understand, or even picture to myself, in one

* A 1968 collection of primary sources edited by Henrietta Midonick.

† To *Awakenings*.

‡ Carl Friedrich Gauss (1777–1855).

state, what I had desired or dreaded or even done in the other." The shape-space of these two states—the state of *poise,* so open and hospitable, with perception of everything in its eachness, its richness, its harmony; and the state of *tension,* so violently warped, like a terrible force-field, and utterly closed, mean, tendentious, and egocentric—their contiguity and their absolute contrast have become much clearer to me since reading Gauss. God knows, one spends so much of one's life in these unbearably but inescapably *closed* inner spaces—like a fly in a Klein bottle;* but gradually, now and then, by a mercy, one does—often quite suddenly—find oneself *outside* the labyrinth, in a divinely spacious *real* space, and able to look down or back at the neurotic state as a sort of transient and local deformation (deformity); but only sometimes—the neurotic state, while it has hegemony, cannot but be felt as absolute, totally confining, and permanent. [. . .]

My present work—now I have left Beth Abraham—is mean and abrasive and unrewarding.† I can scarcely stand it—but it pays the rent (just). I can only stand it by thinking of it as temporary. With my mother's death, and also with leaving (as I had to) the patients at Beth Abraham who were so dear to me, I find I have no real motives/reasons left for staying here in New York: or, rather, only *one* reason, namely my analyst, who means so much to me, and can (as much as anyone) see me through some of the tortures and distortions. I wait eagerly, and with some anxiety, for *Awakenings* to come out. [. . .] Auden (bless him!), to whom I gave a proof-copy the day I saw you, has been magnanimous enough to call it a "masterpiece," which makes me feel good (that is, in those moments when one is *allowed* to feel good).

OS MET RAYMOND GREENE in the summer of 1971, when Greene, a publisher, asked him to write a general book about Parkinson's

* A Klein bottle is somewhat analogous to a three-dimensional Möbius strip, impossible to escape.

† OS is referring here to his work at Bronx Psychiatric Center (also known as Bronx State), where he was frustrated by a bureaucratic and less than humane culture, but perhaps even more to several for-profit nursing homes around the city, where, in the 1970s, living conditions were notoriously cruel and filthy.

disease (as opposed to the rare form of postencephalitic *parkinsonism* that affected his *Awakenings* patients).* Preoccupied with writing *Awakenings,* OS turned in earnest to the more general book only in the spring of 1973, after *Awakenings* had finally gone to press. Over the span of about a week OS wrote two very lengthy letters to Greene, outlining his thoughts on Parkinson's. The following is excerpted from the second of these, which by itself ran to sixteen typed pages.

To Raymond Greene

CHAIRMAN, HEINEMANN MEDICAL BOOKS

APRIL 6, 1973
[NO ADDRESS GIVEN]

Dear Dr. Greene,

[. . .] In the past few days I have been seeing (unless I deceive myself) a clearer and simpler structure to the whole. I may also need to enter the book, as it were, obliquely, and by stealth, through a series of letters which insensibly become chapters. One other preliminary remark: anything I say *now* about a possible plan or design for the book may be ignored when I actually get down to it: I have never found that I could plan or calculate too much in advance without doing violence to my own spontaneity, or to that unconscious planning which is so much deeper and surer than all one's calculations and cogitations, and which one *discovers* in what one has done when one has done it. [. . .]

I find conversation, the writing of letters, and the composition of imaginary dialogues, much easier than "proper" formal writing. And this is true in my reading: for example, I adore Goethe's letters, conversations, aphorisms, *Italian Journey* etc., but can scarcely *move* in his finished works, his *opera.* I have wondered, in relation to this, about writing some "Parkinsonian letters," analogous, perhaps, to the Screwtape letters, or to take a nobler example, Dostoevsky's *Letters*

* Greene (1901–1982) was an older brother of the writer Graham Greene and a renowned figure himself: a publisher, a medical doctor, and an expert mountaineer who had joined the 1933 British expedition to climb Mount Everest.

from the Underworld. I have also wondered about reviving the *dialogue* as an expository instrument: it comes so naturally to me, that I always carry pens of different colours, in order to delineate the imaginary personae who converse in my notebooks. [. . .]

I had not sufficiently grasped at the time of my *Listener* article the necessity of *all* descriptions (pictures, representations) being metaphorical—frameworks-of-reference, ways-of-looking-at etc. which are neither more nor less than *ideas.* For example, I was still hung up with the Newtonian-Lockean idea of "absolute" space, space as a sort of empty container in which things happen: an epistemological assumption which is implicit, for example, in all current neurophysiology, certainly that of the Sherrington and Pavlovian schools. Although I said something about the odd-shaped spaces or fields in which these patients lived (I have one patient who tends to get rather sudden onsets and offsets of severe Parkinsonism: he once said to me, "It's like a magnet being switched on and off in my brain. Suddenly, it's there and pulling me out of shape."), I failed to realize that "space" itself is a metaphor, time equally, and that one may have any number of space-times (or geometries, or metrics) all equally useful, equally elegant, as frames-of-reference, for making pictures of the world. Curiously, although I intellectually perceived the truth of this relativistic principle when I was reading Einstein last year, it only came to me with intense, visual vividness when I read Gauss' incredible papers on curved surfaces of 1825 and 1827; and especially at that marvellous point where the excitement breaks through the rigid formality of his language and he speaks of "a new point of view, where a wide and still wholly uncultivated field is open to investigation." [. . .] I had tormenting dreams in the early part of the night, when (like Alice, perhaps) I found myself before a door fumbling impotently with an enormous bunch of keys, and none of them fitted the door, and I knew it; and then I fell into one of those very deep, dreamless, but so fruitful sleeps one sometimes has (Jonathan Miller, in his entry in the *National Dictionary of Biography* or whatever, lists as his recreation "deep sleep"); and, in the morning again dreamt of finding myself in front of that door, but this time with a single key which I knew was the right key; and it was, and the door opened—and outside was the world, a marvellous feeling of spaciousness and ease. I also dreamt that morning of a hippopotamus as a sort of convoluted, grey, homogeneous, mobile hyper-surface—a dream which was a joke, but a vital one, because in its absurd, idiosyncratic way it fixed the images of curved spaces and surfaces well and truly in my mind). [. . .]

With kind regards,

Postscript (April 8): (I should, perhaps, have posted this letter two days ago when it was mostly written; otherwise it will become one of those interminable Herzogian* monstrosities such as I am always writing and never posting!) [. . .]

One last note, by the way (it gets back to something I have raised again and again). It is a question of style, but I think there is nothing deeper than style: *being* is style, one only has being *in* style. One sees again and again (it would be a lovely subject for an essay) how even the greatest thinkers, if they do not actually do violence to their own impressions and thoughts, seem blind to their nature, or present them in a completely unnatural, artificial, systematic manner which must be utterly different from the vivid, intuitive, living mental processes which produced them. [. . .]

[Newton's] *Principia* (partly because of intrinsic difficulty, but no less because of style) seems to me unreadable: but the Quaeries to the *Opticks* are utterly delightful, and reveal a completely different and very engaging Newton, Newton-at-play, throwing up ideas and catching them, or—in his own death-bed image—playing with pebbles on the seashore). [. . .]

D. H. Lawrence (who is almost the only person I can read at this time)—at his best—comes across with incredible warmth and vividness and aliveness and realness because of the movement of impressions, the continual animal-like immediacy and mobility and the eschewing of the static, the monumental, the systematic. I don't want to write a Treatise, I just want sketches, impressions, "Pansies," Pensées.†

* A reference to the title character in Saul Bellow's *Herzog*, who composes mental letters to friends, family, and famous people, living or dead, apologizing for his own failings or his disappointment in others.

† The two words are etymologically related, as OS undoubtedly knew from his bedtime reading of *The Oxford English Dictionary*.

Not an edifice. Something more like a garden planted with living flowers, ideas germinating and budding, flowering and fruiting, each a different expression or conception of the whole, each a way in which the whole reveals itself—an epiphany, each a door or window opening upon the whole. In this sense, it scarcely matters which flowers (which phenomena) one looks at first: one can wander round the garden any way one wishes, for everything one sees will express or exemplify or present or represent, in its own way, the whole. A book of examples—like the world itself. (The greatest such book I know, and one which for a year or more, has been my only "research" book, is the great *OED,* which I ramble in every day. The *OED* is not a list of definitions—unlike smaller dictionaries, which lack quotations—but a garden of the imagination, an immense word-and-meaning game, a Palomar for gazing into intellectual space. The arbitrary alphabetical order doesn't affect this quality in the least. Indeed, I have wondered about dealing with the phenomena of Parkinsonism etc. *in* alphabetical order. [. . .] When one is dealing with worlds (as opposed to systems)—as in the *OED,* in language, in natural phenomena—there is no "beginning . . . middle . . . end," no logical path, only a progressive differentiation of ideas (and *ideata*) already infinite (or infinitesimal) at their inception. [. . .]

I had best finish a letter which threatens to become interminable, and will tax the patience of the best-disposed man.

I hope that these peculiar effusions can be prolegomena to the book proper.

To the Editor, The Listener*

APRIL 22, 1973
ALBERT EINSTEIN COLLEGE OF MEDICINE, BRONX, NY

Sir:

Donald Broadbent's article on "Alternative States of Mind" not only presented experimental evidence of the first importance, but was deeply fascinating and suggestive with regard to the epistemological, educational, social and political implications of the work he discussed.

That we have, among so many other divisions, a cognitive division in our nature, the possibilities of *two* fundamentally different states-of-mind or modes-of-thought, is an ancient insight, which is continually renewed and renamed in different generations and disciplines. [. . .] Pascal called these modes-of-thought the *analytic* and the *intuitive* [. . .] and devotes the fundamental first section of the *Pensées* to delineating these two modes of perceiving reality, their powers, their limitations, and the relationship between them. He notes that "the one quality can exist without the other: the intellect can be strong and narrow, and can also be comprehensive and weak."

That one can exist without the other, or far more strongly than the other, has become very clear to me from working with children in kindergarten or pre-school age. Thus when asked to draw an object (I usually suggest an animal, sometimes a hippopotamus), some children at once, and, as it were, instinctively, make an outline of the entire animal in a single, swift and continuous stroke; but other children construct the animal serially, feature by feature. And this is perfectly reasonable, for an object (hippopotamus) can *either* be pictured as a continuous, closed, and convoluted grey surface, *or* as an assemblage or association of individual "features." One can also observe a categorical difference in children's approach to numbers: some tend instinctively to cardination and cardinal concepts, and others to ordination and ordinal concepts. But one *also* sees children—and this has

* OS was acquainted with Mary-Kay Wilmers, who edited *The Listener;* she had solicited his article "The Great Awakening." He loved writing letters to the editor and did so often throughout his life, and these were formally addressed, as was customary at the time, to "Sir," regardless of gender.

nothing to do with intellectual calibre—who seem to employ *either* approach with equal facility.

It is essential, I think, to recognize that there exist strong inborn propensities to this mode or that: Holism vs. Serialism, Gestalt vs. Mosaic, Intuition vs. Analysis [. . .], and to respect these, and not deny or violate them in the process of education or subsequently. But it seems to me dangerous to suppose (as Broadbent seems to suggest) that such propensities are based on structural differences in the brain, differences which *because* they are structural would seem to be relatively fixed and unmodifiable, and to foredoom each person to a particular way of thinking. [. . .] One feels that education should be equally hospitable to both approaches—as Goethe suggests in his allegory of the "Pedagogic Province."* There is no doubt that the present Gradgrindian sort of education which usually obtains is hard upon, if not fatal to, the intuitive child. [. . .]

The alternative states of mind, which Broadbent speaks of, are truly and totally alternative; they can neither include nor exclude one another. It is essential, however, that neither is allowed to *occlude* the other, for they are not opposites, not in opposition, but *complementary,* and they come together to picture the world. Nothing whatever—not even a single electron—can be adequately pictured in either way alone; as we need both eyes, we need both states of mind, to give us the depth and reality of the world. [. . .]

Passing from individuals to cultures, we see that a perfectly-balanced doubleness of vision has been achieved only twice, and then all too briefly: first, by the Greeks, in whom (as Auden has emphasized) "these two kinds of activity are inextricably mixed"; and secondly, as Broadbent points out, in the seventeenth century—"the century of genius"—and especially in that middle third of the century which saw the foregathering and foundation of the Royal Society.† Since this time, one feels, our very progress has been a decline, a dissociation of consciousness or sensibility, a torturing split between two states-of-mind, two modes-of-being, which have continually and

* In Goethe's 1821 novel, *Wilhelm Meister's Journeyman Years, or the Renunciants.*

† The founding of the Royal Society in London, in 1660, is often recognized as the beginning of the modern scientific era.

increasingly lost touch with each other. Donald Broadbent's article is a plea for synthesis, for a (re-)unified picture and view of the world. If we can achieve this, we may perhaps expect another century of genius, the third and greatest advance in human culture; if we fail to . . . one remembers the closing lines of the "Dunciad."*

Oliver Sacks

To Brian Southam

EDITORIAL DIRECTOR, ROUTLEDGE & KEGAN PAUL†

MAY 2, 1973
BETH ABRAHAM HOSPITAL, BRONX, NY

Dear Mr. Southam,

I'm sorry—it has been most rude of me not to have replied earlier to your letter of February 21. [. . .] I am afraid I cannot direct you to any particular studies on the subject as such, though doubtless they exist: but I have never done any reading on the subject—nor do I feel tempted to. I think, perhaps, that one of the reasons for my delay in answering is that your question(s) struck a sensitive point. [. . .]

I myself am descended from a long line, indeed several long lines, of Rabbis on both sides of the family. My parents had a deep sense of Jewishness (*Yiddishkeit*)—partly of a noble and generous and real kind, and partly of a mean and hostile and suspicious kind. My (late) mother, I think, never "forgave" my eldest brother for marrying out of the Faith; she wished to, without doubt; all that was motherly and human and individual in her yearned for reconciliation, but

* Alexander Pope's satirical eighteenth-century narrative poem about a goddess, Dulness, and her minions, the Dunces, whose stupidity and greed prevail over the muses of Science, Logic, Wit, Rhetoric, and Morality. Pope writes,

> Thy hand great Dulness! lets the curtain fall,
> *And universal Darkness covers all.*

† Southam raised the idea of a book exploring the topic of the "prominence of Jews in many spheres of cultural and intellectual life" and wondered if there were any serious studies that might be relevant.

something abstract and categorical, imprinted and inexorable, prevented her, in her heart of hearts, from even a partial "acceptance" of my brother's choice. [. . .] My father, as far back as I can remember, has always had a sort of compulsion (the compulsiveness sheathed by a sort of facetiousness or jocularity) to ask of anyone who had signally failed, or signally achieved: "That so-and-so—do you suppose he's a Yiddishe boy?" It may be that my parents suffered a good deal from anti-Semitism in their younger days (certainly *their* parents did, and like so many of their generation were hounded out of Eastern Europe; and many aunts, uncles, etc. were done to death and lost even the possibility of exile). Thus my mother was not only the first Jewish woman (and almost the first woman) to be elected a Fellow of the Royal College of Surgeons, but (I am speaking now of more than fifty years ago) one of the very few Jewish consultants in the whole of England.

You will appreciate, therefore, that my own bringing-up was largely overshadowed by the sense of Jewishness-as-a-special-state (and its subjective and objective corollary, i.e. anti-Semitism as the expected/feared/and unconsciously-desired reaction of THEM—the others—to one's Jewishness, one's claim to special status). *Consciously* I scarcely ever thought about the matter, and thought, indeed, that it was all very exaggerated. It seemed to me that I scarcely ever encountered anti-Semitism as such. Looking back, I am sure that this imperception of mine, in part healthy, was in part a neurotic denial—the exact reverse (and thus the same thing) as my parents' exaggerations. They thought (at least sometimes) that Jewishness and anti-Semitism were the poles of the world; and I pretended that neither existed.

With this sort of background, you see what I mean when I say your letter struck a sensitive point; indeed, it is perhaps only in the last few years that I have been able to transcend some of these neurotic attitudes, and to replace the categorical-abstract sense of US vs THEM, of Pro and Con etc. by a sense of myself, and of others, as *individuals*, unique and uncategorizable.

(A special point relates to the eminence attained by my first cousin Abba Eban. Abba—or Aubrey as we/I know him—is, in fact, the best sort of Jew, with a deep and real sense of Jewish history and culture, and an equally deep sense of its relation to all other history and culture—he has a spaciousness, an openness, a hospitality of mind, which makes him an object of suspicion to the fanatics in Israel, both the fanatics who preach isolation/alienation, and those who preach complete "assimilation" and de-individuation.)

Let me depart now from these personal and autobiographical considerations, which are neither edifying nor pleasant, but which any adequate reply to your letter forces me to review and relate.

I think there are perhaps three "factors" which tend to give Jews an uncommon intellectual and cultural force. All of these, to my mind, are acquired and environmentally-determined (I am no believer in "superior" or "inferior" races, in terms of native endowments or propensities). In order of increasing significance, then:

FIRST—the conditions of ghetto-life, of the Cheder, etc. (which are still recapitulated three or four generations since our escape from the ghettoes) were calculated to produce a sort of intellectual precocity, and a passion for minute and pedantic intellection. This sort of "cleverness" may lead to school-prizes, Firsts, and acclaim and rewards of various sorts—but it seems to me essentially banal and petty; and, more importantly, unrelated to human life and feeling. It is what Goethe called "cleverness *in abstracto,*" i.e., a knack for gaining knowledge which has no inherent relation to one's feeling and being. This is the odious sort of Jewish cleverness, which (like all such parodies of intelligence, in Jews or others) both astonishes and repels, and is doubtless a potent cause of anti-Semitism.

SECOND—Much more important, and rarer, than the disposition/capacity for assimilation and quibbling is the encouragement of a passion for reflection and meditation. One sees in Judaism (as in the Bible) two quite different styles: one dogmatic and legalistic in tone, the other deeply-questioning, allegorical-dramatic, and profoundly conscious of the mystery of things. A beautiful example of this [. . .] is the lovely dialogue or dialectic associated with the Passover dinner, with its great richness of allegory, symbol, ritual, etc. An *essential* part of the Seder night (and one which moves me almost to tears as I think of it, in my own childhood, when my parents were still young and vigorous) is the part-ritual part-extemporare question-and-answer, exploration together of the meaning and significance of the Seder and its symbols, which goes on between parents and children, between *all* the participants. I have no doubt that my own curiosity and love-of-reflection was reinforced by the special meditative quality of the Seder nights, and (in their lesser way) of the ceremonials of Friday nights. [. . .] The dogmatic-doctrinal, legal-scholastic aspects of Judaism always disgusted me, and were a potent factor in driving me *from* Judaism; but the deep, peaceful, contemplative, hospitable mood of the Seder nights, and of Judaism in its lyrical and meditative aspects—this drew me *to* Judaism. If the first inculcated feelings of alienation, the second gave a deep sense of the at-homeness, of a thinking and knowledge essentially connected with one's entire feelings and being.

THIRD—The perennial "specialness" of Jews lies partly in the *doubleness* of their identities and loyalties. (I remember, even when I was five or six, wondering, "What am I—a Jew or an Englishman?" Am I primarily a *Jewish* Englishman, or an *English* Jew? etc.) The Jew is always, to some extent, "a stranger in a strange land." [. . .] He is forced to confront, from a very early age, *the nature of strangeness.* This, if very intensely felt, can (I think) make him or break him as a person. Experienced pathologically, the sense of strangeness becomes a sense of *estrangement*—the feeling that one is above non-Jews, or below them, but hopelessly and unbridgeably *different* from them. Experienced honestly and courageously, the sense of strangeness can give Jews (or anyone) a sense and a feeling which is the opposite of estrangement—i.e., an acute awareness of the beauty of contiguity, of an "otherwise" which is fundamentally different from, yet fundamentally related to, one's nature. A sense [. . .] of hospitality and *xeniality* (xenia, that lovely word, if I am not mistaken, related especially to the mutual appreciation of strangers to strangers, a relation which is that of host and guest): open, friendly, trusting xeniality, as opposed to morbid, lustful-hateful xeno-philia/phobia. Being different—whether one is Jewish, Black, Homosexual, Deformed, exceptionally gifted, or Retarded, or whatever—being different can be a disaster or a blessing: it is a disaster if it makes one defensive-hostile-phobic-manic, etc., but a blessing if one can transcend all this and achieve the almost-divine magnanimity of perfect open-mindedness, a mind always tuned to awe and appreciation, and to a true love of what is strange, of being, of all being. [. . .]

In short, I think it can be either a curse or a blessing to be born, as

Jews are, into a double identity, into paradox, into strangeness. But, paradoxically, if they can transcend their sense of estrangement, and reach sublimity—like Spinoza, Freud or Einstein—then, in a sense, they are no longer Jews, as they are no longer simply Dutchmen, Viennese or Swiss-German. They have become—simply human, as deeply human as a man can become. In this way, it seems to me, that Jewishness (like all distinguishing, and potentially stigmatizing and alienating states) may lead to that so-rare spaciousness and sympathy of mind which is the universal hallmark of greatness in all ages and peoples.

OK, I have said my say, or more than my say. I will be coming to London in late June, for the [. . .] publication of *Awakenings,* and hope to see you then. In the meantime, I would, as always, be delighted to correspond with you,

To Seymour L.

POSTENCEPHALITIC PATIENT

JUNE 11, 1973

BETH ABRAHAM HOSPITAL, BRONX, NY*

My dear old friend Seymour (and it is almost seven years since we first met!),

Thank you very much for your letter which I got today, and for the longer one which came last week. It is always a pleasure, and a compliment, to receive letters from you (and even to be a voice amid your "voices," when you hear voices). I have always had the greatest esteem and affection for you—as has everyone who knows you at Beth Abraham. Your goodness of spirit, your fineness of mind, has never been compromised by disease; perhaps, indeed, it has been forged by it. Indeed, in a sort of way, you have always represented (at least, in my eyes) the conscience of Beth Abraham, the epitome of that unconquerable human spirit which disease, drugs, and the

* Although OS had been fired by BAH, he continued to stay in touch with most of his postencephalitic patients informally. He would resume working at BAH in 1975.

isolation and imprisonment of institutionalization, can never oust or conquer; and it is for this reason that I took the liberty of using your picture—a faceless silhouette, it is true, but a silhouette of profound eloquence, to use in my book, to represent the human spirit: Man—bowed, confined, lamenting, but unbroken. I did *not* however include your "story" among the twenty histories in my new book. I am not sure why I omitted it—for it would have been of the greatest interest. I have learned so much from you—you are among the half-dozen patients who have taught me more about illness, suffering, and the human condition, than all the rest of my patients combined. I should, perhaps, have told your story, but for some reason I didn't. I will (I think) do so, as well as the story of some other patients, should my book run into a second edition.*

This has been a most difficult book to write, and especially to publish. I wrote the original nine histories almost four years ago (that is, in the summer of 1969), and then put them away, feeling that it would be wrong of me to disclose, in such detail, the lives of my patients—of patients who had confided in me, who had trusted me, and who were so dear to me. These scruples and hesitations have been overcome—very slowly, very gradually, and—perhaps even now—only partially. To turn a patient, a friend, into a "character" in a book, seems monstrous in a way; and I would not do so were I not convinced, over-and-above all personal considerations, that your lives and stories (I speak of *all* of you—victims of the sleeping-sickness—whom I have known since 1966 or 1967) are of the deepest interest, and deeply moving, and that they could cast a unique light not only upon the peculiar illness from which you have suffered so long, but upon the nature of human nature, the human spirit, in general.

But, you will understand, I (still) fear to show my book to my patients and "characters," let alone to the staff at Beth Abraham: I do not know whether I will be able to enter the hospital again, once my book has been seen. And this is a very difficult and tense period, awaiting its publication, which (in England) will be in 2–3 weeks. (The American publication will be in the fall.) My feeling is that I should *not* show the book to you—to *anyone*—before it is published, before it is public.

* OS did add two long footnotes about Seymour L. in a later edition of *Awakenings*.

So, I will ask you, as I ask others in Beth Abraham, to be patient. (In a sense, I am glad that the book will first be published in another country; there was even a time when I thought of not allowing it to be published in America, in order to spare the feelings of my colleagues and patients.)

I feel too frail at the moment, in these last days before publication, to show the book, in detail, to *anyone.*[*]

Once it is published, it will be public property, and I will be able to detach myself from such feelings, from identification with it. But this is still a while to go.

I am most happy to see you so well again—most happy, and most astonished. It is a great wonder, almost a miracle, and it shows, once again, the unconquerable spirit which lives in you.

With kindest regards,

To Thom Gunn

JULY 7, 1973
37 MAPESBURY RD., LONDON

Dear Thom,

You are often in my thoughts, and I am always meaning to write to you—but—so little is translated into action.

I fear I never thanked you for your poetic New Year's card—although it resides permanently on my mantelpiece. I gather (from Faber's) that you have changed your address; but I can only visualize you in the surroundings of Filbert.[†] I have asked my publisher to send you a copy of my (so-long-obstructed-and-delayed) new book *Awakenings*—I hope it has arrived, or will do so shortly. First reactions, here, have been very kind. But, fortunately or unfortunately, my mind is so fixed on the *next* book, the *real* one—it will be, I think, a very general one, which I shall call "Station and Motion"—that

* With later work, OS began increasingly to show essays in progress to his subjects.

† Gunn had recently left his house on San Francisco's Filbert Street and moved to Cole Street.

Awakenings, and reactions to it, do not seem quite real. I have also been pondering about *pain*—Faber's have asked me if I would write about it for them (but I fear it would become a theological book); do you remember how we once talked of pain?—it must have been ten or a dozen years ago.

Time: Time the preserver, time the destroyer—I shall be 40 in two day's time. It doesn't upset me as I once thought it would; I suppose I have found the last 5 years easier and pleasanter than the years of often almost hopeless neurosis which preceded them. On the whole now, though it comes and goes, I have a lovely feeling of movement and change, and a preparedness for change whatever it entails: "*Respondeo* etsi *mutabor.*"* The years of neurosis were years of transfixion, of an absolute terror and incapacity for change.

Time. These last few days I have had the strangest rush of ancient memories (some, perhaps, phantasies), most dating from my earliest childhood ('35, '36). I find the old Broad St. train (which I take every day to the Heath) is like a catalyst, a releaser, a Proustian mnemonic; it has become a memory train, conveying me back into the mid 1930's; and if ever I try my hand at an autobiography, I will use it as this, as a train back to childhood.

Images of home, of journeys, seem to occupy me constantly now (I must try to dig up "Travel Happy," that one piece of my former writing you felt was—*real*).

The need for travel, for some sort of pilgrimage (the nature of which will become clear only later) is very intense: I may be going to Italy for a while: two summers ago, I was immensely moved by Goethe's *Italian Journey,* and last summer by Lawrence's Italy-books, (especially *Etruscan Places,* which you mentioned to me in 1960, I think)—And this summer, perhaps, *I* shall have my own Italian Journey.

Do write, Thom, and tell me something of yourself and your plans—and, in particular, whether you have any thoughts of England or New York this year.

Yours,
Oliver

* "I respond although I shall be changed," a phrase used by Eugen Rosenstock-Huessy, author of *I Am an Impure Thinker* (1970).

To Isabelle Rapin

PEDIATRIC NEUROLOGIST

AUGUST 15, 1973
37 MAPESBURY RD., LONDON

Dear Isobel,*

I thank you for your kind letter of July 20. I failed to reply [. . .], perhaps because I have spent most of the past eight weeks in a "paradoxical" depression.

Paradoxical—because *Awakenings* received quite splendid reviews from many of the most significant critics† —at least outside the Neurology scene. As for what my "colleagues" will say, I am not quite sure, but I have been greatly warmed (and, as it were, pre-protected) by a beautiful letter from Luria himself (actually the second of two letters from him), in which he says that he finds the book "delightful," and that I have revived the forgotten and necessary art of clinical observation and biography with great success. And if Luria—the man I most esteem in Neuropsychology—likes *Awakenings,* I will not be so upset by what lesser men say.

The book has also brought me in a flood of correspondence—from poets to philosophers to psychologists to physiologists, and has given me a vivid sense of how many passionately thoughtful, honest, clear-headed people there are, whose sensibilities (like mine) have long been outraged by the lifeless reductionist quality of Science today, and who look forward to, and are creating, the "new" Science-Art of the future. All this makes me realize, in a way, that the *real* thinking of the world is not done in University Departments or Professorial Chairs, but by deeply-reflective individuals here and there, whose significant contacts are personal rather than academic. Certainly our *own* relation—which has always been stimulating and fruitful for me, and I hope (occasionally!) for you—has had this quality, and takes place in a garden or over dinner, etc., and not in the barrenness of

* Although Isabelle was one of OS's closest friends and advisers, he often misspelled her name, as he did with other friends and relatives.

† *Awakenings* was published on July 9, 1973 (OS's fortieth birthday) in the United Kingdom, and a few months later in the United States.

a Conference Hall. These letters, phone calls, seeing people, being asked to talk or partake in small conversational groups, has given me a lovely sense of *being real,* having a place in the world. But—there is another side to it: which is a strong (and not entirely neurotic) tendency to *shun* such contact, because, in some fundamental sense, I have to think out everything for myself, in my own idiosyncratic, roundabout way. I have to nurture my thoughts in something which approaches a total though not hostile solitude—I *feel* this, although I am far from understanding it. [. . .]

To F. Robert Rodman

FELLOW RESIDENT AT UCLA

OCTOBER 27, 1973

11 CENTRAL PARKWAY, MT. VERNON, NY

My dear Bob,

Your letter gave me enormous pleasure—so clear and kind and dear and *you.*

I hope that what you say about *Awakenings* may become true. Texts age, theories become obsolete, but there is (if it is genuine—maybe it is the only test of this) something continually fresh and ageless about real experience and feeling—I *feel* Donne so vividly alive to me, the impact of his experiences and feelings so immediate and relevant—it means nothing that they are "350 years old."

Age, death: so vital that one gives and leaves something—children, books, help, whatever—to compensate for mortality [. . .] And the last month has been mournful too: Auden had become very dear to me in the last 2–3 years (he and my analyst, finally, were my ultimate "reality-bearers")![*] The sense that I—we—will never see his face, hear that extraordinary voice, again gives me (perhaps even more than my mother's death) such a sense of the irreplaceable uniqueness and preciousness of the individual. At least we corresponded fully in his last months, and his letters are so alive, so intimate—more so (at least

* Auden died on September 29, 1973, in Vienna.

to me) than many of his poems—I am re-dedicating *Awakenings* to his memory. [. . .]

You sound so much happier and deeper than you used to be—the transformation of "feel" and tone is unmistakable, and extraordinary: how much you owe to your analysis, and how much to everything else, God only knows. But it is a great mercy and a blessing (and astonishing too, in a way) that one *can* change and develop so much in one's thirties, and later, when so many of one's contemporaries have come to a dead stop, transfixed.

Lovely to hear that you are expressing yourself all sorts of other ways as well—Poems, Carpentry, building, sailing.

Whatever is to be said for an intellectual Jewish upbringing, it is so damnably mental and mentalistic, so blind to the joys and necessities of the body, of *doing* things, the joys of making, of building with one's hands, etc.—I am rediscovering some of these things too, late, but better than never. (Did you ever read Groddeck's* remarkable essay on "Massage"—which Auden, I think, first translated, as a student. He had *such* a sense of the unity of body and soul.) [. . .]

I miss you [. . .]—I must visit you, both of you,† soon. [. . .]

Love,
Oliver

SEVERAL TELEVISION PRODUCERS HAD approached OS to suggest a documentary about the *Awakenings* patients. After meeting Duncan Dallas from Yorkshire Television, OS agreed, hesitantly, knowing that a record on film might be of unique scientific importance. He writes in *On the Move* that he was not sure of the propriety of making such a film, but when he asked the patients how they felt, most of them said, in effect, "Go ahead, tell our story, or it will never be known."

* See footnote on page 405.

† Rodman and his wife, Maria.

To Duncan Dallas

DOCUMENTARY FILMMAKER

NOVEMBER 20, 1973
[NO ADDRESS GIVEN]

Dear Duncan,

I was delighted to get your letter of the 13th. It is very thoughtful of you to let me know that the "raw material," at least, has come out excitingly, and *exists*. [. . .]

It is as well that you filmed when you did, because (as you may have heard) nearly fifty NY Hospitals were struck, stricken might be a better word, by a hateful strike earlier this month.* In many ways my sympathies were (at first, at least) *with* the strikers, who had been very shoddily dealt with by Washington; but the ensuing horrors—which were especially intense in Beth Abraham—rather changed my mind. About half our population had to be transferred, and we did what we could with the remainder. I managed to mobilize some students to help me, and more or less acted as an orderly-lifter-maid-of-all-work etc. for ten days. During those ten days 22 of our patients died (including seven Parkinsonians), as a more or less direct effect of the strike, not being properly fed, turned, toileted, etc., and the final cost—in terms of the ultimate effects of trauma and strain—is not to be calculated. Really, a murderous business. On the last evening, enraged picketers smashed my car up, as well as those of most of the others who cared for the patients "in defiance" of them. It was the *mood* of the strike, the strikers, the whole atmosphere of the discourteous, nothing *reasonable,* anywhere, just an awful murderous battle with x thousands of patients as pawns, and their lives at stake.

I can imagine that the editing [of the film] will take you a full month and more. Although *I* keep thinking of all the things which were missed; for example, Gussie S. "woke up" in a most extraordinary way *during* the strike—in a way which she hadn't since 1969—and was talking nineteen-to-the-dozen, very clearly, very fascinating. And, and, and—so much else. But, for *you,* I imagine, the problem is

* Some thirty thousand hospital workers went on strike for a cost-of-living increase, but since hospital strikes were illegal according to New York State law, penalties against the union were very stiff and the workers went back after eight days, having gained very little ground. They would go on strike again in 1976.

almost the reverse: of having too much material, and all the dilemmas of selecting and cutting. I *think* I might come to London around Xmas myself [. . .] and wondered whether perhaps things might be ready, or in some sort of rough shape, for me to look at with you then? [. . .]

To Kenneth McCormick

EDITOR IN CHIEF, DOUBLEDAY & CO.*

NOVEMBER 24, 1973

[NO ADDRESS GIVEN]

Dear Mr. McCormick,

Many thanks for your two letters of the 20th. [. . .]

Be reassured!—there are no more footnotes concealed deep down in the envelope, and my paroxysm of footnotery (my friend Jonathan Miller speaks of my propensity for making comments on comments on comments etc. "commentorrhoea") seems to have exhausted itself. I am what I am: like many of my patients, I suffer from an "obstructive-explosive" sort of character, so that at times I *cannot* think or write to save my life, and at other times I cannot stop doing so. And here, you, as my editor, are forced to take a sort of therapeutic role, as moderator or umpire, as someone who says with authority "Enough!" "Time's Up!" etc. Sometimes I need encouraging, and sometimes I need stopping—like my patients: I am sorry.† I will await the copyedited manuscript (?galleys), and perhaps we can then decide together whether or not it would be worthwhile to insert these last *addenda*.

I am busy now with "Tics & Spasms"‡—I have just been rushing

* McCormick (1906–1997), a longtime editor in chief at Doubleday, had bought the rights to publish an American edition of *Awakenings*.

† OS's comparison of himself to his parkinsonian patients is quite apt. Throughout his life as a writer, he needed an editor at close range to keep him from his frequent "too-muchness." Perhaps this innate "obstructive-explosive" quality indeed is what drew him so powerfully to his two major neurological interests: parkinsonism and Tourette's syndrome.

‡ Another working title for an essay or book inspired by OS's patients that was never completed.

into it without too clear an idea of what form it will assume; but I find I do violence to my spontaneity if I try to plan ahead too minutely.

Many thanks for everything again,
Yours sincerely,

To Thom Gunn

NOVEMBER 29, 1973
11 CENTRAL PARKWAY, MT. VERNON, NY

Dear Thom,

I was most affected by your letter of—almost two months ago; I appreciated it—its feeling and thought and care—more, I think, than any other letter anyone has written me about *Awakenings;* and if I have been slow in answering (and if this answer is laboured or stilted), it is not because I have been unmindful or ungrateful for your letter, but, rather, that it obsessed me in a way, and gave rise to such a range of reflection and retrospection, and such conflict, that I wasn't sure how to write back.*

First, I thank you for your sympathy about Wystan's death, and I feel too the loss this must be to you, and so many others. I originally met him four or five years ago, and we approached each other very slowly as people—we were both intensely shy in a way. In the past year he had become very dear to me—I really came to love him in all sorts of ways. In particular he stood for *trust,* an immense solidity and substance and depth, with a beautiful candour and transparency:

* In a letter two weeks later to Ken McCormick, OS wrote:

> I have been moved by a letter from another poet, Thom Gunn, whom I used to know a bit in the early sixties when I was in San Francisco, when I occasionally burdened him with various writings of mine. He says: ". . . I found you so talented, but so deficient in one quality—call it humanity, or sympathy, or something like that. And, frankly, I despaired of your ever becoming a good writer because I didn't see how one could be taught such a quality . . . What I didn't know was that the growth of sympathies is something frequently delayed till one's thirties. What was deficient in those (earlier) writings is now the supreme organizer of *Awakenings,* and wonderfully so. It is literally the organizer of your style, and is what enables it to be so inclusive, so receptive, and so varied."

so that the impulse to lie or distort or conceal or falsify (etc.) went away in his presence, and in some fundamental way he taught me how to be myself, by being *himself.* You speculate as to what has happened to/with me, what has brought out some sort of humanity or sympathy which seemed absent or dormant in the early sixties. It has been a slow process, and God knows one fraught with the continual danger of lapses and regressions—when something hateful takes possession of me, and I again find myself writing (though not acting out) things and situations of the "Doctor Kindly" genre. You are surely right when you say that one cannot be *taught* sympathy, humanity, etc—in the sense that it is not amenable to injection, instructional instillation, etc. I always had a sort of empathic passion for "Nature," but, bye and large, it was a "Nature" which more or less excluded human beings. I could feel at ease with a theorem, or a poem, or a landscape, but almost never with a person, and almost never with myself. I think this *has* changed a good deal over the past few years, but I really don't know what have been the chief determinants. I have found working and loving increasingly synonymous; I cannot understand anything, I cannot *approach* it intellectually, except as a relation, in a sort of devotion or intimacy; and I had the blessing of a marvellous group of patients, who were at once as singular as the fauna of the Galapagos, and yet intensely real to me as human beings; I think they made a decisive difference to me, as (perhaps) I did to them: and this, in a way, has been a sort of love-affair.

The two individuals who have meant most to me in the last five years have been my analyst—and Wystan, among the "living living"; they have stood for trust, and love, or at least the sense of the *possibility* of these, in a way I could scarcely have imagined five or six years ago. These, and the "living dead," the voices and teachers from the past, who have spoken to my condition, who have induced and carried me through various crises, and who have felt so palpably present, at times, as to give me an almost physical joy and warmth. There have been a dozen or more such teachers, out of the past: and the one who has most moved me, who has seemed to me richest in reality and suggestion and aesthetic delight, has been Leibniz, who I "discovered" in April of last year. It is odd, I found I had read quite [a lot] of him twenty years ago, as a student, but had "forgotten" it all; it left no impact; I wasn't ready; and then suddenly—indeed it was sudden, like a revelation, at 4.30 on April 18 of last year!—I "remembered" him again, and plunged back with a sense of continually-growing delight and awareness, and above all, that lovely sense of recapturing the past, one's past, the human past, my childhood—a sort of marvellous Platonic anamnesis. I have suffered very much from losing the past, losing my childhood; I have, unlike you, lacked continuity and roots in this absolutely fundamental sort of way; it has given rise to all sorts of nostalgias, genuine and phoney (drugs, on the whole, seemed to me—or at least the way *I* used them, which was mischievous and narcissistic, to give me only a phoney past, a regression, instead of the awareness and continuity I sought); but various people—patients, students, friends, Wystan, and my analyst—have helped me to recover some of what I had thought to be irredeemably lost, and to become—if only intermittently—a real person.

But no, alas!, there has been no "falling in love" with anyone, in the more literal sense, in the sense of a deep mutuality and respect and wish to share, etc; my "infatuations" and fatuous fetishes etc. have grown weaker, and less satisfactory as love-substitutes, as substitutes for anything; but they have left something of a vacuum, erotic, spiritual, or whatever; and I often feel intensely and genuinely lonely now in a way in which I never used to. I would love to love and be loved, and have a lover, and be a lover; but I am very isolated, I see very few people, and perhaps (ironically, now I am perhaps mature enough to maintain such a relation) I am now too old. At least I *feel* painfully old, at the moment; I have since the death of my mother, last November. [. . .]

I'm sorry—this is a rambling, and maybe very egotistical sort of letter. As I said at the start, I have not known how to reply. When you criticized my writing (and, by implication, I felt, *me*) a dozen years ago, I used to feel annihilated—incidentally, I think I *did* write a few decent things (like "Travel Happy"); it was not all virulent or phoney like "Doctor Kindly"—and now you express appreciation of *Awakenings*, it gives me immense pleasure, and a sense and sort of existing (which I lose all too easily).

I have been thinking of going to the West Coast for a few days around Xmas–New Year: will you be around then? I would love to see you—and, again, many many thanks for your exact and generous letter.

Love,
Oliver

PS: I think—indeed I am sure—you are wrong about Wystan not liking you. The two of you had not met too often—and, as I say, personal feelings were slow to grow and mature in him. But I know that he liked and admired your *poetry* immensely: after meeting you last summer (was it a poetry reading?) he said, "Thom writes magnificently—but he does *dress* very strangely!"*

To Thom Gunn

DECEMBER 31, 1973
37 MAPESBURY RD., LONDON

Dear Thom,

Thank you for your friendly postcard—I much appreciated it, having felt that I had sent you a letter somehow *wrong*. Indeed I continue to feel somewhat obsessed by your letter (which I tend to carry with me), and have the increasing feeling that my own—like, perhaps, so many of my own writings and feelings—was less than candid, if not positively disingenuous; wanting in candour, in penetration, and perhaps too in that sort of sympathy and humanity which unites me so naturally to my patients and students, but which I find so hard to extend to my friends (or myself).

Other things make me think of you. It is a marvellous morning, of a sort one remembers from childhood, but which seems to be exceedingly rare, these days, in London: an utterly clear and cloudless sky, low low December sun, everything frosty, intense, alive, bathed in a lucid, almost Arctic sort of sunlight—the sort of sunlight which

* Gunn was fond of wearing leather jackets and boots, unusual among a literary crowd in the 1970s, while Auden favored rumpled suits.

reminds me of "Sunlight"*—sharp, unsparing, yet extraordinarily gentle. And yesterday (in the train, coming back from the North) I reread *Etruscan Places,*† and *that* made me think of you—I remember how warmly you spoke of it soon after we met.

And (it being a bit below freezing) I have on my favourite (now rather ancient) leather jacket and mittens, in which the fetish is sublimated into proper role and friendly presence; and sitting in the garden, in the frost, in the sunlight, mitted and kitted up to my eyebrows—this *also* makes me think of you, what we share and variously apprehend, and of my dismissive dishonesty, in my previous letter, in speaking of "fatuous fetishes" as something behind me, done-away-with, forgotten.

What a complex business, these "emblems of identity." Easy to be overcome and infatuated; easy to be dismissive and sneering; but *so* difficult to prehend or present in its full complexity (and full simplicity), without reduction or caricature. I cannot pretend to understand—although I understand enough to see the partial folly (and the partial truth) of earlier feelings, earlier writings. You, I think, always had a humour, an irony, an art—the power of detachment, of simultaneously *being* and *seeing*—being both inside and outside a certain frame-of-reference. This is, in effect, what you said to me the first time we met, *viz.* "I am *me*—not my parents, or the subject of my poems. You mustn't confuse the writer and his subject"—or something of the sort. You always had, I suspect, this amalgam of genuine feeling from the "inside" with an artistic seeing from the "outside," a humorous mastery which prevented any sort of total identification or engulfment: the doubleness which transforms one from passive (pulsive) to active, patient to agent, victim to—I don't know how to say it, but a *sort* of mastery, without doubt. It is the sort of doubleness which I try to encourage in the patients I see, and which my analyst tries to encourage in me, in which one is both inside *and* outside the state of "patienthood" (forgive horrid word!), inside and outside what one suffers, and *is;* so that one is not dominated and devoured by a part of oneself.

* A poem by Gunn.

† By D. H. Lawrence.

I lacked that sort of doubleness, detachment, humour, mastery, for many years; doubtless (though less) the lack is still there. There was a sort of ferocity and totality of identification (with some fierce, fantastic, impregnable "Wolfhood," as opposed to what I felt as my soft, unformed, vulnerable Oliverity!), which at once promised identity and a loss of identity, a sort of enthralling (and fatal) death and rebirth. Perhaps I gave way to such ideas especially between '63 and '67, after I had left San Francisco (and relative sanity), and moved into a labyrinth, a mire, of phantastica and drugs.

I cannot tell you (though there is no need)* the intensity of *lostness* and *foundness* (but a false "foundness") I traversed in those years. Nor can I assess, as yet (if ever it will be possible) the contribution of those years of almost-total immersion in drugs and phantasies to what I am now, or to be in the future. To call them "regressive," to write them off, to disown them, to deny them—as perhaps I tended to do in my previous letter—is certainly a foolish reduction and fraud, and I feel ashamed of the dismissiveness, or contempt, which I may have expressed. But you will understand better, and forgive, if I make clear to you the degree of *engulfment* which I have been through in the past (and which, perhaps, in varied forms, *perpetually* faces me). It is a sensitive area for both of us, complex, exciting, dodgy, as any. The relation of one's poses and postures and "pseudo" identity to one's "real" and constant, ever-growing richly-aspectual identity is far beyond my understanding at the moment: clearly, one's identity is in no sense *simply* the sum of one's identifications; but nor is it something pure and Platonic, or completely transcendent. Auden (like the Red Queen)† used to say, "Always remember who you are." And I think he had a very good idea of who (what,

* Gunn used prodigious amounts of speed, LSD, and other drugs, but even he was impressed by OS's consumption. Lawrence Weschler, in his book *And How Are* You, *Dr. Sacks?*, quotes Gunn as saying, "He did do a lot of chemical experimentation, I mean, outrageously extreme, far more than anyone else I knew."

† From Lewis Carroll's *Alice's Adventures in Wonderland*. In a 1965 letter to Jenö Vincze, OS wrote, "It has been my bedside reading for 20 yrs; under the veneer of its Victorian prose (manners, heroine, morals etc.) it is Beckett, Sartre, Joyce + Wittgenstein; the human mind in puns and parables; the book of a mathematician, incapable of, yet also beyond, poetry; the final reduction of epistemology; a joke, a dream, a game once more."

why) he was. *My* idea of self is much fainter (or given to vanishing) and more clouded with conflicts and contradictory identifications. And *yet,* it is (or is it?) (at least some of) these very conflicts which provide the possibility of fruitful reconciliations and renewals in oneself.

7

Astronomer of the Inward

1974–1975

THE RECEPTION OF *AWAKENINGS* was mixed. OS's medical colleagues greeted the book with a resounding silence, and in the United States it received only a single review (by Peter Prescott in *Newsweek*). Yet it was noticed by the literary world, at least in the United Kingdom, and would win the Hawthornden Prize, placing OS in the company of such previous winners as Vita Sackville-West, Graham Greene, V. S. Naipaul, and Michael Frayn.

Colin and Anna Haycraft made OS a guest of honor at their 1973 Christmas party. As he recounted in *On the Move*, Jonathan Miller said to him after that party, "You're famous now."

To Thom Gunn

JANUARY 13, 1974
11 CENTRAL PARKWAY, MT. VERNON, NY

[Postscript to the letter to Gunn from New Year's Eve 1973]

To resume this odd letter (if "letter" is the right word, perhaps it is more a soliloquy, or an apologia, or—I don't know): to resume it, then, after a fortnight, I *almost* phoned you on New Year's eve—I tore myself out of a party and went restlessly from one bar to another, thinking of the happy-acid-New Year's eve which you had spent, which I had hoped *we* might spend, four years ago, 1969/70.

My 17 days in England were marvellous, marvellous for self-focus and self-esteem: if the first 50 weeks of 1973 were among the most miserable of my life, the last two weeks, mercifully, extraordinarily, have been among the happiest. *You* knew recognition (recognition of your true self and powers and achievement, etc) at 25 or so; I have had to wait till the age of 40. But it is a sweet feeling, especially after so many years of non-recognition, rejection, getting-nowhere, being-nothing, etc. The *sort* of recognition is interesting, but perhaps not entirely surprising: I write a book about the central problems in Medicine (at least, these are as close to the book's centre as anything could be in an essentially centreless book), and everyone seems to be fascinated by it except my own profession, my colleagues, my "set," who have managed to cold-shoulder it (and me) entirely. [. . .]

Trying to digest all of this, I have the feeling that it represents, somehow, an "identity-crisis," or, rather, the resolution of one (which is normally accomplished in adolescence or early manhood, but which in my own case was protracted for a quarter of a century). It has become clearer to me who and what I am, and what I must do. And very necessary, I now realize, is the act of confessing but also transmuting the experiences and changes of the last few years in the form of an autobiography—an autobiography which neither omits nor simplifies, neither pleads nor extenuates; candid, but without self pity or contempt; funny, maybe picaresque: something which will render (to the best of my ability) suffering, perversion, inauthenticity, cruelty, etc in the mode of sympathy, and, above all, *comedy;* which will expose the fruitfulness of conflicts and contradictions; the core of authentic passion and enquiry which has always lain "behind" the fetishes, fixations, etc . . . I will regard this autobiography as a fuller attempt to answer the questions you posed in your frighteningly-penetrating letter of last October. [. . .]

I am sorry I did not get a chance to see you over Xmas; but I had a sudden, peremptory feeling that I needed to go to London. I think, however, that I will come to the West Coast in Spring or early Summer—

Please write, and tell me how *you* are. Reading between the lines of your letter I find myself somewhat concerned, and would be reassured to hear fully from you.

Love,
Oliver

To Martin Snellgrove

CORRESPONDENT

FEBRUARY 9, 1974
[NO ADDRESS GIVEN]

Dear Martin Snellgrove,

I am most grateful to you for your letter and your Poems. I had an immediate empathic response, and much admiration for your skills and sensibilities. [. . .]

It seems to me—as perhaps to you—that words are our greatest blessing (and privilege) and our greatest curse. They are so powerful—and so impotent. Twenty years ago, as a student, I saved up for months to buy the big OED, thinking then that words could give one a precision in a *pointing* sense for conveying one's thoughts and feelings. I still love the OED, but in a completely different sort of way now. I ignore the definitions and find delight in the enormous variety of meanings which words have, their evocative precision, as opposed to their defining or pointing quality. I think of all experience as metaphorical, and I suppose I think of the world too as metaphorical—at least, as inaccessible and unconveyable in any mode *but* that of metaphor. I don't mean in any sense that I deny *objects;* merely that the only way in which we can know the world—or in which the world can know and compose itself—is in terms of an infinite richness of presentations and representations. [. . .]

I don't see why you should say of yourself that you are employing poetry "as a vehicle for something essentially extraneous to it"; because nothing is extraneous to poetry, or, putting it another way, that *everything* is essentially poetic and symbolic. I think the dissociation of, say, poetry and mathematics is quite artificial. (One feels that really great mathematics and mathematicians are also great poets, and essentially so, in their way, using *their* vehicles of election. One sees this beautifully, for example, in Gauss' extraordinary papers on the curvature of space and possible spaces, on his insistence that "space" itself is only a way of describing experience and action, that it is a metaphor, or, rather, a plenum of metaphors all of which are interesting, relevant, useful.)

I love the richness of concrete imagery in your poems. The only thing which disturbs me—once in a while—is when you sometimes seem to speak of "pure" thought, or Aethers, or a sort of structureless Oneness, or Nothingness, or "holes." (Similarly, in Thom Gunn's poems, which I so love, I sometimes get disturbed by references to a sort of abstract Will, rather than the orchestration of an infinite number and variety of wills.) I mean, I think there is always Order, infinite Order, in all real thoughts and in reality itself, infinite differentiation, infinite integration, infinite simplicity, infinite complexity (as opposed to fracture and fractionation, or addition and concatenation, or simplification and complication). And I cannot distinguish "soul" from "body," or Thought from Reality: it seems to me that everything is Thought, but thought which is incarnate and *enactive*—deeds. [. . .]

With kind regards,

To Mel Erpelding

FEBRUARY 10, 1974

[NO ADDRESS GIVEN]

Dear Mel,

[. . .] It is a bitterly cold night—for New York, cruelly cold, and we are low on heating-oil in the house. And so I have wrapped myself up, I have put on a thick woolly jersey and my old and faithful leather jerkin; and (my fingers being freezing after a couple of hours of shovelling snow outside) I am wearing an old and faithful pair of gloves. I am encased, enclosed, contained, in warm and protective clothes—skins and tunics. And now, after hours of feeling frozen and drenched and miserable, I feel deliciously warm and *cosy.* I feel like a naked fledgling protected in its nest—almost like a foetus in the womb. I feel *safe*—I feel *at home.*

And your words echo and re-echo in my mind. I am thinking of your words about the Barbour suit,* your sense of safety and protection, and almost of invulnerability. [. . .] The sense of being safely,

* In the 1930s, Barbour introduced waterproof "Barbour suits," for motorcycling in wet weather.

warmly, beautifully *enclosed* is, surely, one which must go back much farther than the time we first met; it must go back to earliest childhood, and perhaps still earlier; and it is not peculiar to us, it is something universal, something deep and mysterious in Nature itself, which each of us (*all* of us) remember and find in our own special ways. And it is *this* thought/feeling which you touch on in your letter (over and above whatever sense of warmth, safety, protection, etc. we ourselves shared in our gear, our kit, on our bikes, in our travels, in our own and personal togetherness), and this thought, to which I shall try now to give a more general form. (Incidentally, am I to take it that you still have your bike, or another one? And, I also wonder, do you still Scuba-dive sometimes, in that other and allied protective suit?) As I write, my eyes shift every so often from the paper before me, to a lovely pearly Nautilus which rests in my study; and I think of the Nautilus, naked and vulnerable, in its pearly chambered suit it has made for itself. [. . .]

The idea of *enclosure* is surely as elemental as any—whether it is the nucleus of an atom enclosed within its electron-shells, tiny animals and plants in their cysts and spores, chicklets in eggs, embryos in wombs, birds in nests, the nesting of words within phrases (which Chomsky so beautifully analyses), knights in their armour, animals in their skins, or *you,* Mel, in your Barbour suit. There was a time, for many years, when I thought of bike-leathers and other sorts of "kit" as mere fetishes or fixations, something deeply compelling and compulsive, but also entirely neurotic and regressive, something *shameful,* designed not only to protect but to *hide,* to hide behind, to be dishonest behind . . . And, no doubt, in a way, it was—that is, when I thought of "kit" in this excluding-and-concealing way, and used it as such. I now see that when I thought and behaved in such a fashion, I emphasized the neurotic aspect to the exclusion of other aspects—aspects which were healthier and more natural, universals of Nature itself. [. . .]

I do *not* think that one can properly speak of men, of any beings, of Nature, as having an "inside" as opposed to an "outside," as having a "core" as opposed to superficies. There are only—phenomena. But the phenomena of the world have an infinite plication: there is continuity, and *yet* there are layers; one's skin is infolded to become the membranes and mucosae of one's organs and viscera; their membranes are infolded to become the membranes of one's cells; their membranes are infolded to become the membranes of molecules; molecules are infolded to become atoms enclosed in electronic shells; or, going "outwards," one's skin is "exfolded" to become the (in)vestments one wears on one's clad body, one's surface, one's skin, extends and coextends into the skin of the world, that infinite membrane which is Einsteinian spacetime. One is naked, one is clad, one is necessarily both. A man in his Barbour suit is as natural a phenomenon as a squid in its mantle or a penis in its foreskin. The mystery of the world is a mystery of *folding*—one sees why Leonardo, why all the great artists, have had such a passion for draperies and folds, those mysterious multi-plications (omni-plications) which show us the variety and unity of Nature. The earth itself, clad in mountains and valleys, infolded, outfolded, is like a man in a suit. [. . .]

There is nobody else in the world to whom I could have written (would have *dared* to try writing!) so peculiar a letter. I hope that it neither offends nor upsets you. It is not my wish (as it once was) to vex or to claim you, but to see what we share in this odd business of living, of being fellow-travellers in the mysterious world of reality.

Dear Mel, Brother-Animal, I love you, your thoughts, your body, your Barbour suit, and all. (If it makes sense) I even love the atoms which compose you. And love—real love—doesn't make claims or possess; it gives reality, recognition, to oneself and the world. Write when you write; come when you come; the very thought of your presence, however separated we are in space and time, gives me a selfless joy and the loving-lovely feeling of sharing the world.

To Samuel Sacks

MARCH 3, 1974

11 CENTRAL PARKWAY, MT. VERNON, NY

Dear Pop,

I tried to phone you this morning, but your (silly) Answering Service said you were away for the weekend—I spoke, briefly, to Auntie Len

and Michael; and David phoned me twice, reassuringly, last week.*
[...]

I was also *particularly* pleased by that Editorial in the *British Clinical Journal* (do you know who the editor is, by the way?). It has been the *first* generous and genuine response from the profession; I wrote to tell him how grateful I was. Because there is no doubt that *Awakenings* has been received by my colleagues, on the whole, with a mixture of hostility, bewilderment and mean-ness—just the mixture which has so *confined* me since Arnold Friedman, 1967, and even years before this. It is paradoxical that the *last* people to grant one a proper recognition are so often one's colleagues and fellow-workers in a subject. Anyhow, I see intimations, as doubtless you do also, that my stature and nature (however idiosyncratic!) is coming to be properly recognized at last.

Medicine has fallen on evil days, since the days of Henry Head, Kinnier Wilson, etc. The very notion of *Care* has practically vanished, and been replaced by a pedantic passion for new drugs, procedures, methods, techniques. I used to wonder why, almost 60 years ago, you chose to relinquish a potential career as a Specialist, and to become a General Practitioner instead. I *now* see that you instinctively and intuitively made the right move; you have loved and cared for tens of thousands of patients, as they have loved and cared for you in return; I think your professional life has been infinitely more significant and rewarding, more *genuine*, than the half-lives of most hospital Consultants and Specialists.

I think good General Practitioners are the salt of the earth; and it is a major tragedy and symptom of our times that they—and the whole concept of care, the whole Art of Healing—have all but vanished from contemporary life.

I have made this point rather central in a piece I have written (a requiem) for Auden.† *His* father was a general practitioner, and Auden himself, throughout his life, was deeply concerned with the nature of doctors, healing and care. In his last book of poems, no

---

* Only fifteen months after Elsie's death, Sam Sacks's brother, Bennie, had also died, and his sister Alida was quite ill.

† "Dear Mr. A. . . . ," published in a 1975 tribute volume edited by Auden's close friend Stephen Spender and later adapted for *On the Move*.

less than four are dedicated to beloved physicians—the last of these to me; and in his very last letter to me, this summer, he quoted some words of Osler which *his* father used to quote him: "Care for the individual patient, more than for the disease from which he appears to be suffering."

Both you and Ma (whatever else was the case) showed a lifelong passion for the care of individual patients. Medicine, for both of you, was a dedication and a calling; and I am deeply happy and grateful that you have both transmitted this great tradition of compassion to me. And equally (again, despite anything else) to Marcus and David, and Caroline as well.*

I am well and active and happy. I have just written—spontaneously, unintendedly—ten short essays on various aspects of *Cares* and perhaps these, rather than anything else, can be dovetailed together to form my next book. I expect to come home for Pesach, though briefly, and look forward to being with you—you all—at that time.

Love,
Oliver

EARLY-TWENTIETH-CENTURY PSYCHOLOGY was largely dominated by the behaviorist work of Pavlov and Skinner, which simplified human behavior into stimulus and response, but this began to change with the work of A. R. Luria and his colleague Lev Vygotsky. These two Soviet psychologists created the new discipline of neuropsychology, working from their deep investigations of linguistic and cultural factors, which they were convinced were equally important to brain function. As a medical student, OS had heard Luria lecture in London, and he had read Luria's monumental *Higher Cortical Functions in Man* and *Human Brain and Psychological Processes* shortly after moving to New York. He also greatly admired Luria's two very detailed narrative case histories: *The Mind of a Mnemonist,* about a man with a highly unusual autobiographical memory—he seemed to

* OS's brothers Marcus and David, as well as David's daughter Caroline, were general practitioners.

forget nothing that had happened in his life, and *The Man with a Shattered World,* about a young soldier with severe traumatic brain injury and his struggles to put his fragmented world back together. These works had become models for OS in writing about his postencephalitic patients. Although they would never meet—it was difficult for Luria to travel, given the restrictions of the Soviet Union—they carried on an extended correspondence, and he became a lodestar for OS.

To A. R. Luria

NEUROPSYCHOLOGIST

APRIL 13, 1974
11 CENTRAL PARKWAY, MT. VERNON, NY

My dear Professor Luria,*

Thank you very much for your last letter—my heart always misses a beat or two when I see in my letterbox one of your delightful, rare, prized letters, with their great Russian stamps and their unmistakable handwriting! [. . .]

I just had occasion to see a youngish man with an extraordinarily severe Korsakoff's syndrome†—his memory-slate for immediate traces seems to get wiped clean in about five seconds. What makes his predicament so striking, so pathetic, so *challenging,* is the fact that he is in no sense "psychotic," and though he "confabulates" (as it were, automatically) he is neither delusional, mendacious, nor lacking in insight. Indeed, he seems to me altogether a fine man—fine in intelligence, fine in character and courage. He wrings my heart (as he challenges my intellect); and in the hour or two since seeing him, I have been continually asking myself *how* he might be helped, what *sort* of therapeutic programme or activities could *conceivably* help him in his—his chaos. And chaos it is, in a sense, for he is like a man outside History, absolutely unmoored and disoriented, in a constant

* Luria and OS had begun corresponding in 1973, after OS reviewed Luria's work in *The Listener.*

† Korsakov's (also spelled Korsakoff's) syndrome can occur with alcoholism or deficiency of vitamin B1; it damages parts of the brain's memory systems.

movement of ever-changing, bewildering, *meaningless* moments. He is enveloped in immediate amnesia; but beyond this amnesic circle, so to speak, his general memory and associational nexus seem to be entirely intact.* How can we help him to leap or bridge this intense amnesic gap? How can we restore him to continuity? Of course he has continuity, in innumerable ways: his use of language, for example, is eloquent, perfect. I haven't started to investigate him or his problems, and it is precipitate and premature to mention him/them now. But I am wondering what your own experience has been with such patients, how *you* would approach/assist him, etc. I constantly have such thoughts when I see patients: I am always saying to myself "Now what would Luria think? How would *he* formulate or approach the (phenomenal or procedural) problems here?" So forgive me if I take the liberty, this once, of actually asking, where I so often ask in imagination. [. . .]

DURING THE FIRST FEW MONTHS of 1974, as he mentions in the previous letter to his father, OS wrote more than a dozen essays about the patients he was seeing at Bronx State, a public psychiatric hospital—mostly people with autism, schizophrenia, or severe learning disabilities. He had tried to engage with them in ways no doctors had before: he obtained a piano for them, and a Ping-Pong table; he even took one patient, a nonverbal artist long immured in the hospital, to the nearby New York Botanical Garden (where the young man uttered his first word: "dandelion"). But when OS objected to the institution's use of "behavior modification"—i.e., punishment—his bosses retaliated by insinuating that there were rumors that OS had "abused" his patients. Horrified, and in a fit of rage (typically, directed at himself as much as anyone else), in mid-April he impulsively destroyed his essays, trying to erase the whole traumatic experience.

* This forty-nine-year-old man, whom OS would later write about in "The Lost Mariner") had retrograde amnesia such that all of his memories after the age of nineteen had been wiped out, and he was incapable of forming new memories. But he could still recall his early life.

In the meantime, Bob Rodman, the psychiatrist who was OS's closest friend from his UCLA days, was consumed by a crisis of his own: his thirty-eight-year-old wife was suffering from debilitating pain in her abdomen, and he feared the worst. By June, Rodman was writing long letters to OS almost daily, sharing his agony and uncertainty as he and his wife visited numerous medical specialists—none of whom could agree on a diagnosis or, indeed, even whether to tell Maria of their findings. He wondered whether to take the family to Maria's homeland, Sweden, for what seemed likely to be her final days. How could he continue treating his own patients when he was so distraught? How could he face the prospect of telling their young daughters, aged nine and eleven? He remarked in a letter to OS that he could not afford the "luxury" of suicide because of them.

To F. Robert Rodman

JUNE 25, 1974
11 CENTRAL PARKWAY, MT. VERNON, NY

My dear Bob,

I was painfully surprised and distressed by your letter and wept as I read it, for you and with you. I did not quite know, however, how to "respond": I have thought of you many times in the last few days, picked up the phone receiver and put it down, started letters and stopped them. I wish I had not been so blind before. I *did* see that you were in the throes of crisis, but I failed to see the real and agonising events behind it and sort of took you at your own word, that it was merely a sort of age-generation-generativity thing.

It sounds as if the two of you have had a frightening time these last few months—no, frightening is too mild a word, horror and panic, which you use, are nearer the mark. And any sort of coming-to-terms precluded by *uncertainty* as to what is going on. I gather that Maria will be seeing the lung-man around now—and perhaps he will offer you the knowledge, if not the reassurance, you so need; or perhaps not. You *have* to know what's happening, whatever it is (I pray that it turns out to be no more than a tardily resolving post-surgical pneumonia). Because if you don't know, you are in a quicksand of anxiety, everything keeps giving way beneath you, and you are prevented from achieving that certitude, and still worse, that *fortitude* which you so much need . . . and of which you have so much, when you are not being undone, or unmanned, by extraordinary circumstances.

I think too that you *must* have someone to talk to at your end, someone who is both physically and emotionally close to you; you are maddening yourself by being silent, and secretive, on so grave a matter, and when you feel as you do; whether it is your former analyst, or another psychiatrist, or a friend, you *must* disburden yourself and find a sympathetic (and practical, but above all sympathetic) ear and heart. You *must* have at least one such friend near you in Los Angeles. It is folly, and a sort of pride, to bear this cross alone.

Your strength will be tested enough, and more than enough, by facing what you have been facing, and may have to face, even given sympathetic understanding and support; to deny yourself this is surely an act of condemnation of some sort, condemning yourself to suffering alone, in gratuitous secrecy and silence.

As sickness is the greatest misery, so the greatest misery of sickness is *solitude.* Solitude is a torment which is not threatened in hell itself.

I know very well those forces which shut one up and in, which confine one in an *extra hell,* when one most needs to reach out towards others, towards one another. You must fight these forces, and let your friends deliver you to some extent: this is, ultimately, what they are for. I am physically too far away, alas! to do more than write or phone, on the whole. You need the actual, physical, living presence of people, of someone, who can commiserate, comfort, sympathize, advise.

Wittgenstein says somewhere, about suicide (*he* was always thinking of it), that a man never commits suicide unless *he rushes his defences.* I am relieved that you feel you must not consider this last luxury because you have your girls, who depend on you; but you make no mention of the feeling, and perhaps it is lost at the moment (in which case it must be revived), that you have no right to suicide because of your duty to yourself, a most gifted man, in the prime of life, and the prime of his powers, who has so much to give and live for, so much to fulfil, *even if the worst should happen.* It is the sense of being annihilated, of having lost the grounds and spring of your *own* sense of life, which strikes an ominous (and morbid) note in your letter. Or perhaps not morbid: of course, one is diminished, at times almost extinguished, by the illness, let alone death, of those one most loves; the soul shrinks back into itself, and says "Me? Let me die too! What right do I have to live now? For whom, for what, should I live?" and the answer "For yourself, for others" carries almost no force. You must sit quiet, however much grief and terror wash over you (by sit quiet, I mean not give way, not rush your defences). I am sure that seeing your patients, however painful it is, is as vital for you as it is for them: you must hold on, and gather your courage, and—above all—have someone *near* you on whom you can lean (and it is *no shame* to lean on someone at such a time: we are all leaners and leaned-on, anaclitic, at desperate times).

I hope more than I can say that you have painted too grim a picture, and that things are already taking a turn for the better. Do write again, or phone, Bob; and perhaps we can meet in the summer, soon; *don't* despair—it's a sin, and don't forget that there are *many* who love, esteem and need you.

Love,
Oliver

To F. Robert Rodman

JULY 22, 1974
37 MAPESBURY ROAD, LONDON

Dear Bob,

[. . .] Yes, I *have* received all your letters, with their impressive and frightening range and depth of feeling, and their beauty of expression, which never deserts you however tortured you are. [. . .]

I feared for you from the tone of some of your later letters, because some sort of "breakdown" seemed to be—hanging over you; you were close to the edge (although there are people, and you are perhaps one of them, who *can* go close to the edge, and even over the edge, and may have to, in order to *traverse* an experience entirely); and I wanted to say—but couldn't: "*don't break down now, when she needs all your sympathy, support, and above all, sanity; break down—if you must—later, but attend to her, not yourself, in the few weeks that remain.*" I don't know whether this would have been the right thing to say—how you would have taken it; whether I should say it now; or whether, indeed, either saying or not-saying such things make any difference at all.

Let me add (but you know this already from your knowledge of me): that I see nothing shameful or disreputable or improper about a "breakdown"—whether it be paranoiac or of any other form; if one is to taste and test one's emotions to the full, to go through an experience of horror undenied, one almost *has* to become psychotic, or at least to enter the realm of "as-if" psychosis; one has to do this, to submerge completely, and then re-emerge, purified of the horror. But it is a dangerous process, an internal Odyssey before which the strongest may quail unless they have some human life-line to "reality." For my patients, perhaps, whom I was constantly with, and whom I could treat with that special sort of absolutely intimate, yet dispassionate, concern (which is the proper relation of physician to patient, or Artist to Art), some such guidance was possible; as again it has been possible between my analyst and myself; but I don't know how possible it is between *friends*—especially if they have been separated, and are at a (physical) distance from one another. I was discussing this very point with Alvarez* last summer: you remember how he opens his book, *The Savage God*, with a beautifully-analysed yet agonized memory of Sylvia Plath's suicide and final days, and the tormenting question as to whether *he*—as her friend, her confidante, her neighbour—could or *should* have known more clearly what was going on, and whether he might not have *acted* to save her: these questions of Insight and Responsibility were still on his mind, fourteen years later.

I am sure that you are both doing the right thing in returning to Sweden for your wife's last days: one should die, as one should be born, among one's own, and *at home*. (My poor mother said to me, many times, that when her time came she wanted to die at home, in the garden, at Mapesbury: and one of the things I found most distressing was the thought that she had been denied this, that she had died in Israel, on some idiot tour, under the burning heat of an alien

---

* Al Alvarez (1929–2019) was a literary critic who served as *The Observer*'s poetry editor from 1956 to 1966. A poet himself, he also published many nonfiction books, including *The Savage God: A Study of Suicide* (1971).

sun.) I think the *metaphysics* of dying (if this is the right word) are of the profoundest importance: by which I mean—not any sort of deathbed conversion, repentance, etc.—but a final coming-to-terms with one's life and one's death, with everything one has been, and the—the nothingness, darkness, mystery (who knows?) before one. This is the time for deathwork, *her* death-work, in which you must be her closest companion: *not* the time for neurosis of your own, which you have every right to, but *not now* . . . later. [. . .]

It is not given to us to die like oak-trees, or like animals (do you know Tolstoy's story "Three Deaths"?), but I think that death-anxiety, and *death-fuss,* are the great *spoilers* of death. Even if one has been tormented by death-fears all one's life, one's death, when it comes, can be, if not welcome, at least a deeply peaceful and serene acquiescence (one sees this very strikingly, as Boswell describes it, in Johnson's last days, his last weeks of life, when he was relieved from the intense death-fears which had obsessed him all his life).*

Obviously this is much more difficult to achieve when one is struck down in one's prime—the sense of *anger,* of *outrage,* the continuing *violence* of one's will to live, may make such an acquiescence tremendously difficult; but it *must* be achieved, if one is to die decently and properly, without a fierce peace-depriving agony. One's death-work has to be the greatest and hardest work of one's life; and, perhaps, as the philosophers say, all life, in a sense, is such a preparation for death.

Enough—or too much—said. All I can do is to reiterate once more my deepest sympathy to you both in this—this farewell situation; and to assure you that I am not far away, and am within easy writing, phoning, or coming distance.

Your friend,
Oliver

* This is one of many letters that foreshadow OS's own farewell letters in 2015. He was well versed in the vagaries of the end of life—both in life and in literature—and often related stories of how a great poet or scientist had died, their last words, their final acts.

IN AUGUST, STILL DEEPLY SHAKEN by his treatment at Bronx State, and facing Shengold's annual month of holiday, OS took off for a few weeks in London and Norway.

To Al Alvarez

POET, WRITER

AUGUST 20, 1974
HOTEL ULLENSVANG, LOFTHUS, NORWAY

Dear Al,

I am sorry not to have seen you since last summer: some wretched spirit of perversion makes me keep away from people I like and admire.

My particular reason for writing *now* is because just before coming here I lighted on your book on Beckett (it is the *only* book I have with me here in the fjords!). I think it's uncannily penetrating—frighteningly so; I was mesmerized by *Molloy* and *Malone Dies* twenty-plus years ago, when they came out, but have not read any Beckett since; your book rouses my appetite—but alas! there isn't a Beckett-containing bookshop within a hundred miles of here! To paraphrase something you said in yr review of *Awakenings, your* book on Beckett shows how deeply you understand his world of nothingness—eternal sloth and nothingness and boredom, nothing to say with the obligation to say it. I think your book is an extraordinary act of sympathy or empathy, entering into another's almost-unimaginable imagination, *and* conveying it. And it fills one with admiration for Beckett's *resolution,* that he has kept going, *and said*—incomparably, and *beautifully*—what is almost too awful to say. It also fills me with admiration for *you,* that you keep so active and productive, and so affable and benevolent, when (it is clear) you yourself have to contend with and *contain* a nihilistic spirit of perhaps comparable depth and power.

I wish I could say the same of myself. The book I *have* to write now (I keep destroying what I write) is *also,* in a way, a book about nothingness, hopelessness, futurelessness, situationlessness, vacuity, boredom, monotony, sloth. You had suggested, last summer, that I should perhaps try to write a book on Institutions, and it is—especially—a devastating experience *since* then (in a madhouse) which on the one hand almost did me in (I compromised with suicide by literary infanticide—destroying a book of essays I had written in February–March: pity, crime, because I think it was perhaps the best thing I'd ever done), but on the other has given me, and focussed for me, how appalling and *perverting* institutional life can be. I shall call it "Ward 23" (it was Ward 23, in Bronx State Hospital: for youngsters doubly labelled "autistic" and "retarded," and in most cases abandoned by their parents, and institutionalized since the age of 5 or 6). I had originally thought of doing it in quasi-fictional form, like Chekhov's "Ward 6" or Solzhenitsyn's "One day in . . ."[*] but now incline, for various reasons, to write it in the form of *essays,* interspersed with anecdotes (illuminations, epiphanies, sudden brilliant, revelatory *Kairoi,*[†] standing out against timeless *Chronos,* the dullness and darkness and nothingness of the rest). What so intrigued and appalled me, and what needs to be conveyed, is the amazing *potential* richness and vividness of mental life, especially moral and authentic sensitivity and experience, even in those who are virtually devoid of "intelligence" and "Reason" (as usually understood); the hopefulness of the patients, despite their dire histories, as opposed to the hopelessness and stupidity and *cruelty* of the Institution: the awful resonance of passivity and confusion and nihilism *between* the patients, the place, and the shadowy, sadistic *System* in the background. Ward 23 was run by a complex, strange, and rather terrible man, a Buddhist Hobbesian (!)—a mixture almost as lethal as the Jewish Jesuit in *The Magic Mountain.*

My experiences on Ward 23 were awful in a way "Mount Carmel"[‡] occasionally hinted at, but never really approached. They brought back, with overwhelming force, my own childhood experiences in a concentration-camp-like boarding school, between the ages of 6 and 9.[§] I wonder if something similar gripped Orwell all his life: I

* *One Day in the Life of Ivan Denisovich.*

† A Greek term for "correct moments."

‡ "Mount Carmel" was the pseudonym OS used in *Awakenings* for Beth Abraham Hospital.

§ While this comparison is obviously exaggerated, OS remembered being beaten and starved at Braefield, and he often drew the comparison between his experience and George Orwell's.

am struck by the similarities between "Such, such were the joys" and *1984*—it is essentially the same, or same *sort* of, nightmare in both.

Have you read Bettelheim's* most recent book (*A Home for the Heart,* or some such title)? It's rather laborious, but the *genuineness* of the man and the place stand out—an ideal sort of institution—one cannot doubt for a moment the power of such a therapeutic universe even for the most "hopelessly" ill. An *essential* clue to Bettelheim's insight and achievement, it seems to me, were his *own* years confined in concentration camps; knowing from *his own* experience, the depth of indignity and hopelessness to be faced, and that which could (*perhaps*) allow a man to survive—similarly with Viktor Frankl, who greatly impresses me; *his* understanding could not have been achieved without *his* having endured a concentration camp.†

I wonder—you don't really touch on it in yr book—what sort of early experience tilted Beckett into *his* savage darkness; and I *also* wonder (impertinently) what sort of experience(s) gave you your very exceptional power to appreciate and convey some of the worst things in the world. And again, I wonder about Thomas Mann—I reread *The Magic Mountain* last month; I hadn't read it since I was eighteen or so—and was stunned (among so much else) by his wonderful depiction of *empty time,* all the paradoxes of vacuous, actionless, pastless/futureless time, endless and empty, slothful and restless; how easily and terribly it can devour a man's life; and how insidiously and easily, without really knowing it, some people can be beguiled, bewitched, into timeless nonentity.

I'm sorry for this sudden, rambling, peculiar letter; I hope it doesn't offend you. But I felt I *had* to write to you, because I have been so moved by yr book on Beckett, and—and—all sorts of things. I spent my days here swimming further and further out in the fjords; I swim (I suspect) as you climb (how marvellous "Snow" is in *The Magic Mountain*). [. . .]

Again, please forgive this wordy, illegible, and perhaps impertinent

* Bruno Bettelheim's work was generally well regarded during the 1960s and 1970s. It was only later that revelations of his plagiarism, misrepresentations, and bullying of patients led to his discrediting.

† Frankl, a neurologist and psychiatrist, was imprisoned at Theresienstadt concentration camp, where his father died of starvation and pneumonia. In 1946 he published *Man's Search for Meaning,* about his experiences in the camp.

letter; and, again, congratulations on your excellent book—and thank you for—all sorts of things.

My best regards to you both,
Oliver

A FEW DAYS LATER, as he was hiking alone up a Norwegian mountain, OS fell and broke a leg; the accident would have been fatal had he not been discovered, just before nightfall, by two reindeer hunters. He was taken to a local hospital, then flown back to London and operated on. The leg was immobilized in a full-length cast for several weeks—and, in a deeply uncanny way, soon ceased to seem attached to his body. He could not will it to move, and he could no longer even imagine doing so. This experience ultimately led to a very different book than the one he had in mind; it was a book not only about institutions but about the passivity and helplessness of patienthood.*

To F. Robert Rodman

AUGUST 30, 1974
MEYERSTEIN WARD, MIDDLESEX HOSPITAL, LONDON

Dear Bob,

Thank you for your brave and sensitive letters of July 30 (from Sweden) and August 15 (from Hannover). I am going to have to make this a very brief, and inadequate, and rather "inappropriate" reply, as you will see.

I'm sorry I didn't reply to your first letter, which arrived about three

* One wonders whether OS's rage over Ward 23 infected the Leg book, which would ultimately take a decade to complete. He rewrote the book many times over, stymied in part by the lack of a clear neurological explanation, in those days, for his experience. But he also attributed the blockage to the anger he felt when his surgeon dismissed his questions about proprioception and body image as unimportant. Once again, he was being ignored by his medical colleagues. And it is not difficult to imagine that such a sense of powerlessness and self-hatred (not to mention an inability to deal with authority) derived from his childhood at Braefield—certainly that is what both he and Shengold felt.

weeks ago—I was rather depressed and distracted myself then, as my father was ill (he is better now), also I was not sure *how* to reply, altho I strongly shared your feeling that Sweden was the right place now for [Maria]—for both of you—and that whatever you were doing to strengthen links (emotional, cultural, etc.) with Sweden could only be wholly advantageous to you—to you *all.*

It may have been the strength of this feeling, plus some confused quasi-identification, which made me feel—quite suddenly—around Aug 10, that I wanted to go, that I *had* to go, to Norway, which I did. Once there I many times lifted the phone, and wondered about "dropping over," so to speak; but each time some inhibition made itself, along with the sense that I should do best to keep myself in "the middle distance." On my last day in Norway, I had an idiotic accident (if "accidents" exist as distinct from "incidents"), and took a baddish fall on a mountainside, breaking my left leg, completely rupturing the quadriceps tendon, etc. Mercifully, because it was high and remote, I was found and rescued before nightfall, and returned to England, invidiously and self-consciously, feeling like a fool but also very glad to be alive, last Sunday. I was operated on at once, and today has been my first day sufficiently clear of pain and the obfuscating effects of medications to be able to write any letters.

I will be immobilized here in plaster for 6 weeks or so, and *only then* can I start progressive weight bearing, exercise, ambulation, etc. It looks, therefore, as if I am likely to be in England until November, possibly December.

Let us keep in touch; take the Middlesex Hospital as my address or forwarding orders until I know where I stand.

And no! You *never* sound "maudlin"; you have shown extraordinary resource of every sort since you first wrote to me in the beginning.

I will write again soon.

My love to you both,
Oliver

APART FROM HIS IMPATIENCE to recover the use of his leg and his general fear of passivity and lack of control, OS needed to understand, intellectually, what had happened. How could the injured leg completely disappear so rapidly from his awareness of his own body, and why did his surgeons seem so unconcerned by this? What neurological theory could explain the brain's ability to "forget" a whole appendage almost immediately? He needed to know why this experience, this rupture in his own body image, had been so terrifying, so annihilating. And why his doctors, by and large, had rebuffed any attempts to discuss this existential problem.

OS's main interlocutor at this time was A. R. Luria. Following the accident, OS dispatched a number of huge letters to him—some running to forty-five pages—ruminating on these questions. Some of the following letters to Luria have been greatly condensed.

To A. R. Luria

SEPTEMBER 13, 1974
MEYERSTEIN WARD, MIDDLESEX HOSPITAL, LONDON

Dear Professor Luria,

It is a while since we have been in touch. I have been meaning to write to you for some time, but am *especially* prompted to do so now following some strange neurological experiences of my own!

Three weeks ago, while climbing in Norway, I fell, and tore the tendon of my left quadriceps completely off the patella. I was alone, in a scantily-populated part of the country; it was a great good fortune that I was found before the Arctic nightfall, because otherwise I would not have survived to tell the story.

I was operated on August 26th—seemingly a straight-forward injury operation, with no significant peripheral nerve damage (although, I suppose, all sorts of nerve endings would be destroyed or stimulated in the torn tendon muscle).

Thus it is now my 18th post-operative day. What seems to me of great interest—but perhaps it is a commonplace finding (I confess that I am ignorant of the matter)—and what I found terrifying at times was the *central* disorder associated with injury. Thus, for the prior 13 days after the operation (or 15 days after injury) I found it not only impossible to elicit the least contraction of the quadriceps, but *unthinkable.* When asked to "tighten the muscle" I would go red with effort—but nothing would happen. I felt my effort diffuse uselessly—that it had *no point of reference or application;* and that, indeed, it was not a real exertion of "will" because there was *no proper intention or idea* (image) *involved.* I found this profoundly disquieting, the more so as I did not expect it. My surgeon spoke of "reflex inhibition" (at spinal or peripheral level), but this was obviously inadequate as description. I wondered fearfully whether I was harbouring an obsessive or hysterical inhibition (or amnesia); and whether the use of the leg would ever return. I found reassurance useless in face of the intensity and strangeness of my own experience.

When I was told to "try to raise the leg," *I could not think what was required;* I found it impossible to imagine any *procedure* by which I could raise it. I felt I was deprived of *the idea of the action,* the idea of "how." I felt like a man with a stroke—or so I imagine—who is faced with some bizarre *apraxia* or amnesia of action. On the 13th post-op. day I discovered, by accident, how to twitch my quadriceps and pull the patella, but I could not at this time make any flexion motion at the hip, or *imagine* doing so. On the 15th post-op day this became thinkable and doable, for the first time; this was a great relief to me, and felt like a miracle. The power to do this (and the simultaneous *idea* and *intention* of doing so) were possible on this day, and came and went in an unpredictable way. Thought I might make the flexion-motion, but find that I had then "forgotten" the movement and could not make it a fourth time; a minute later, perhaps, I would "remember" the movement, and be able to do it—or perhaps not. I was reminded of aphasiac patients who can *sometimes* get a word, sometimes not—who find the word/idea *continually slipping through their grasp.*

On the 16th post-operative day I was got up on crutches for the first time. I found this bizarre and terrifying. When I looked straight ahead, I had *no idea where my left leg was,* nor indeed any definite feeling of its existence. [. . .] Whenever I looked down I found my visual perception strangely unstable: I had momentary difficulty in recognizing an "object" next to my right foot as my left foot: It did not seem to "belong" to me in any way. When I accepted it as my foot, it seemed too heavy, too far away, tilted to one side, in the wrong place, etc.—really a sort of central visual-kinaesthetic dis-orientation, analogous, I felt, to the loss of "motor images." [. . .]

One speaks of having to "relearn" actions and motor patterns after they have fallen into disuse for a while—but I confess that I had no real anticipation whatever of the sort of difficulties I would encounter, and the mode of their recovery.

I have been intensely put in mind of patients with stroke, and of the loss and processes of learning, recover and discover, adaptation, innovation, etc. you relate in *The Man with a Shattered World.* I can now appreciate the beauty and accuracy of the descriptions in there, as never before.

Strange business! I would be most interested in your comments. I have to wear the plaster another 3½ weeks, then will have to spend a period strengthening the leg (it is so wasted and weak already, after 2½ weeks in plaster), and learning to walk *alone* (the final difficulty, I fear). I do not see at the moment how I could extend my knee or hold it extended. I cannot imagine walking without the plaster. So I will stay in London another 6 weeks or so before I can return to my work in New York.

The experience of *being a patient* (the first time for me) is quite extraordinary; and I am tempted to write up this whole experience as a little *journal* or book. [. . .]

I hope that you are in good health, and have been having an enjoyable summer.

I spent a delightful day with our friend Richard Gregory about 3 weeks ago. Have you seen his latest book, an anthology of papers which Colin Haycraft has just brought out?* Beautiful work.

I should very much enjoy a letter from you. I feel so *isolated* in my hospital room, and visitors and letters are the breath of life.

Yours cordially,
Oliver Sacks

* *Concepts and Mechanisms of Perception.*

To Richard Gregory

EXPERIMENTAL PSYCHOLOGIST[*]

SEPTEMBER 14, 1974

BOND ST. WARD, MIDDLESEX HOSPITAL, LONDON

Dear Richard,

Thank you so much for your book and your letter. Your book "hit the spot" perfectly—I was reading too much, but I was utterly *starved* of visual nourishment and stimulation, and found myself drinking in the delights of these eagerly from the book.

This experience (which is not finished yet!) has been an astonishing one in so many ways—so much so that it is now clear to me that this is the "material" for my next book. I have no doubt that my experience has been basically similar to that of all patients who are suddenly and unexpectedly injured, rendered helpless and dependent, immobilized and isolated; in my own "case" (I suspect) all sorts of symptoms and phenomena have been magnified by the magnifying glass of my intelligence and imagination, coupled with obsessiveness and fears. As an astronomer peering at a new planet gradually "sees" or "makes out" more and more features, which become clearer and clearer as he "concentrates" upon them; so I have felt like an "internal astronomer," peering and gazing at landscapes of symptoms, and by the act of continual exploration increasing the symptoms and becoming enveloped ("wrapped up") in them.[†] (There are dangers in this sort of introspective astronomy, especially when it concerns itself with morbid phenomena.) However, whatever intensification or prolongation of various strange disorders I may have caused, I am *now* getting better—visibly, palpably, with each passing day, almost each hour—and have as a "trophy" an exceptional penetration into the "central" aspects of all this (injury, operation, enclosing and immobilizing part of the body, suffering a *hiatus* in normal experience) and *this*, I think, is very important to write up. [. . .]

* Gregory's work explored optical illusions and other visual phenomena, and he conceived the idea of "perception as a hypothesis": that the brain does not simply register a scene as a given, but must use sensory input and actively try to construct a scene from it.

† OS uses this metaphor knowing that Richard Gregory's father was a distinguished astronomer.

The first walk was grotesque: I would find myself mysteriously unable to proceed, and *discover* that one foot was stuck behind the other, or in some wild unusable position. [. . .] The 2nd and 3rd walks were each better by many orders of (imaginative and practical) magnitude. The 6th walk (yesterday) I experienced/enacted a true "kinetic melody" for the first time: i.e. I did not have to think "First I do this; *now* this; *now* this, etc." but "it went by itself" melodically and naturally; then suddenly [. . .] I "lost the hang of it," and had to go back to my graceless, conscious, cautious, unmusical and unnatural pseudo-gait.

Something else fascinating. I was confined for 20 days and nights in an almost-featureless oblong room 7′ x 9′—(the window was blocked by scaffolding, and I could neither see through it, nor out through the door). When I was moved into a larger room, yesterday, which showed a complete prospect through both door and window, I noticed a sharp and complete stopping of stereo depth perception at a point *within* my present room; and looking outside, across the corridor, a pipe on the opposite wall appeared to be in the same plane as my own door, despite a gross parallactic difference observable between each eye. [. . .]

This rectified *within a few minutes*—as I gazed and visually "palpated" the new situation, I felt my visual space becoming larger, until I could clearly *see* the depth between them.* [. . .]

I *also* experienced abnormalities, instabilities, uncertainties of scale and perspective. The garden looked *minute* (and idyllic)—almost as if seen thru the wrong end of a telescope. The right angle of the rectangular lens (which I suppose subtends about 135 degrees on the retina) appeared too long for about a minute, becoming more obtuse or acute. The rather grotesque view above the garden—the hospital building dominated by the ludicrous, *meaningless* (here ambiguous) spoke of the Post Office tower (looking like an *axle* of the earth, if anything at all)—all of this looked grossly *fore-shortened,* as if I were looking through a 300 mm. telephoto lens. Again it has come to look more "normal" by degrees.

* OS had been fascinated by stereoscopic vision and other visual phenomena since boyhood.

So, it is clear how profoundly one can lose one's depth and distance perception and judgement when "starved" of space for a matter of days; this seems *exactly* analogous to the loss of sensory and motor images, procedural "know-how," intentionality etc. which I experienced with the immobilized, invisible, de-afferented leg.

Forgive this rambling letter; but I am bursting at the strangeness of all this, and *you* are someone who might be interested. I will be in London either here in the hospital, or in the Convalescent Home (Athlone House, on Hampstead Heath) until the end of the month. Do *please* visit me if you are in town on the 23rd.

My best regards to Freja and yourself,
Oliver

To F. Robert Rodman

SEPTEMBER 30, 1974
ATHLONE HOUSE, HAMPSTEAD LANE, LONDON

My dear Bob,

I have just received your letter (there is a great mail hold-up in London at present). I was much saddened to hear of Maria's death, and offer you all my sympathy in your loss. I did not know her well, but I knew her well enough to feel her rare and special quality—and *this,* of course, is irreplaceable. I hope that her last weeks were peaceful, and not fraught with too much pain or fear.

I know what a frightful, limitlessly-harrowing experience this has been for you (and perhaps, tho' less, for your children also). I know too that the *real* loss, as opposed to the anticipation of loss, starts *now.*

I am heartened to see that your letter is full of clarity and courage; it is the letter of a broken-hearted man, but also of a man who has—beneath or beyond the depths of his grief—still greater depths of strength and faith to carry him through his affliction. You may not be conscious of this, but it shines through your letter, and it is *there.* It will especially be this ultimate and unconscious strength and resolve which will support you in the agonizing months ahead, and which will gradually—without any denial or obliteration of reality—start to heal the unbearable wounds. And it is good and important that you have your practice to return to—whatever your later decisions or moves. Also, you will understand your patients as never before (I don't mean that your understanding and sympathy weren't deep before; but that they will be still deeper now you yourself have plumbed the depths of painful experience and emotion).

I forget exactly what I wrote to you from hospital—it was a difficult time then, with a lot of pain, uncertainty, fear. The *last* thing I would want to suggest, or *seem* to suggest, is that there was any significant connection between my accident and your situation. Such things are always, of course, multiply determined and "over-determined," and this particular accident, I think, especially so. No! I had many excellent "reasons" for contriving (or facilitating, or *allowing*) such an accident, which had reference to pressing parts of my life (in particular, an ugly situation in New York, which led me to a whole series of destructive acts from the burning of my book on April 19 to my fall in the mountains on August 24); no, many many reasons not bearing on you. And, I will add, not all of them entirely destructive or "bizarre." I have found the entire business absolutely extraordinary: the shock of sudden crippling; the multiple aspects of "patienthood," helplessness and dependence; excruciating sensitivity to the behaviour (and especially the motives) of those round one; the certainty that one will never recover (I had a severe "negative phantom" for almost 3 weeks, and could neither move, perceive, nor imagine my leg; I felt acutely amputated, bereft of my leg); and then the unexpected, incredible steps of recovery, which seemed like a miracle (the wisdom of Nature, the grace of God; but certainly the sense of nature as a definitely wise and healing force, nature as provider and Providence for all of her creatures). In my lesser way, I too have plumbed the extremes of *this* experience and feeling, and feel definitely and indeed radically altered by it. I also shall see my own patients in a more compassionate and comprehending light, having *been through* patienthood myself.

I will also write (have *already* roughed out 40,000 words of) *a book* on the whole experience, with all its implications + ramifications. It is, I suppose, the book on Care (and "anti Care") which I was planning to write before, but which has now been given a far deeper reality and authenticity through being a patient myself. I see that I *needed* injury and patienthood *of my own* to deepen my insight and write a better book. The experience is both *the source* and *the cost* of all genuine "material."

I wonder whether sometime—later, not now—it may help to assuage your grief, clarify your soul, set things in new significances + perspectives, if you were to "transform" some of your recent and present experiences into Art, Communication, Insight—I don't know what word to use. But I don't think one can (or *should*) try to do this while one is still *overwhelmed* by feeling. One cannot feel to the utmost and examine one's feelings *simultaneously;* but certainly one can (and *must*) first experience and *then* explore.

It is inconceivable to me (as I am sure it would be to anyone who knows you) that you should become "metalled" or mechanical or "dead inside" as a result of your loss. You *will* feel, you *are* feeling, a merciful *numbing,* which one *has* to have when the wound is still so fresh and new; this is like the temporary scab which staunches a flow of blood, and which is now gently staunching what would otherwise be a spiritual haemorrhage. But a *carapace*? No, Bob, this you will never develop. I am sure you need (as do your children) every sort of sympathy + support at this time; probing and examining—I think *not now* but later.

My friend, I am not the one who needs visiting; I wish I could visit *you.* But I am still in a cast (tho' quite active on crutches), and I am told that I must expect not less than 3 months from time of operation (Aug 26) to functioning adequate for self-sufficient existence. So, willy-nilly, I cannot expect to return to New York until the start of December, or thereabouts.

I hope I may visit you when I can move fairly easily, perhaps early in the New Year. And there *will* be new tomorrows, new years, new life, awaiting you, the other side of the valley.

Yours,
OLIVER

To A. R. Luria

OCTOBER 19, 1974
ATHLONE HOUSE, MIDDLESEX HOSPITAL
CONVALESCENT HOME, LONDON*

Dear Professor Luria,

It is a month since I wrote you an outrageously long letter, and I have been eagerly awaiting a reply. But the mails are very slow now—especially at the London end; no doubt a letter from you will arrive in a few days' time!

I am sitting in the warm October sun as I write; it is a most perfect Autumn day, extraordinary, a miracle (the Convalescent Home is set on a hill, amid woodlands, a little Paradise in the middle of London); and everything round me—the varied colours of the October leaves, the scents of new-mown grass and flowers, the singing of birds, the heaven-blue sky, the pure, sweet, transparent air, the noonday glory of the sun itself—everything in a wonderful and infinite harmony, Nature herself a living song of praise.

And *I* feel well, wonderfully well, not only with my leg, but in the whole of myself—is not recovery a *wonder,* an absolute *miracle*? I keep hearing in my mind some words of Nietzsche (I think they preface *The Gay Science—Die Fröhliche Wissenschaft*): "Gratitude pours forth continually, as if the unexpected had just happened . . . for *convalescence* was unexpected." [. . .]

My surgeons, my specialists, did a beautiful job, joining the torn tissues, restoring anatomical connection; but *they* could not teach me to walk again, to *feel* my legs, to jump in the air. [. . .]

The *real* human beings who have helped me to recover, to move, to live, have done so, essentially, *by being themselves;* not by their words, but by their actions and deeds, by communicating to me their own existence, and then communicating to me the actuality of existence, and even its *possibilities,* which I had all but forgotten.

Thus, in Norway, where I spent the night in a little cottage hospital following my accident, there came in the next morning a nice country doctor with comfortable old clothes and twinkling eyes. I opened

* In *On the Move,* OS refers to the same place as Caenwood, using an older name for this grand mansion, which was converted to a convalescent home in 1955.

my mouth to speak, but he put a finger to his lips; and then, without warning, without a word, he suddenly *jumped* onto my bedside table (at least a four-foot standing jump!); then he jumped down, and, without a word, placed my hands on his legs, just above the knees. And there I felt—but you have already guessed!—I felt the scars of bilaterally-sutured quadriceps tendons. "You see," he said, "it happened to me. Both legs, while skiing. You see, I have made a perfect recovery." I did indeed; his words were needless. What he did, his *deed,* was in itself communication enough, communicating the *hope* of recovery as no words could.

Similarly with a young doctor here, who seemed to *understand* everything much better than the highly-academic specialists at the University Hospital. One day I asked him, "How come? What gives you your *understanding*?"; and he said (as I had already guessed!) "Experience. I broke *my* leg. I spent weeks in a cast. *I know* some of the problems from my own experience."

Similarly a young orderly at the convalescent-home, who some years before had smashed his pelvis and legs in a parachute-jump, and been in plaster from the waist down for more than three months. He talked to me—fascinating; and still more valuably, he *walked* with me, and he promised we will play football together by the time I leave here!

And my dear, wryly-humorous, admirable physiotherapist, who has helped me out of the labyrinth (of weaknesses, fears, disallowals, disbeliefs) more than any other single person.* She could not have done this by the exercise of skill and will alone: she too had been through distresses and disabilities, hers dissimilar to my own. *Only* through experience can one learn—or teach. [. . .]

* See OS's letter to the physiotherapist, Mrs. Miller, dated November 27, 1978.

To A. R. Luria

JUNE 7, 1975
[NO ADDRESS GIVEN]

Dear Professor Luria,

I was quite overwhelmed by your letter of May 12. You are far too complimentary, and your praise makes it difficult for me to find the right thing to say.*

You speak of my "deducing a whole philosophical system" from my injury, and later, as being "an outstandingly gifted man"—in being able to make deductions from "events which the ordinary man does not even see." I feel I must at once disclaim and concede what you say! I think I *do,* perhaps, have certain gifts, but they are not gifts of intellect or reason as such (I am not particularly "clever," and my formal abilities are very limited—they certainly would not allow me to deduce a whole philosophical system). But I think I do have a certain gift of "seeing," of attention, and an obstinate power of holding things in my mind, for days and weeks and years if necessary, until they begin to yield me their secret. In particular, my gift is for detecting *inward* changes, minute shifts in perception and feeling which the ordinary man doubtless experiences, but which in the ordinary course of events he is unlikely to be concerned with or pay much attention to. My special proclivity is to detect such changes when they are scarcely perceptible, and *to not ignore them or dismiss them from my mind;* but, on the contrary (and this could be construed as a sort of contrariness!) to *fasten* on them and *focus* on them, until they become clearer and clearer—and in so doing, of course, I *magnify* them, so that they come to fill my whole field of vision.

When I was ten, I developed an absolute passion for astronomy, and it is this *sort* of passion, I think, which still dominates me today (although it has been transformed and turned inward, so to speak, so that I now find myself bending a minute and ceaseless and rapt

* Luria had begun this letter by saying, "Thank you so much for your long and absolutely remarkable philosophical letter! It is astonishing that you should construct a whole philosophical system from the injury you had."

attention upon the inner firmament today); but there has been a change of *direction,* towards subjectivity and inwardness, but with a continuance (I hope!) of the original astronomical precision and passion. Over the years, and almost against my will, I have become a sort of astronomer of the inward, focussing with extreme introspection on my own state-of-mind, but doing this (I like to think!) in order to learn something of other minds, of Mind, and not as a mere egoistic indulgence: for if I am unique—as everyone is unique—I am also very similar to everyone else, and what I observe to occur in myself, must surely occur in others as well. This, at least, is my (the) Credo. [. . .]

By a curious coincidence, rummaging this evening through some oddments of Leibniz, I came upon a strangely similar image, contained in his description of a half-Platonic, half-scientific *dream:* "I was no longer in the cave, I no longer saw a vault above me, but found myself on a high mountain which revealed to me the face of the earth. I saw in this distance everything that I wished to look at . . . but when I considered a particular place fixedly, immediately it grew, and in order to see it as if close at hand, *I needed no other telescope than my attention.*" My own special gift, then (and how often it seems like a burden or curse!) is far from being analytic and deductive, and perhaps equally far from being empirical and inductive; it is—*existential:* it starts from the most concrete, particular and individual experience, and then burrows downwards and inwards, towards the most general determinants of existence and action. Thus, I do not feel that I am *ascending* to deductions, nor do I feel that I am *extending* to deductions; I feel that I am *descending,* deeper and deeper, towards the bedrock, the foundation, of my experiences. I think that deductions and inductions involve a *departure,* a moving-away from the grounds of experience, whereas my own "method" is to *stand still* on the same spot, the *ground* of my experience, until I "discover," explicitly, what sort of ground it is:

And the end of all our exploring
Will be to arrive where we started
*And know the place for the first time . . .**

[. . .]

Yours sincerely,
Oliver Sacks

To A. R. Luria

JUNE 14, 1975
11 CENTRAL PARKWAY, MT. VERNON, NY

Dear Professor Luria,

I fear I have been a nuisance with my enormous letters and precipitate telegrams—I cannot thank you enough for the unfailing understanding, good nature, kindness and tact with which you have received, and at once responded to, my various (and perhaps, sometimes, half-mad!) communications.

I often wonder what one man can really do for another in this odd world of ours. It is clear what he *cannot* do (and should never attempt to do): he cannot *be* another person, nor can he tell another *how* to be, or how to think or feel or act; and he cannot coerce, or *cajole* another without compromising the proper relation of mutuality. What then *can* he do? *He can be himself* (quietly embodying, without ostentation, the cardinal qualities of curiosity and courage, of a bold and honest and persistent *striving* in the face of all difficulties (and doubts) which assail him). And he can *convey* some of his qualities to another human being: that is, he can *inspire* him, *entrust* him, and—above all—*encourage* him. And it is exactly this which you showed me, and conveyed to me, in your magnanimous letter of May 12, and [. . .] it is what you have *been,* and *done,* since we first corresponded. [. . .]

There is one specific point (or a *nexus* of points) which you bring

* From T. S. Eliot's poem "Little Gidding."

up in your letter of May 12, and which I have been thinking on very intensely, because it touches on the deepest parts, the *mainspring*, of my character and thought; namely, what is the sort of *passion* which drives one, and what is the sort of *gift* one possesses?

I think there are good images in the analogy of certain "Adventure" and "Detective" stories (to which I am addicted—and to which, I sometimes suspect, *you* also turn occasionally as well!). And one of the things which delights me in Kierkegaard is that *he* often refers to himself as a sort of detective—a detective of inwardness, a detective of the soul. And surely many of the images here—the sense of MYSTERY (and *suspicion*), the finding of CLUES, the way in which one TRACES events to their origin (however impossible or paradoxical they may seem!)—surely these images have a great (and comic) relevance to the endeavours of Science: to the Psychology and the Structure of INVESTIGATION as such. What is so entertaining with regard to the "classic" Detectives—Poe's "Dupin," Conan Doyle's "Sherlock Holmes" et al.—is their wilful (or whimsical) *misunderstanding* of the methods they actually use. Thus both of these pretend to be "entirely scientific," "rational," etc.—and when they come to expatiate on their methods they portray themselves as engaging in long trains of systematic and *deductive* thought, or as sifting and "analysing" a great mass of evidence, calculating, comparing, computing probabilities, and *then* proceeding to inductive conclusions. But is this not utter nonsense, *post factum* rationalization? It is not *in the least* the way they actually work—anymore than it is the way creative scientists work! (Sherlock Holmes, at work, is like an excited *dog* "on the trail"!)

Do you know the "Father Brown" stories of G. K. Chesterton? These give, to my mind, a much funnier (and more penetrating, and truer) picture of the motives and methods of the Master-Detective, and the deeply irrational and paradoxical (sometimes crazy!) way he works.

You speak in your letter of my "gift" for seeing "events which an ordinary man does not (may not) even see," and my going on to ponder on these. Such a "gift" could be a genuine and more-than-ordinary sensitivity or it could be, as they say, "an over-active imagination"—*"invention," "illusion," seeing what's not there:* Such an ambiguity envelops all such terms as "a gift," "a sixth sense," "a third ear," etc.

What *you* say here in admiring belief, my surgeon said in amazed disbelief: he said "Sacks—you're unique! I've never heard anything like this before. You see things which none of my other patients ever mentioned to me!" I didn't know whether to feel flattered or insulted, and after a few seconds I replied to him: "I *may* feel certain things abnormally acutely, but this doesn't mean that others don't *feel* them, because they don't mention them, at all. I refuse to believe that I am constitutionally unique! I bet *other* patients go through similar things." [. . .]

I had many similar experiences when I tried to describe and publish some of the extraordinary phenomena shown and experienced by my post-encephalitic patients, such as some of the things I describe in *Awakenings.* In 1969 I saw things which I myself could scarcely believe, and when I started to describe some of them the following year I met with an almost-universal distrust and disbelief—at least among my neurological colleagues. A number of them expressed their incredulity in print; to those who did this I extended invitations to visit "Mount Carmel." "Come and see for yourselves," I wrote to them: none of them ever came; apparently, *they didn't want to see*—at least this was the conclusion to which I was forced. [. . .]

And yet: I could demonstrate the "incredible" to my students every week; *furthermore,* I could demonstrate them (and *did*) to lots of "ordinary" people—relatives of the patients, volunteers who came in from the neighborhood to help, the hospital electrician, the hospital gardener. I found again and again, by a singular paradox, that "plain," "simple," "ordinary" people *believed their eyes,* and were much more receptive, than many highly-trained and "cerebral" colleagues (who seemed blinded by their own language and methods and mental habits and assumptions—so much so as to be *incapable of experience,* or listening with an open mind to the experience of others).

This is why, finally, I addressed *Awakenings* to "ordinary" people, and why it is "ordinary" people, on the whole, who have responded to it with appreciation and interest. [. . .] And this is why Kierkegaard, who himself was such an extraordinary man, such an oddity and genius, spent hours every day talking to "ordinary" people in the streets of Copenhagen; and why in the very last thing he ever wrote (which, sadly, was only published after his death), he said: "Thou plain man! I have not separated my life from thine; thou knowest it,

I have lived in the street, am known to all . . . if I belong anywhere, I must belong to thee, thou plain man . . . I was always of your company"; and, in his final words, speaking of the profound and absolute paradox of faith, he brings the greatest and the plainest together: "It is infinitely high . . . but it is possible to all." I desire and believe the same of myself: I have been given certain gifts—less of intellect than of *seeing,* and of *saying* what it is I have seen; but I feel I am of the company of plain men—and my gifts are *theirs,* if in higher degree; I am not *set-apart,* but am *united,* by my gifts; I see what he sees, but with clearer eyes; *I experience what he experiences, and I say it for him.* If I have been superiorly endowed, my endowments are employed for the sake of the plain man. I don't give a damn for "intellectuals," "specialists," "experts," etc.; I seek to give voice to the universals of experience, to see and speak clearly what everyone feels.*

And yet—I cannot pretend that I am simply and merely "the man in the street"; I am "in the street," in the world, existent, but I am also a *thinker,* which most people are not; and, in particular, I am a thinker who thinks (*tries* to think) of existence. [. . .]

I need to make clear the extreme *practical* importance of an "inward" approach, the *power-to-help* which comes from understanding precisely what patients are "going through"—the inward determinants of their outward behaviour. If we can get "inside" the patient, *a new sort of understanding* is rendered possible, and along with this—*a new sort of therapy.* [. . .] With regard to the phenomena of negated and alienated parts of the body, an enormous amount of suffering and disability might be avoided, or reduced, if only some understanding were extended to such things. [. . .]

Again, for example, with my Parkinsonian patients, by patiently trying to get "inside" these patients, I have learned that a great many of the external phenomena of Parkinsonism—movements which are too fast, too slow, too big, or too small—arise from the *inwardness of the Parkinsonian experience,* in particular from peculiar distortions of space and time-sense, *illusions about the "scale" of movement.* [. . .]

Once one grasps that such Parkinsonian phenomena have this "inward" foundation, then, and only then, can one help patients to

* This paragraph is obviously both true and not. OS cared a great deal about certain intellectuals and experts—sometimes ones who had received great recognition for their achievements: Nobel laureates and the like. But he saved his highest regard for those with "spacious" minds, capable of understanding things in a new way.

regulate their irregular movements by providing them with non-pathological metrics or frames-of-reference. [. . .]

Yesterday I was talking with a cousin of mine, a highly gifted (and verbal) woman who three years ago was rendered aphasic by a stroke, but mercifully made a perfect (or near-perfect) recovery—I asked her yesterday *how she got better:* whether language seemed to come back *of its own* "spontaneously"—or whether she had to struggle to regain her words.*

"Both," she said, "I had to struggle for words to come, but they would only come if they were 'ready' to come. But *their* readiness-to-come was connected with my readiness-to-fight." She continued, "My recovery was like a miracle—I experienced in myself the miraculous rebirth of language."

She added, "It reminded me, in a way, of giving birth to my children. I had to struggle for them to be born, but they could not be born 'til they were *ready* for birth."

This, in the most general terms, is "the continued miracle" (in our own person, in the world) of which Hume so ironically, but so accurately, speaks. The miracle of birth, the miracle of becoming, the infinite mystery of coming-into-existence: and the *fight* of the unborn to be realized, to be born, the *struggle* of the possible to become the actual. [. . .]

Oliver Sacks

* The cousin OS is referring to, Madeline Gardner, had a catastrophic stroke when she was in her fifties, affecting movement as well as speech. She did not have formal speech therapy, but once she gained enough strength and stamina, she would force herself to sit up in bed and call various friends and relatives to talk on the phone. In essence, she devised her own speech therapy.

8

Atavisms

1975–1977

To Gunther Stent[*]

MOLECULAR BIOLOGIST

OCTOBER 26, 1975

11 CENTRAL PARKWAY, MT. VERNON, NY

Dear Professor Stent,

I am moved to write to you through a chain of associations—and a recollection.

Yesterday I had lunch with Stephen Spender and Ed Mendelson (W. H. Auden's literary executor), and our conversation turned on Auden's scientific reading and interests, and the extent to which these continued during the later years of his life. I mentioned that he was a sedulous reader of the *Scientific American,* and often discussed various articles in it with great interest and animation.

This has suddenly brought back to my mind the last time I saw Auden, when I visited him at Christchurch in February of '73. While we were at lunch he suddenly spoke of your article, and we talked about it, and the issues you raised, for most of the afternoon. He was extraordinarily fascinated by it—and suddenly I find myself wondering whether he knew you, or wrote to you at length personally. [. . .]

I think he was often exercised by the notion of "prematurity." Back in 1969, when I showed him the MS of my first book (*Migraine*—he later reviewed it when it came out in '71), he said that he found Part V of the book very startling, and "premature" (this was the word he

* Stent wrote an influential article entitled "Prematurity and Uniqueness in Scientific Discovery," published in *Scientific American* in 1972.

used); he felt that it might seem incongruous with the earlier parts of the book, and prove unintelligible or offensive to would-be readers (it was partly in consequence of his saying this that I completely excised Part V of the book. Indeed I completely *forgot* what I had written, and only "re-discovered" some of the notions several years later, when they came to me in a wholly different context. So maybe in '69 they were also too premature for me!). [. . .]

When I showed Auden the proofs of *Awakenings,* in February '73, he again warned me about "prematurity," but added, "Go ahead, and be damned, and publish!" He added, "Remember about prematurity, and don't take it *personally* if it doesn't meet with recognition," and he cited your discussion of Michael Polanyi.*

I am hesitating and hovering over another book at present (based on a very strange experience of my own as a patient). It was *so* strange that I found it incredible myself, and kept pushing it away from consciousness. I mentioned it in a letter to A. R. Luria of Moscow, who wrote "You *must* publish it . . . it is very important." Luria added that he thought such a publication *might* provoke a major alteration in certain established and canonical medical approaches, but also that I should be prepared for it to have no effect whatever, because *it might not be time yet* for such a reconsideration to occur. Paradoxical business! I can't help wondering whether, in some sense, *every* creative formulation is "premature" (although, in another sense, precisely appropriate—for how can anything be discovered unless it is *time* to discover it?).

Do forgive this rather rambling and not-too-coherent letter, but I couldn't refrain from writing to you.

ISRAEL SHENKER, WHO IN 1969 had written a *New York Times* article about the postencephalitics at Beth Abraham, had followed this up with a 1971 article about the tics such patients often had. As a result, OS began receiving letters from people with other conditions that produced tics, especially Tourette's syndrome. One of these correspondents was the man he would

* Polanyi (1891–1976) was a Hungarian-British philosopher.

eventually describe in "Witty Ticcy Ray"; another was a man with an extreme form of Tourette's (OS sometimes thought of this as "super-Tourette's"), and he wrote to Luria about this case.

To A. R. Luria

DECEMBER 24, 1975
[NO ADDRESS GIVEN]

Dear Professor Luria,

[. . .] There has been something [. . .] which has excited me intensely in the last ten weeks—namely, working with a most extraordinary patient, an extraordinary individual with an extraordinary disorder (Gilles de la Tourette syndrome). I have spent about forty hours talking and working with him, and very much more than this thinking about him. I don't know whether you have encountered any of these patients in depth: I feel very strongly (and I have felt this since 1970, if not earlier) that there doesn't exist, that there *could* not exist, any other neuropsychological disorder of comparable complexity, challenge, and fundamental importance. [. . .] There is an extraordinary *diffraction* of attention (somewhat reminiscent of what one sees in "hyperkinetic" kids, and also in some ways suggestive of, yet also quite different from, what one sees in a distracted obsessive or schizophrenic patient). Within a few seconds, on entering a strange room, he will touch, smell, handle, and "test" dozens of objects in his immediate environment—a *violence* of inquisitiveness which can scarcely be conveyed (I use "inquisitiveness" here in a sense antithetical to a spirit of *enquiry*). He shows myriad, continual, involuntary, "automatic" imitations, impersonations, *incorporations* of others, and these others, these brilliantly eidetic and histrionic images, continually jostle in his stream of thought, and are continually *manifest* in a bewildering bursting-forth of different "faces," "voices," "tics" and "noises." The complexity and speed of this "dissociation" is such that if I close my eyes, sometimes, when he is telling a story (he is a *marvelous* story-teller, and extremely witty, and *becomes* all the people he describes as he talks) I may get the impression of half-a-dozen people in the room (*uncanny* his powers of conveying, of *being*, many others—one may pick up several "personae," or hieroglyphic noises and motions characteristic of them, all in the course of a single sentence).

Some sort of barrier, all sorts of barriers, seem to be abnormally permeable in him (and have been, I think, since his earliest days). The filtering, the selection, the discrimination etc. which *has* to develop with the growth of the self, which has to develop if the self is to have any differentiation and boundaries, and not to be *overwhelmed* by a thousand stimuli at once—this sort of boundary or barrier seems missing or at least strangely deficient [. . .]. So, in a way which is uncanny, and almost incredible, one can see and hear the very genesis of thought, the rudiments, the beginnings, of perception/thought/action. It is like looking through a sort of magical and transparent window, into the very heart of thinking and the creative process. [. . .]

Despite his grotesque and severe disorder, he is a most intelligent man, and holds a highly responsible (though uncreative) job. Also—and this seems to me an *essential* aspect of his cerebral and existential "style," and I have seen it in varying degrees, in all other patients with Tourette's I have seen—he is inventive, playful, imaginative, witty and surprising and spontaneous, to a remarkable degree. Typically the drug haloperidol (which inhibits dopamine-action), whilst curing the over-reactiveness, the impulsiveness, the tics, etc. *also* "cures" the inventiveness, the playfulness, the imaginativeness, the wit. So, we have a pretty problem, with this pandemoniac but in its way profoundly productive and creative disease. *How can we effect a re-organization of his remarkable but chaotic powers so as to produce a potent and creative person*—as opposed to a pharmacologically dulled "zombie," who has been rendered impervious to all the sources and roots of creation (as impermeable as the patients were originally over-permeable). I know your great fondness for puzzles and riddles, and I think of you (amongst so many other things) as a great Detective who is, equally and necessarily, a great Reconstructor. I think this man is as complex as the Rosetta stone (or far more complex than it): and that if one could "crack" all his enigmatically disguised expressions, and all their determinants and co-relations, one could accomplish a fascinating and fundamental neuropsychological task, as well as help to reconstitute a deeply disabled and suffering fellow-creature.

To F. Robert Rodman

CHRISTMAS DAY, 1975

11 CENTRAL PARKWAY, MT. VERNON, NY

My dear Bob,

It was very good hearing from you a while back—(both your phone call(s) and your good letter of Nov 24) and, especially, hearing that you are at last finding it possible to pour out, and articulate, some of your feelings on paper.* The terrors, the anguish, the loss, the bereavement, are probably the deepest and most far-reaching experience you have ever had to go through. I know how close it came to "unhinging" you, and to knocking away from under your feet the most basic securities, the very desire to live. I'm sure, in a sense, that one *never* recovers from such experiences as these: but then, it is not a question of "recovery"—but of surviving an abyss, a complete rupture of one's deepest securities, of that which gave life most of its meaning and value. Lots of people are broken by such breaches; you haven't been, and won't be, but assuredly you will henceforth be a very different person. Nobody who has known the Exile, the Abyss, can ever be the same again: the simple, innocent immediacies are gone, and they can *never* be regained (except in regression and phantasy—or in a sort of spurious and sentimental Art which is, essentially, evasive and nostalgic). I sense, from the *tone* of your voice, and the *quality* of your attitude to what you are doing, that you are facing and making art of a fundamental reality—and not fuzzing it up with false sentiment and rhetoric. I think that the whole experience, and your writing of it (and it may not be relevant whether you publish or not—it is the writing, the externalization, the honesty, which counts) *has* to form the most important turning-point and crisis in your life.

I don't know that such experiences make one "better," but assuredly (if one lets them) they make one *more profound.*† One grows up with them, one sheds one's first innocence, one faces the profoundly questionable quality of existence. Suddenly, so vividly I thought I had

* Rodman's wife, Maria, had died in September 1974, and he had begun to write a memoir about her illness. The book, *Not Dying,* was published in 1977.

† OS added this footnote to his letter: "I realize, re-reading this, that I am unconsciously paraphrasing part of the Preface to Nietzsche's *Fröhliche Wissenschaft.*"

left the radio on, I hear a Brahms song playing in my mind: one of the Opus 92 songs, I think, I'm not sure, but is so obviously written, and sung, by men who have known death, and who have somehow transmuted the anguish into rapture, but without losing so much as a dot of the pain: and suddenly, on the heels of this musical association, a couple of lines of Auden come to me, something like:

Remember, no metaphor can express
A real historical unhappiness.[*]

No, no metaphor can. Perhaps certain symbols can—but only in the *universal* way which symbols do. In particular I am thinking of the profound and tragic symbols of Loss-and-Restoration, Exile-and-Return, etc. (*Galuth* and *Tikkun*) explored by the extraordinary Safed Kabbalists of the mid-sixteenth century, above all by the profound poet and mystic, Isaac Luria of Safed.[†] I have been dipping into this virtually-forgotten and almost-always-misrepresented beautiful and mysterious development in Jewish sensibility and thought. (Have you read any of the remarkable works of Gershom Scholem?)[‡] For the first time in my life, I find myself clearly able to delineate what previously I could only vaguely and blunderingly intuit: the *genuine* roots of Jewish consciousness, in particular history and universal myth, as opposed to the senseless, dehumanized, legalisms and pedantries which—for the most part—constituted the pseudo-religious atmosphere in which I was brought up, and to which I have spent a lifetime submitting and rebelling; but never, *never*, finding any real nourishment or meaning. But in the beautiful and profound symbols of the desolate soul, the soul-in-exile, the soul which has been shattered and cut-off from its origins, and the vision and practice of healing the soul, one's soul, and others'. By right activity, by *Tikkun*—here, for the first time, I find something which corresponds to my own

* The actual wording is "No metaphor, remember, can express / A real historical unhappiness." From Auden's poem "The Truest Poetry Is the Most Feigning."

† Isaac ben Solomon Luria (1534–1572) reinterpreted and systematized the mystical doctrines of Kabbalah.

‡ Gershom Scholem (1897–1982) was a German-Israeli philosopher whose writings in the 1960s brought the history of Jewish mysticism to broader awareness.

experience and also, equally, to my clinical experience. I think I shall be drawing heavily on this astonishing richness and depth of these Lurianic ideas (which, in a sense, have been forgotten for almost four centuries) in *A Leg to Stand On,* and also in another book which is taking shape in my mind, based on a recent and current exploration and relation with an *incredible* patient (with Gilles de la Tourette "disease"). I won't say anything about him now, but will simply indicate that I *think* (this may be the most outrageous claim or phantasy) that I am discovering things about him, about the development and constitution and sicknesses of the Self, comparable in importance to Freud's work on Dreams (and also his wonderful, and *funny,* and so rarely-read book on Jokes). [. . .]

One cannot hope (usually) for more than a very little recognition in one's lifetime—one must have *some,* and I *have* had some, from those whose calibre makes their recognition double, trebly, crucially important. Being an original, and a sort of genius, in a field as hidebound, as *demented,* as "Medicine," (so brainwashed and "officialized" that it is absolutely nothing—neither Art, nor Science, nor Popular, nor Profound); it is a very difficult and isolating and anomalous position, especially for someone as unaggressive as myself; it comes to me, more and more clearly, that I must simply proceed in my own odd way (and my "oddness" is an essential aspect of my powers, and my feeling towards my work, my deepest sense of the mysterious, the enigmatic, and the essentially funny, and almost "crazy" quality of Nature and its "laws"—for, increasingly, I see the logic of Nature as like that of Alice, of Dreams, of Jokes, of Absurdity, of Fun: I think of Nature as *essentially* symbolic and playful, and not at all as a superficial and "square" Order-of-"Reason"): that I must proceed then, in my own very odd, idiosyncratic, and inimitable way, hope that I shall encounter a few understanding spirits, and inure myself to the predictable pig-headedness and bewilderment of my colleagues. And this, I think, is not really paranoid: I mean, it is not an attitude which stems from contempt, fear, or hate; but rather, the sad (but in some ways rather pleasant) realization that one has *outgrown* the majority of one's colleagues and contemporaries, that one has transcended their powers of understanding or acceptance, and that (even if they mean well) it is no longer in their power to follow or recognize one in the higher and more recondite flights of one's thought. Recognizing this, one is bound to feel an essential isolation and loneliness, but perhaps a diminution of bitterness and rage.

I have rattled on irresponsibly, egotistically, far too long.

I am dead-tired inside: I wish I could take about three months off—but damn it! I have no money, no freedom. I *need* about three months away from this horrible winter (there has been a gale and a snowstorm going on all this week—I feel it will last, without interruption, till the Spring). I long for some comfort, some release from the drudgery, I want to be comfortable, to recover my health, and to devote myself to the two books which are burning me up. And damn it! I can't afford to. I can't afford to, and yet I can't afford *not* to. (I had some of this desperate feeling last year, and it was undoubtedly one of the reasons why I contrived to injure myself and have a creative "holiday" in hospital; but one can't make a habit of that sort of thing—and besides, it's regressive, it's a cowardly way out.)

Oh, dear Bob, what does one do? I wish I had children, progeny, like you. My *biological* sterility and isolation dismays me. But then I have students—and readers.

Anyhow, my warmest best wishes to you for the Coming Year—I think it will be a *good* year for you, a year such as you deserve and need, and haven't known for a very long time.

Love,
Oliver

THE FOLLOWING CORRESPONDENT, Bill Pearl, was a founder of the Tourette Syndrome Association, a group that had coalesced around a kitchen table in 1972 and consisted of perhaps a dozen parents of children with TS. The condition was then considered very rare—less than one hundred cases, it was thought, in the United States; and it was usually treated (unsuccessfully) with psychoanalysis. But, having observed similar repetitive tics and automatisms in his postencephalitic patients (and even "prehuman," almost simian behaviors), OS intuited an organic basis for Tourette's syndrome—and he soon observed that it was thousands of times more common than previously thought.

Tourette's ranges widely in symptoms and severity, and the cases OS was seeing tended to be at the more severe end of the spectrum. This was particularly so for the patient with "super-Tourette's" whom he had described to Luria. (OS referred to this patient in *On the Move* as "John P.," so he is identified thus here.)

To Bill Pearl

TOURETTE SYNDROME ASSOCIATION FOUNDER

FEBRUARY 2, 1976
[NO ADDRESS GIVEN]

Dear Mr. Pearl,

You [. . .] are lucky not to be in New York today: the worst storm in umpteen years, roads iced, traffic stalled, all life and business brought to a standstill. Mercifully, this is my one free day, so I can stay at home and catch up on my writing, my letters, etc.

[. . .] I think I have seen [. . .] three "paths" in Tourette's: "low-level" tics which are felt as ego-alien, and show no tendency to be mannerized, or moulded into and by the personality; extensive "linguistic" transformations of tics, to peculiar and almost hieroglyphic "private languages," but without shivering [*sic*] the integrity of the personality as a whole [. . .]; and, in some patients—though *singularly few* (as you yourself could not stress enough)—a massive breakdown of the whole personality. [. . .] What is so remarkable is that more Tourette patients *hold together* despite the force and strangeness of their illness, and do not go mad, or shatter, or suicide. [. . .]

I think one must allow, *if only as a hypothesis,* that in some patients the involuntary motions and noises and words may serve as "carriers" of meaning and feeling, may be pressed into use as peculiar symbolic languages, in which the symbols are movements, actions, *enactments* (as opposed, for example, to the phantasmagoric *images* of a dream). [. . .]

I will say, in parenthesis (but again, you understand, this is only a hypothesis, or a thought—one of many such which pass through my mind) that such a business of extraordinarily rapid and vivid symbolization, with extreme condensation and multiple determination of the symbols, and extraordinarily facile (and, at times, jocular and playful) association, is extremely reminiscent of what Freud delineates as characteristic of *dreams.* Of course *here,* in contrast to the dreamer, the *ticqueur* is awake and not enclosed in sleep, he is reacting constantly to people and things and stimuli all round him as well as to constantly-erupting intrapsychic contents, and his "images" are not merely inner perceptions, but actions, enactments, which are open to view. And yet, I think the analogies to dreaming, and also to joking, are certainly useful, and perhaps fundamental. [. . .]

We must also ask: is there *anything else* significant or strange, in the motions and noises and behaviour of ticqueurs, besides tantalizing similarities to the ordering of dreams and riddles and jokes and charades? [. . .]

Going over some of the tapes of [John P.] last weekend, [. . .] something which is scarcely credible emerges. We were playing over this bit of tape, Kennie Gospodinoff (Scheflen's pupil) and myself, when Scheflen* himself walked into the room. He exclaimed "Holy Mackerel!" and his mouth fell open; he felt, we all felt, that *what we were seeing, in this brief interaction, was not human behaviour of any known sort, was not human at all, but essentially—simian.* There have been "hints," I have had "hunches," but some of the tapes which I got last weekend, and some of [John's] reactions and recollections, (when very delicately, very carefully, very tentatively, I brought this up in our session on Saturday—for this is not exactly the sort of thing which comes up, which has *ever* come up, with a patient), provide at least partial verification of this almost mind-boggling hypothesis: *that there may remain in the human brain functional "traces" or "residues" of pre-human behaviours, and that these may become activated as atavisms under certain conditions.* [. . .] If one could demonstrate beyond doubt that such atavisms or "reversions" of behaviour can occur, and to pre-human (phylogenetically primitive) as opposed to infantile (ontogenetically primitive) patterns, this would be of momentous importance in various ways, from the standpoint of evolutionary theory and cerebral organization on the one hand, to questions of identity and social response and recognition, on the other (I will expand on this a little bit later). As I write I hear Darwin's words in my mind, at the end of his famous chapter on "Reversion": "The

* Albert Scheflen, a professor of psychiatry at Albert Einstein College of Medicine.

fertilized germ of one of the higher animals . . . is perhaps the most wonderful object in nature. But on the doctrine of reversion . . . the germ becomes a far more marvellous object, for, besides the visible changes which it undergoes, we must believe that it is crowded with invisible characters, proper to both sexes, to both the right and the left side of the body, and to a long line of male and female ancestors separated by hundreds or even thousands of generations from the present time; and these characters, like those written on paper with invisible ink, lie ready to be evolved whenever the organization is disturbed by certain known or unknown conditions." [. . .]

I have only had this feeling strongly once before, namely—in the summer of 1969, when my post-encephalitic patients were reacting so strongly, and in some cases so strangely, to the administration of l-DOPA. Jelliffe,* fifty years ago, spoke of "menagerie" noises (barks, grunts, howls, etc.) as being among the strangest characteristics of certain patients in the context of an acute encephalitis or its aftermath. People who came to "Mount Carmel," in that almost-incredible summer and fall of '69, would cock up their ears when they heard certain sounds from our post-encephalitic ward above: "You keep animals up there?", they would say "You have experimental animals? Or maybe pets? . . . Or is that the sound-track of a jungle movie?" [. . .]

You may say: "But all this is very theoretical, you know—what bearing does it have on the *lives* of these patients?" I wish I could have shown you the videotape of [John P.] during certain "touchings" and "interactions" with his girl-friend in the room (I myself was out of the room, but the camera "saw," the tape "remembered"). They have problems, like any young couple, but there are certain *sorts* of problems which would not occur with any other couple, problems whose occurrence has specifically to do with Tourette's—or, more precisely, with the ever-changing composite picture produced by the dynamic interaction of this man with his illness. Thus some of the touchings and the "interactions" *go wrong*, because his speed, and over-reactivity, put the two of them "out of synchronization," (like two pianists playing a duet at different speeds); some go wrong, secondly, because he suddenly presents a different "face," a different

* Smith Ely Jelliffe, an early observer of encephalitis lethargica.

"*persona,*" to her (for example, a sudden, rough "Marlon Brando" face-and-voice, a sudden crude "Waterfront" *persona*) in the middle of an otherwise consistent mutuality; the final, and most disquieting incongruity, is when he is suddenly seized by an apeish impulse, and *presents an ape's and not a man's face.* As you may imagine, these incongruities cause sudden bewilderment, laughter—and terror; and not only in her but in others as well; one cannot explore the phenomena here in *isolation* from their social and cultural context; one has to see the Tourette patient *in society,* to see how he reacts to others, and others to him.

And by others I mean not only *human* others: it is fascinating, for example, that [John P.] is always being attacked from behind by *dogs;* but that, on the other hand—*and I must film this*—he may cause an uproar when he goes to the Zoo, especially when he visits the Monkeys or Great Apes. When he goes into the Ape House, he feels very much "at home" in a mysterious way; involuntarily he "apes" the Apes, except it is not Imitation, but "Memory"; he *instinctively* knows, as it were, how an ape behaves, and communicates, and feels; and, acting apeishly, he excites the Great Apes, and *they* are suddenly taken aback, and gaze at him with a mixture of "recognition" and shock, as if sensing in him a long-lost "brother." His descriptions of this are quite plausible and convincing; it would certainly be most important to record such an interaction on tape or film; for to deceive a human audience by animal imitations is one thing, but to deceive the animal species itself into a sort of spurious "recognition"—why, this is incredible, if so, and must be submitted to critical confirmation (or refutation). [. . .]

TOP: OS, ca. early 1960s.

ABOVE: Sidney's Cafe, Santa Monica. *Photo by OS.*

LEFT: California Highway Patrol officer ticketing a motorcyclist. *Photo by OS.*

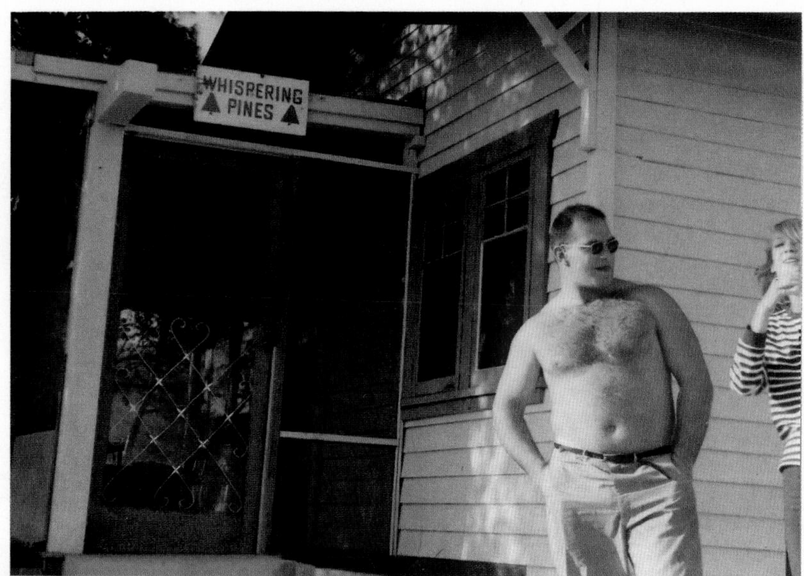

ABOVE: OS with friend outside his Topanga Canyon house, ca. 1963.

LEFT: An official UCLA portrait of OS as a medical resident, ca. 1962.

BELOW: Lifting weights in the early 1960s.

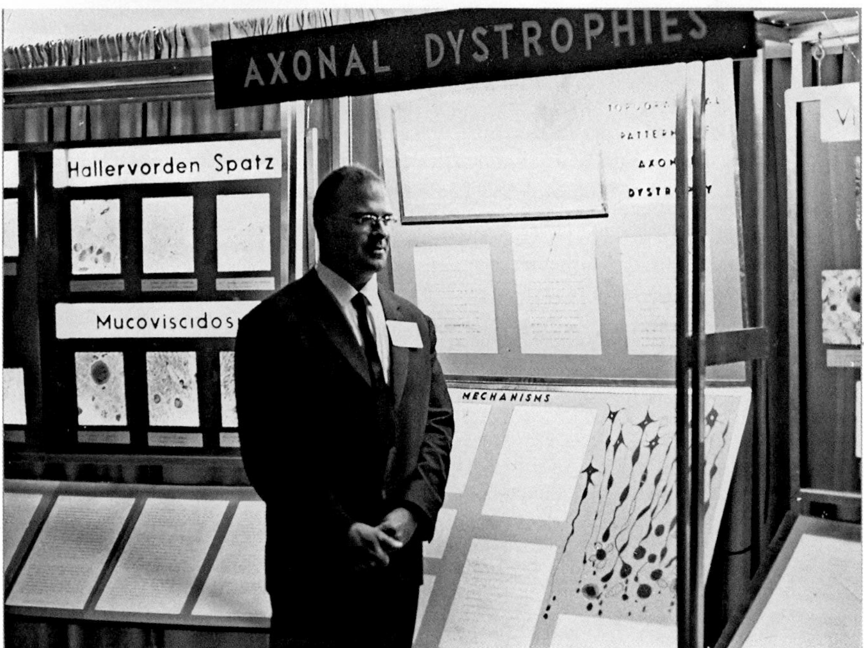

TOP: With colleagues at Mount Zion Hospital, ca. 1961.

BOTTOM: With his poster exhibit at an American Academy of Neurology meeting, 1964.

ABOVE: Thom Gunn in Hyde Park, London, 1959.

RIGHT: W. H. Auden, in a portrait that hung over OS's desk for many years.

RIGHT: Jenö Vinzce, early 1960s.

BELOW: OS on his BMW, ca. 1970.

ABOVE: OS presenting some of his thoughts about the *Awakenings* patients, ca. 1970.

RIGHT: Aunt Len (*left*), polar bear, and friend.

LEFT: Sam and Elsie Sacks in the garden at 37 Mapesbury, at their 1972 golden wedding celebration.

BELOW: OS and his three brothers. *From left:* Oliver, David, Marcus, and Michael Sacks.

LEFT: Bob Rodman, around the time he met OS at UCLA in the early 1960s.

BELOW: Typing on the deck of a friend's house (note the dictionary atop the bench).

*To Jonathan Cole**

MEDICAL STUDENT

MARCH 21, 1976

11 CENTRAL PARKWAY, MT. VERNON, NY

Dear Dr. Cole,

I thank you for your letter (your letters!) of February 27, and am sorry to be so long in replying.

I very much appreciate your generous response to my books, and feel grateful (and amazed!) that you should want to work with me.

My delay in replying is because I don't know what to reply. But here, roughly, is my "situation":

a) I don't *have* a Department.
b) I am not *in* a Department.
c) I am a "gypsy," and survive—rather marginally and precariously—on "odd jobs" here and there.

When I worked full-time at Beth Abraham (the "Mount Carmel" of *Awakenings*), I often had students spend some weeks with me for their "electives"—and this was an experience we would always find very pleasant and rewarding—I have the happiest memories of those far-off days.

But now I am, as it were, without any "position" or "base" or "home," but peripatetic here and there, I can't possibly offer any *formal* sort of teaching—or anything which could be formally "accredited" to you.

"Informally" (I sometimes think) I see and learn and do a great deal, with the extremely varied patients I see in various clinics and Homes, and every seeing-and-learning-and-doing situation is, *eo ipso*, a teaching situation. I find every patient I see, everywhere, vividly alive, interesting and rewarding; I have never seen a patient who didn't teach me something new, or stir in me new feelings and new trains-of-thought; and I think that those who are with me in these

* This letter was also published in slightly condensed form in *On the Move*. Cole did come to observe OS's work with patients for several weeks in 1977, and he became a lifelong close friend and colleague.

situations share in, and contribute to, this sense of Adventure. (I regard all Neurology, *everything,* as a sort of Adventure!)

But it's all entirely unorganized and impromptu—unorganized, that is, in any *curricular* sense. If you lived in New York, and could casually wander round with me here and there, you *might* find it an interesting and novel experience (or, then again, you might find it frankly intolerable!). But I couldn't possibly *advise* you to come from London to go on Neurological tramps with a Tramp like me—and I *am* a Tramp, a Gypsy, a Nomad, and not (as you seem to think) a Head of Department! If you worked with someone else (who did have a Department and all that goes with this) then you might "drop over" from time to time.

But, you see, I have all the disadvantages, as I have all the advantages, of being a Non-Establishment, Non-Established, person—I rove freely and widely; I spend as long as I want with any patient I want; but I have no Department, no courses, no colleagues, no help, no "position" (and for good measure no "security," and almost no "means").

There! I have said much too much, but it will do no harm (and may amuse you) if I acknowledge and "confess" how things are; it is doubtless indiscreet—on the other hand, it is candid.

Enough (or, most probably, much too much) said—

I'm sorry I didn't reply before, but (as you see) any candid reply had to involve, amongst other things, a certain embarrassment.

Do write and let me know how things work out with you—once again, I would be delighted to see you in an informal, casual, peripatetic way, but I am in no sense "set up" for any formal teaching whatever.

With best wishes—and thanks,
Oliver Sacks

To Paul S. Papavasiliou[*]

PHYSICIAN

MARCH 30, 1976

11 CENTRAL PARKWAY, MT. VERNON, NY

Dear Dr. Papavasiliou,

[. . .] I am (finally!) writing to you this morning, in the aftermath of a phone-call from a patient of mine, a very nice woman who was almost "terminally" Parkinsonian when I saw her in October, but is now doing absolutely beautifully on Sinemet,[†] and enjoying such a *renewal* of life as would have been absolutely inconceivable in pre-DOPA days.

First and foremost, I owe you both a belated apology, and a belated expression of appreciation and admiration. [. . .]

You said, very candidly and disarmingly, that you could scarcely proceed with your own bold work if you were beset by the sort of *doubts* which I had; that you *had* to be optimistic in order to be—a Conquistador. I appreciate this perfectly, and I *do* think of you (and your colleagues) as Conquistadors, patient, ingenious, dauntless Adventurers into the realm of the great chemical Unknown. I appreciate this if only because I am *also* a Conquistador—altho' a different *sort* of Adventurer in a different *sort* of Unknown. The Conquistador *cannot* be plagued by doubts—his Endeavour, his Enterprise, would be undermined by these. But *doubt is needed,* no question of this: constant, critical, sceptical questioning, questioning of all reported (purported) "observations," all modes of measurement and observation, all the specific assumptions which are made, and all the general, underlying, intellectual and emotional attitudes. *Someone* has to question these, vigilantly, disinterestedly, someone who has no special "interest" or "*animus*" in the matter; and, willy-nilly, (the matter was not of my choosing) it has partly fallen to me to be such a Questioner. I think you recognized this when we were able to speak, more genially, over dinner, when you said (the highest compliment from

* Dr. Papavasiliou worked in the lab of George Cotzias, who pioneered the use of L-dopa in people with Parkinson's disease. *The New England Journal of Medicine* called their work "the most important contribution to medical therapy of neurological disease in the past 50 years."

† A refined version of L-dopa.

a practical man), "I see you are also a useful man, Dr. Sacks." *This* pleased me more than anything else!

We are, in a way, different *species* of Scientist—you represent the bold, questing, enterprising Adventurer, and I the cautious, careful, critical Doubter; and yet we are also two sides of the same coin—for the very words "Quest" (question) and "Wonder" are two-faced; questing and wondering are adventurous and affirmative, but they also turn inwards, and look at themselves, testing the foundation on which they are reared.

If I was unduly sharp or rude that evening, I apologise; and I must ask you to remember, in extenuation, that I am someone who has been *silenced* for the past seven years, by the pressures and powers of professional disapproval. I am a consummate observer, and an unprejudiced one (I have what I might call "a pure love for *phenomena*," for every manifestation of Nature, whatever it is—Curiosity lies at the very core of my being). I have been privileged to see as much, or more, of Parkinsonian patients (and patients with all other forms of movement- and impulse-disorder) as anyone living; I can support my observations with films and tapes (you saw, that evening, how I walk round with a video-camera, because I want to catch as much as I can, including all sorts of things which I may find unintelligible or disquieting); I am honest, I am brilliant, I have made the best observations—and yet, damn it!—I find myself, I *have* found myself, these last seven years, virtually "unpublishable" in the Professional Press. By 1971, it was clear to me that almost anything I wrote would be rejected out-of-hand by most Medical Editors. In December 1970 I had found myself "attacked"—by [many colleagues]: attacked in a way which impugned my veracity and sanity. I wrote personally, and civilly, to all of these colleagues, inviting them to see me, and my patients, at "Mount Carmel"—none of them even had the courtesy to reply. [. . .] A single brave medical editor—just one!—representing (and this is significant) general practitioners and not specialists, wrote a remarkable editorial* in which he spoke of "the strange *mutism*" of the Profession in response to a book of outstanding importance, and one which was aimed and lodged at the very Heart of Medicine. [. . .]

You, in your way, and I, in my way, wish to *fortify* our patients, to

* In the *British Clinical Journal.*

strengthen them against the insidious invasion of an ugly disease, and the special vicissitudes and dangers to which this exposes them. [. . .]

Our approaches are wholly different, yet wholly complementary—What I took exception to, what I *reacted to,* on Tuesday, was your "passing-over" any and all non-pharmaceutical methods in the treatment and management of Parkinsonism: your speaking of Parkinson's disease *as if it were a purely chemical disorder, wholly treatable (at least in principle) by purely chemical means.* As a clinician, as a former neuropathologist, and as a former investigator of the neurophysiology of the nature of "denervation sensitivity," I *know* that Parkinson's is *not* a purely chemical disorder; and you know it too. We *all* know it, and the knowledge is unpleasant, and uncomfortable. There are *incorrigibly-damaged "functional systems"* (to use Luria's term), devastated microsomes, synapses, whatever. There is incorrigible damage to the fine fabric of the brain—and neither Dopamine, nor Apomorphine, nor any other amine is going to be able to repair this structural damage.

Whatever other considerations have relevance here, there seems to me a stark, inescapable *"economic" dilemma.* If the "functional" reserves (and not merely the chemical reserves) of the patient are depleted, then supplementing these with l-DOPA (or *any* dopaminergic agent or adjuvant) must obviously be of *limited* use. What is surprising, and significant, is that we repeatedly see seemingly-complete restorations-to-normality. But then one's finances are "restored to normal" by a loan from the bank. One *seems* to be solvent, one has money to spend, but there is a hidden snag—an "occult" debt. What subsequently happens (and, to my mind, is very inadequately, and *misleadingly,* dismissed as incidental—as "side-effects"), as far as I can see, is that l-DOPA *starts* as a "supplement" but *turns into*—a "stimulant." When this happens—and I think there is a threshold, a critical point, which may vary very greatly in different patients—I think there is an "*overloading*" and a *stressing* of already-damaged and thus somewhat labile functional systems, and that this can only eventuate in their recalcitrance—or breakdown.

I see "ON/OFF effects" as expressions of an intermittent (and typically *sudden*) system-breakdown—and I regard it as the most *ominous* of "side-effects" (because the least calculable). I agree with you that *gross* ON/OFF effects are (mercifully!) rare, but I think that something *qualitatively* similar (even if very mild, and scarcely symptomatic) is likely to occur in 100% of patients—that is, *all* patients who show a spectacular initial response to l-DOPA—I think that all those who initially do very well *must,* ipso facto, "pay the price" later; because this "doing well," in itself, is physiologically *costly,* and involves the patient in an ever-mounting physiological *debt. I see no pharmacological way out of this "bind."*

What then? *What should one do?* Should one recommend l-DOPA (or Sinemet, or AMP etc) to Parkinsonian patients "without reservation"? These are the very words you used. I think that one needs to have *many* reservations—altho' it is a further dilemma (both therapeutic and ethical) as to how much one should vouchsafe these to patients one is treating.

Let me give you two individual examples. [. . .]

Last weekend I got an anguished, bewildered, and angry phone-call from the son of an old man I had been seeing for a while. The old man (he was 86, but remarkably vigorous and alert for his age) had had Parkinson's for fifteen years. It was obviously a very slow, "benign" sort of illness, for at 85 he was still able to live alone, to climb stairs, to shop, etc. (and, on social occasions, being a merry old soul, to sing and dance with people a third of his age). Certainly he had—*problems,* but *he knew where he was;* he knew his powers and limits, what he could do and couldn't do. In short, he was *managing OK* without pharmacological help (at least, only a little Cogentin for rigidity). A year ago he was started on l-DOPA by Dr. X at Einstein. He was told, in the most glowing terms, that this "would make a new man" out of him; that he would feel "just great," and have "no complications." The old man and his son had various questions; these were waved aside as time wasting, irrelevant. "I recommend l-DOPA to you *without reservation*"—these were the very words which Dr. X used! The old man did famously for about six months, and *then* he started to run into trouble. His present situation, seemingly, is this: that he has become severely *hypotonic** with the continued use of l-DOPA [. . .], so much so that he has difficulty maintaining any postures—he is as limp as a rag-doll, or a nerveless "floppy baby"; but if there is the

* Having reduced muscle tone.

slightest reduction in his l-DOPA dosage, he immediately becomes disablingly (and rather dangerously) *rigid.* I *cannot* find any "right dose" any more—he is damnably hypotonic, or damnably rigid: *he is damned, apparently, whatever we do.* And this is precisely what his son said to me. He said, "You've got my old man *hooked* on l-DOPA. He can't take it anymore, but he can't take *not* taking it—l-DOPA has put him in one hell of a mess." [. . .] You will say "What's one case-history? What's one individual? What does one man matter against all the figures I can show you?" I say it does matter, and matters intensely; suppose that this man was your father—or you.

Now for the other case-history—or allegory. This was a woman in her mid-fifties, with a rather rapidly-progressive form of Parkinson's. She had managed, with ever- and rapidly-increasing difficulty, to keep house etc. until last summer, and then, in a catastrophic sort of way, everything got "too much" for her, and within a few days she collapsed into a state of cripplingly severe Parkinsonism allied with a reactive (and perhaps also partly "organic") depression of suicidal and nihilistic intensity. When I saw her first, last October, she was drooling, rigid, unable to stand, unable to talk intelligibly, bedfast, unable to turn over in bed, to feed herself, etc. She also felt (not unnaturally) that she was "finished," that this was "the end of the line"; and she could hope for nothing but that God would have mercy on her soul.

But "recover" she did—and has—with the use of Sinemet (l-DOPA caused intolerable peripheral effects), and all other sorts of help and support. She had a period of severe "side-effects," but now, in a mysterious way (which I don't pretend to understand) she has achieved an *Accommodation*—a temporary one, at least. Seeing her walk down the street, running for a bus, working again (she is a registered nurse), cheerful and realistic, enjoying life to the full with her family and friends, one would never guess that she has, that she *had,* cripplingly severe Parkinson's disease, and that her seeming "normality" is but a brief and merciful interlude in a deadly situation. *She* knows this, and accepts it with a wonderful equanimity, courage, gratitude—and humour. "I have been granted a *reprieve,*" she says, "and whatever the future holds, I shall never forget the happiness, the *goodness,* of this special time. It may be 'borrowed time,' but I shall live it to the full. And when Payment is demanded, I shall pay without demur; I was saved by Sinemet from the jaws of Death. I would have been dead by now if you hadn't given it to me. My only feeling is gratitude—and 'Thank you, thank you' to the good doctors who developed the drug, to the good people who have understood and helped me in every way, and to the good Lord who allows such miracles to happen."

There! There are two case-histories for you. It seems to me clear that the old man should *not* have been put on l-DOPA; and equally clear that l-DOPA/Sinemet *was* the right treatment for the woman. Can one generalize? Can one say in advance, or at any point, what constitutes "right action," and what not? This letter is already absurdly long—and I shall therefore leave all these questions open. The important point is that they are seen *as* open questions.

You are a bold and brilliant man, and you have provided many answers. But every "answer" you provide opens vistas of new questions—technical, chemical, ethical, existential—and if someone like myself has any use in the world it is this: that I am a *Questioner,* I question incessantly, I question everyone and everything; and I shall have plenty of questions for the Lord God when (if) I ever meet Him face-to-face!

Let me conclude by

a) enclosing a copy of the new, enlarged edition of *Awakenings* (it will come out here in a few weeks)
b) Sundry short articles + papers I have written on and around the subject.
c) Thanking you for being a courteous, a gallant, and an honorable adversary—a man with whom I profoundly disagree, but whom I cannot help admiring and liking
d) Hoping that we can meet, and talk at length, and perhaps see some patients together, some time in the future—

With kindest regards,
Oliver Sacks

To Michael Sacks

APRIL 5, 1976

11 CENTRAL PARKWAY, MT. VERNON, NY

Dear Michael,

I just got a letter from Pop today, in which he mentioned that Butch had died.

I know how much Butch meant to you (and you to him)—you often said you felt he was your "only friend," and I remember the sense of wonder, rapture almost, with which you would sometimes exclaim "*Beautiful* Dog!" I think the loss of such a beloved animal, a true friend, can be very hard to bear—or believe—and I feel I must at once write you a letter of sympathy and fellow-feeling.

I don't know whether you ever read any of J. R. Ackerley's books—among them he wrote a particularly fine one about his dog Tulip, and later (in *My Father and Myself*) he was to write of Tulip: "She offered me what I had never found before—single-hearted, incorruptible, uncritical devotion . . . From the moment she established herself in my heart and my home, my obsessions . . . entirely fell away . . . I sang with joy at the thought of seeing her . . . The fifteen years she lived with me were the happiest of my life."

Freud, you know, was deeply devoted to, and loved by, a succession of dogs (chows), and wrote very movingly of "their fidelity, their affection, entirely without ambivalence." He would be heartbroken each time a beloved dog would die—it is so sad that a dog's life-span is only a fifth of a man's. Perhaps—although such a thought would scarcely be bearable to you now—you might consider, in the fullness of time, getting yourself another Dog, another "Butch." (Not that Butch, an *individual,* whose life was inseparably shared with yours, can ever truly be "replaced.") Mourn for him, lament—for you need, one needs, the true feeling of Grief (and loss and parting), as one needs the Joys of meeting and finding. He was a lovely animal, indeed, and I think he lived a good and full life—for a dog—and enjoyed it to the full to (for a dog) a ripe old age.

My landlord and landlady, who are really friends now, have just been bereaved of their little dog—a dear creature whom they got as a puppy, fifteen years ago.

There comes to me, as I write, the poignant grief I felt at the death of Greta,[*] when she was run over in front of the house; and I remember digging her a little grave under the old Magnolias in the garden.

By the way, *when*, exactly (and *how*) did Butch die? I had a curious experience a short time back, because I got a sudden sense of his dying or death. By association I connected this with Bounce, [Alexander] Pope's dog, and *his* sadness at losing his old companion. I even found myself, "out of the blue," quoting one of Pope's last letters and verses (copy enclosed), in which he enquired after, celebrated, and lamented his dear Beast.

I *may* be back for a few days at Seder; otherwise, most probably, later this month. Look after yourself, and after our old Dad—

Love,
Oliver

To Samuel Sacks

APRIL 6, 1976
11 CENTRAL PARKWAY, MT. VERNON, NY

Dear Pop,

[. . .] Magdalen [*sic*] Gardner-Capp phoned up one Sunday and asked if I'd come to lunch. I said, "No, please forgive me, but I am in a frantic writing mood, and would hate to interrupt it. Another time maybe?" She said, "Please make it today, if you can, because *Aubrey* is coming, and he particularly wants to see you, he said." So I went along, and a very fascinating encounter it was. I have had a number of people say to me that I looked, or talked, or "moved" like Aubrey, without their having any idea we were blood-relations. I'd always assumed that such comments, if they had any substance at all, were based on superficial and acquired characteristics *e.g.* a certain sort of English (Oxbridge) accent, etc. But when I did meet Aubrey—and this was really the first *proper* meeting between us in our lives (previously he has just been a lofty public presence, and myself a withdrawn, silent figure on the few occasions we have been in the same

* A dachshund, a family pet in OS's youth.

room), I was startled to find (and he was equally startled!) that we were almost-uncannily similar in many ways. It startled us, and it caused much merriment in the whole company: for our gestures, our manner-of-speaking, our motor idiosyncrasies and turns-of-speech, etc. were almost as alike as those of identical twins! The similarity went *deep:* we kept finishing each other's sentences, inadvertently, we tended to make the same, sudden, unexpected associations. I have never had such a strange sense of "a double" in my life, and I said this to him: "Aubrey," I said, "although I scarcely know you, and our paths have scarcely crossed, and our lives have been utterly different, I have the feeling of an *innate* similarity between us—I think you are 'constitutionally' or genetically closer to me than any of my brothers." He agreed about this, entirely; and made a very interesting comment. He said, "*I* always felt, as others did, that *I* was very similar to my grandfather, Eliahu; and when you walked in the room today," he continued, "I got a shock, because *you* seem extremely similar to him in all sorts of ways, although, of course, he died before you were born." Curious, isn't it? *Atavus* means grandfather, and by this curious freak of atavism (because I think one must see our odd genetic similarity as precisely this), I got a very strong sense of being truly a Sacks—in the sense of being my grandfather's grandson! What do *you* think of this? It is a curious example of the interactions of Nature and Nurture: Aubrey and I shared the feeling, I think, that *each of us might have been the other,* had our "upbringings" etc. been reversed! [. . .]

You mention that there has been a *mild* flu epidemic in England. There has been a most *severe,* and *malignant* epidemic here—really a coming-together of several different strains of flu. The worst has caused (I have seen this in two patients) a very nasty sort of *encephalitis* with severe, and, one suspects, permanent brain-damage. There has been an identification of *this* virus with the "Swine flu," or "Spanish flu" virus of 1918–19. As you may have read, there is sufficient alarm over here for [President] Ford and Congress to have rushed through emergency legislation to immunize the whole country. This malignant sort of influenza, and the incidence of *encephalitis lethargica*–like complications, may give a special and dreadful topicality to the story of *Awakenings.* [. . .]

This has been a long, gossipy letter—but it is to make up, in part, for my naughtiness in not writing for so long.

Take care of yourself, Pop—there's nobody like you!

Love,
Oliver

To Eleanor Pyle[*]

JANUARY 15, 1977
11 CENTRAL PARKWAY, MT. VERNON, NY

Dear Eleanor,

It was very nice to get your good letter—I too am sorry that you hasted (felt you *had* to haste) back to Boston straightaway, and sorry too that I monologued so much, and didn't allow you—or Bob—to speak enough! I can only say that after nine months of "shut-up-ness" I had such a sense of liberation in my brief holiday on the West Coast—some of this surely from the delight of companionship with Bob, a Bob who has deepened and grown so much with the years—that I could hardly *stop* talking! And perhaps you, by your very silence, your deeply listening and feeling silence, somehow drew me out.

Thank you for your emboldening words about standing one's ground before Editors. What you say, and what Bob said, has greatly helped to release me from a sort of *abjection* of posture, and to recover a needed sort of erection and erectness. I had a talk with Bob Silvers[†] just yesterday, and will be giving him a slightly clarified and more forceful, but essentially unchanged, version of my original article on Monday. If he doesn't like it, he can lump it, and I shall take it

* Eleanor Bronson Pyle was a friend of Bob Rodman's.

† Robert Silvers, a founding editor of *The New York Review of Books,* began reading drafts of OS's work in the early 1970s (though it is not clear which essay OS is referring to in this letter). OS's attitude of bravado here ("he can lump it") contrasts with the deep respect he felt for Silvers. For his part, Silvers spent years getting to know OS (as he must have done with hundreds of other promising writers), encouraging him, sending him books of interest, and giving detailed critiques of works in progress. It was not until 1984, though, that the *NYRB* first published OS, with "The Lost Mariner."

elsewhere: it is up to the NY Review (or whoever) to expand for me, not for me to contract to them! [. . .]

I am delighted—and amused—by your reference to Starship Enterprise; and it set me thinking. I had thought of this as just a harmless fad, a sort of relaxation I could look forward to after the toil of the day. But your letter makes me see it somewhat differently: I think that it is neither accidental nor superficial that I have grown so fond of *Star-Trek,*[*] and it must be because it resonates with my own sense of being an Explorer in Deep Space, continually setting out to discover new worlds, and somehow to see the "alien" as *dear* (without in the least denying its otherness). Although *Star-Trek* often degenerates into an improper xenophobia or xenophilia, at best it deals with that right relation to the Strange for which the only proper word is (that very rare word!) XENIAL: grasping, and being grasped, by the strange as dear, welcoming the strange(r), making it (him) at home, and feeling oneself at home in the strange; the mysterious ultimates of home and hospitality, host-ship, guest-ship, the relation between them. Somehow, at best, I think this is what Starship Enterprise is about; and perhaps what any pioneer, or any properly, creaturely, enterprise is about. Anyhow, in some very deep way, I feel absolutely *regenerated* by my visit to the West Coast, and my Starship Enterprise is winging once more, either to the Rim or the Hub of the World, or into that good Space and Spaciousness whose "Centre is everywhere and Circumference nowhere."[†]

Cordially,
Oliver

* OS's fondness for *Star Trek* led him to call the nurse in *A Leg to Stand On* "Nurse Sulu."

† "God is an intelligible sphere whose center is everywhere and circumference nowhere." Dating back at least to the twelfth-century booklet known as *The Book of the Twenty-four Philosophers,* this image has been revisited by the Kabbalists, Pascal, Voltaire, and Borges, among others.

To Jonathan Kurtis[*]

FORMER STUDENT

JANUARY 15, 1977
11 CENTRAL PARKWAY, MT. VERNON, NY

My dear Jon,

[. . .] I gather that you must just have started your year in Greenville. I suppose it's rather an uprooting from Duke—but perhaps not really, because now you carry your own roots and centre with you, *within* you . . . and as for the future, [. . .] I think that one's nominal or external or titular "position" matters less and less as one begins to find oneself and one's *real* position.

This came up, amusingly, and paradoxically, a few days ago, when I met some of my old professors at UCLA, who quizzed me closely on "where" I *was* (*what* I was, *who* I was, etc.). When I mentioned my peripatetic, nomadic life—I have become a real free-lance now, driving from borough to borough, even State to State, with nothing but my good head and a portable videotape, one of them said: "But, Oliver, this is all very—eccentric. You mean you don't really have any position at your age?" I said, Yes, I did have a position. "What position?" he asked. "You have no centre, no base."

"It is very simple," I answered. "I have the only real position there is, the only tenable one, the only one with real tenure. My position is precisely at the centre of Medicine—taking in the view from all sides, sitting serenely, securely, at the very *heart* of Medicine." Maybe it sounded arrogant, or crazy—but at least it made him pause, and took that look (half-sneer, half-concern) off his face. What a passion people, at least one's colleagues, have to *label* one. I was also asked: "You're not doing regular, *orthodox* neurology—what *are* you doing? Some sort of zany, *fringe* activity?" So I invented a word. I said, "I make a precise structural analysis of all sorts of neurological phenomena, in a way which nobody else has done. It's Neuro-Analysis. I'm the first neuro-analyst, and my present patient the first neuro-analysand, and I have the first neuro-analysis in its entirety on tape."

They obviously don't know what to think: they both understand and misunderstand; they see that I'm authentic, and have something

* A student of OS's at AECOM, now embarked on his surgical residency.

to offer, something *new;* and they've asked me back to give some talks; but they imagine it's something "up my sleeve," sort of a new gimmick, or method, or technique, or theory—when really what I "offer" is none of these, isn't a "thing" at all, but merely an example of a total concern, and a sort of transparency which lets everything through; and this, which should be the commonest, simplest, most *elemental* "position" in Medicine, is apparently rare, suspect, and unintelligible to the narrow, doctrinaire, *opaque* majority. Extraordinary. I shouldn't let off steam like this, but I know that you are also simple and transparent, in the best possible way, and I imagine that *you* may run into odd difficulties with your colleagues, if not now, then later, as you achieve your autonomy, that sort of autonomy which is so "dangerous," such a stumbling-block, to the "heteronomy" of the group. It's just like the old *odium theologicum,* where simple honesty and autonomy are seen as heresy. [. . .]

I found my brief, sudden, impulsive visit to the West Coast (mostly the Beach Fringe in Venice and Santa Monica) absolutely fascinating. It's so innocent, and sweet—and silly, and corrupt. I see that the picaresque Beach novel I sketched out a dozen years ago, and imagined to be entirely "dated" now, is in fact as relevant as ever, because the whole "climate" there is absolutely dateless, timeless, unHistorical, *against* History. I see what I had forgotten—though it was this which, romantically, first drew me to California, and which later led to my leaving it in fear: life is so *easy,* so comfortable, so lacking in problems and challenges of all sorts, that one could easily be a child for ever, living in a childlike, childish, ingenuous Eden, an "endless summer" of body and spirit, never taking on oneself the responsibilities and pains (and joys) of adulthood. I went to California, in my twenties, because I wanted to "escape"—escape the rigidities and repressions of my (English) life hitherto; and I think it was the right decision. Then I left California, five years later, because I felt myself adrift in structureless emptiness, a rather comfortable "situationlessness" which everyone there shared. Now, I must confess, after having been bruised and battered into adulthood [in New York], being centred in myself (I mean creatively centred), being a going concern and a Worker (whether I like it or no), I wouldn't mind going back to California now. I could no longer be tempted, but life would certainly be easier and sweeter, infinitely less stressful, in a physical sense. And increasingly I am beginning to feel the vulnerability of the flesh, the need to preserve oneself from unnecessary stress, and to give oneself an atmosphere of unhurried peace. [. . .] I will look forward to hearing from you when you get settled down. And, of course, *seeing* you, on one of your occasional visits to New York (unless I make one of my even-more-occasional visits to South Carolina!).

Cordially,
Oliver

To David Goodacre[*]

VICAR

JANUARY 17, 1977
11 CENTRAL PARKWAY, MT. VERNON, NY

Dear Mr. Goodacre,

I have just received your review of *Awakenings* with real delight—delight that anyone should be moved to read the book with such passion and penetration, and great gratitude for your thoughtfulness and generosity.

Work—and writing—isn't done in a vacuum (although sometimes one gets the feeling that it is!); an author *needs* his readers and reviewers, especially ones as perceptive as yourself; because it is only through *you* that a real sense of communion is established, the feeling that one exists (and one only exists *for* others), of a good intercourse with the world. And it is only through the transparency and courage of people like yourself that one gets a real idea of what one is doing—your good review makes me see *Awakenings*, myself, and my work, in a new light.

I am particularly happy that such appreciation comes from a "spiritual" quarter—because I suppose now (I am not sure that I realized this, consciously, when I wrote it) that *Awakenings is* a sort of theological myth—of Exile and Redemption, Crucifixion and Resurrection,

* Goodacre, a vicar in the north of England, reviewed *Awakenings* for a tiny publication called *Carême*, a quarterly review of books on health and spirituality.

or what you will. And I suppose, increasingly (but it is especially your response and review which makes me think in these terms) that what I try to grasp, and what grasps me, are those intersections of history and myth—clinical situations which are parables or incarnations of the human predicament—of our wretchedness and uprootedness, and the great promise which is given to us all. What is so instructive about disease, like disaster, is that it *shakes the foundation* of our previous securities, of everything we have previously taken for granted; abysses open beneath our feet; helplessly, hopelessly, we fall in, engulfed, with a sense of absolute lostness and despair; and then "miraculously" or "providentially," *perhaps,* we find firm ground, *new* ground, in the depths of the abyss, and this is a *radical* healing or redemption. The lure, the danger, the temptation of l-DOPA was that it *seemed* to provide (and, for a short while, so to speak, provided the conditions *for* providing) a radical hope where none existed before: this mythical-theological sense was explicitly felt and formulated by some patients—thus "Leonard L.", with his characteristic humour and irony, called dopamine "resurrectamine" as soon he had heard of it, long before he had actually taken it! And he also called Cotzias "a Chemical Messiah." [. . .]

Now, in 1977, almost all of the patients are still alive; they are not very "well," physically speaking, but their experiences, their *bearing,* has made them profound human beings. I find their courage and faith—inexplicable, and inspiring; and it makes me sure of some-*thing,* or rather some transcendent *power,* to which one turns, to which one *must* turn, when the resources of the finite are exhausted. I have a book coming out, which is (perhaps) as eschatological as it is medical: it concerns a patient, a patient I know well: the patient was myself when I spent several months in hospital in 1974, and came very close to losing a leg. I *had* lost it, subjectively, and could discern no basis, no way, in which it might recover or return. That it did seemed scarcely short of a miracle, and has given me an enduring sense of the great, suprapersonal powers in whose hands we lie, and before whom we must give up our calculations and arrogance, and give ourselves totally, freely, in faith. This will be a sort of neuropsychological-theological book—an experiment, if you will, in "clinical ontology"; and since I can't resist a play on words, I am calling it *A Leg to Stand On.*

Really it is a great privilege, and destiny, to be a physician—*or* a patient. It makes one pause, and question what one has taken for granted before. It shakes one's foundations, and it makes one deeper. And, ultimately, it gives one a gratitude and an awe, which the healthy and happy have perhaps never known. [. . .]

I would not *wish* or *prescribe* sickness or injury or calamity for anyone; but *if* it should happen, a new and special destiny may be found, or created, by one—and perhaps it is THIS which I am ultimately interested in: the way in which people are "made" or "broken" by disease and misfortune, and this, ultimately, depending on whether one finds any ground beneath one's feet, after ill-fortune has thrown one in the abyss. I think disease, in this sense, is an ultimate testing-ground of a man's *manhood,* the wounded creature losing and perhaps (re-) finding its foundations.

I am sorry I have written a rather verbose, tangential, uncalculated letter—and one which is perhaps not too *directly* related to *Awakenings* or your review. But it is about attitudes, orientations, which are central to us both—to us both as caretakers of the Body and Spirit.

With many thanks and my kindest regards,

Yours sincerely,
Oliver Sacks, M.D.

ALTHOUGH OS WAS STILL STRUGGLING with the book that would become *A Leg to Stand On,* he was also continuing to accumulate more case histories. In this letter to his old boss at UCLA, he tells a story that a few years later would become "Cupid's Disease" in *The Man Who Mistook His Wife for a Hat.*

To Augustus S. Rose

FEBRUARY 11, 1977
[NO ADDRESS GIVEN]

Dear Augustus (if I may!),

It was very good of you to write me; I am sorry there has been so little contact over the years (for I have always thought of you as the teacher, above all, who taught me how to *observe*—although you may now feel a bit disquieted at the *sort* of observations I make!), and did hope to see you in my brief (unplanned, impulsive) visit to LA. [. . .]

Something which often makes me think of you, and wish I had a hundredth of your clinical experience in the area, is seeing (a quite astonishing amount of) neurosyphilis. It is very odd—perhaps it is chance—but just in the last year or two I have seen *dozens* of cases, where I scarcely saw any in the preceding decade. It really is a "wonderful" disease (I'm afraid this sounds a very outrageous word to use, but you, if nobody else, will know what I mean) with respect to the astonishing variety and complexity and *strangeness* of the syndromes it produces. Two weeks ago I saw a woman of ninety, accompanied by her hale husband of ninety-three: the history was of "a change in personality" coming on at the age of 88, when the old lady—habitually prim and staid, characteristics accentuated by age, started to become more and more "animated," and distinctly disinhibited, rather flirtatious and jocular. I found a remarkable psychomotor excitement, with (as yet) no real dementia. She herself made the diagnosis—she said, "Would it have caught up with me after sixty-five years?" I asked, "What?" rather dazedly. She said, "The squiggly things, Cupid's disease," and went on to tell (she was an Armenian) how she had been impressed by the Turks into a brothel in Salonika, in her early twenties, contracted syphilis, but was given "heavy metals." [. . .] She wondered if it had persisted, latent, all these years, and was now the cause of her unnatural animation. We did a spinal tap: it was—astonishing story.

I do hope you are keeping well and active—it sounds very much as if you are; and I will look forward to seeing you in May. Again, my warmest thanks for your very kind letter,

To Josephine De Luccia

CORRESPONDENT

FEBRUARY 22, 1977

11 CENTRAL PARKWAY, MT. VERNON, NY

Dear Mrs. De Luccia,

Thank you very much for your letter of the 19th. I offer you all my sympathy—I think *nothing* is more dreadful, or tragic, than the development of a pre-senile dementia in one's spouse—but I am not sure that I can offer you anything more.

If the diagnosis was made at Einstein, and called "Alzheimer's Disease," then, I fear, so far as current knowledge goes, nothing can be done—nothing, that is, in the way of any "miraculous" Medicine or Surgery. A great deal can be done, of course, in the more general matter of *care*—by which I mean everything which invests life with meaning and dignity. One can have a relentless dementing disease, be it Alzheimer's or Huntington's chorea, without it being felt as a torment, or an indignity. It *is,* of course, a catastrophe—but catastrophes can be borne with dignity and calm—indeed there *is* no other way to bear them.

And *you,* you have to do the bearing too—and perhaps, as the disease progresses, it may be harder for you than for your husband to bear. When misfortune or calamity strikes one member of a family, everyone else is affected too: the suffering is multiplied—but also it is *shared.* I think that the *proper* sharing of suffering—sharing in the mode of concern and compassion, as opposed to any *improper* worrying and fussing—is ultimately, perhaps, as much as one can do to ease another's lot: for the worst horror of all is being-left-alone, left alone, abandoned, to sicken and die.

Having said this, I hope that you do not misunderstand me. I think that you need to "support" your husband, to be with him, as it were, as the darkness closes in—this is the least, and the most, another human being can do; but you must *not* sacrifice your own life in the process, or feel that you have to be with him every minute of the day. He himself, I am sure, would want you to live a full life yourself, and not anguish or fret over him in a way which is profitless and can only torment you. In particular, I think that you are on dangerous and foolish ground if you are (in your own words) "looking for a miracle"—a miracle-drug, or anything similar. And you sound as if you are obsessing for "a miracle," constantly reading everything on senility which you can lay your hands on. There *are,* in some sense, "miracle drugs": cortisone, as you say, was/is in some ways "miraculous," and so was/is l-DOPA which I describe in my book. But no such "miracle" is available for Alzheimer's—nor, I must add, is one likely to be forthcoming. You ask me to be "kind" and "considerate," and I fear that you may find my words harsh; I don't mean to be harsh, I only mean to be truthful, and to try and spare you further futility, when I say that yearning and searching for "miracles" here is profitless alike to you and him, is provocative of anguish, and stands in the way of a serene and dignified acceptance of reality. If the reality *could* be altered, the matter would be different; but if it can't, then any hopes are illusions, and one *has* to accept what one cannot alter. This is the law of Fate—and though it seems strict, it is merciful as well.*

Yours sincerely,
Oliver Sacks, M.D.

To Mel Erpelding

MAY 14, 1977
11 CENTRAL PARKWAY, MT. VERNON

Dear Mel,

Thanks very much for the package—and your letter. [. . .]

I found my brief visit to Oregon deeply happy, and reconciling—and, perhaps, in some part of me, a bit sad. It was very good to see you, to see you *both,* so happy and settled: and, for the first time, I didn't see Pilar† as an enemy or "rival," didn't feel any envy or hostility,

* De Luccia replied to OS, thanking him for his "prompt and honest" response. She wrote: "You chose your words very, very carefully, and I know that this dreadful illness can never be reversed or cured in the immediate future. I will continue to care for and love my husband as best I can. [. . .] Your letter was received on a dark and rainy day with my mood to match, but it was written with gentle thoughtfulness and compassion. I am going to save your letter and when my spirits are low and depression hits me, I will read it over and over again."

† Erpelding's wife.

didn't want to *bisect* you with her and I think *she* felt this, this new acceptance in me, and the long-standing tensions between us ceased. I no longer want to "*possess*" you, and the disappearance of this phantasy, after so long, gives me a serenity and a relief. But perhaps also—and this is why I spoke of sadness—the renunciation of a long-held phantasy is a sort of bereavement; but it's a *sane* renunciation, and a *healthy* sadness—such as go with an acceptance of the unalterables of life.

On the other hand, with the disappearance of these old possessive-competitive tensions, I—we—found ourselves freer and more at-ease. It was a *deep* (and almost mystical) delight climbing up Mount Hood with you—the immensity, the spaciousness, the *goodness* of it all (and I love your new sensitivity, and communion with nature, whether it is feeling the great trees as friendly angel-spirits, or spotting new lichens, and knowing what's edible. I have the feeling that you could live off the land, *with* the land, in a sort of peaceful, holy communion with it). And those lovely ancient baths, fed by generous Hot Springs: how right you are—it was indeed a Special Place, a Secret Place, a Holy Place, a Healing. You saw (how much you see!) the wound in my centre, my central maiming, which has been so overwhelming since that Norwegian experience: you saw what had happened, and what I needed: a true healer. And taking me with you, on that path on Mount Hood, was a deeply healing symbolic act—of a sort Kabbalists call *Tikkun;* somehow repeating the experience of a mountain path, but this time not alone but with an old and dear friend, and not destructively but humbly and holily, and not with a Bull but in the presence of friendly Trees. All this somehow *exorcized* the horror which has held me for three years, and broke the trauma, the ragged nightmare. I have felt much better, and specifically, *surer-footed* since that Monday evening—the terror, of slipping, of falling, of self-killing, I felt it going away in our walk.

There is some strange current between us—nothing we can ever (or even need to) "act on": it has always been there, and will probably endure. I don't even know what sort of words to use—it is a feeling beyond words, which is liable, perhaps, to get spoilt by words: it is certainly a sort of love, and a sort of communion, and it has something sexual about it, but in that mystic and sublime sense which Dante uses when he speaks of "the holy and glorious flesh," and the flesh as Mystery, as the Word made Flesh. Something physical, something mystical, a sort of mystical sexuality. I cannot formulate it, I have already tried too much: suffice that it's there—somehow mysterious and ordained—and that it makes for a mysterious flowing and contact, and can do so without tempting or teasing or tormenting us into anything merely and profanely physical which would be damaging, and somehow unnecessary as well. Perhaps I shouldn't have said this—but I feel the need to—it's part of the clear-sight, the reconciliation, the quietness, the acceptance, which mingles joy and sadness strangely. [. . .]

I had a lovely experience in Toronto—it started well with a symbolic deed: a young man came up to me and said, "Oliver Sacks? I have been wanting to hug you!" and he did so forthwith. (He was chairman of the film committee of the APA,* and said he couldn't read *Awakenings,* or see the film, without wanting to hug me!). It was a lovely, genuine, spontaneous reaction, and—especially after the hateful coldness at UCLA, where I felt anything but hugged—it gave me a wonderful feeling of warmth and welcome, and this carried on into the meeting, where, in turn, feeling loved, I loved the audience back, and there was a marvellous discussion which was like an embrace. [. . .]

Love,
Oliver

* The American Psychiatric Association had its annual meeting that year in Toronto; the program included a showing of the 1974 Yorkshire Television documentary of the *Awakenings* patients.

*To R. E. Case**

CORRESPONDENT

JULY 25, 1977
11 CENTRAL PARKWAY, MT. VERNON, NY

Dear Dr. Case,

I have just received your most generous and thoughtful and thought-provoking letter [. . .].

Questions of outlook and orientation in regard to medical education, no less than treatment, have been exercising me greatly at the present time—and it is a curious coincidence that your own letter should arrive right now. My feelings are at once passionate, ambiguous, and not fully resolved—and therefore [. . .] I find it difficult to write on this matter, above all to write in a critical way. For this reason, among others, when I was recently asked to contribute a memoir or reminiscence to be included in a book called "My Medical School" after some weeks of vexation I had to decline, because I found Medical School *deadly*—a real murdering of the perceptions, the heart, and the mind—and I could have expressed and explored nothing except negative feelings. I feel that my Medical School (and it was one of the best in England) retarded and deformed my education and development, as a scientist, as a physician, and as a human being—and that it was this (among other things) which delayed my finding my own feet, defining myself as a physician and a person (and I think the *person,* the personal qualities, are pre-requisite for a physician, though perhaps not for a physicist) until I had "escaped" from Medical School, and the systematic idiocy which it imposed upon me—on all my contemporaries. There was a lack of intellectual excitement and wonder, and a lack of human feeling and concern: the nerve of astonishment—and compassion—was not nourished; sometimes, perhaps, it was almost killed. The sheer *dullness* of doctors—their lack of intellectual curiosity and passion, and their resistance to entering fully into the feelings of their patients, seems to me very much a reflection of miseducation—an education which reduces everything

* Case, a scholar, wrote that he wanted "to find a way to show the urgent need to introduce matters philosophical into the education and training of scientists [. . .] especially those in medicine, psychology and social work."

to "information" and "skills." I think it is difficult, and rare, for physicians to surmount the mischiefs built into their training, the false ideas and emotions on which so much "Modern Medicine" is reared. I stress "modern" because I think that this is partly or largely a reflection of our age: unbridled technology and growth of information, and a loss of the leisure, the human dimensions, the sense of *apprenticeship,* which characterized medical education fifty years ago (for example, the education of my parents, who both became physicians during the First World War). I think there has been a rupture of a most ancient tradition—call it the Hippocratic tradition if you will—in which epistemology and ethics and education are inseparable. I by no means advocate a mere going-back to the past: our world has changed irreversibly this century; what I envisage as an almost desperate necessity is bringing all our new knowledge and concepts and skills into a human and humane tradition or, as Buber puts it, "humanizing technology before it dehumanizes us."* [. . .]

For physicians, there must continually be a sort of "double" orientation, an exact and dispassionate grasp of the *phenomena* involved (which, in a sense, are impersonal, supra-personal, or sub-personal)—and this constitutes the *Science* of Medicine); and an empathic and compassionate grasp of the personal *predicament* involved—the intelligence of the heart no less than the mind. [. . .]

No, I am afraid I cannot send you any articles etc. because I no longer write them—I only write *letters* (like this!), and—occasionally—*books* (I think of a book as *a letter to everyone*—at least to the sort of people I would like to correspond with). By the same token I fight shy of any public lectures or involvement, although I deeply appreciate your generosity in suggesting my name. But I am essentially a solitary physician-thinker-writer, and I have to work and publish in my own rather private and personal way. Your own series of lectures sounds profound and exciting.

Enough said. My deepest thanks, and best regards,

[. . .]

Oliver Sacks

* OS frequently invoked the philosopher Martin Buber (1878–1965), often in reference to Buber's concept of the I-Thou relationship and the I-It relationship.

To Richard Lindenbaum

SEPTEMBER 6, 1977

11 CENTRAL PARKWAY, MT. VERNON, NY

Dear Dick,

[. . .] Such a sadness, and a release, has come from the death of Luria, which I heard about ten days ago (it was reported here, initially, but not in England). In a way which must sound absurd if not unintelligible to you, *my* "idealizations" are for "good fathers"—like Auden and Luria—who are set at a safe distance from reality. My in-a-way very intimate correspondence with Luria, over the last four years, was the most deeply affirmative experience of my life; and I would feel (as I think Gorki says of Tolstoy) "While this old man is here on earth I shall not be an orphan" (or something like this). I *never* saw this old man in Moscow whom I loved, and after some initial yearning and pining, no longer wanted to, and felt (as I think he did also) that we could give of our best, our most "ideal," given (and *only* given) a distance between us. But, of course, it is absurd to press the comparison further; I wanted communication, even communion, but not conjugation. I knew, of course, that he was old and ill, but he was so vividly alive, imaginatively and emotionally, so vividly present to me, every day, many times a day (I would continually be saying to myself, at work: "Now what would Luria say of that? Must write to him instantly! Send him a tape!") that I was utterly stunned by his death, and spent a couple of days absolutely *howling,* in a way I haven't done since my mother's death—and, to a much less extent, Auden's. (I had become closer to Luria than I ever was to Auden, perhaps, partly, because we had so much to *share.*)

His death "unlocked" me, cleared my heart, cleared my mind—and suddenly I was able to go, to flow. When I found that there had been no news, or obituary, in England, I instantly felt that writing an obituary was the right thing to do, and, without the least hesitation, even a second thought, I sent a telegram to the Editor of *The Times.* When I *tried* to write the Obituary, my feelings overflowed—"feelings" is too narrow or sentimental a word; I flowed into an extended biographic-critical-lyrical essay, 20,000 words, almost a little book; and they poured out in a single sitting—I scarcely stirred from my chair until I was "finished." I think it is probably good, because it came out in this way—very much the sort of way in which *Awakenings* came out (at least the impassioned part which I wrote straight after my mother's death)—with a transparency and purity of feeling: *genuine* in feeling and thought, through and through (as opposed to my conflicted, ambiguous piece in '73).[*] It was an Act, an existential Act: an Act of Mourning, Remembering, Thanking, Farewell; and, in its way, an Act of Love. Whether this makes it essentially private I am not sure—I think, probably, it doesn't, because so much of what I say is not just "my feelings," but Truth, or at least a deep understanding, which others can share, of a rare and very significant human being, whom I could not help, again and again, comparing to Freud (and "neuro-analysis" to psycho-analysis, etc.). Only *when* I had done this could I write the Obituary. Eric[†] phoned me this morning, to tell me it was printed—a bit cut—in yesterday's *Times* (Sept 5) and I have just been down to the Foreign Papers Shop to get it. I am glad to see it. I have done the right thing. Most of my life is dominated by a feeling that I have done/am doing/may do the *wrong* thing, an almost perpetual sense of Crime and Sin. So the certainty of doing the right thing is very important—indeed it is the *only* time I can ever make a move. [. . .]

To Samuel Sacks

SEPTEMBER 24, 1977
11 CENTRAL PARKWAY, MT. VERNON, NY

Dear Pop,

[. . .] Thank you for your lovely New Year's telegram—and I hope you got mine. And that you had a good, festal New Year, and will really enjoy this New Year to come. [. . .]

We had a very nice ("Sacks") get-together at Rosh Hashanah here—the Capps, and their ramifications; and Jonathan,[‡] and Carmel[§]—

* In 1973, OS wrote an essay about Luria for *The Listener.*

† Korn.

‡ Jonathan Sacks, OS's nephew and David's son, who was then going to graduate school in New York.

§ Carmel Eban Ross, Aubrey's sister.

and Aubrey. It was odd and moving to see a Sacks Face, and hear a Sacks Voice, in so many different inflections and modulations. Perhaps I have been more aware of the Landau Face and Voice, and of *these* determining me, *in* me; but I had a vivid feeling, on Rosh Hashanah, of the other half, the Sacks patrimony in me. This was strengthened by a long talk with Aubrey, in which he spoke particularly of our grandfather, your father; and, afterwards, I borrowed a proof-copy of his autobiography,* which so riveted me that I read it at one sitting (not easy, because it is about 700 pages long!). Beautifully, brilliantly, vividly written, it never loses its tone and pace, gives one a tremendous feeling of History—History *lived.* The only note of sadness—never explicit, but somehow present throughout, especially in the opening chapter on Boyhood—is a certain sense that he himself never had much say in the matter: that he was "chosen," or "fated," really from childhood, and that, to some extent, *personal* feelings and development, etc. had to be sacrificed to a certain Public role—although a very noble one, as Witness, Voice, Architect, and Historian—of age-old Jewish yearnings and plight, and the brand-new, yet so-long-awaited, State of Israel. But that's how it is when one has a "Mission": one's personal life *is* sacrificed in a sense, but consecrated too. And now I think of it, these two words have the same root, indeed they are synonyms which have split-apart in meaning. The other pity—and something which must be exercising Aubrey now—is where he belongs in the *present* State of Israel, which, alas! has lost some of the selfless splendour of its first days, and is so riven by petty party politics and squabbles, which are so below, so inconceivable, to someone like Aubrey. One feels he is almost *too big* for present-day Israel. [. . .]

I see quite a lot of Carmel—really very fond of her—please give my love to the indomitable Yitzhak, and of course to Lina, and Florrie, etc.

Love,
OLIVER

* *Abba Eban: An Autobiography,* published in 1977.

To Harvey Shapiro

MEDICAL DIRECTOR, BETH ABRAHAM HOSPITAL

NOVEMBER 10, 1977*

11 CENTRAL PARKWAY, MT. VERNON

Dear Dr. Shapiro,

I was somewhat distressed by our talk this afternoon, and I could see that you were as well. There is, indeed, no matter as emotionally charged—for anyone—as all questions relating to death. [. . .]

As a person, as a physician, as a biologist, and as a man of religious consciousness, I have, of course, a reverence for life—indeed, it is the deepest feeling I have, and the root and spring of all other feelings. Moreover, it is not simply theoretical: as I mentioned to you, I have a power of seeing and bringing out life, potential, hope, in the most severely brain-damaged patients whom others have long since given up. Time and time again people are dumbfounded by this power, and say "Dr. Sacks, we never heard her talk before. Look she's smiling! We never knew she had any feelings. It's amazing how you bring people to life, etc." Last Friday, I saved a man's life by doing a tracheotomy with a table-knife—he had a seizure while eating, and aspirated a chicken-leg down his larynx. A couple of weeks ago, in a nursing home where I work, a patient had a coronary and immediately went into massive pulmonary oedema and shock—I worked on her, desperately, for two hours, brought her through at last, and now she is recovering quietly. I am a practising physician, and I do not speak theoretically of life and death. *I am entirely for life, on the side of life—true life, personal life—whenever I see it.*

I think you were hasty in comparing me to executioners in the death-camp, as you did, if only by implication; but I know that here, *you* were carried away somewhat, and—forgot yourself, and who you were speaking to; but realizing this, and the obvious excitement and anxiety which led you to invoke such a comparison, I overlook it. You forget, perhaps, that I lost most of my "extended" family in the death

* A note in OS's hand at the top of this letter says, "NOT SENT. This man is already too 'threatened.'"

camps—as, very likely, you did as well. Neither of us could possibly be potential Eichmanns!

This is not the issue, at all. We are not speaking about killing the living, the truly living, the personal, the human . . . which, of course, is the ultimate atrocity.

What we *are* talking about is a rather terrible twilight zone which has only come into existence in the last thirty years, and which has no precedent in the entire history of humanity. Namely, our ability to maintain, even for decades, those who have the most severe and *seemingly* incorrigible brain damage—for example, those who are apparently cerebrally "dead," and have flat EEG's, etc. Such patients could *not* be "maintained" a few years ago, and the problem of what "Life" meant, in relation to them, did not arise. It is only in the last few years that we have developed the ability to keep such patients "alive"—if the term "alive" can legitimately be used of someone who has no detectable cerebral activity—or cerebral activity which does no more than maintain autonomic functions, and is *apparently* incapable of improvement.

Here the problem is not that of killing a conscious and personally-alive human being, as of the sense, and morality, of maintaining, by artificial means, a sort of "zombie"—someone who may be "essentially" dead, but who can be kept going as a sort of living corpse. This is a horrible and uncanny region—because one is dealing with what it means to be "half-alive" and "half-dead." You extended my attitude, as you saw it, to the Death-Camps; I could extend your attitude, with similar unfairness, into a vision of a world where, if a man or a dog has been decapitated by injury, we might keep their hands (and bodies) alive, by perfusion. I do not say whether we should or should not—simply that this is *a region which has no precedent in human experience,* and which faces us all with problems which have no precedent, and where we cannot but be confused as to meaning and application of the classical and biblical injunctions, because no such situation was envisaged when they came into historical being. One can imagine a hospital entirely full of decerebrate patients maintained artificially—one can imagine a hospital or a city, or a world: this is the other side of the Death Camp vision. And one which some might consider scarcely less terrible, unholy, and obscene. These are very deep and frightening waters, and it is far from clear what course one should steer. But I *do* think that such matters do need discussion and "agonizing reappraisal." But, of course, I fully realize that such matters are not for the chart!*

I wish we lived in a simpler and more innocent world!

* I.e., the patient's medical chart.

9

Coming to Terms

1978–1979

To Innes Rose

LITERARY AGENT, JOHN FARQUHARSON, LTD.

MARCH 13, 1978
[NO ADDRESS GIVEN]

Dear Innes,

[. . .] I think that David Lan* is a gifted and responsible young artist, who has a real and deep interest in *Awakenings*. I think that he would spare no effort, as he intimated when he first approached me, to create a play which would be "faithful in substance and spirit" to *Awakenings*. I think he would probably be successful in so doing. In answer to a very nice letter he sent me about three weeks ago, I said that I would be happy to show him my patients, notes, films, and other "material" pertaining to *Awakenings*. So far, so good.

What *worries* me are the following considerations:

1) The propriety, or otherwise, of making a play about patients and persons still living. As you know I *agonize* constantly about "exposing" my patients—or myself—to the Public: *divulging* what, in some sense, I feel as confidential or private. This is the alpha and omega of my *own* inhibitions—and the reason why I have only published two rather than twenty books. [. . .]

A book is already at an imaginative remove. And a play—inspired by a book—is at *two* removes. Nonetheless, I cannot avoid a shrinking and fear—and trauma to my patients, and trauma to me. But then

* Lan became an acclaimed playwright and director and served as artistic director of the Young Vic from 2000 to 2018.

again, I know that many of these feelings are neurotic, and that they are likely to become assuaged with time. Thus NOW, five years later, seeing my poor patients, or those who remain, de-differentiate once more into the greyness from which they came, I am unashamedly glad and grateful that I *did* write *Awakenings,* and let a [documentary] film be made: for otherwise, their story, their situation; something phenomenologically, historically, medically, humanly, dramatically, and poetically unique would never have known the light of day. [. . .]

If I have observed faithfully or truly in my book, I *want* it—or any part of it—to be the subject of scientific research, responsible *investigation.* Do I want it to be subject of an artistic presentation? Here I hesitate. I do—and I don't. I do have a strong feeling of *Awakenings,* and all my work, as representing a rather rare sort of genuine Art-Science; I have the strongest feeling that all sorts of things, which are not amenable to scientific analysis, *are* depictable by artistic presentation—and, indeed, in no other way.

2) [. . .] My first point, essentially, has to do with *propriety;* and my second, essentially, revolves round the concept of *property* (it occurs to me that both words form an etymological doublet!). It is true that I am the sole author of the book called *Awakenings*—but to what extent, or in what way, is the subject "mine"? To what extent is *anything* mine—to give another exclusive rights to? [. . .] If *Awakenings* were entirely a work of fiction, then perhaps I could more justly lay claim to it; and obviously, in some sense it *is*—in the sense that almost everything has been transformed by being refracted through *me.* But I *also* feel it is a piece of History, and a piece of Science, which belongs to everyone, is in the public domain, and public property (if it can be called "property" at all). I would not, for example, feel that I had any right to stake a claim in Physiology: for example, to claim that "Parkinsonism" or Migraine was *mine.* [. . .]

There is obviously some complex ambivalence in myself about having anything to do with other people, in any sort of collaboration. Of course here, things should be much easier, because I am in no sense *directly* involved. I would make my work and myself *available* to David Lan—and, so to speak, bless his endeavours; but his endeavours would be wholly and solely his own, as my own are wholly and solely my own; what he made of *Awakenings* would be HIS business, and not mine—and yet . . .

How damnably complicated and involuted all of this! Or is it?

If he wrote a *wonderful* play, then—of course—I would be delighted and grateful. But supposing he wrote an *awful* play; or (to avoid value judgements) a play which aroused my passionate *distaste;* or one which I felt was *not* faithful to the substance and spirit of *Awakenings,* but, indeed, in violation of them at every possible point—what then?

I do not want to be put in a *critical* position. I am a creator, not a critic or censor.

I don't want to be put in a position where I might feel compromised, or embarrassed, or somehow "responsible."

I do enough agonizing and obsessing about *my own* work, and don't want to do it with anyone else's. [. . .]

A last point. This relates to the *exclusiveness* of rights. And this in turn relates to my ambiguous feelings about property. Supposing *someone else* wished to make a play about/from *Awakenings;* they would be *prevented* from doing so by the terms of the contract. But, by the same token, David Lan would be *protected,* as he *should* be, in a way, if he devotes much time and effort and spirit to the work.

For example: Alan Bennett, it so happened, expressed an interest in a dramatization of *Awakenings* the moment it was published, in 1973; and *re*-expressed this, when I saw him in New York in December. I told him, at the time, that I had met and discussed things with David Lan, and had been much impressed by Lan's sensibility.* Alan said something about "not committing oneself," adding, "Why shouldn't twenty of us make dramatizations if we wish?" Why not indeed. Should I prevent Alan Bennett, say, to protect David Lan? Have I the *right* (or the wish) to do so? Is "prevention" in fact involved. Or precisely *what* is being prevented? Obviously *plagiarism* should be prevented (as when a certain Dr. X† published several chapters of my *Migraine* under his own name); but should "free enterprise" or "competition"? Again, this all comes back to the vexed notion of property, my sense that I am not the Owner of a (literary) Property, with the right/power to grant "Rights" of one sort and another. And yet, in a

* OS never granted exclusivity to either playwright. In 1982, he received a package from Harold Pinter containing the script of a one-act play, *A Kind of Alaska,* based on *Awakenings,* which he gave his blessing to.

† In *On the Move,* OS tells the story of how Arnold P. Friedman, his boss at the headache clinic, tried to appropriate his work.

way, I suppose I am—I do not have a strong sense of, say, Copyright. [. . .]

Sorry about this mad letter. Give me some sensible sober Scottish advice!

To Harvey Shapiro

MAY 7, 1978

11 CENTRAL PARKWAY, MT. VERNON, NY

Dear Dr. Shapiro,

I take the liberty of enclosing this fascinating and important article from this month's *Epilepsia*.

I found it particularly interesting, and *relevant,* because I have been finding an astonishing incidence of precisely such syndromes in the geriatric population at Bronx State—and in the elderly non-psychotic, but often very confused, patients I see elsewhere.

The importance, obviously, is this: that we have here an *eminently treatable form of "organic brain syndrome,"* which may be overlooked and untreated indefinitely—unless one is vigilantly aware of its existence, and prepared to do careful EEG monitoring of patients.

I now have quite a series of patients who had been labelled (often for years) as "senile," "confused," "arteriosclerotic," etc. who have shown striking improvements in clinical state with the use of anticonvulsants.

I bring this up, as I could bring up a dozen further reasons and indications, because I feel that *not having an EEG service* at Beth Abraham—one which is instantly available, and causes no stress or inconvenience to patients or others, and can be read by someone who is acutely and intelligently aware of the patient's syndromes and situations (myself!)—is definitely *detrimental* to patient care, and may deprive some patients of beneficial treatment.

I have waited more than three months now for "Administration" to respond to my offering such a service. It was obvious, indeed, speaking to them, that they were placing ludicrous considerations of a "political" and "commercial" sort *before* the primary consideration of patient care. I say "ludicrous" because such considerations do not in fact arise at all. Can *YOU* do anything to expedite matters? It may be, as happens increasingly with *everything* at BAH, that far from *opposing* my suggestion, Administration has simply *forgotten* it. It is paradoxical that the more people (and computers etc.) they have, the more inefficient they become!

Yours sincerely,
Oliver

To Barbara Beasley

COLLEAGUE

MAY 13, 1978
11 CENTRAL PARKWAY, MT. VERNON, NY

Dear Barbara,

Nice talking to you this afternoon—I hope all goes well down at NIH,* and I will do my best to run the clinic smoothly on Wednesday. [. . .]

I got a funny feeling from the Movement Disorder Group—mostly at Columbia.† I have been trying to define what it is that I find "funny"—the feeling is very strong, but not easy to define. I think it's this: that most of them seemed to have a "professional interest" in the subject, as opposed to a real interest—and by a "real" interest I mean a real love for the subject, a feeling and wonder for *the phenomena themselves*—as opposed to an itch to label them, classify them (in a purely nosological, not phenomenological, way), and—get rid of them. I have pretty much the same feeling now about the Tourette Group, which has professionalized (and also commercialized) itself so much. And I did, years ago, with the Migraine Group—the people who "run" Migraine in this country (somewhat, perhaps, as General Motors is "run"), and who regard the subject as "*theirs.*" I think there is something *essentially* wrong and suspect about such Groups, such Attitudes, such a proprietary feeling towards part of

* The National Institutes of Health, in Washington, D.C.

† A group of physicians concerned with movement disorders (Parkinson's, Tourette's, Huntington's, etc.) who met monthly to keep abreast of new work in the field.

Nature. Something quite horrible—it gives me the creeps—and something utterly different to the "amateur"* amative approach, which distinguishes Purdon Martin,† and You, and Me. But it is not easy to *survive*—professionally, financially, etc.—when everything is increasingly taken over by Medical Business & Professionalism (it is in an attempt to survive, I imagine, that you are going to NIH). They kept talking about "data" and "data processing" last week [. . .] much as they do at the Tourette Association. I wanted to make a protest—and say that though I had nothing against "data," of course, I felt that essential sorts of *understanding* were being ignored, but I think that such a protest would have been *unintelligible*. Anyhow [. . .] that evening sharpened my feeling that I (and a few fellow-spirits) constituted a separate *species*, and that we must struggle at all times to maintain individuality, and to minimize contact with professional "groups" (I almost wrote "gangs"—Friedman et al. certainly formed a Migraine Gang!). And, by the same token, not form (even defensively!) a "Group" ourselves. There is something most dangerous about Groups *as such*—they can pool all sorts of resources and powers, but they are likely to do so by doing away with individual freedom. And yet, one can be so limited and cut-off if one reacts to the other extreme, and become a total Solitary or Hermit. I must confess, for myself, that I find the whole question of *association* (personal, professional, creative, or whatever) extraordinarily difficult—and yet unavoidable!

Best,
Oliver

* OS loved to use the word "amateur" in its original sense, as referring to someone who might be quite knowledgeable but is motivated primarily by love for the subject rather than professional advancement. His book *Oaxaca Journal* (2002) chronicled his travels in Mexico with a group of amateur fern enthusiasts.

† James Purdon Martin, a neurologist specializing in movement disorders.

To Colin Haycraft

SEPTEMBER 15, 1978
[NEW YORK]

Dear Colin,

[. . .] I will be coming to London on October 12 and staying for 2–3 weeks (my old man is having some surgery—not intrinsically serious, but he is old, somewhat ailing, and not the best surgical risk). I expect to bring with me a completed manuscript of the Leg book—at least, I pray to God, now, that I can go on to complete it, with no more inhibitions, or *gross* errors of propriety and tone: hold the whole thing in the right mood and perspective, and I will look forward to seeing lots of you then. *Specifically,* something I'd love to go over with you is a fascinating notebook (*De Motu Locali Animalium*) which Harvey* put together in 1627 (the year before he published *De Motu Cordis*). The MS was "lost" for three hundred years (in the British Museum, or some such place), but published, belatedly, at the *tercentenary* of his death (1959). I found reading it, even in the obviously paraphrased and sometimes misleading translation, an astonishing experience—I cannot tell you what it meant to me, but something of the feeling will be conveyed if I say that there was such an intimacy and affinity of feeling, that I felt almost as if this half-written, half-conceived masterpiece of 1627 had been waiting, three and a half centuries, for *me* to finish it! But I would very much like to go over the original with you, to *mull over* its meaning(s)—and I hope you may not be too busy to spare an hour or two on this. *You* might find it interesting as well—he is so Aristotelian, in a way; but also, obviously, writing in reaction to Galileo, and, as such, dwelling on *living* motion, "inner" motion, generating motion, joyful motion; as opposed to the outer, "impressed" motion of Galilean projectiles and bodies. And, again, he is so clearly a physician—and experimenter. And, again, an artist who is intensely sensitive of the *art* of movement: the last part is a wondering wonderful mass of images and metaphors dealing, especially, with the *music(ality) of animal motion,* once all its mechanical (and "cybernetic") aspects have been discussed, quite incredibly, considering the

* William Harvey, best known for his delineation of the circulation of blood in the human body.

time; then Harvey, says, This is not enough. It is not just a procedure or technique. Not just a process, or series of actions. It has more than Gravity—it has Grace. It has the lightness and beauty and wholeness of Art. And especially that Art which is the closest to life, and not dependent on Images, "Meaning," Representations—namely Music. This is *so* close to my own feelings and approach to the theme, that I have the book as a most *dear* companion.

Interesting too that Harvey could never complete or publish it, even though he lived another thirty-plus years, and retained his extraordinary powers to the end. It is perhaps essential in the *subject* that he (one) *could* discover and articulate the Circulation, a closed Circuit; but not Animal Motion, which, whatever circuits and nerve-circles it *contains,* is ultimately open, and "inspired," and stamped by grace—which is not the case with an automatic, largely-mechanical motion like the heart's.

Talking about Latin, I find in *Chambers Common Quotations* at the end: "*Solvitur ambulando:* the problem of reality of motion is solved by walking, by actual performance." This could almost be the motto of my book—I have felt this, again, quite overwhelmingly this week, where the problems involved in the nature of motion, and ambulation, when I took my first walk on Sept 11, '74 were solved (at least, in many fundamentals) precisely four years later—by a recollection of that first walk as vivid as the original: in general it is only by such re-plays, "playing" the experience over and over again, that I can and do solve problems. I *must* be faithful to the original experience and, if necessary, *repeat* it, as nearly as I can, again and again—otherwise I am apt to get lost in abstraction. Can you tell me *where* this tag comes from—who said it, and in what context, etc.?

In the same week as I was reading Harvey, about a month ago, I was also reading a book by Babinski*—and a whole lot of Simone Weil†—which would *seem* very heterogeneous and arbitrary, but made a lot of inner sense for me. Babinski was an extremely eminent, world-renowned neurologist, which makes the whole business even curiouser. This was a book stemming from his experiences

* Joseph Babinski (1857–1932), a French-Polish neurologist.

† A French philosopher who died in 1943 at the age of thirty-four, Weil was the author of *The Need for Roots* and *Gravity and Grace*.

in WW I—and is entirely devoted to delineating a "*third* realm of disorder—neither attributable to organic lesions nor to hysteria," what he calls the "*syndrome physiopathique,*" as a not-uncommon, often devastating, but (*provided* it is properly understood and treated) essentially benign neuropsychological syndrome arising from shock and reflex changes in the spinal cord and brain, following peripheral injury on the battlefield. The original library label was pasted in the book. I observed that it had been taken out in 1918 (when it was first published), 1937, 1955, and by myself, 1978—just four times in sixty years! Here is a readable, at times *racy* book, on an extremely important subject, by a world-renowned neurologist, and it seemingly became, from the moment of publication, one of the *least* read books in the world! This is *exactly* what happened with Weir Mitchell's great account in the Civil War (1864), and Leont'ev/Luria in WW II (1944). I mentioned this to you last November, and finding Babinski's book—not easy because there were no references, no *clues,* to lead me to it—has completed my "series" of these bizarre "overlookings." Perhaps there are others, from other wards: but it seems to need *an extraordinary situation*—with large numbers of emergencies and wounded people, and a rare atmosphere of passionate honesty and concern—to break the barrier, and allow some seeing, some insight. But then, as soon as the situation, the War, the atmosphere of rare concern is over, *everything is forgotten—as if it never happened.* I cannot help compare this with the situation in *epidemics,* for example, that of encephalitis lethargica: the huge scale of this, the insights achieved, the sense of pressing and fundamental importance and then, everything forgotten within five years. My discoveries, my detections, are not of anything *new*—but of what is always around, but nearly always overlooked. [. . .]

ADDENDUM Sept 22.

I wrote this letter *last* Friday and perhaps should have posted it then (or, alternatively, have destroyed it, as extravagant). I have felt miserable this last week, *undone,* and have been quite unable to write anything (and want to undo everything—letters, book-writing, etc. previously done, a temptation or compulsion I will do my best to resist). What has shaken me has been one real calamity (which perhaps has had neurotic resonances), and one nasty neurotic mischief (of a sort to which I am particularly vulnerable). The real sadness is over a cousin of mine here, about my own age, who is very dear to me*—an immensely gifted and articulate woman who had a stroke last Friday night, and is now completely mute and aphasic.† Absolutely clear and lucid in all other ways, but completely without any language or speech. It is too early to say whether there will be recovery, or how much. She had a sort of forewarning, quite explicit, and said to me a couple of weeks ago, "I'm headed for a stroke. These headaches are *real*—there's something going on in my head," and is now so frantic, and anguished, and angry, that she yells all day, like a damned soul in Hell. I think there is *nothing* more unbearable than aphasia for someone greatly gifted in language, who lives in a language-world. Perhaps the neurotic element is the sense of *identification* with her (so I keep having nightmares that *I* am aphasic, the terrible sort of identification which happens with family relationships, and though she was only a cousin, she was very close indeed—in effect, my "family" in New York), and an awful feeling of *accusation* (which, quite unfairly, gets inflected inwards), because I feel that her headaches, which *were* real, all too real, and due (one now sees) to the stretching of blood vessels in a cerebral angioma, were dismissed as "nothing" by her doctors, her neurologists. So the feeling that this whole thing is partly the result of *neglect,* and might have been avoided by proper attention, the feeling of the "bad" negligent, mindless-heartless doctor, and the feeling of the wretched, hopeless, patient "victim" *both* enter and alternate in my mind, even though I am *not* her doctor, and even though I went to much trouble to try and get the best neurologists for her as soon as she mentioned the headaches to me. But by a series of mishaps, the "good" doctors were all away, or had left New York, and she fell into the hands of some "bad" ones.

I suppose it is just this theme, of feeling-falling helplessly or

* Madeline Gardner, who had also had a stroke six or seven years earlier, again regained a good deal of speech afterward, this time with the help of a speech therapist.

† There are two main forms of aphasia: receptive aphasia, in which people are unable to understand the speech of others, and expressive aphasia, in which they are unable to express themselves (if both forms are present, the disorder is termed "global aphasia"). Madeline had expressive aphasia but learned, over the years, how to communicate effectively, with the patience and help of those around her.

neurotically into the hands of "bad" doctors,* which has clouded my feelings about the Leg book and experience: the feeling which *must* be transcended, or distanced, if I am to write a good and graceful-grateful-gracious book, and not a silly-hateful-vicious one. But I have been unable to write anything for the last week, because the hateful feelings have been re-activated, partly by what has happened with Madeline. [. . .]

To Robert Katzman

NEUROLOGIST†

OCTOBER 9, 1978
[NO ADDRESS GIVEN]

Dear Bob,

In view of the strike last week I saw [Mrs. M., a patient, and her husband] at length here—in my place.

I feel I should see her briefly, frequently, and regularly—say weekly: briefly, because as her central functions fade, they become more and more labile and fatiguable, and as I saw very clearly, her state changes radically, even catastrophically, with fatigue and stress: there is the most *enormous* difference between her "fresh" and "fatigued" (perhaps there is some chemical basis for this); frequently, because she is losing ground so rapidly: she was markedly worse, almost frighteningly so, since I had gone over her at the end of July and the *march* of the disease, in her case, is so rapid that it is detectable almost every week; and regularly, because only with this can the requisite support and centreing be provided—if, indeed, it can be provided at all.

The phrase "losing one's mind" takes on a fearful and literal significance here—as the mental machinery falls apart, fades, dissolves,

* OS often felt that the reason it took him the better part of a decade to finish *A Leg to Stand On* was his ambivalence and anger at his surgeons, and his inability to articulate that.

† Katzman, a pioneer in the study of Alzheimer's, published an editorial in 1976 suggesting that Alzheimer's was the most common form of dementia. Until then, most physicians were still using the diagnosis of "organic brain syndrome," "dementia," or simply "senility" to describe such patients.

before one's eyes. It comes to me [. . .] that I may, so to speak, have to make a study of dying, a meticulous investigation of the dissolution of every cerebromental function, a most relentless and rapid disintegration. I have the sense that one could *learn* an enormous amount this way—and, so to speak, in a most *accelerated* way, seeing the immense elaboration of cognitive and mental functions reduced from intactness to non-sentience in little more than a year—as I think happens in the more rapid and malignant forms of "Alzheimer's."

As to what can be done by way of maintaining, restoring, substituting, or strengthening such rapidly-dissolving functions and functional systems, I am not too sanguine. Even in the last six weeks, the loss of abstract thinking, of logic, and language, and "quasi-spatial" schemata has been so great in this patient that I am not sure that *any* "algorithmic" approach would hold up. By contrast, her "*concrete*" powers and judgements—getting the "feel" of a situation or picture, its "mood," its "meaning," its intrinsic-expressive significance as opposed to the deciphering of its details and signs, are very well preserved. It seems to me particularly important to observe what *sorts* of things she can recognize and "hold," and what sorts of things she is so rapidly losing; and *how* a sort of intuitive-aesthetic-feeling cognition is preserved, while her analytic-abstract (re)cognitions are so rapidly eroded.

The gentlest *attempts* to examine abstract functions—not really a formal "testing" at all—tended to provoke or bring out immediate failures, which she felt as *mortifying*, and which provoked extreme anguish and anxiety. At the end of July, she was able to put the numbers—or some numbers—in a clock,* although their alignment was poor: on Friday, with the clock in front of her, she made *smidges* instead of figures, apparently not realizing what she was doing. When I drew her attention to this, she was profoundly shocked, mortified; she became quite ashen, and trembled, and said, "Am I that far gone?" I couldn't help *sharing* her horror to some degree.

The consciousness of mental failure, loss, dissolution, impotence, of an ever-greater and hopeless sort, seems to me, at least with this

* As mentioned in chapter 1, OS frequently used this classic neurological test of drawing a circle and asking the patient to write the appropriate "clock face" numbers in it.

patient (and her husband) to carry an affect of horror beyond that of any "physical" illness. The sense of humiliation, of degradation, is much greater—especially as this was a woman of superior abilities, who prided herself on her wit, her judgement, her good memory, etc., and who was outstandingly self-reliant and independent.

I am rambling—I suppose I am not sure of the point. I think I want to say that this may be an extremely grim, and even *ghastly,* sort of study. This is not to say that a minute study of the dying of a mind may not be of the greatest value and interest. But it is a study of something terrible—and there is terror *entailed.* I am not sure how far one can be "objective" and "unconcerned," how far one can separate the cognitive and the affective, any more than the patient can.

For myself, without being "hard," I *have* been watching mental decline and dissolution, in the largely geriatric populations I work with, for the last dozen years—and can "take" the feelings, combining concern with detachment. But this may be very hard on the patients, and even more on their families—and one scarcely knows whether to encourage a sort of "fighting-back," a fighting-to-hold-on despite the certainty of steady losing and defeat—or to encourage a sort of "abdication" and resignation. The *indifference* (and/or denial) which Joshua* shows seems to me atypical. I do *not* think that Alzheimer-patients are (in general) protected by a merciful blindness or oblivion, any more than most MS patients become "euphoric." But perhaps I am over-reacting to one patient, and obviously I need to see at least a dozen more.

Mrs. M. will be seeing Jack Wilder† this week, and he will undertake—I gather—whatever psychiatric support and other measures he feels needed. I have to find some sort of *balance* myself between the coolness of an Investigator and the warmth of a Physician. I hope I can. Her baseline is changing so rapidly that choline, if it works at all, will have to be rather fast and dramatic. [. . .]

Forgive an overlong, and not too coherent letter.

---

* Another patient known to Katzman.

† A psychiatrist.

To Mrs. Miller

PHYSIOTHERAPIST[*]

NOVEMBER 27, 1978

11 CENTRAL PARKWAY, MT. VERNON, NY

Dear Mrs. Miller,

I cannot refrain from dropping you a short note to say what a special pleasure it was seeing you again this month.

It is a curious—and significant—feeling re-encountering, at a different time, and in a quite different capacity, someone who was most significant to one at an earlier time in one's life. There is no doubt—I cannot put my appreciation strongly enough—that you helped (pushed, bullied!) me through a most important time of disability, regression and dependence to independent function—and, no less, independence of spirit again. There is an erectness of spirit which goes with erectness of gait, as an abjection of spirit goes with an abjection of gait. It is this combined physical-moral correspondence which I hope specially to bring out in my book (*A Leg to Stand On*). The experience has *continued* to be of the utmost importance—really, my whole orientation has been changed since the experience: I *listen* to patients as I never did before, try to enter into their feelings and predicaments, and feel Rehabilitation is the most glorious thing in the world. I have little sympathy or contact with my colleagues, bye and large, but have found great pleasure (and spectacular results!) from working closely in these last four years with all my colleagues in Rehabilitation: physiotherapists, occupational therapists, speech therapists, music therapists, dance therapists, drama therapists—you name it. I think that you—all of you—have a depth of understanding about what the patient goes through and *must* go through, which (on the whole) is denied to physicians (except some rare ones, like Mr. Scott, who had himself been "through it"). One is never the same after one has been a patient oneself. It teaches one as nothing else can: at least if one *lets* it.

As I mentioned, I have seen dozens, scores, literally *hundreds* of patients since with various problems of limb "neglect" and dis/mis-

* Miller had been OS's physical therapist at the convalescent home in London where he stayed following his leg injury in 1974.

use, sometimes on central, sometimes on peripheral bases, as well as scores of amputees. I could not have *begun* to understand their problems, and the challenges of Rehabilitating them, had it not been for my own experience—and the crucial work with you. [. . .]

I hope you can visit New York when you are here—if not professionally, then just on a visit. I would love to take you round some of the patients I see—and introduce you to some of the fine therapists and "physiatrists" I know and work with here. So, a letter of encouragement—and, again I meant to write two lines, and have written two pages! Brevity has never been a quality of mine! Do write, please, at your convenience, and let me know what is happening with you. [. . .] Again, my deepest thanks for all you did for and with me in '74.

With kindest regards,
Oliver Sacks, M.D.

To F. Robert Rodman

DECEMBER 19, 1978
11 CENTRAL PARKWAY, MT. VERNON, NY

Dear Bob,

[. . .] I shouldn't have started to write—perhaps I am not in a (letter) writing mood: I feel rather ill and sickish and distracted and distressed, especially after a phone call from my father this evening [. . .] saying that my aunt* is back in hospital, and sinking. She never recovered after *her* operation, which was a terrible evisceration almost—which would have taxed a youthful constitution, and was murder, in a sense, in a frail body of 87, and, damnably, was a consequence of the hospital strike, which postponed her surgery (and that of countless thousand others) at the time when it was elective and proper, [until] she had to have a partial gastrectomy, splenectomy and hemicolectomy, evisceration. I suppose I am mourning, in advance: I had concealed from myself the seriousness of all this; the *feelings* broke out,

* OS's beloved Aunt Len.

before I got the news of her readmission to hospital, when I heard the "Lamentations" of Zelenka last week—and perhaps some (all?) of what I said about Lamenting and Resenting had more reference to *this* (and *my*) "situation"—rather than yours. What could be more different than losing an aged aunt—and a young woman whom one loves?* Between the going-away of Death, and the going-away of Abandonment. And yet, they are united (surely) in the dynamics, the universals, of Love and Loss—of Grief and Grievance, in an elevated and noble Mourning, *and* a lowering and "ignoble" ranting "How *can* she leave me like this? How *can* she go away? How *dare* she! *I shan't allow it!!*" (etc.) All what you say about that enraged, outraged, sense of omnipotence.

In a very important, a *decisive* way, I think my aunt was sanity-saving, life-saving, during the deathliest period of my life, between the ages of six and nine, when I was "sent away" by my parents, into a prison-like school. The affect and symbolism of this "abandonment," and this "imprisonment," have dominated much of my (unconscious) life, perhaps in a way both creative and destructive. At this time, when I must have approached, if not entered, an almost-autistic state, my aunt was the only one who unmistakably cared, and whose love was practical, and embedded in action: in particular, her opening to me, during the holidays, *her* school†—a remarkable one which she had founded and built, from first to last; and which was centred around the Symbols and realities of "Home," "Openness" (it lay in a forest, open to the forest in all directions, and yet not exposed, but protected and cosy; but with great windows which showed vistas in every direction, and doors that gave free ingress and egress). Do you know that lovely definition of "Home" which Auden quotes: "Home, as you may or may not know, is the only place where you go out and in. There are places you can go into, and places you can go out of; but the one place, if you do but find it, where you may go out and in both, is home." Home, Openness-Spaciousness—and *Enclosure* (so close to,

* A reference to Rodman's wife, Maria, who had died in 1974.

† Helena Landau did not actually found the Jewish Fresh Air Home and School for Delicate Children, but she was its head of school from its founding in 1921 to her retirement in 1959.

yet absolutely different from, Closed-in-ness or Shut-up-ness). The enclosures were little *gardens*—a little square given fully and freely to each pupil, in which he could grow and culture and cultivate what he wished (always, of course, within the limits of Nature and Necessity, and a domain of limited area). These little gardens were playthings, play-grounds, meeting-grounds, and—potentially—works of art, where infinitudes could be explored within the finitude of a few square feet. [. . .] What you say about "that overlapping area between oneself and others, the location of Cultural Experience," brings poignantly to my mind the Garden-of-gardens* in Delamere: where we were, at once, individuals yet together, not fused, not crowded, but in that wonder-realm, play-realm,† which can accommodate us all, commodiousness enough for every creature in the Universe, and yet so intimate that in it they become *fellow* creatures. [. . .]

I cannot think (acknowledge!) that I ever spoke of "poetic symmetry." It doesn't *sound* like my sort of phrase!! For myself, in some sense, I *live* for—for surprises, for constant disturbing newness—and renewal. My life is unfinished and untidy and asymmetrical in the extreme! [. . .] *I think symmetry is Death*—whatever its beauty: it is mathematical beauty, conceptual beauty, the beauty of snow-crystals; the beauty of the lifeless, deathless, *inorganic*. [. . .] I think Life is *essentially* asymmetrical and "odd"—and at the molecular no less than the "poetic" level. Is it not of great moment, here, that Pasteur, that most poetic of biologists, found all life-molecules to be asymmetrical? Not a nice racemic balance—not a symmetry of right *and* left. It is this which sets aside *bio*chemistry from inorganic. I felt this as a sort of revelation when I found it: this "perverse" triumphant *asymmetry* in Nature, life insisting on asymmetry from the start, asymmetry a prime and *prius* of life: as, much earlier, in childhood, I had a special *personal* feeling for numbers, but—prime numbers, irreducible, individual, "asymmetric." I thought of prime numbers as *the first individuals,* and, in a strange metaphor (though metaphor is the wrong

* OS, ever the etymologist, was fond of pointing out that the word "paradise" originally meant a protected walled garden.

† D. W. Winnicott (1896–1971)—an influential psychologist as well as a pediatrician and psychoanalyst—was deeply concerned with the role of play in child development, and here OS, knowing that Rodman was a Winnicott scholar, is invoking that work.

word), I used to imagine *a garden of primes,* also a Community and Order of Primes. Sorry I am being superfluous and prolix, but I was stung by your calling me (not that you did exactly) an Apostle of Symmetry.

Lamenting, Resenting—Love and Loss—Past and Present. Another personal resonance (what could be so different, and yet there may be something in common) has been a sense of Loss (and Estrangement) from my friend Jonathan Miller, whom I have known and loved for thirty-three years. I hope this is only partial and temporary—but, it is important that I recognize a certain reality (I could not bring it to consciousness, thought I was *beset* by it, until Shengold "delivered" it). The sadness and anger has especially to do with his current film-and-book, *The Body in Question.* He came to New York for a week and more, and talked with and filmed many of my patients; and we had countless discussions etc. on the issues he raises. I find these patients, some of them, in the film and some of the discussions; and clinical stories I told him in the book, and yet never a breath of acknowledgement anywhere (though acknowledgement is made to many others). I shared freely and unthinkingly—it never occurred to me to do otherwise: it is what I have *always* done in relation to him—what one *does* do in the Play-Space, the Love-Space, of friends. It had never occurred to me that in any sense he might see me as a rival, and ambivalently *take,* instead of "receiving" (although I felt something lacking in the filming in New York—a certain lack of personal access; a lack or withdrawal of his usual generosity and grace—that circle of the three Graces, giving-receiving-returning). For myself, I *love* to acknowledge, to say "Thank You." It never occurs to me to—to *appropriate:* I lack (for better or worse) any sense of "property"—at least in the love-realms of Thought, Art, and Science. Indeed it is so precisely *because* I feel these realms are so free to all, so free of proprietorial or territorial sense, that they allow and form for me *an exemption from neurosis,* neurosis which so masochistically-unfreely centres round being "owned." Art-Space, Thought-Space, Play-Space, for me, is incompatible with any sense of ownership or claim, and I am *astounded* when I find this is not so for everyone. [. . .]

This incredible business of *capitalism* in Science! [. . .] (To quote wonderful Simone Weil, who as physicist-cum-saint, is peculiarly equipped to speak in these matters): "[. . .] The Spirit of truth can dwell in Science only on condition that the motive prompting the savant is *the love of the object* that forms the stuff of his investigations. That object is the universe in which we live. What can we find to love about it, if it isn't its beauty? . . . The true definition of science is this: the study of the beauty of the world."

It is this *love,* this *warmth,* this *beauty,* which is missing—along with the sense of wonder and mystery. Why should Jonathan, who shows and has (in so many ways) *so much* love and warmth and beauty in him, write a book* so loveless, cold, and ugly (well, not ugly, but lacking in beauty, lacking in the sense of himself: depersonalized, abstract). [. . .]

I find the door closed in Jonathan's book; something inaccessible; he denies others entrance, as he denies himself entrance—the same cold, inhibitory, forbidding lack of access, and inner hospitality and generosity, as chilled me (and others) when we met him in June. I hope this is only something passing: that he has not—in this special sense—been "wrecked by success" (the success of the *Image,* in a depthless televisual world). In some sense, I think, he is gifted too much, with too much facility, and perhaps he knew success too early. He was polished and finished, perhaps, at thirty. Whereas you and I—we are very late developers, but we *do* develop, constantly, painfully, stammering and stumbling, hurt by experience, to an ever-deeper reality, searching for new ways to express that reality—and this, this development, this surprise, this incalculable capacity to hurt, be hurt, blunder and grow, is what makes us so alive, so creatively alive, and (as you are experiencing with such intensity now), so *vulnerably* so. But I think *wounds,* if they don't kill, are the true *path* to life—at least, life past the animal, the lyrical, the pastoral level. Not that I despise any of these in the least: and to show I don't, I enclose a postcard of the most peaceful, uncomplex, unconflicted, pastoral scene—it gives only the faintest idea of the wonderful, elegiac steeped light of the original.

My love and best wishes for the New Year, to you *both.*†

Oliver

* *The Body in Question,* a companion book to the thirteen-part television series.

† Rodman had recently remarried.

To Lance Lee

POET, PLAYWRIGHT, AND NOVELIST*

DECEMBER 22, 1978

11 CENTRAL PARKWAY, MT. VERNON, NY

Dear Lance,

(As you have discovered) I am a very erratic letter-writer—capable, equally, of year-long silences and sudden 10,000-word effusions! [. . .]

Essentially, and if nothing else, I want to say how much I was moved and impressed by your review of Thom Gunn (and Ted Hughes)†—and, specifically, how closely your interpretation of Thom's "*poetic*" attitudes and development over the last twenty years corresponds with what I have seen, felt and known of his *personal* life. I don't know whether this is "relevant" or interesting; a creator is to be judged by what he creates, not what he "is"—but there is a special beauty in the event of a close consonance, and this is something which is very important *to me,* because I am less interested in Art-as-such, than Art as an expression of a passionately experienced, investigated, attested, and articulated life: Art as a sort of (auto) biography. [. . .]

I think much of the "violence" etc. [in Gunn's early poetry] was *assumed* (so to speak) for purposes of exploration—I don't imply any pretense or duplicity, of course. Thom used to speak in many voices, and freely spoke and wrote, at this period, of "donning" identities, like clothes—"emblems to recall identity," etc.—the savagery of the imagery in startling contrast to the extreme, sometimes almost excessive, *formality,* altho' I don't think the latter can *entirely* be seen as a defence against being carried-away, ego-dissolution, etc. It was also and equally the exactitude of the word-smith, the linkage of thoughts and words linked and knit as closely as chainmail (cf. "Byrnies").‡ I think I may have sabotaged my own thought here by ending up with "Byrnies"! Anyhow, he spoke in many voices, and some of these had peculiar (perhaps perverse) "romantic" resonances which, as you say, tended to give him a certain, intense, limited "appeal." I have to

* Lee was a neighbor and friend of Bob Rodman's.

† Lee's review, of Gunn's *Jack Straw's Castle* and Hughes's *Gaudete,* was titled "The Roots of Violence" and appeared in the spring 1977 *Chicago Review.*

‡ Byrnies are a type of chain mail shirt that Gunn made the subject of a poem.

say that (I think) there was/is much more to this (his early) erotic-violent "identity" than apparent at a first reading. In particular, the landscape of fairy tales and sagas, especially Norse myth, seem to me to suffuse many of these early poems, and his sensibility at that time. Auden, introducing a selection of G. K. Chesterton, cites G.K. on such "inscapes":

> There is at the back of every artist's mind something like a pattern or a type of architecture. The original quality in any man of imagination is imagery. It is a thing like the landscape of his dreams;* the sort of world he would wish to make or in which he would wish to wander; the strange flora and fauna of his own secret planet; the sort of thing he likes to think about.

Thom liked—likes—to think of Beowulf, the Eddas, etc. He finds their atmosphere particularly congenial. Auden too was brought up on these, and, curiously (for a Jewish boy of good bourgeois antecedents!) my own imagery and desires etc. were largely formed by such Northern (and unclassical and unSemitic) myths and "landscapes" (no doubt, parenthetically, this is why I "chose" Norway to have my nearly-fatal accident in).

Then, by degrees, and yet radically, Thom started to speak in his own voice, and not in ("other") voices. As you poignantly remark, the emergence of such a genuine, "quiet" individuality, and being faithful to it, requires great courage—and may *lose* one many former readers who cared only for the donned identities. His own voice is that of "the Individual"; the other voices were those of *playful identities being employed for "indirect communication"*—or this is my own "interpretation." Obviously I am making an explicit comparison between Kierkegaard's early "polyonomy," the countless partial-personalities, and pseudonyms, and the assumption of his own name, the finding of his full individuality, in his later works—but this may well be a tendentious comparison or projection.

When *Awakenings* came out Thom wrote me a letter which obsessed me; he said that back in 1960–1, when we had last seen a lot of each

---

* OS later used this phrase of Chesterton's, "the landscape of his dreams," as the title of a chapter in *An Anthropologist on Mars.*

other, I had seemed enormously gifted—"gifted with every gift except one, precisely the most important, call it sympathy or humanity, or what you will." And that he had despaired, then, of me and my writing. But [he asked,] "What happened? What was most lacking in 1960–1 is now at the very heart of your sensibility and writing—was it ——, or ——, or ——, or" (Falling in love; drugs; analysis; yr patients etc)?

I thought about this—for months—and said to him, finally, that I didn't know, and/or couldn't say, and could only explore or expose or explain in an autobiography, sometime.

I wondered then, and am certain now, that when Thom saw a profound change in me, and questioned it; it was, in part, because he was making such a passage *himself*—and questioning *it*. And one of the significant "coincidences" of his current creative renewal, after 2 years of total inner as well as outer silence, is this: that he himself is composing an autobiography at this time.

Incidentally, one of the best memories of 1974 I have was spending a day with Thom wandering around Hampstead Heath, having a beer at Jack Straw's Castle, etc.—and finding that this was the *actual* landscape in which we had both grown up as boys, and that we both knew, had "discovered" all sorts of "secret" places and crannies associated with our having used Hampstead Heath, over and above its *literal* delights, as the backdrop for our dreamscapes, and "the strange flora and fauna of our own secret planet."

It was a very singular conjunction of literal and symbolic, past and present, etc., and paradoxical, in a way, that while I had travelled 6000 miles, in 1960, to meet Thom in his "adopted" Californian Home; I only *really* met him, in a sense, 14 years later, on his own (*our* own) indigenous Hampstead Heath, where we had both spent the secret, formative childhood years.

Anyhow, my congratulations on a very fine review—and your dramaturgic activities—and I hope to *see* you, again, in 3–4 weeks, when I expect to be in LA.

My very best wishes for the coming New Year,
Best,
Oliver

To Helena Landau

DECEMBER 30, 1978
[NO ADDRESS GIVEN]

Dearest Len,

We have all of us been hoping so intensely that this month would see your return to health; but, alas! this was not to be.

My heart is torn when I hear of your weakness, your misery—and, now, your longing to die.

You, who have always loved life, and been such a source of strength and life to so many, can face death, even choose it, with serenity and courage, mixed, of course, with the grief of all passing. We, I, can much less bear the thought of losing you. You have been as dear to me as anyone in this world.

I shall hope against hope that you may weather this misery, and be restored again to the joy of full living. But if this is not to be, I must thank you—thank you, once again, and for the last time, for living—for being *you*—and bid you a final farewell in this life.

God bless you; And if your time has come, receive you into the Long Home where there is no more parting.

I am, dearest Len,
Most lovingly, most sadly,
Your nephew, 'Bol'*

P.S. You know that I can be with you, in a few hours, if you wish—you have only to give the word. In the meantime, you are continually in my thoughts and my prayers—
Jan 1/79: I phoned Hospital this morning—and though you were too weak to come to the phone, Staff Nurse told me you seemed to be rallying—the happiest news for the start of New Year. *Do* get well, and live—*fight* to get well, you can still have much good life before you yet.

I dearly hope that this New Year sees a renewal of your energies, bodily and spiritual, a leaving-behind-you the deathly crisis of last year.

* Bol was Lennie's pet name for OS.

So my best wishes for this New Year—may it see you healthy and happy once more![*]

To F. Robert Rodman

JANUARY 5, 1979

11 CENTRAL PARKWAY, MT. VERNON, NY

Dear Bob,

I received your good letter with much pleasure. [. . .]

Your good letter made me feel what a *bad* letter I wrote you—I mean, bad-tempered, bad-mannered, and (I fear) bad-willed, and I will make this short in case any further badness comes out! It is a very bad time, this, with my aunt not exactly dying, but *hovering around death,* one which raises the (inexcusable) thoughts: "Why is she taking so long about it? Why can't she get better—or die—*something,* NOW?" Thoughts she has been having herself, and which have caused her to turn towards my father—us—with piteous pleas that she be put out of her misery: bad time; *dangerous* time; I feel savage thoughts rising, and must be careful not to act on them. Not to project them. Not to express them. Looking back through what I wrote, I feel that whatever truth may have been contained, there was a *vindictive* note which was uncalled-for (doubtless imported from somewhere different): and I am ashamed of it. Fortunately, I have not expressed any vindictiveness to Jonathan himself, and perhaps I passed on a few toxic thoughts to you because I feel you have a large capacity to *detoxify* them.

Again (and again and again and again!) I have to remember that others are—others. I have great difficulty, sometimes, distinguishing them from myself: or "projecting" myself into them, without knowing that I'm doing it. Jonathan is a man of extraordinary gifts—of heart as well as mind—but he is *extremely* different from myself (even though we sometimes enter the same play-area). I keep forgetting, or denying, this difference. I have no right to reproach him

* Lennie died on January 22, 1979.

for—being himself, with all the virtues (and defects) of his brilliance. I am not brilliant myself, I don't sparkle, glitter, I lack facets, but I have something different—a deep, concentrated, original force—and I must hold onto *this,* and remember it, and honour it, and express it, and not concern myself too much with what anyone else does, unless they also have an original force, whose workings and products I can admire . . . at a distance. [. . .]

Love,
Oliver

PS I am forgetting! I have been *delighted* by the Kierkegaard you sent me—many, many thanks for a most perfect and thoughtful gift. It found its way instantly to the very select few books I keep on my bedside table and established the illusion that it had *always* been there! It instantly became as familiar as an old friend, and I had no sense that I had *just* received it! Thank you again—

To F. Robert Rodman

AUGUST 18, 1979
11 CENTRAL PARKWAY, MT. VERNON, NY

Dear Bob,

It was very good, as always, talking to you last week—I am sorry that I let *months* elapse, and appreciated your occasional kindly calls in the interval, even though I didn't respond to them. [. . .]

I thought of you last week, on waking from a fascinating dream (which bore on some things you had said about your running, and I about my swimming). I dreamt I was on the Moon—it was not clear how or why, nor did it seem to matter!—and taking my first Moon-Walk. At first, with the unprecedentedly altered gravity, and thus the unsuitability of my own gravity-sensors etc., I could not calibrate my own movements, alternated between absurd, wild overshootings and underestimates, etc., and had to take extraordinary pains to count, calculate, cautiously, consciously, to work out my own trajectories, as if I were a missile, or a lesson in ballistics—a tremendously difficult (and somehow absurd) trigonometrical exercise applied to "myself" (or, rather, this body which had become so altered, almost alienated, because of the alteration of gravity, and the inappropriacy of its customary, gravity-oriented responses).* But—by degrees—I acquired skill: indeed became *quite* skillful, without, however, being able for a moment to relax my conscious cautious calculating and counting; thus I acquired competence, and skill, in the total absence of Ease or Grace, and it all felt very artificial and unnatural to me, and dichotomizing into a ballistic calculator (my "mind") and a ballistic object (my "body"). And then, *all of a sudden,* I got the hang, I got the feel, of a wonderful new motion, a most free and joyous sort of leaping-walking, body and spirit completely joined in a most free and easy, natural-graceful motion. I said to myself in the dream, "Ballistics has become Ballet, Metrication Music," and then I woke up—and wrote it all down!

"Wrote it all down!" Ah, if only I could!! So many thoughts, teeming, resolving, ceaseless, I don't know whether they are a joy or torment: rather they are joy to *entertain,* and a torment when they cannot be given outer, manifest, *public* form. It is joy to fill my "Notebooks," write notes on patients, write letters to friends—and this is a million-and-a-half words a year—but somehow, or *sometimes,* an evasion of the one thing I *must* do. The Book of Job has almost everything in it, including precisely *this* feeling, when Job cries out:

O that my words were now written!
O that they were printed in a book!
That they were graven with an iron pen and lead
In the rock for ever!

These feelings, always with me, have reached a sort of climax today, because I am going away tomorrow, to the North once again, with its dangerous allures—in the half-desperate hopes of writing my maddening leg-book, so intolerably inhibited and postponed. The

* This dream has obvious resonances to OS's leg-injury experiences and learning how to walk again. But it also evokes his earlier descriptions of parkinsonian patients, as well as a patient he had recently seen and would write about in "The Disembodied Lady."

absolute need to *write it out,* if only so that I do not, once again, *act it out,* and I cannot help some fears and inner quakings, when I think on my last journey North in August '74.

So—I have packed the trunk of the car with typewriters (3) and fishing rods, and will leave for Canada tomorrow. It doesn't have to be Canada—there are plenty of nice, restful lakes upstate in New York—but I associate Canada with a sense of freedom, and "home" (without the special dangers of England, at the moment). I hope the febrile intensity of feeling goes down, because I have a sense of some *danger* in this mood—and continually overlook the angry and destructive impulses down below, which I can scarcely feel—until they suddenly *erupt.* It is a funny *mixed* mood, both creative and destructive; and I must be careful not to *push* it too much—I am very exhausted, bone-weary, inside—and I am not under sentence of death if I don't write now (although I tend to cultivate this feeling). Perhaps I can, or should, just "relax," as they say, and have 2–3 weeks, if I can, of vacation from outer (and especially inner) obligations. It is the *obligations* that are dangerous—and the *freedom* that I need.

Some mixed feelings of needing to take care, and make provision, etc. led me to write a long letter to my lawyer, last night (I was very restless, too restless to sleep), in lieu of a Will—or, in effect, a "literary" Will. It dwelt especially on the enormous, the *bizarre* volume of unpublished but perhaps, occasionally, publishable writing; my enormous but "secret" (or secretive) productivity, which has filled the apartment with (fearfully destructible and combustible) Notebooks and manuscripts. Anyhow, I indicated that in the event of something unfortunate happening, that I wished my Assets, or Estate, such as it is (it is very little) to be used on behalf of this "literary" legacy—for it is all I leave, and I have no dependents. [. . .] I feel somewhat easier in my mind having written it—but it should not be seen as expressive of a death-seeking impulse (I hope!), so much as a sort of *protective* measure (both internally and externally).

Tricky time: I miss Shengold's presence—perhaps I have become too dependent on him. Perhaps I am writing to you, in part, in lieu of him; you must use your "third ear" and decide! [. . .] I should be back here the week of Labour Day—perhaps around the 5th or 6th—though I *might* stay longer if I fall into a writing streak, and things are doing particularly well. I *hope* I can bring back a beautiful book;

failing this, possibly a fish—or a moose; and, I pray, not a broken leg or neck. It is possible, of course, that it will all be uneventful—even to boredom. What I want is that it should be uneventful, but *quiet;* a good, peaceful, living quiet

"The quietude
At the centre of the movement"[*]

I will look forward to seeing you when you come to New York in October. And, please, look after *yourself*—I know this year has been difficult and distressing for you too.

Love,
OLIVER

To Jonathan Cole

OCTOBER 9, 1979
11 CENTRAL PARKWAY, MT. VERNON, NY

Dear Jonathan,

I'm sorry I've been such a lousy letter-writer myself—I always enjoy yours, which are full of life and interest, when they come, and the sense of good expansion (and crises!) they show.

I thought of you especially today, because I saw Sister (now Mother) Genevieve,[†] whom I had not seen since she left New York, and she asked about you with great affection, and asked to be remembered to you.

Your [. . .] job at the Whittington in geriatrics and neurology sounds ideal—I mean, for someone like yourself who will *bring*

* From Octavio Paz's poem "Solo for Two Voices."

† Sister Genevieve was a member of the Little Sisters of the Poor, a religious order founded in nineteenth-century France. The nuns serve the elderly poor by running residences around the world, and many of them rotate to different residences every few years. Thus, Sister Genevieve had left her post in one of the New York residences to become a Mother Superior elsewhere. OS worked at several LSP homes in the New York area for almost forty years, and he kept in touch with many of the Sisters as they moved about the country.

appreciation and concern to it. To someone *un*giving, it would probably be a thundering bore. Clever and skilful doctors are common (and, so to speak, cheap): what is lovely, and rare, is to see the Three Graces—Giving, Receiving, Returning—*as well as* intelligence, knowledge, talent, skill—and, pre-eminently, this is what I (and clearly many others) see in *you.* [. . .]

I have an increasing, an almost constant, feeling that one is lost without grace—and I use the word in a way which covers everything from the grace of a squirrel to the grace of a saint. The lovely *rightness* of Mother Genevieve, which so moved me when I saw her again yesterday, is because she is guided by grace. (I thought of my favourite title, Simone Weil's *Gravity and Grace*). The obvious *wrongness* of Parkinsonian movement is because it is *not* guided by grace: this came out very strikingly in one of my Parkinsonian patients whom Jonathan Miller interviewed when he was here for some days filming for his series: she said that she was painfully conscious of how "wooden, and mechanical, and robot-like" her walking had become, and how it now lacked grace—and life. But, she added, Nature provided a cure, in that, increasingly, she was apt to have music come to her—music she used to dance to, with a young man she once loved—and when the music came, she recovered her grace. [. . .]

I felt tense and miserable for the first half of the year—which especially had to do with the most cruel and wretched death of my poor aunt (whom you met), after a series of appalling post-operative complications [. . .] it all happened at the time of the damnable Strike, and what should have been fairly minor elective surgery became very major emergency surgery—practically evisceration, which would have taxed someone young and strong, and was virtually murder in an old body of 86. And there were a series of post-operative infections of a sort scarcely seen since the Crimean War—and I could not help associating these with the fact that the Hospital had no sterilizers, and had run out of sterile packs of all kinds. But she had a very good, very long, full life; and, in a way which aunts, uncles, godparents, grandparents can sometimes do—when parents cannot (simply because they are too *close,* and forced to play all the roles which Authority and Dependency force upon one)—she stood for Health, Sanity, Humour, etc. almost for Life, in the darkest days of my childhood—and ever after. [. . .]

A turning point for me was going away to a lovely island in Lake Huron, Manitoulin [. . .]. I have never known a place of such beauty and quietude—and I felt my torn, irritable, nerves ("heart") healing with every minute and hour, partly in the immense *natural* beauty and peace of the place, and partly in the remarkable feeling of *community* I had—of very strong, simple, kindly souls, wisely innocent and incorruptible, a very remarkable island community altogether. Unprecedently, for me, I found myself talking, easily, and deeply, with everyone I met: absorbing, equally, the social and natural history of the place, its ecology and economy, and how it had *survived,* so full and free and spacious and fertile, with all the dangers of isolation and desolation, on the one hand, and of spoilage and corruption, on the other. [. . .]

I had an immense feeling of mildness, forgiveness, the absence of censoriousness, anger, accusation, on the island. I don't know how much was inhibition and introjection from a very rare and lovely place: and how much projection of an eased heart, which, after months of tormenting, had reached some sort of unaccusing, unpersecuting, forgiving *inner* peace. After about ten days on the island, I felt a sudden "quickening" of my book within me, and had written 50,000 words in ten days. I did hope that I might complete it there, but fifteen hours of typing a day started to do my back and cervical roots in, and it became apparent (and has since become more apparent) that—at least in *this* writing—I have a far vaster task on my hands than I imagined. I had thought of a Bagatelle, a Divertimento, of about 30,000 words (*Migraine* and *Awakenings* are about 70,000 each), but am instead facing a Symphony of about 250,000 (a single day—admittedly a crucial one—in the Hospital, occupies 70,000 words, and may be the longest day in literature since *Ulysses*!). So I find myself, willy-nilly, not intending it, in a sort of *Marathon,* and despite a reasonably steady 4-5000 words a day, may find myself writing till Xmas—and *then* have my publishers (or public) say "Yes, it's marvellous, each page is full of interest—only you can't expect us to read three *thousand* pages!" It has extended itself, you will gather, very much beyond a case-history, into a sort of novelistic autobiography, and—and—

"Expand not into boundless desires and designs" says Browne.

"I thought all I should have to say on this matter would have been contained in one sheet of paper," writes Locke, in his *Epistle to the Reader,* "but the further I went, the larger prospect I had; new discoveries led me still on, and so it grew insensibly to the bulk it now appears in." He adds, "I will not deny, but possibly it might be reduced to a narrower compass than it is, and that some parts of it might be contracted. But, to confess the truth, I am now too lazy, or too busy, to make it shorter." Anyhow, for myself, I will at present let it take the shape and scale it seems to flow out in, in defiance of all my programmes and plans (because, I suppose, it is *genitum non factum**—or, better, *factum* cum *genitum*)—and only cast a critical-contracting eye on it when it is done. I hope I can go back to Manitoulin, for it was there that the magic circle was completed, the threads were gathered together, and the needed sense of spaciousness and mildness came upon me—and I would like to set an "Epilogue" there.

I must stop now—perhaps in future I should use airmail forms—otherwise *you* are likely to receive a Treatise rather than a letter. I hope to come to London, MS in hand, sometime later in the Autumn, God willing! My thanks again for your very appreciated and cordial letters—and my regrets at being so erratic (and eccentric) in reply.

Warmest regards,
Oliver

* Begotten, not made.

10

Clinical Tales

1980–1984

To Robert Katzman

JANUARY 14, 1980
11 CENTRAL PARKWAY, MT. VERNON, NY

Dear Bob:

Just today, at one of the Little Sisters [homes], by chance, I had occasion to see one of the not-so-rare sorts of patients I mentioned to you on Friday, with a fairly rapidly-progressive *concurrent* Parkinsonism and dementia. I herewith enclose copies of my notes etc. since I first saw him in September '75 (there are also CAT scans and EEG's—but, of course, no brain biopsy!), and you will see how in little more than four years he has passed from a very minimal syndrome, clinically, to an almost terminal one. It didn't occur to me to *link* the Parkinsonism with the dementia etc.—and perhaps they are not linked. I found his course sufficiently puzzling to send him in to Labe* last year for a diagnostic work-up (he came back, I think, with the usual *double* diagnosis of "Parkinson's disease" and "Alzheimer's disease")—and also to Bronx Lebanon for a work-up back in the winter of '76–77, when he presented a *transient* appearance of what was later to become the continuing picture. Some of these strange and sudden "flare-ups," which I have seen in other such patients, have been (mis)interpreted as "strokes"—but there is neither good clinical nor other evidence that he *has* had strokes. The course has really been continuously and relentlessly downhill, though with the odd *lability* (see, for example, the great change in intellectual function—ability to write—within

* Labe Scheinberg, a neurologist at AECOM.

a *five minute* period in test sheet of Jan 30/79), of a clinical picture which I have seen in so many. He also shows a very characteristic *excessive sensitivity to* (and intolerance of) both antiparkinsonian and tranquillizing *drugs*—which I also think may be characteristic.*

As you will see, I didn't *think* to see his dementia and Parkinsonism as a single clinical/aetiological entity until—today. So, my notes have no bias in this direction! However I could provide similar case-histories, more or less, for every week in the year.

I don't say this *is* an entity, but I do say it is different in various ways from either "senile dementia" or "Parkinson's disease," in that these tend to run slow courses of 15 years or so, and usually without much admixture (at most a *little* dementia in the ordinary Parkinsonians, or a *little* Parkinsonism in the "ordinary" dements). This rapid (2–5 year) course of Parkinsonism-and-dementia conjoined sets apart a sub-group (perhaps several sub-groups) of patients; as does their extreme lability of function, and pharmacologic hypersensitivity. In one such patient, a former dancing teacher around 50, the course from first symptoms to profound and terminal dementia and Parkinsonism was no more than eighteen months (she was among my first L DOPA patients in 1969).

Would you be interested in further work-up/biopsy, on this chap? I *think* the family and Sisters etc. would go along. Alas! As I mentioned to you, autopsies have virtually ceased at Beth Abraham—it happens however that we did have *one* autopsy recently on a man I had followed closely with this dementia-Parkinsonism syndrome, who was full of tangles, etc. and said by neuropath. to be "typical of Alzheimer's."

Oliver

* These careful observations are an early recognition of a rare and malign form of Parkinson's with dementia that is now known as Lewy body disease, formally identified only decades later.

To Robert Katzman

FEBRUARY 5, 1980
11 CENTRAL PARKWAY, MT. VERNON, NY

Dear Bob:

I seem to be inundating you with letters etc.—I get these paroxysms from time to time—but this will be fairly brief, and the last of the quartet.

Obviously some of the *specific* points which arose have forced me to consider more *general* issues—with regard to possible approaches to these patients.

Specifically, this has been in my mind today with regard to the Alzheimer patient (or *couple*) I have been seeing, because it *is* such a family situation. You had originally, I think, envisaged me as an "investigator," and not a "physician," with regard to these patients. And I myself would have *preferred* this to be so, and to avoid any "involvement," distress, etc. But I have *had* to "get involved," or be concerned, willy nilly, because I am dealing with a *person*—and not a "preparation." [. . .]

In the Preface to my *Migraine*, twelve years ago, I wrote "I was . . . delighted at the complexity of the histories I received . . . Every patient . . . opened out, as it were, into an entire encyclopedia of neurology. I was recalled from my neurological preoccupation by the suffering of my patients and their appeals for help." And so it has been with these Alzheimer patients. I (one) *cannot* be just an "investigator" and not a physician: at least *I* cannot; if they were faceless members of a series, having standardized tests of one sort and another, I could; and obviously this sort of quantitative investigation must be done. My own sort of investigation is essentially *qualitative* (this does not make it any less precise—only, more delicate, and complex): a qualitative neuropsychology, with kinships to Luria's; but, again, one which I *cannot* detach from human responsibilities and concerns.

Basically, perhaps, I am "a naturalist"—nice, old-fashioned term—and all my other interests (in psychology, neurology, physiology, etc.) are subordinated to the Naturalist in me. As a child, I read Bates* on

* Henry Walter Bates (1825–1892) wrote about his explorations of the Amazon rainforests with Alfred Russel Wallace.

the Amazon, Speke[*] on the Nile—I was imbued by the spirit of adventure, exploration (of a sort *at once* scientific and romantic). This spirit has transferred itself to my clinical work: I regard every patient as an Amazon, as a Nile—I see *landscapes* of disease, physiognomic, geographic. I do not know in advance what is "relevant"—which makes it difficult or impossible to draw up programmes or projects of work beforehand. I have to ramble, to meander, to explore—not knowing in advance what I may find (though, hopefully, *prepared* and open for finding(s)). Exploring a realm, one cannot confine one's attention to this or that (although, obviously, there *must* be a deep, perhaps unconscious, selectivity). One doesn't know in advance how all the fauna and flora will relate, how everything will fit into an immense "eco-system" (I see I am unconsciously paraphrasing, at least, the title of Gregory Bateson's *Ecology of Disease*[†]—I know the title, but not the book!) [. . .]

One must see (or, at least, be in a position to see) *everything* which is relevant in/to the life of the patient—if one is to be either a naturalist or physician. I do not see how this is possible in the context of a *clinic*—where, so to speak, a patient is torn out of their natural setting, and seen only in a narrow and artificial way. One needs to see the patient in their natural setting—"at home."

I found this very strikingly with that *most* fascinating patient Hans H. (with the object agnosia, etc.—the old musician).[‡] I *could not* make sense of what was going on when I examined him at [the clinic]—also, I *could not* examine him adequately (for example, I needed a tape-recorder and a piano! And I needed to see how he conducted himself in his own surroundings—at home). *Only then* could I begin to comprehend what life might be like, both for the patient, and for everyone else, for a rather gifted man with a most profound visual agnosia. [. . .]

I am sensible of *how poor* my observations of [this patient and

* John Speke (1827–1864) made three expeditions to Africa to search for the source of the Nile and became the first European to see the great African lake now known as Lake Victoria.

† Bateson's (1972) book was called *Steps to an Ecology of Mind.*

‡ He later described Hans H., using the pseudonym "Dr. P.," in the title chapter of *The Man Who Mistook His Wife for a Hat.*

her husband] are (in a way), because my information has been restricted to seeing them in the Clinic, and phone calls from the husband. [. . .] It seems to me that I have learned less of this patient than of somewhat comparable patients whom I have seen in Institutions. [. . .] Because, in a way, *an Institution is like a little world,* and, especially in one like Beth Abraham, where I know every patient, every nurse, every therapist, every current and nuance of life which goes on, I *can* be something of a Naturalist—and Physician: I am able to observe, and (to some extent) influence, a whole world—although (of course) it is a miniature world, a special world, a tragic world, and an *exiled* world, no longer entirely *in* the large world.

It is partly this cut-off quality of institutional life, the little institutional world, this sense (sometimes) of working in a leper-colony, which has made me want to see some patients who are still in the great world; who have not yet been "put away," restricted, confined.

But, I have found, *if* I can generalize from this experience with one patient, that I cannot learn as much in a clinic setting, as in an institutional setting (no doubt Hughlings Jackson found the same, which is why his most valuable observations were made in—Asylums, the West Riding asylum, etc.). On the other hand, if I confine myself to Institutions, I see only the most advanced cases, the sickest, the worst—I have no possibility of seeing what it is like in the early stages, how the person copes, and how their family, friends, everyone, *copes with them.*

I think one can *only* see this if one has some domiciliary practice—if one goes into their Homes, as an observer, at least; and, perhaps, as their family physician—at least the Family Physician *with respect to the Alzheimers.* I think that most essential observations may fail to be made in a clinic, and cannot even *in principle* be made in a Clinic. I think experience of such patients at home, and perhaps in day-hospitals, and various programmes, is absolutely necessary to getting a full picture, to seeing the ecology, and not just the neuropsychology and physiology, of this disease.*

* OS writes in *Uncle Tungsten* of how, as a boy, he used to accompany his father, a general practitioner, on his house calls.

I better stop now—or I will run onto another page. I will keep my promise, and make this the last of the four!

With kind regards,
Oliver

IN NOVEMBER 1980, OS BOUGHT a small house on City Island, an island off the Bronx that is home to a number of boatyards and boatbuilders. There he kept a rowboat, and he could swim or row in the waters of Eastchester Bay or ride his bicycle up and down City Island Avenue. In addition, he had a dining room that would provide space, after his father's death in 1990, for the Bechstein grand piano from Mapesbury Road. He hired a handyman to build bookshelves in every room, including the kitchen and basement.

To Stanley Fahn

NEUROLOGIST

APRIL 14, 1981
119 HORTON ST., CITY ISLAND, NY

Dear Stanley,

[...] I was going to phone you—for help—last week, in regard to a patient at the Little Sisters Home ("Holy Family") in Brooklyn, but she seems to have settled a little. An intelligent (perhaps psychotic) old lady with the most tremendous Parkinsonian "drive" I have ever seen—at least, in a non-DOPA, non-post-encephalitic patient. She was *constantly* running—if not into walls, then into exhaustion, so much so that I feared for her life. She could not sit—every two seconds or so one felt this sudden surge or urge take hold of her. DOPA/sinemet seemed to make her worse, and haldol stupefied her . . . I had hoped St. Barnabas which is my first port-of-call, whenever I run into neurological problems at the Little Sisters, could help—but she was no better when she returned from them. Mercifully, we have found she has a great passion, and a great flair, for *sewing*—she sews with incredible speed, and perfect precision, and when she is sewing she is free from the blind, mad force of the subcortex: the moment she stops sewing, she reverts into violent, almost suicidal, Parkinsonism. At present, we are tearing up everything in the House, so that she can sew it together again! However, this may be only a temporary measure: and I may have to send her to you—her name is Rose T. [. . .]

With best personal regards,
Oliver

To A. K. Brenner

CORRESPONDENT

APRIL 28, 1981
[NO ADDRESS GIVEN]

Dear Ms. Brenner:

Thank you for your interesting letter—and your kind comments about my work, in the brief glimpse you saw of it on the Dick Cavett Show.* You must *read* "Awakenings"—and, equally, see the film† (which has often been shown on public television in England, alas! never here).

I mention the film particularly because this is (amongst much else) *about* the unique power of music and dance in these patients—we are, indeed, one of the few such facilities in the State to have a full-time music-and-dance therapist.‡

I regard all living movement, all "life," as dance—of a sort (in the universal sense of Havelock Ellis' book *The Dance of Life*); and, specifically and practically, I think the *only* way of gaining or regaining propriety and grace of movement, after this has been disturbed by some movement-disorder (as a neurologist, my concern is with

* OS's cousin Carmel Ross, an actress and publicist, had arranged for him to be interviewed for the well-known talk show, and Cavett dedicated two shows to his conversations with OS. This considerably raised OS's profile in the United States, and he received a number of letters from viewers.

† The 1974 documentary.

‡ At Beth Abraham Hospital.

disorders on an "organic" basis), is through dance. My conviction of this was confirmed by a personal experience—the leg injury I spoke of briefly in the (second) show, which deprived me for three weeks of the ability, and even the *idea,* of walking. That I did walk again, and recover my motility, I owe no less to music (it happened to be Mendelssohn!) than to neurological recovery.

I do not think of dance as an "additional" treatment-tool, but an essential, and quite central, one! [. . .]

So, you see, I am entirely in sympathy with your ends and means; and do what I can to promote these in the tragic but inspiring world of Asylums, Institutions, Chronic Hospitals, etc. where I spend my life working among the incorrigibly damaged.

With my best wishes,
Oliver Sacks, M.D.

To Vanya Franck

ACTRESS

MAY 10, 1981
119 HORTON ST., CITY ISLAND, NY

Dear Ms. Franck:

Thank you very much for your kind and most interesting letter—I was quite inundated by letters after the Dick Cavett show, and can only answer some of them—belatedly, and inadequately. [. . .]

I am sure, as you say, that one of the special distresses of Parkinsonians is the frustration of normal emotional expressions—even the basic outlets of laughter and tears (not that they *lack* deep emotions; it is specifically their *expression* which is blocked). And certainly laughter and tears (for them, as for everybody) when they are possible, discharge countless tensions. I am entirely for any, every, form of therapy which *releases*—I was talking about this the other day to a "clown-therapist." I see art and play as particularly vital—and *active* (in a way which "support" is not).

The sheer weight of tragedy—both the existential tragedy of a confined and ruined life, and the tragedy of incorrigible neurological disease—is colossal in these patients of mine. What astounds me, finally, is that so many of them *survive*—and as vivid, real, unembittered human beings. They have taught me, above all, what courage really is.

Needless to say, we provide what forms of therapy, counselling, etc. are available to us—but our resources, in all ways, are painfully limited. But I do thank you for the generous impulses which prompted your letter—and yes, it is curious, but there *can* be a sort of intimacy about television—though only if one keeps one's real face, and avoids a sort of public face ("Private faces in public places/are wiser and nicer than/public faces in private places").*

With my thanks and best wishes,
Oliver Sacks, M.D.

To Walter Parkes

FILM PRODUCER†

JULY 14, 1981
[NO ADDRESS GIVEN]

Dear Walter,

[. . .] I feel—perhaps—I put you off, in various ways, without meaning to, or, perhaps, realizing what I was doing. I *forget*, for example, what a shock Beth Abraham must be to an "outsider"—and yet I see this, constantly, in every student or resident or anyone who first comes: they get overwhelmed by horror, disgust, disquiet, despair etc.; and it is only *later*, perhaps after a couple of weeks, that a mysterious something else takes hold: a sense of a strange serenity, beauty, poetry; of humour, of courage—of the most inspiring order. One can see the Sickness, the Misery, the Affliction, straightaway—it takes time to see the mysterious Strength, the immense (and unexpected) reserves of health and Life. Partly for this reason [. . .] I feel I may have been wrong to *show* you Beth Abraham (especially in its present deplorable

* A quote from Auden's work *The Orators: An English Study.*

† Parkes, along with his coproducer Larry Lasker, were interested in making a feature film based on *Awakenings*, and in 1981, OS took him to meet some of the postencephalitic patients at Beth Abraham.

state—so different from how it used to be, at the time *Awakenings* was acted-out and written). [. . .]

One can pay a visit to Beth Abraham and see it as—purely awful. But (even now, in its present state) it does constitute an entire *World,* where passionate (and often comic) dramas are played out. A World—and an Underworld—and all that I say in *Awakenings,* and more. [. . .]

I *hope* I haven't "put you off"; on the other hand, it will have done no harm to show you a genuine complexity—one that will have to be purified and simplified, of course, by imaginative and romantic vision (as it was, already, in the book of *Awakenings*). [. . .]

Which brings me, finally, to the start of my letter! The salient matter, the only one that matters, is your own sensibility, and your generous interest in my work and "vision"—and the sort of story (stories) I have tried to relate. That you should *wish* to present, or re-present, this story, these stories, is a great compliment, and a happiness, to me.

The *real* horror is lostness, oblivion; the real *need* to remember and represent—and compared to this, everything else shrinks into insignificance.

I felt this, quite overwhelmingly, when Duncan Dallas, who made the [documentary] film of *Awakenings,* revisited Beth Abraham in 1977—3½ years after he had made the film there. He was most movingly welcomed by the patients who were still living—they all remembered him, and the whole experience, with gratitude and affection. Duncan said then: "They seem to be de-differentiating, a lot of them going back into the Limbo from which they came." In the moment that he said this, I felt it too, and felt overwhelmingly (and unreservedly) grateful and glad that I had written *Awakenings,* and that we had made a film too—that we had *rescued,* from Oblivion, a remarkable, almost fabulous, and yet *real* situation. It came to me—but this, indeed, is precisely what the patients themselves said, when they saw me obsessed with hesitation—that if I had *not* written *Awakenings,* permitted the film, etc. *their lives would never be known*—and, to this extent, with their help, I brought them into "history," and made commemorations of, memorials to, their strange, unique lives.

By the same token, I am *elementally* glad and grateful for *your own* interest, because I see this as yet another way of celebrating life—and by remembering and celebrating an extraordinary life, the collision (or coalition) of a great soul in a wrecked body, so akin to the theme you are drawn to in "The Genius."*

Such *reservations* as I have are relatively minor—relative, that is, to the salient fact of your creative interest in *Awakenings.* One fears vulgarization, sensationalism, distortion, etc.—always; but it was very clear from meeting you that your own integrity and sensibility, should you embark on the matter, would prevent any gross distortions of this sort. Of course, as you were the first to say, others (directors, etc.) might overrule you.

There are any number of "authentic" renderings—all of which could present an essential "truth" (even if they departed rather widely from "the facts"). After ruminating on the matter for some weeks since we met, I doubt if I am as "obsessed" or as vulnerable, or as impossibly demanding—critical, as, perhaps, seemed the case at first. If things *should* work out well, it would be a delight for us all; if they don't, I doubt if much harm could be done. So, for myself, at this juncture, I want to say that I am much more grateful than fearful; I feel *open*—and wonder if this is the case with you too? [. . .] How are you both? How is LA? Is the end of the strike† in sight? Please write or phone.

To Mother Genevieve

LITTLE SISTERS OF THE POOR

NOVEMBER 2, 1981

119 HORTON ST., CITY ISLAND, NY

Dear Mother Genevieve,

I am so sorry I missed you (Sister Winifred looked after me very kindly)—I thought I would drive down by little backroads—and I got well and truly lost on them, so that my journey (though beautiful) lasted seven hours instead of three!

---

* A working title for a film that Parkes and Lasker had written, released in 1983 as *WarGames.*

† Of the Writers Guild of America.

I found it quite heartbreaking to see poor Sr. Geraldine,* so damaged, yet so gallantly herself; I thought I would burst into tears when I saw her, and I wanted to hug her when I left. I disallowed myself—but I am sure she felt my feelings. She showed such concern for others, and the Little Sisters—asked me keenly about Jeanne Jugan Residence in the Bronx, and was delighted when I said that it was finally finding itself, and becoming a real Home. She showed no self pity, or resentment, or protest—not, at least, at the deepest level; but an acceptance (not without humour) of her hard fate.

Sr. Mary Geraldine had said to me (this was ten days earlier) that she could or would not talk, and she seemed "withdrawn and unresponsive"—presumably depressed. This, then, was what I expected. The reality I found was utterly different: she was vividly responsive, and luminously herself—and I could not help feeling that the "unresponsiveness" etc. of the preceding week(s) must have been partly organic, a direct consequence of thalamic and hypothalamic depression, and, as such, lifted with the use of steroids.

This can only be temporary in the face of this relentlessly-advancing pathology—I think one must thank Medicine (and God) that this final remission *has* been granted to her; and that her mother (and all of you) will find her so alive, and real, and present, in order to make her last peace and farewells.

I had the sense, as I think she had, of a last (and valedictory) *reprieve:* a last deep and grateful drinking-in of the world, and all of the goodness and beauties of life, before the final taking-away. She was so gracious, so grateful, for everything when I saw her: I thought of the last line of Auden's last poem, "Let your last thinks all be thanks."† It seemed to me that this was precisely her mood—a final and gentle and comprehensive Thanksgiving—and, as such, the only mood, the *right* mood, in which to approach Death. Doubtless there are *also* inner protests and angers and fears—but at the deepest level, there was a profound calm. It is a mercy that she is not suffering much physical pain—nor, I think, will she; but as the tumor

* A relatively young sister who was dying of a brain tumor.

† "Lullaby" was the final poem in Auden's last book, *Thank You, Fog*—but it was not the last poem he wrote. And "let your last thinks all be thanks" is not its final line.

grows, and the steroids fail to hold, there will be a gradual and painless easing-away.

I cannot judge (nor is it my place to) how many weeks or months she has left; but clearly she is dying—and, as clearly, with great grace.

If I can switch from sadness to gladness. On my last day in England, this summer, I went to Jonathan Cole's wedding, in a tiny country church. It was a wonderful ceremony, and Eternity was there. He (they) looked wonderfully happy—there was a palpable sense of a Blessing upon them, blessing descending with the September sunlight, raying in and enfolding them through the ancient church windows. Jonathan asked specially to be remembered to you, and I take the liberty of enclosing the prayer-and-hymn sheet they had in the church.

I hope *you* are well, and in good spirits, Mother; and that I may be able to see you when I come to Philadelphia again soon.

Yours sincerely,
Oliver Sacks

*To Harold Pinter**

PLAYWRIGHT

MARCH 9, 1982
119 HORTON ST., CITY ISLAND

Dear Mr. Pinter,

I have received A KIND OF ALASKA, and read it with very great pleasure. I was rapt, completely. You have entered the world of AWAKENINGS, and of Rose R. in particular, with uncanny success—or, rather, you have reconceived it in your own unique way, but a way which is utterly faithful to the themes and spirit of AWAKENINGS. It is absolutely authentic, and marvellously real—and if Rose R. were still alive

* Playwright (and later Nobel laureate) Harold Pinter. His one-act play inspired by *Awakenings* was published along with two others in *Other Places: Three Plays* (*A Kind of Alaska, Victoria Station,* and *Family Voices*).

(she died two years ago) she would be deeply moved and exclaim wonderingly "Yes, that's *me*! That's how it was . . ." There is an extraordinary sort of sympathy or resonance between the two works—I feel your work as a great compliment (and also complement) to my own, a wonderful transformation of it, and I am delighted you did it. What do you have in mind for it now?

AWAKENINGS itself achieved a lot of good notice when it came out in 1973, and still, I think, attracts attention. It was republished, enlarged, in both the States and the U.K., in 1976, and is to come out in a rather radically revised edition this coming autumn or winter. In this new edition I continue the story of these extraordinary patients, whom I have now known, and loved, and worked with for sixteen years. In my "follow-up" of Rose I wrote ". . . we never saw anything like the events of 1969 again . . . She had been inconceivably, inaccessibly and incommunicably ill since . . . 1926." I often used to sit with her, for hours at a time, trying to fathom what sort of state she was in, *where* she was; and I think you have divined it in A KIND OF ALASKA.

I hope we can discuss further the other points you raise in your letter, and I will look forward to hearing from you. I am sorry the original package went astray—it must have been distressing to have two months without word—but now you have my right address (and telephone number) for further communications.

Yours sincerely,
Oliver Sacks

To Thom Gunn[*]

APRIL 1, 1982
119 HORTON ST., CITY ISLAND, NY

Dear Thom,

I am so sorry, I only just received your letter—when I moved here (a little nautical house near the water) 18 months ago, I didn't realize that the P.O. forwarded letters for a year only—I was (but this is better than not getting them at all) just sent a batch of letters, including yours, by my former landlords, who found them accumulated on their return from wintering in Florida. [. . .]

Yes, it does seem a long time since our ferry-ride to Staten Island.

I live—alone, as always (but "*nunquam minus solum esse quam cum solus esset*")[†]—in a nice old frame house, with lots of books and a piano to keep me company. But I am not misanthropic or isolated, even if I am solitary, and love entertaining my friends in the house—especially in the garden or verandah.[‡] I also have no money—I will have to borrow for Tax Day—but have found a *sort* of equanimity—though I desperately wish for more leisure.

I have just made a third—probably final—revision of *Awakenings*, for my poor patients are vanishing fast; I have worked with them, lived with them, loved them, for sixteen years—a third of my life, which is now coming to its end. My final revision is a sort of Thanksgiving and Farewell. I have a funny sort of feeling that my (22 year long) "American Adventure" is drawing to a close—and am not quite sure what the future holds.

I have explored depths, heights and depths, I could not have imagined even five years ago; but all this has been in a strange sort of

* OS had continued to stay in touch with Gunn, and they occasionally visited each other in San Francisco or New York. In a letter from 1984 to his friend Douglas Chambers, Gunn described OS: "I did see Oliver, on one of his lightning visits here, and he was bubbling over with his peculiar kind of worried energy which I love so much in him but could so easily find irritating." (From *The Letters of Thom Gunn*, edited by Michael Nott, August Kleinzahler, and Clive Wilmer, which contains a number of wonderful anecdotes about OS.)

† "Never less alone than when alone."

‡ This seems like a rather grand, Kiplingesque word to describe the modest front porch of OS's City Island house, though it did have room for a small dining table and a few chairs.

secrecy and silence—I have not been exactly "blocked," but *mute* (or shut-up) in some sort of way for—oh dear! nine years now . . . But, liberatingly, I have just written an explosion of "Essays" (I don't know *what* to call them—maybe "Meditations"), and finally feel a very overdue book quickening inside me (it will be called *Quickening(s)*[*]—and am starting to reach for the typewriter, instead of my endless, secret "Notebooks"—that endless Journal (?to an ideal friend) which has absorbed a million words a year since *Awakenings.*

I do congratulate you on your magnificent "Collected" Poems—you bring a whole life into poetic focus; and would love to receive your new poems in the summer. [. . .]

Best,
Oliver

To Edwyn Silberling

ATTORNEY

APRIL 28, 1982
119 HORTON ST., CITY ISLAND, NY

Dear Edwyn:

[. . .] I received *this*[†] today, which is identical with the form-letter of three-and-a-half months ago. If an *organism* functioned like Medicaid it would not survive a day! What sort of headless, mindless Thing is this which has no internal communication, no organization, no unity-of-action whatever? No wonder most of my colleagues refuse to see Medicaid patients—the patients themselves are made to suffer from the defects of the System. Can *you* do something about this, please? Is there no-one who *runs* the show? Perhaps it is as decapitate as it appears. [. . .]

* This provisional title for the book later became the chapter title "Quickening" in *A Leg to Stand On.*

† A query from Medicaid, the government insurance program that covered most of OS's nursing home patients. A few years later, OS would give up billing Medicaid altogether (though still seeing his patients), since that income barely met the expense of submitting bills.

Incidentally—I hear this from many colleagues—the incidence of (clerical/computer) errors in their own Remittance-Statements and processing of claims has now risen from its previous 25–30% to about 70%. Two thirds of everything they do is wrong—this must be some sort of record!

I hope you are keeping well and enjoying the Spring.

Yours sincerely,
Oliver

To Lawrence Weschler[*]

WRITER

AUGUST 18, 1982
119 HORTON ST., CITY ISLAND, NY

Dear Ren,

I was delighted that you were able to come out to the Island today, and, as always, enjoyed the visit hugely. Alas! (as I feared) I excruciated my back with the rowing, for the third time in two weeks—it was already half "out" in the morning. I had difficulty bending down to put on my shoes, and at such times I have to be especially careful. [. . .] I probably shouldn't write now, because I am half-stoned (on some percodan), which makes the pain more bearable, but makes me loose and effusive.

But there was a particular point you raised. [. . .] I spoke of my early scientific passion, and its disappearance aet[†] 14–5. You asked why. I said, inadequately, that I thought it due to the rise of sexual and neurotic distraughtness. I think this is a very partial truth (if, indeed, a truth at all).

* Weschler, a young staff writer at *The New Yorker*, had proposed to write a profile of OS for the magazine. They developed a close friendship, but once it became clear that Weschler felt he could not write the profile without discussing OS's homosexuality, OS asked him not to proceed. Weschler recounts many of their conversations from this period in his 2019 book, *And How Are* You, *Dr. Sacks?*

† OS liked to use the rather archaic "aet" as an abbreviation for "at the ages of."

In early childhood, aet 3–5, I certainly had some illuminations and raptures of a "scientific" or "cosmic" type—the clearest I recollect being connected with sunflowers and primes (finding their whorls were derivatives of these, and getting an overwhelmed sense, therefore, of the mathematical order and beauty of the world). In the years of intense misery during my Evacuation (aet 6–9½) I ceased to have such experiences. And then, on my return to London, aet 10–10½, I had them again—indeed they dominated my life, they *saved my life* (I felt then, and still feel) in these years. Why then did they go away?

I think the reason may be altogether more "human," and less pathological, and to do with the fundamentally different character of the "scientific" and the personal. I think I managed to *exclude* the personal—personal feeling, personal involvement, etc—during these charmed years of scientific absorption—but if I excluded its anguishes, I excluded its warmth as well. There is, indeed, a scientific love, but it is pure, cold, austere, impersonal—it can be radiant, radiant, as dazzling as the sun, but it is not *warm,* not warm in the least. I had illumination, I had light, but very little warmth—and, in the nature of things, or the nature of *my* nature, this could not last. I think it was not neurosis or sexuality as such, but the need for warmth, for "life," which broke my absorption.

But is this any answer? Why (you will say) could one not have *both*? Are they not, finally, complementary, to be joined? Why an "*either/or,*" and not a rich *both*?

Do you know Darwin's "Biography of an Infant"—his observations on his first-born son, George, in the first months of his life (c 1840)? What is so marvellous—and so natural (and so, *so,* rare)—about these is that they *do,* most perfectly, combine the scientific and the personal, scientific detachment and fatherly love are perfectly fused. Much later, this is what Darwin himself had to say, in a passage of great candour and pathos:

> In one respect . . . my mind has changed during the last 20 or 30 years . . . Formerly, pictures gave me considerable, and music very intense delight. But now . . . I have almost lost my taste for pictures or music . . . My mind seems to have become a sort of machine for grinding general laws out of large collections of fact . . . The loss of these tastes, this curious and lamentable loss

> of the higher aesthetic tastes, is a loss of happiness, and may possibly be injurious to the intellect . . . by enfeebling the emotional part of our nature.

And yet, one would say, are not *all* Darwin's observations drenched in love (and humour)—right to the end? Whether he describes orchids, or earthworms, he does so with love; there is *always* the love of phenomena, the love of Nature, what Spinoza (and perhaps Einstein) would call "the intellectual love of God."

The question is: whether *this* love—the love of phenomena (Einstein speaks here of "*einfuhlung*"), the love of order, the love of nature, *amor intellectus Dei*—whether these forms of *intellectual* love can, in fact, "go" with a more *personal* sort of love, or whether they are, in some sense, incompatible, even hostile? [. . .]

To Benedict Nightingale

DRAMA CRITIC, *THE NEW STATESMAN*

OCTOBER 23, 1982

119 HORTON ST., CITY ISLAND, NY

Dear Mr. Nightingale,

I want to express my appreciation for your penetrating review in the *Statesman,* and especially your words about *Awakenings.*

I was intensely surprised myself when, a few months ago, the post brought me *A Kind of Alaska,* with a lovely letter from Pinter about how he had read *Awakenings* many years ago; how it had "suggested a world" to him, which then sunk from consciousness; and how, last summer, he had woken with an image of the woman lying in bed, and the first words of the play in his mind, and this strange "world," re-erupting, became the play in a few days.

Like you I wondered: Why dig up this bizarre case-history from the archive (though, of course, having lived and worked with these patients for sixteen years now, they are living realities to me, not bizarre histories in an archive)? But as I read the play I had an extraordinary sense of consonance and rightness—he *had,* as he hoped, entered this strange world, and in a way which seemed to me uncanny, almost clairvoyant. I felt he had been with me, known my thoughts and feelings, known the patient, etc. When I showed the play to a colleague of mine who had known Sylvie ("Rose R.") intimately, she said, "He has caught Sylvie's essence. It is not like 'Pinter,' it is just like the truth"—a wonderful tribute to the essential truth, the *transparency*, of Pinter's art (I wish I had quoted this to him when I met him last week).

I had long feared *Awakenings* was forgotten, and *had* become no more than an "archive," and I was overjoyed, of course, that Pinter *had* seen it, remembered it, incubated it, and, finally, reconceived it in his own unique way. And, as you say, it is perhaps, finally, not so surprising, since Pinter is intensely concerned with the depths, the submerged nine-tenths—and so am I. My work always seems to bring me to the intersection of fact and fable, and to provide me with metaphors or parables of being human—extreme metaphors, as you say, or (perhaps more accurately), metaphors of human beings in extremity. All my patients (in the "locked ward," the tragic underworld I work in)—and not just the postencephalitic ones—have extraordinary stories to tell and be told—and I think their stories *are* relevant to more ordinary and universal feelings and situations. I do, in this sense, think of myself as a sort of neurological novelist—though, alas! one who has published all too little. But the powers of a Dramatist are much finer and rarer—"Alaska" (and Judi Dench)* convey the essence with incomparable economy: and make it *real*, which words alone cannot do.

Thank you, again, for remembering *Awakenings*—I have just finished the proofs of a new edition which carries my patients' stories up to the present, and will take the liberty of sending you a copy when it comes out in the Spring. (By the way, I am English, and *not* American, though I live and work in the States.)

* Dench originated the role of the patient, Deborah, in *A Kind of Alaska*.

To F. Robert Rodman

NOVEMBER 13, 1982

119 HORTON ST., CITY ISLAND, NY

Dear Bob,

A letter—why not? The rumble of train wheels, the passing scenery, the sense of a Journey—I am *on my way*—to Boston. [. . .] Eric* has a bookfair in Boston this weekend—somehow he gives me "permission" to travel. [. . .]

And, in general, these last days, I have had a sense of *release*—after the endless (the 8 yr.) imprisonment of "The Leg." My editor likes what he can read†—and, whatever revisions I may (have to/want to) make, *in some sense* the bloody thing is over. I feel no joy (as after *Migraine*)—just a great sense of release, and the eager-fearful, *alive* question of "What now? What *next*?" (There has been no "Next" for the longest time, for the Leg book *obstructed* everything else). [. . .]

Sense of farewell. My mother died today, in 1972. I had forgotten, consciously, until I put the date on this letter—and thought "Christ, this is it. The very day. Incredible." *What* is "incredible" I don't quite know (I think Freud is wrong—that a mother's death hits one harder than a father's, is altogether more mysterious and fundamental). Eric was greatly, even decisively, changed by *his* mother's death, and speaking to Thom Gunn, whom I spent yesterday afternoon with, it was clear what an *immense* effect—as Loss, as Liberation, as incitement to movement and creation—came with the death of *his* mother (but he was only 15). Deaths of mothers . . .

As I wrote the date 11/11 at hospital a couple of days ago, I suddenly thought of a particularly dear patient (Lillian W—she appears in a footnote in *Awakenings*)—whose birthdate, unforgettably, was 11/11/11. When she died, earlier this year, I felt it very keenly—I wept on and off for a couple of days—and felt her death as the death of my patients, the postencephalitics to whom I made a sort of vow 16 years ago, that I would be with them "till Death do us part"—but now

* Eric Korn, who frequently traveled to the United States for bookfairs.

† His American editor, James H. Silberman at Summit Books, had to make his way through dozens of pages of OS's often illegible handwriting.

Death has; 9/10th of them are dead (tho' those who are alive are *very* much alive); and, with this, somehow feel my "work" here is done. The Epilogue, which I wrote in February, after Lillian's death, has this sense of "Thank-you," Farewell, Requiem. (I hope I am not distressing you—I know *your* mother is postencephalitic.)

In some sense, Shengold now is the only clear reason for remaining in New York—there are no longer specific (groups of) patients, or clinical opportunities. What work I do—if it *is* "work"—I suspect I could do *anywhere* at this point. And I have—quietly, and quite persistently—*hated* New York; or, to put it less hatefully, have never *loved* it as, in such different ways, I have loved the West Coast, and England—and even Canada. So—some *stirring*, some sense of movement, some sense of "A Move," is afoot.

I ache for the West (Coast)—Thom's visit this week [. . .] sharpened the ache—the Pacific, the Desert, the sense of immense *Space*. I cannot write, cannot think, cannot feel properly real and alive, unless there is (this sense of) Space—and a (physical) world I can love. That I should have had to go to the *Adirondacks* to complete the leg book—this (increasingly) is typical: I *need* the grandeur (the quietude, the immense otherness) of Nature—the sense of the Good Mother, Mother Earth perhaps—to give me a sense of reality, to *root* me, sufficiently to write. (I think, wistfully, of your own lovely place, the garden, the hill, overlooking the Pacific—how *rightly* you chose. I hope, amid the fret of life, you *allow* yourself, give yourself liberty, and leisure, to enjoy it.)

Forgive—if you can read—this ranting (and it may be, self centered) letter; phone (or write) me soon. [. . .]

A hug to the girls, and all my love to you,

Oliver

OS CONTINUED TO BE DEEPLY interested in visual illusions, such as the sort his friend Richard Gregory studied, along with the mechanisms of stereovision, color vision, motion vision, and so on. He was struck especially by the "strobe" or "wagon wheel" effect; sometimes, while biking up and down City Island Avenue at night, he'd catch sight of his bike spokes, which appeared to be stationary or even rotating in the wrong direction. He would return to these questions often throughout his life.

During this period, he often discussed visual questions with Gregory or with his friend Bob Wasserman, an ophthalmologist he met soon after moving to New York.

To Robert Wasserman

OPHTHALMOLOGIST

DECEMBER 13, 1982
119 HORTON ST., CITY ISLAND, NY

Dear Bob:

A brief "follow-up," from my end, on the STROBE effects.

I think you must be right about mercury lights etc.—certainly the effects are *most* striking when I ride at night under these. Only then may I see the ridged pattern absolutely motionless for half a minute or so, or slowly revolving, or jerking, forwards or back, according to my speed.

However, I *also* get some "strobe" effects, though only momentarily, in uniform and steady illumination, in daylight. It is less easy to make observations here, for the "strobe" effect may last only a fraction of a second. Indeed I was not quite sure of it till today, when I saw it *repeatedly* under rather special conditions—viz. cycling along a tree-lined road in late afternoon, with the oblique rays of the sun (perhaps) being intermitted by the trees. But I also see it, more rarely, when there is no intermittency to the illumination. I also observe this with my bicycle chain—moments of absolute arrest. I incline now to think this is purely physical, and that the tread of the tire, or the links of the chain, *create* an intermittency of illumination; but I am not absolutely sure. In particular, it seems to me that there may be an odd "off" effect, so that in the moment of looking AWAY from the tire or chain, I suddenly see it motionless, arrested. Or, occasionally, if I just *glance* at it, rather than *gaze*. Perhaps a glance CAN "cinematize," and fracture the continuity of perception.

Whatever goes on, it makes one realize what an achievement, what a synthesis, perceptually, "motion" is. One takes the sense of movement for granted, and it is only through these strange aberrant moments, these strobe effects, that one experiences perception stripped of any sense of motion and duration, and reduced to the raw physical or physiological—which has no true motion, or duration, but only (a succession of) motionless configurations. Perception is normally a stream—a sort of perceptual melody. It is fascinating—and frightening—to see this stream stopped. I suspect this often happens, on a purely physiological basis, with some of my post-encephalitic patients.

Anyhow, what do you observe from *your* end?

Hope to see you soon—that last weekend was a delight.

Love,

To Richard Gregory

JANUARY 11, 1983
119 HORTON ST., CITY ISLAND, NY

Dear Richard,

Lovely talking to you just now on the phone—how few people one can, one would *dare* to, phone in this way! But you were wonderfully friendly, and instantly acute—to anticipating my implied thought of perceptual flashes or frames. This matter has been in my mind for four weeks or so, and *hits* me again every time I go out on the bike (as I had just done, a few minutes before I phoned you). I wrote to an ophthalmologist friend about it (I take the liberty of enclosing a copy of my letter) and he too had thought of saccades, but could only guess, and had no experience (I don't suppose these phenomena are "symptomatic" in the ordinary course of life, though since noticing them a month ago, I *catch* them, now, all the time.)

I am both sorry (and relieved) that there is no evidence for perceptual discontinuity or "frames." If this is the case, I am going to have to confront a massive task in *re-interpreting* all sorts of clinical data which come to me, which would seem to suggest a sort of "atomized" perception—atomized in time, space, or both (what I have tended to call "cinematic" and "mosaic" perception). And, more generally, the quality of *delirium,* which is (described as, and seems to me) essentially discontinuous, a sort of perceptual/conceptual *flutter.* And the sort of thing I have called "freeze-frames," "stills," "perceptual arrest," etc. in *Awakenings.* It may be that there are discontinuities of *attention,* not perception; or some incoherence in *aspect* perception, so that instead of seeing something steady and whole, one confabulates or fancies or hypothesizes *scenes* "based" on some fragmentary or figmentary (mis)perception. Such perceptual deliria—I especially think of the productive visual paragnosias seen with right parietal occipital lesions—beside their quality of superficiality and flutter, are curiously devoid of affect or *tone.* [. . .]

For me "kinetic melody" is not just a turn-of-phrase, but a precise and irreduceable characterization of animal motion. Similarly I want to talk of "*perceptual* melody"; and I am very conscious of the melodious movement of my *thought* (that is, when I can or do think!). In short, I cannot dissociate the idea of ACTION from music (and, specifically, melody). [. . .]

Whenever I *do* anything—ride a bicycle, or think—I am proceeding in a way which is beyond procedure: that is, beyond any definable or describable procedures. I cannot *tell* anyone HOW to do anything—even something as simple as riding a bike: I can only say "The way to do it is to do it." I can, indeed, give him information, instructions, suggestions, procedures—but these do not come together until he suddenly "gets the feel," "gets the hang," "gets the idea," and then what he gets surpasses all procedure: or, rather, reflects a *personal* proceeding or procedure which he cannot describe ("What CAN be shown CANNOT be said," etc.).* [. . .]

What especially challenges me in patients is to evoke the power of, the stream of, Doing, despite severe impairments of perceptual/conceptual procedures e.g. getting an apraxic† patient into a complex continuous task, even though he has fundamental difficulties in programming and sequencing. My "method" here is Play—or Art, which, I feel, can *embed* procedures. Again, my method is to engage the Person, the Actor, the active acting Self, which can achieve a unity of action, even where there is severe neurological disunity.

* He is quoting Wittgenstein.

† Apraxic people have an inability to execute fine movements—for example, with posture or with speech.

Leibniz, comparing Animals to Clocks, speaks of "an active Principle of Unity" (this strikes me as an anticipation of Kant). Hobbes, on the other hand, at least in the metaphor of Leviathan, is concerned with "artificiall Life" and Automata—a sort of seventeenth-century anticipation of Cybernetics. I know how interested you are in the "A.I. enterprise." [. . .] I sometimes feel you are describing artificiall Life, Man, or Intelligence—and not "Natural" (I feel my *own* function, in a sort of dialectic or dialogic complementarity to yours, is to study the nature of Naturalness). [. . .]

Do reply—it is so long since we've written. And I will hope to see you in March.

Love,

To Gerald Stern

NEUROLOGIST

SEPTEMBER 14, 1983
119 HORTON ST., CITY ISLAND, NY

Dear Gerald,

I have just got your good letter of the 7th, for which I thank you immensely. It sharpened my sense of regret and annoyance, however, that I didn't manage to *see* you when I was in London last week. It was a very frantic week. [. . .]

I am glad you liked the "Hat" piece*—there is a special sort of pleasure in writing such things (they bring back the image of a patient in affectionate and vivid detail, and there is a sort of fidelity, and rightness of feeling, as are built in to the clinical contact), and, hopefully, in reading them—at least to "old-fashioned" clinicians like ourselves. I have just written (but not even titled) a Korsakov piece,† about a

* "The Man Who Mistook His Wife for a Hat" was published in the May 19, 1983, issue of the *London Review of Books* and would become the title story in the book of the same name.

† This, too, would eventually appear in *The Man Who Mistook His Wife for a Hat*, as "The Lost Mariner."

delightful intelligent man with not only the most devastating anterograde amnesia, but a retrograde one which has erased him back to 1945—so he is very much marooned in the past (not, of course, that *he* feels it as "past"—his early memories are quite startlingly "immediate" and distinct). I hope it will get published here, and will send you a copy when/if it is. (Have *you* seen such Korsakov patients? Are they common? The *syndrome,* including gross retrograde amnesias, is described by Luria, though most of his patients, of course, have midline tumours, rather than "ordinary" Korsakov's. For myself, curiously, though I see a lot of alcoholics, I have not too often seen classic Korsakov's.)

Quite *why* I wrote this piece—or the "hat" one for that matter—I don't know: they just came to mind. I think, however, that almost everything which just "comes to mind" now has something to do with organic alterations of identity; and perhaps the body-image book I was thinking of will include histories like this, and illustrate (what I suppose is becoming my central interest, in a way)—what one might call "The Neurology of Identity."

The frenzy in London, and my long (and I fear rude) silence for many months both had to do with *A Leg to Stand On,* or whatever I shall call the "Leg" book. I was driven half-crazy by my editor-publisher in England taking *six months* to edit and retype the book, when I had thought it pretty much done in March, and expected to get the retyped version in April, and to see it published this month. Although I am a terrible procrastinator, actually it was a *very* difficult (and tricky) editing job, because, among other things, it was necessary to reduce the length from 240,000 to 70,000 words—nothing like this was done with *Awakenings.* But I think now, finally, that there is a taut narrative—which won't make impossible demands on the reader, and a nice Epilogue, which really amounts to a sort of intellectual history and critique of Neurology. I only got the Epilogue right on my last day in London. *Now,* at last, I feel comfortable with the book—comfortable, not exuberant (it has wearied and worried me for far too long to allow this)—and somehow, comfortable, or resolved, in relation to the clinical experience, and in relation to Neurology and Neuropsychology. Feeling "reactionary" or "counter-enlightenment" is *not* the right feeling; what I hope came out is the feeling of a great (and necessary) sweep a century and a quarter long—from "preconceptual" neurology (exemplified by Weir Mitchell),[*] to "classical" ("Sherringtonian") neurology (exemplified by Head),[†] to neuropsychology (exemplified by Luria and Leont'ev) to . . . to an "Existential" neurology, or neurology of identity, which seems to me now to be demanded and due in describing syndromes and situations which defeat any description in terms of Function or System, and *demand* description in terms of (altered) Self or Identity. This was the case with "Dr. P."[‡] and with so many of the right-hemisphere syndromes one sees; it is the case with my Korsakov patient; it was the case with my own body-image, body-ego, disturbance. I am sure we *contain* circuits, puppets, etc, and that one *has* to have a neurology of these; but I am sure we are not *just* puppets—at least in health (though pathology may make us so; as described in the fascinating self-account you sent, in which Schooling[§] speaks of periods in which his limbs behave "as if controlled by a drunken marionette master"), and that one needs a neurology of freedom, of health, of a Self, *transcending* a neurology of reflexes, systems, cybernetics and puppets. [. . .]

I *am* verbose—this letter is too long—but it is partly because there *is* so much to be said! Anyhow, thank you, sorry I missed you, hope to see you soon.

Yrs,
Oliver

* Silas Weir Mitchell, a Civil War–era physician and writer.

† Neurologist Henry Head (1861–1940).

‡ The subject of "The Man Who Mistook His Wife for a Hat."

§ A patient of Stern's who had described living with Parkinson's.

To Lawrence Weschler

[SEPTEMBER 1983]
[BLUE MOUNTAIN CENTER]*

Dear Ren,

I should be writing "the book"†—but my only impulse is to write *a letter*—yet the two feelings are always very close for me. I have a vivid sense of readers as *individuals*—I wish I could address them, each and every one, as Kierkegaard does, as "Thou" (I always felt that Kierkegaard was actually writing *to me,* such is the authentic magic of his "Thou"). Conversely, every time I *do* address myself to a "Thou"—a particular individual—I tend to find myself writing a book! At least, if there is *a subject* to write on, something which vitally interests us both—Indeed these are my favourite sort of letter. [. . .]

I love the "*scientific*" or *philosophic* letter—this was the sort of correspondence I had with Luria; there was a medical or scientific or philosophic *subject,* often lots of them; yet the letters were also *between us,* personal. I like letters to and from "patients" or readers—patients, a long way away, who've read my books; or readers who, if not patients, respond with a peculiar resonance of their own. Auden liked to "draw me out," liked me to write long letters about my work—he had an inexhaustible interest in *patients,* and strange clinical and human experiences—as seen through the patient's or doctor's mind and heart. Much of what later became *Awakenings* started as conversations, and correspondence, with him. He was also interested in my "philosophic" reflections on these—tho' less so than Hannah (Arendt), whom I knew a bit. There were times when it was *to her* that I especially wanted to write—and I regret that I was always too diffident to do so. And now they're all dead—these good "parents" whom I loved—and now I too am older,‡ and am less turned to "parents," and more turned to contemporaries, friends, and sometimes "sons." We have become close in the two years we have known each other—and yet, in an important way, we are quite different, we are

* A literary colony in the Adirondacks.

† *A Leg to Stand On.*

‡ OS had just turned fifty.

"strangers"; and I find, involuntarily, that my pen turns, and my mind and heart open, to *you.*

I love letters which deal with investigations and problems; but I also love narrative and narration in letters; and what I love most is to have them combined—a problem, problems, *embedded* in a tale—a *narrative* Philosophy, a *narrative* Science, in which the Riddle and the Tale are one. I suppose this *is* what one means by "case-history"—and, in its universal form, what Luria meant when he spoke (as he frequently did) of "Romantic Science"—his case-histories, or Freud's, for example, *are* studies, and investigations, presented as tales (and perhaps they could be presented no other way). This, very clearly, is the case with my "Leg" book—and my piece ("The Leg") in the *London Review.*

But what I have written, and what I have published, and what I have spoken of to you, only go as far as the first walk*—*I have said nothing yet of the rest of the story.*

And it is *this,* now, which floods into my mind (partly because this is where I have "*got to*")—not least because of these idyllic surroundings, the whole *feel* of this place (an old Manse, in a village, surrounded by lovely country) which reminds me so intensely of the Convalescent House in '74. And it is precisely the same time of year—a wonderful Indian summer of September–October, with the long, warm, lucid golden light, and the leaves turning and carpeting the ground.

Above all, the identity of *spiritual* feeling—the sense of a most precious and beautiful interim—a Sabbath—in which Mind and Body find perfect repose, but find this in the perfect stillness, the mysterious still, of action. So one is not lazy, one is intensely alive—it is life intensified into a pure and perfect "Now"—just as it was, when I was convalescing, eight years ago—

. . . For this special interim,
So restful, yet so festive,
Thank you, Thank you, Thank you, Fog.†

* I.e., the moment when he was first able to walk following his injury.

† From Auden's poem "Thank You, Fog."

For Auden it was being fog-bound, at Christmas in "an ancient manor-house" in the company of old and dear friends—a brief "time-out" of "lenient days" before having to re-enter the World, "the world of work and money."

External circumstances—especially if gratuitous, imposed by necessity—can provide this special sweet Interim; but, equally, it is a need, the most profound need, of the Mind. And no-one analyses this more beautifully than (his friend) Hannah Arendt, in her last book (*Thinking*)* which we both so profoundly admire, when she speaks of the Gap between Past and Future—"the quiet of the Now in the time-pressed, time-tossed Existence of Man."

IN 1979, MARY-KAY WILMERS, the editor who had earlier invited OS to contribute to *The Listener,* cofounded the *London Review of Books,* a sister publication to *The New York Review of Books.* Between 1981 and 1984, she published four of his "clinical tales" in the *LRB*—including "The Man Who Mistook His Wife for a Hat."

OS had also been in touch, for a decade, with Robert Silvers, the editor of *The New York Review of Books;* in January 1984, Silvers published OS's debut essay in that paper, "The Lost Mariner." Silvers had a remarkable instinct for selecting newly published books to send to his writers, hoping to instigate an essay review. It is likely that Silvers sent Steven B. Smith's 1983 book, *The Great Mental Calculators,* to OS, leading to the following reply.

* The first part of Arendt's last book, *The Life of the Mind,* was subtitled *Volume 1: Thinking.*

To Robert B. Silvers

EDITOR, *THE NEW YORK REVIEW OF BOOKS*

MAY 24, 1984

119 HORTON ST., CITY ISLAND, NY

Dear Bob,

[. . .] During a brief "release" in the Botanic Garden this week I found myself thinking again of the Finn twins*—the calculating identical idiot savants we spoke of—but in a much more general way. What seems to me centrally important about them, (and so many of *The Great Mental Calculators,* a book I feel to be as wrong in its conclusions as fascinating in its material), is that it is not, in fact, "calculation" at all, not a series of mental operations, but an instantaneous and intuitive *seeing*—as Dase† would see, at a glance, if someone threw down a handful of peas, that there were 117 of them. He did not *count,* he instantly "saw" the 117-ness, as a unique, familiar numerical physiognomy or landscape. This is obscured in Steven Smith's book, I think, but clear from F.W.H. Myers' account a century ago, though it is indicated anecdotally. [. . .] The Finn twins are referred to (pp. 29–30), and their "methods" ascribed to the formation of unconscious algorithms. I don't think this is the case, and I have known them for eighteen years. I think they live, so to speak, in a romantic landscape of numbers—and that this is the *only* landscape, the only passion, in their poor narrow lives. My own calculations, as a child—and I *was* quite a calculator, and can speak from experience—were far more akin to music, or language, than computation: everything was suffused with feeling and meaning—there were *never* strings of figures, but number-words and poems.

And this makes me think of, and in an article might turn to, other, higher forms of "calculation"—for example Mendeleef's calculation of the existence and properties of missing elements (which I also did,

* Charles and George Finn, the twin autistic savants OS wrote about in his essay "The Twins."

† Zacharias Dase was a famous nineteenth-century calculator; though of limited intelligence and unable to understand mathematical principles, he was able to "calculate" tables of prime numbers into seven digits. Steven Smith wrote about Dase in his book, and F.W.H. Myers also wrote about him in his 1903 *Human Personality and Its Survival of Bodily Death.*

in my chemistry days, before I had heard of Mendeleef). I would carry with me lists of the properties of elements, but what was *perceived* was not lists, but "characters," "physiognomies." I felt the elements as my friends, I got to know all their faces, and after a while one just *knew* what the missing faces would be like—the actual computations, the quantitative, was entirely secondary.

And I have been very excited by rereading Helmholtz* (*Sensations of Tone*), and his speaking of tones as perceived, instantly, "synthetically" as qualities, even though they might be broken down by (a Fourier† or other) analysis.

So a consideration of the Finn twins, hopefully, could lead to a general discussion of synthetic perception, "judgement," "intuition," and it would be especially nice if one could in fact show that the most mechanical sorts of thought, seemingly, and in the narrow heads of morons, was as romantic and intuitive as anything in Helmholtz. Anyhow, this is just a thought for a future piece. [. . .]

I am more grateful than I can say for your generosity to me this year. "The Lost Mariner" made me happier than any piece I have ever had published—and (I have received many further letters since we spoke) has had a wider impact and resonance; and it is marvellous of you to "risk" yourself, and the *NY Review,* with the opening chapter or prologue of my book. Thank you, thank you, thank you, Bob!

Will contact you on my return—

Oliver

EARLY IN 1966, WHILE ON a visit to New York, Jonathan Miller had introduced OS to his friend Susan Sontag, but OS was too shy to stay in touch with her. (He wrote to Miller, "I liked Susan Sontag immediately, which is almost without precedent for a human female. I will *try* to keep in touch with her, although I feel intimidated both by her calibre and her sex. You, of course,

* Hermann von Helmholtz (1821–1894) was a pioneer in the study of human vision and hearing, among other fields.

† Joseph Fourier (1768–1830), a French mathematician.

feel exhilarated by both, which is only right and proper. Give me a year with a good analyst, and I'll feel the same way.") It would be nearly twenty years later that the two of them began spending time together. Sontag was one of the founders of the New York Institute for the Humanities at NYU, which hosted Friday luncheon-seminars for its fellows, and she admired OS's work. In an August 30, 1984, letter to him, she wrote: "So much of what I read, and I am (alas) a compulsive reader, makes the space in which I feel and write seem smaller. You make it seem larger, and more accessible." OS was elected to membership in 1984 and began seeing Sontag more often.

To Susan Sontag

WRITER, AUTHOR OF *AGAINST INTERPRETATION* AND *ILLNESS AS METAPHOR*

MAY 24, 1984
119 HORTON ST., CITY ISLAND, NY

Dear Susan,

I meant to write to you straightaway after that (for me, at least, immensely enjoyable) evening—and your quite outrageous, but much appreciated, introduction of me—but the intention has somehow got caught with the general chaos and inhibition of these last, difficult weeks—I am very sorry.

I enclose "The Man Who Mistook his Wife for a Hat"*—I hope you enjoy it. I think the *issues* involved in such cases are quite fundamental, and at the heart of what is involved in any *living* recognition, in "life." The article was fairly reduced from the expansive (10,000 word) MS I gave to the LRB, but I hope I can bring back some of the deleted portions if I can collect this (and other pieces) in a book, as I want to. I had a piece in the LRB† earlier this month, which was *so* cut as to be rendered completely meaningless and reduced to the level of anecdote or curio—I am sort of furious about this, and am

* The essay.

† "Musical Ears."

not sending [you] this (indeed I hope that no one sees it!). The issues which arose here were also deep—it may be that sometimes things just occur to me, first, *as* "stories," and only then get charged with imaginative depth and power. I do thank you for your sensitivity in bringing out the "allegorical" aspect—perhaps not too clear in these articles, but absolutely central in *A Leg to Stand On,* my own journey to a neurological Underworld and back—which is published (in England) next week and here (with quite a different ending!) in July. I am quite nervous about this—it is so long since I have brought out a book, and it is autobiographical (and so "naked") in a way I have never attempted before. This is one of the things which has been paralysing me these last few weeks. I hope I may take the liberty of sending you a copy of this.

I found Friday, at NYU,* unexpectedly delightful—and am sorry only that *you* were not there. I did not realize that I would be expected to re-present what I had said to a quite different, and perhaps terrifyingly critical, audience. But in fact there was a lovely thoughtfulness and rapport, a sense of thinking *community* I suspect I am starving for. I felt the ridiculousness of my *isolation* almost violently after the lunch, and perhaps I can start to emerge from it, in part.

I had admired you, as you must know, for many years—since I met you (with Jonathan, one breakfast-morning in the early sixties, and before), and hope we may perhaps see more of each other now.

With kindest regards, and deep gratitude,
Oliver

IN 1984, *A LEG TO STAND ON* was published in England, and soon after, Mary-Kay Wilmers published Michael Neve's very caustic review of it in the *London Review of Books.* (OS often wondered whether Wilmers had used Neve's review to "punish" him, since, he thought, she had felt "betrayed" by his offering "The Lost Mariner" to Robert Silvers for *The New York Review of Books.*)

* At one of the New York Institute for the Humanities luncheons.

To Lawrence Weschler

JULY 2, 1984

119 HORTON ST., CITY ISLAND, NY

Dear Ren,

[. . .] it was very nice to get *your* letter, with its friendliness, and its business-like questions. I am relieved to hear there is some (internal and external) consistency in what I have said (and what has been said about me) in the past three years—the more so as a disquieting lead-review in the *London Review,* in addition to everything else, accuses me of confabulation—the words "disbelief," "unbelievable" occur seven or eight times. Similar imputations of my veracity/sanity appeared wholesale in that dreadful issue of JAMA,* of which I don't *think* that I have a copy (but I could get one, I suppose, if you so wished). ITS date was about Dec 15 (1970), and its poisonous effect lasted through January into February, obnubilating any pleasure I might have got from the publication of *Migraine* (which, though dated and slated for 1970, didn't appear till January '71—looking up my album† I find Jan 25. 71). It was greeted that day, Publication-Day, by a very *nice* thing in the *Times,*‡ unlike ALTSO§ which was greeted, on publication-day, by a very *nasty* thing in the *Times* [. . .]. Eric tells me I must be more philosophical and less "vain," but I think that vulnerability, rather than vanity, is the problem—though perhaps, they go together, or at least share too great a dependence on the reactions of others. At the *deepest* level, of course, there is no such dependence, and I judge myself and my thoughts evenly and justly. [. . .]

Fondly,
Oliver

* *The Journal of the American Medical Association* had published a number of letters criticizing OS's letter about the side effects of L-dopa—without giving him the opportunity to reply. See his unsent letter dated August 28, 1971.

† OS kept an album of press clippings relating to *Migraine.*

‡ The London *Times.*

§ *A Leg to Stand On.*

JULY 5, 1984

P.S. I had better get this in the post this morning—otherwise, with my gift/weakness for deferment, I will *never* get it off. Isabelle* cooled me down, made me feel saner, less self-indulgently "annihilated," about the LRB review etc. Nevertheless, I fear I may just have to "work through" a summer of bad feelings. [. . .] Glad Shengold is around, so I don't translate my negative feelings into actions, At least, not too much. [. . .]

Thank God for Helmholtz! He stands for a better world, the good of the intellect, integrity, balance, and ceaseless creativity. He is the only salve for my wounds. Also, specifically, he is the pointer for so much. I have especially been reading him on stereoscopy, and think I may write something on this [. . .] about the perception, the feeling, the *idea* of depth, and how fundamental it is to one's sense of the world. [. . .] My only recreation these past weeks has been playing with the stereoscope I bought at the book fair.

When I was in England, I spent almost a whole weekend with Duncan† (dear Duncan!) talking about stereoscopy, and little else! I said it was one of the earliest and most constant interests I had; he wondered if this was because I regarded stereopsis, binocular fusion, unconsciously at least, as "a metaphor for a mental act." I think he's dead right—I do; and I find Helmholtz does, and I think this must be why it was one of *his* most constant interests (his papers on it come out over a thirty year period). I don't know whether I ever talked to you about my early passion for stereophotography, making stereoscopes, hyperstereoscopes, pseudoscopes, etc. and also, though less, color vision & photography. But now, strangely, almost forty years later, in this period of disquiet, my mind goes back strongly to this. It is the only thing which gives me peace.

I also want to write a piece on color. I have some patients, and can find others, with partial/total congenital/acquired absolute lack of color vision i.e. only black-and-white vision. I wonder what *their* idea of colour is like. You will recall that this (in a particular example:

* Rapin.

† Dallas.

as to whether a man who had never seen a particular shade of spectral blue could *imagine* it) is discussed by Hume.

Must post!

To Chaim F. Shatan

PSYCHOANALYST

JULY 7, 1984
119 HORTON ST., CITY ISLAND, NY

Dear Dr. Shatan,

I must at once acknowledge, though I cannot fully "reply" to, your very kind letter and important, fascinating reprint. [. . .]

This power of music to articulate, organize, animate perception and movement is something I constantly see in neurological patients—perhaps especially Parkinsonians (and as such is something I repeatedly advert to in *Awakenings*). Whatever occurs is eminently *demonstrable*—sufficiently so to convince the hard-boiled State to allow us a full-time music therapist at Beth Abraham (the "Mount Carmel" of *Awakenings* . . .). The music, as you imply, has to *affect* the person, to "touch" them, to "move" them—I also see this quite amazingly in patients with cerebral palsy, patients practically doubled up with spasticity and athetosis, incapable "normally" of anything but the most limited and pathological movement, who can "relax" and *dance* quite beautifully. I have patients who say, "When I hear music, when I feel music, I *forget* my cerebral palsy, my Parkinsonism, etc." I think there can be a Parkinsonism of the mind (rigidity, inertia, difficulty starting and stopping, perseveration, etc.) in Parkinsonians, abetted by a sort of inturning and obsessing of the attention, that, in this way, they "think Parkinsonian" but, quite radically, can get out of this, and find a mental and affectomotor freedom in art, most especially a kinetic or time-art.

As you indicate and imply there seems to be a limitless transparency between body and mind—the somato-psychic no less than the psycho-somatic [. . .]. So that not only does every state of mind find expression in bodily tone and posture, but vice-versa. It is typical, as you say, that all our most evocative terms ("rigid," etc.) have precisely this double connotation. [. . .]

Parkinsonians, and especially post-encephalitics, may show an extreme acceleration or retardation of movement, thought, perception, time-judgement—so much so as to make them almost unintelligible, through being so "out-of-synch" with normal people around them. [. . .]

Anyhow, thank you again for your letter and reprint. [. . .] I am sure we are both dealing with important realities, however difficult they may be to conceptualize.

With kind regards,
Oliver Sacks

To Lawrence Weschler

JULY 18, 1984
119 HORTON ST., CITY ISLAND, NY

Dear Ren,

It was indeed nice getting your Sunday phone-call—I am sorry I forgot to congratulate you on your fine Hockney piece which I thought full of interest. [. . .]

I was just in at Beth Abraham, seeing the situation today. The picketers were marching up-and-down, chanting, "Want, want, what do we want? We want more money, that's what we want." Inside, *at this early stage,* it is still felt as a bit of a lark—fun joining in, volunteering (like my firefighting was, in B.C. back in '60). But, already, the twelve-hour shifts are taking their toll; it is not quite fun as it was on Day 1. Nurses, doctors, secretaries, volunteers, etc. are all looking a bit strained and exhausted. Many patients, who are normally got up, and turned, and exercised, are left to lie in bed all day. This too is "fun," like a holiday, for a day or two. Then the skin breaks down, bedsores start; breathing flat isn't enough, pneumonia sets in; joints and muscles start to stiffen, there are contractures in a week—*this* is the deadly effect of a strike which lasts for more than two days.

And, of course, only the *physical* effects: unstimulated, left in bed, apathy, confusion, disorientation—and fear; the demented patients no longer know what's happening, or where they are—they start to feel punished, abandoned, in prison, in hell.

Perhaps you and a couple of colleagues from *The New Yorker* should act as volunteers—and document in detail what happens to EVERYONE in these strikes (the picketers, the Staff, no less than the patients). There are mendacious (at least, superficial) reports put out to reassure, telling us that the level of Care is being maintained. You should see for yourself *exactly* what happens—physically, morally, to everyone concerned. Had I not worked round the clock with my students for 11 days in 1974, there would have been a fearful morbidity and mortality among our poor "Enkies":* as it was, the bonds between us—the patients and myself—were perhaps strengthened by the experience. But it is a *fearful* experience, horrible, dangerous, after the first day or two. I again feel that *such a cost* in human suffering, *such* a bargaining with vulnerable lives, should never occur whatever the grievance. [. . .]

I met Wystan first at Orlan's apartment—I don't think you ever met Orlan Fox. (He was Wystan's closest friend, I would think, for [. . .] the last 15 years of Wystan's life; and a *fairly* close friend of mine from the time we first met, in November '65.) I am afraid I cannot quite date it† (something to date it may occur to me)—it was either in '67 or '68.

I had seen him before, in June '56, when he gave his first lecture, his Inaugural, as Professor of Poetry at Oxford. [. . . But] I had certainly never seen him at close quarters before. I was petrified, mute, with fear and awe, that first time. I was fascinated by his furrowed, Jurassic face—I had never seen a face which so resembled a geological landscape—at the same time trying not to stare. I was fascinated by his memory and wit—which reminded me, strongly, of Erik‡ (indeed I never ceased to feel, here, an almost physiological resemblance between them). But, and here he differed sharply from Erik (and

* In the United Kingdom, "encephalitis" is pronounced with a hard "c"; thus this nickname for the encephalitic patients.

† I.e., his introduction to Auden.

‡ Eric Korn, whose name OS often misspelled as Erik.

perhaps I felt there was something more akin to myself), the flow of wit, the gossip (he loved gossip!) might suddenly cease, if something *deep* chanced to be brought up. The memory, the wit, were prodigious, phenomenal; but what really moved me, what excited awe, was to see Wystan *brooding*, to see him suddenly silent, *arrested* by thought (I often saw this later, I loved to see him ponder; I saw it first, but forget the precise occasion, that evening at Orlan's in '67 or '68). Erik doesn't ponder—he's quick as a flash. He's as clever as Wystan—but he lacks (or disallows, or is disallowed) that strange depth. And Wystan himself had to struggle to find it—he often railed at "cleverness," especially his own, how this would tempt one to clever instantaneous solutions, how it stood in the way of genuine thought; and how he himself, in his early work (so he thought) was often "clever," but at the expense of genuine poetic depth. I do not remember—Orlan might—that I exchanged a single word with him that evening, other than saying "Pleased to meet you" and "Goodbye." But Wystan noticed me—despite silence and shyness (he was himself often painfully shy, and knew very well how to be patient, to wait, for the shy).

It may have been a year or more before I ever met him *alone*—I'm inclined to put this sometime in '69. He got me to speak of my *Migraine*, then in press, and was fascinated that I had given Groddeck* so central a place. He then brought out something he had translated, but never published—a work of Groddeck on massage (Groddeck's father ran a sort of gymnasium-sanatorium), with unexpected insights on the somatopsychic effects of this, and the somatopsychic and psychosomatic in general. It was in relation to this Body/Mind theme that I lost my own silence and shyness, and found myself talking, for the first time, quite freely with him and, in particular—he was endlessly, though tactfully, curious—being "pumped" on clinical and personal experiences. I spoke freely of the clinical, reluctantly of the personal: Wystan himself seemed to equate them—and I saw then what perhaps I had only sensed in certain poems, how profoundly, even essentially, "clinical" he himself was. But not in a glib sense (though he was sometimes glib); in a *deep* sense, which combined mind and heart.

* Georg Groddeck (1866–1934), a colleague of Freud's, was a physician and pioneer of psychosomatic medicine.

So, I met him as a friend of Orlan's; and then, so to speak, as myself. Curiously—this is often the case with me—there seemed to be (until close to the end) more *letters* between us than actual meetings. And a sort of decorum was there for—for years. He was always "Mr. Auden," I was always "Dr. Sacks"—I am not sure that we used first names until '70–'71. [. . .]

Love,
Oliver

To Cyrus J. Stow

DENTIST

SEPTEMBER 7, 1984
[NO ADDRESS GIVEN]

Dear Dr. Stow,

I am sorry not to have thanked you before for your very generous letter—though I would scarcely care to present myself, or be regarded, as any sort of "model" for anyone! It seems to me the basic role of the physician is—to listen, to *attend*, whatever special skills and techniques and instruments he may then use. The danger is that all the technology may overwhelm him, overwhelm everybody, and that the still small voice, and the quietude of listening, get lost.

It is this *loss* of quiet listening and attention which seems so much the casualty in our huge, bustling, technologized, bureaucratized world—and it is tragically easy for the young and innocent to become victims. And then the crumbling of all heritages and traditions. But Man is a hardy creature, and a wise one—despite his follies!—and I think we will return to our senses, and the dignity and sanity of life, before it is too late.

Again, my thanks, and kind regards,

Oliver Sacks, M.D.

To Isabelle Rapin

SEPTEMBER 11, 1984
119 HORTON ST., CITY ISLAND, NY

Dear Isabelle,

How nice to hear from you tonight—I'm sorry I have been a bit boorishly out-of-touch since we last met, mostly feeling low-ish, but now perking up since a refreshing visit to the West Coast (it was very nice being welcomed at my old hospital in San Francisco, Mount Zion—I was surprised at being remembered. [. . .] And seeing old friends, familiar, dear faces and places, so it was a good visit altogether; a *nostalgic* one; I half want to go back!).

I enclosed a somewhat reduced copy of Jerome Bruner's* lovely, generous, *deep* review of my book†—along with a funny cartoon (which I like—the *Times* one made me squirm!). To have appreciation from a man of Bruner's calibre and integrity makes me feel much better, has done something to boost my collapsed self-image and self-respect, which sunk to a low after that horrible thing in the LRB. I think now I am beginning to feel more philosophic. One must be, and not react to reactions, but quietly, firmly, confidently, modestly, go one's own way. Easier said than done! I was also surprised and happy to be offered a Fellowship at the N.Y. Institute of the Humanities (the one which sponsored my "Neurology and the Soul" lecture, and has a weekly or fortnightly luncheon-discussion group, a real mix of artists, scientists, historians, philosophers, etc. presided over by Bruner). I think/hope it may offer the sort of intellectual *community* I have felt starved of, and need—because I am more isolated than I should be: but, of course, some of this is necessary solitude, and some defensive withdrawal, so I am not quite sure *how* I will respond. But it makes me feel less like an outcast, a pariah. [. . .]

Love,
Oliver

* Jerome Bruner, a psychologist, was involved in research into cognitive and educational psychology. His many books were very influential, and he later became a good friend of OS's.

† In *The New York Review of Books.*

To Galen Strawson

PHILOSOPHER AND CRITIC

NOVEMBER 30, 1984
119 HORTON ST., CITY ISLAND, NY

Dear Galen Strawson,

I just got sent a bunch of reviews, including yours, which I found of especial interest, and for which I thank you. I will, if I may—but this may be more a talking-to-myself—take up one point which you take up, and which, we both agree, is central.

I am often asked "Why don't you write fiction?" and the answer is, I'm too interested in fact. Equally often I am asked "Why don't you stick to the facts?" and this is a deeper and more difficult question, for it has to do with the *status* of "facts." I am not sure whether there *are* any "brute facts," "mere facts," "raw facts," etc. It seems to me that facts are always *embedded* in experience, and that this embedding—this enormous connectedness—is both theoretical and dramatic; and what is meant, what Wollheim* means, by "leading a life."

A pure case-history, a pure consecution, of pure facts, would be something like: "First there was fever. Then spots appeared. Then the urine became bloody . . . Finally the heart stopped." Curiously, when I first drafted *The Leg* for the LRB, my editor said, "The phenomena are fascinating, but something is missing." When I said "What?" she said "*You,* the experiencer." It is ironic that when, after endless hesitation and inhibition, I introduce the experiencer—and this is all I meant when I spoke of "a neurological novel"—I get rounded on with extraordinary vehemence (not least by the LRB itself). [. . .]

I do *not* regard a neurological "Tale" or "Novel" as a *fiction*—and yet, clearly, it is more than a case-history or consecution. It is a "Tale" because it has a *centre*—a Person, a "Who"—but a neurological tale because it deals with the impact of a "What," a neurological "What." Everything I write deals with relation of a Who to a What—in particular, the patient's *experience* of a neurological disorder (which will include all its resonances and implications for him, the way it alters his life and feelings and thinking, etc.)—how he is altered by neurological disorder (and how, perhaps, he alters it). This

* Richard Wollheim (1923–2003), a philosopher and Freud scholar.

constitutes—a "Story," and a story is *essentially* dramatic. I hope I am not using "dramatic enhancement"—if I do, on occasion, it is an aberration to be pruned; but I must and do bring out the *actual* drama, and it is this which makes my "Tales" dramatic and yet wholly factual. I do *not*, consciously, introduce any "fiction(s)," nor do I feel I have "licence" to do so; nor, I think, do I "make up" or "fabulate"—though I think clinical stories, stories of people dealing with disease and disorder, may *have* a fabulous, or fablelike, quality. It is not a question of "fact" vs. "fiction," but of how to convey a human reality, a reality at once clinical, phenomenal, historical. [. . .]

11

Going Deeper

1985–1988

I (KE) FIRST ENCOUNTERED OS IN 1983, when I was asked to type a section of *A Leg to Stand On.* It was handwritten and covered with water stains, and thus hard to decipher. (He had, I later discovered, written parts of it during breaks from swimming in the Catskills, where he often spent weekends.)

I began traveling to his house on City Island once or twice a week, helping him sort papers and deal with his correspondence, and acting as a sounding board for the new set of case histories he was writing. In time they would become a collection named after the first story, *The Man Who Mistook His Wife for a Hat.*

To Kate Edgar

EDITOR

FEBRUARY 17, 1985
CHANCELLOR HOTEL, SAN FRANCISCO

Dear Kate (if I may),

I hope you [. . .] have not been overwhelmed by the *tidal wave* of my manuscripts—it is far more than any of us (including myself!) anticipated.

I have been enjoying SF immensely. Such a joy, as a start, to be out of wintry New York—This has not stopped me, I'm afraid, from *adding* a bit—just a bit!—to my writing. [. . .] Last night, insomniac, I wrote two more (quite short) pieces—"Murder" and "The Dog Beneath the Skin," as well as 2 or 3 odds-and-ends, 2 more additions to the Preface (not having the original I feel confused where they might go, and will have to see the originals, and now these (in all) *three* addenda, typed out, to see how they fit). Also a footnote to (the title of) the "Hat" piece—quite forgetting whether or not I had already said this. These nocturnal additions, then, should reach Jim* on Tuesday, and I hope may be messengered to you then.

THIS IS IT!! And to make sure I stop adding and fiddling and obsessing, and because it seems a time to rest and take off, I am flying to *Australia* tonight—and will be there, or in the South Seas, until March 4 or so. By then, hopefully, all will be typed, and Jim and I will be able to set all in order, and have ready for Press, by the 15th ("deadline"!).

I enclose another list which indicates the new pieces.

Thanks, more than I can say, for the Marathon task you've embarked on.

Yrs,
Again, thanks, thanks, thanks, thanks!!
O.

To Kate Edgar

MARCH 15, 1985
119 HORTON ST., CITY ISLAND, NY

Dear Kate,

I was meaning to write to you—I had the impulse as soon as I received and read your *spectacularly* beautiful and intelligent typescript, and the way in which you pulled everything together, so that it has the *feel* of a book. I cannot imagine anyone doing this as well as you—and as animatedly and devotedly. And it is very sweet of you to say (after burning so much midnight oil!) that you actually *enjoyed* it.

I always enjoy writing myself—at least if I am in the right mood and mode, and genuinely carried along, and not *forcing* it.

* Jim Silberman, editor in chief of Summit Books, who had also published the American edition of *A Leg to Stand On*.

I value your taste, your judgement, as much as your typing—and greatly appreciate your suggestion of an extra pair of hands (in your case, with your versatility, *several* pairs of hands) for the editing. I am not quite *sure*, at this point, what will be involved.

Jim is reading the ms. this weekend, brooding on it, and will be meeting me on Wednesday afternoon (the 20th).

I have some impulses, and see some necessities, myself. The *impulses* are to do with a few more footnotes—where I think I can enrich the pieces, and/or their connections with one another. The *necessities:* some redrafting here and there. [. . .] Like you, Jim likes, or at least prefers, Preface A, and has told me to stop "proliferating" Prefaces. [. . .]

ONCE THIS BOOK IS OVER I want to launch into another one—on Tourette's.* I would almost *prefer* to write it (for various reasons) if you could type it. I would hope to start *immediately* the "Tales" is out of our hands and off to the printers, and to proceed as fast as physically possible (for this is a book which is *at least 10* years overdue.) [. . .]

I herewith enclose my check [. . .] with more than gratitude. I *know* you did far more than appears in your itemization!

To Orlan Fox

MARCH 23, 1985

119 HORTON ST., CITY ISLAND, NY

Dear Orlan,

I went, impulsively, to Australia—an amazing experience (Darwin called it "a second creation")—practically had my last day in a remote rainforest (a partial reprise of Norway)—but am now back and in one piece . . . with a *mountain* of mail (a quite extraordinary amount from my piece in the NYRB "The Twins"). I have another, "Portrait of the Autist" (illustrated by the Autist) in the coming issue; and yes, of

* Tourette's syndrome, which had come to OS's attention as he worked with the *Awakenings* patients, remained an active interest for the rest of his life, though he never devoted an entire book to the subject.

course, I know "The Possessed" is by Dostoyevsky (perhaps I should call my idiot-autist piece "The Idiot"). And, for that matter, I have called another piece "A Passage to India" [about] a girl dying of a brain tumor, with visions, hallucinations—cerebral? spiritual?—of her birthland. So I am grabbing at titles anywhere and everywhere. Which reminds me, I used an Auden title for a piece I wrote in San Francisco, en route to Australia: "The Dog Beneath the Skin." Which reminds me that Thom Gunn, who has tender and affectionate feelings for you, sends his special love.

I'm afraid I *don't* feel like a perpetual student. Indeed, alas!, my dear, rather geriatric. (But this may be because I reinjured myself in Australia, am 100 lb overweight, and puffing and perhaps looking, like G. K. Chesterton. GKC, that is, with a Bakunin-like beard. And I can't get back, yet, to my own velocipede (tho' I am very glad you have acquired one—it can be a great delight, and it is certainly a health). And as for Romance? It can always come if one is open to it (you, my dear, are; I'm an old celibate). Except when, as this morning, I wake up yearning from a dream, and realize my libido is still there, hungering, waiting, making up little romances, in the wings.

I have now collected and ordered my "Tales"—there are about 25 of them, very different lengths, moods, themes, but I think with *some* sort of unity (if only one of sensibility) holding them together—and will go back to my original title, *The Man Who Mistook His Wife for a Hat . . . and Other Clinical Tales.* Final editing is this week, and I hope the book will come out in September/October. [. . .] And *now,* I have my book on Tourette's, which I shall call *The Possessed* (this was also Tourette's own title for one of his studies in 1885, so I want to honor the centenary).

Let us, at least, write—keep up a "correspondence," even if we don't see each other too often. Friends become more important . . .

Love,
Oliver

To Evelyn Fox Keller

PHYSICIST AND BIOGRAPHER OF GENETICIST BARBARA MCCLINTOCK*

APRIL 12, 1985
119 HORTON ST., CITY ISLAND, NY

Dear Dr. Keller,

I was again away (in London), and again found a mass of mail on my return. But I think I have to write to you before anyone else, and express my deep appreciation of your letter and your book. I have just finished reading *A Feeling for the Organism*—I stayed up most of the night—and I find it almost as remarkable as its subject: Chapter 9 goes so deep, and the last chapter is just beautiful, beautiful.

I have always loved biographies of scientists and naturalists—when I was a child my mother used to read me the biography of Marie Curie (by her daughter)—my mother herself had met Marie Curie twice, and was one of the founder members of the Marie Curie Hospital—but I have very rarely seen the *kind* of biography you here create, in which the character of the thought itself, the feeling which underlies it, its language, and its poignant incomprehensibility and isolation, are so powerfully brought out. It is a magnificent portrait of an extraordinary thinker and person.

I found it especially reassuring, after being chilled, even scared, by Joshua Lederberg† three weeks ago, when he spoke to a group of us about "Discontinuities" and "prematurities" in Science. I thought he spoke of McClintock in slighting (but ambivalent) terms—in an odd sense he could not *get away* from her—and of Science itself in frighteningly dogmatic terms. I am not surprised she found him "arrogant"; I came back with a sort of horror, feeling "And is *this*—Lederberg's 'programme'—all there is to science? Is *this* the beginning and end of the scientific life?" It precipitated a sort of crisis in me—in particular, if you will, a crisis of *love.* For (even though what I do can hardly be called "science") yet it is—I think, at least hope—nearly

* McClintock, who studied the ability of genes to change position on the chromosome, using corn as her experimental model, received a Nobel Prize for her work.

† A molecular biologist and Nobel laureate.

always informed by love, and a product of love; and if I also have had a life with few intense personal relationships, yet it has been warmed, since childhood, by "a love affair with the world." I found Lederberg's "philosophy" profoundly disquieting—but not as disquieting as the lovelessness underlying it. *And yet* (although I know too little of it, and I wished he had talked of *his own* work, instead of sniping at Barbara McClintock), he may himself, perhaps *must* himself, in the days of his own creativeness, have had some intense, quite passionate, feeling for the organism—for the sexuality, (as it were) the *ardour,* of bacteria. Or perhaps not: perhaps he felt himself "mastering" or "torturing" Nature. (I get an unpleasant feeling from *The Double Helix,** and though the achievement was/is so great, so productive, yet the *attitude* to Nature seems to me wrong: perhaps not "wrong," but one in which sympathy or reverence or love plays no part. I sometimes wonder whether one should distinguish "engineering" from "visionary" science.)

I wish I had more *personal* recollection of the matters you describe—I should have, I would have, were it not for "the tunnel" (a sort of dark period I had for almost twenty years, between the opening love-affair with the world, the pubescent romance with science, mostly physics and chemistry, when I was twelvish—and a sort of "awakening," from a long greyness, when I was thirty two or three). In that dark period I completed school, got scholarships, went to University, Medical School, came to America, trained in neurology—but none of it was too real: reality only came back when I really turned to *patients* (in '66).

So all my memories, my feelings, are very attenuated for this time. By the time I came to do genetics at school, I had already lost most of my feeling for science—and the genetics of the mid-late forties did not, in any case, seem too interesting. I *did* hear (though not directly) of Barbara McClintock in '51—it was my first year in Physiology and Biochemistry School at Oxford—and had a sort of feeling of a hydrogen bomb, far off, in the world of intellect. But then, it seems to me, the feeling was forgotten: or, perhaps, more accurately, *replaced* by

* James Watson's account of the discovery of the structure of DNA.

the molecular enthusiasm of those days. It was especially Beadle's* one gene: one enzyme [hypothesis], then. Then, of course, there *was* the hydrogen bomb of Crick and Watson. I have a faint memory of meeting Crick—some meeting in '57—and asking him about Barbara McClintock; and of his answer—a shrug of the shoulders.

The *coldness* of Joshua Lederberg (which especially scares me when I think of the power he wields) reminded me of a man I met years ago—extremely clever, cold, and precise—and of his saying to me: "'Dear'? What do you mean 'dear'? What do you mean saying your patients are 'dear' to you? What is 'dear'? What is 'dearness'? It doesn't exist in the world!" This produced a radical shivering, an *algor,* of the spirit until the next morning, when I had a letter from Luria, which started "My dear Dr. Sacks." And then, I was reminded, I remembered, I again felt, that "dearness" existed: I knew I was "dear" to Luria, and, like him, found my patients "dear" (and, perhaps, was sometimes "dear" to them).

Although (as with my mother, indeed both my parents) the world of "being" is chiefly given to me by patients, there remains a love for other forms of life, especially plants. When I was in Australia, a few weeks ago, I *had* to go to the rain forest—I have an especial love of ferns, tree-ferns, I don't know why—and got an overwhelming sense of why the young Darwin, seeing it briefly, was overcome with wonder, and spoke of "a second Creation." I was especially moved by your descriptions of Barbara McClintock when she spoke about plants (pp. 199–200), and reminded of what Darwin often said about plants: how enormously they were *underestimated,* especially by biologists (and my favourite Darwin books, and the ones which to my mind show most intensely *his own* "love-affair," are the botanical ones, above all *The Fertilization of Orchids,* which is sometimes like lyric poetry).

Nor does it seem to me that this love, this "feeling for the organism" is necessarily confined to the "organic" (unless, of course, the whole world is seen as "organic"). I had an extreme love—in those far-off days of scientific pubescence—for the *elements:* I thought constantly of their properties, carried them with me when I could

* George Beadle, a geneticist and Nobel laureate.

(my friends still remember how my pockets were full of bismuth and manganese, how I raved about lithium, had a spoon made of indium). I had the sense of a personal relationship—they became familiar to me, "friends." It seemed to me, sometimes, that I could enter their world (I love your description of how the chromosomes enlarged for McClintock, how she felt she was entering *their* world)—that I could actually feel their "feelings." And as Goethe took the phrase "elective affinities" from chemical to human relationships, I found myself taking it back to chemical relationships. Specifically, I could not believe the inert gases were *wholly* inert: they *had* to have possibilities of some relationships, I felt; and this conviction allowed me to "predict" the fluorides and oxyfluorides of xenon. Many years later, in the early sixties, when Neil Bartlett produced some of these, some of my friends said "But you were *talking* of these right back in '45." Similarly, without having heard of Mendeleev, I predicted, as he did, the various "eka" and "dwi" elements: "predicted" is too pallid a word—they *had* to exist (if only ideally)—because the nature of the chemical family, chemical relationships, demanded it.

About a year ago, moved by some nostalgia, I bought a copy of Mendeleev—the wonderful seventh edition he published, as an old man, in 1902, so rich in footnotes, and personal annotations (and welcoming the new wonder of radioactivity). I found this enchanting—it became my bed-reading for weeks. And there is absolutely no doubt of *his* "love-affair" with the elements, with *all* the chemicals, the "plants" (and the "genomes") of chemistry. This recurred to me when I came to write about "The Twins."

(And, here again, there was something personal—a sort of nostalgia—involved: for before the elements, when I was a child, I too had a strange love for numbers. They were, for me, the *first* precursors of some sort of "cosmic" feeling, but also personal friends. But I don't think I'm "projecting" in seeing this in The Twins: rather, my own sense of such feelings "sensitized" me to *their* sense.)

I meant to say something, perhaps to "analogize," about being a physician, but I have exhausted myself (and doubtless you!) by going on about other things: but your book was so exciting, I *had* to respond.

Certainly, for a physician, "a Feeling for the Organism" *has* to be central. And yet so often, frighteningly, these days, it is not. A majority of my colleagues, I sometimes fear, have come to think like molecular biologists. Not that I in any way "disapprove" of molecular biology (how could I? I am as excited as anyone else); what frightens me is not the acquisition of new viewpoints, but the *loss* of that ancient and fundamental "feeling for the organism." This feeling *imbues* the best XIXth century Medicine, all "naturalistic" Medicine, but has become, or can become, a total casualty with "Science" or, at least a certain, narrow sort of scientific method and approach which, as McClintock says (p. 201) cannot give us "real understanding." What I found marvellous in the last chapter of your book, is *her* holding to this fundamental, naturalistic "feeling" or "vision," but *renewing* it, deepening it, with a unique and new insight so that Nature and Science are totally fused. It is this which seems to make her so extraordinary a figure—at once "old-fashioned," Victorian, and at the same time a "prophet," at least precursor, of a scarcely-imaginable future: at once XIXth and XXIth century. This is the *kind* of neuro-science which I dream of—and which I think was also dreamed of by William James (the epigraph to my epilogue comes from his marvellous essay on Agassiz).*

I have said *much* too much—sorry—but I was affected deeply. [. . .]

With my best regards,
Oliver Sacks

* What OS calls here his epilogue is actually the final chapter of *A Leg to Stand On;* the epigraph by William James reads:

> The truth of things is after all their living fulness, and some day, from a more commanding point of view than was possible to any one in [a previous] generation, our descendants, enriched with the spoils of all our analytic investigations, will get round to that higher and simpler way of looking at nature.

To Humphry Osmond

PSYCHIATRIST[*]

MAY 11, 1985
119 HORTON ST., CITY ISLAND, NY

Dear Humphry (if I may),

Many many thanks for your card about The Autist Artist,[†] and the fascinating Sergeant's Story, with your comments, and the paper on "Mood Pain." I will type, because I lack your calligraphy; and reply too briefly, because I'm a bit "down." But if nothing else I want to say how much I *appreciate* everything you say or send me. [. . .]

I was very interested by your observations on *mood pain:* I have long known what Johnson[‡] said, and always felt it myself when I got melancholic (not infrequent!). Whether one would *actually* feel happier to find one's spirits restored, but a limb amputated, is another matter: but this is certainly how one *feels,* and it is certainly a feeling which can impel one towards suicide.

In my own case, I should or should not add, it is not suicide but (do they call it?) "parasuicide" which is the danger, (unconsciously) provoking accidents and dangers. I was in a very violent mood at the time of my accident: I had had two or three potentially serious "accidents" earlier the same week—swimming to exhaustion in the icy fjord; managing to lose an oar while rowing, foolishly, in a storm; and *then* the mountain. As soon as the accident happened, and its seriousness came upon me, I *heard* an inner voice (? my super-ego) say: "O dear, Ollie, I didn't want you to go *this* far. You're exonerated from everything—just try and get out of here!" There was an instant lifting of a terrible "mood pain" in the instant of the actual injury. (Auden, in his review of my *Migraine,* quoted something of Coleridge's about physical pain being more tolerable than mental.)

Having smashed up both legs moderately badly—I did the same

* In the late 1950s and the 1960s, Osmond investigated the use of hallucinogens, particularly LSD, in treating alcoholism and schizophrenia. He invented the word "psychedelic" and was the first to give hallucinogens to Aldous Huxley.

† This essay had just been published in *The New York Review of Books.*

‡ Samuel Johnson (1709–1784), lexicographer and writer, famously said, "He who makes a beast of himself gets rid of the pain of being a man."

last year, along with my right arm, in a Bronx gutter,* not a Norwegian mountain—I myself certainly find "purely physical" pain much easier to bear. But then physical pain—especially if it is gnawing or recurrent—can erode one's morale and *call out* "mood pain" (as you say in your paper). Thus I have had a grumbling, infernal, unstable "disc" for 25 years . . . and *this* can seriously undo my morale; so that I think "I'd give a limb to be rid of *this* pain, this chronic hateful brittleness, vulnerability"—I never found an acute injury or pain "hateful" in this sort of way. I have also known in myself and heard from others, how hateful a *migraine* can be: Freud said of his (migraines): "the essential man is unaffected," but this may have been more an expression of *his* strength of will, or "ego." I think "the essential man" can be almost swept away, and one may have to "*hold on*" through a migraine's fury, as through a psychosis (though this, admittedly, is the other extreme).

I have written more—and more "confessionally"—than I intended. I take the liberty of sending you my *Migraine,* of which a new edition comes out this month (but it adds little to, and may be worse than, the original 1970 one).

With kindest regards,
Oliver

To Kate Edgar

AUGUST 30, 1985
119 HORTON ST., CITY ISLAND, NY

Dear Kate,

I hope you are well. I have been meaning to get in touch with you for a longish time, but have felt . . . inhibited, hesitant, uncertain, whatever . . .

One (perhaps minor) thing which bothered me—but I should hate to be seen as ungrateful—is this: that Colin decided to cut *all* Editorial Acknowledgements out of the book (including, therefore, your

* OS had slipped on black ice, rupturing his right quadriceps (an injury oddly similar to that sustained in the Norwegian accident) and injuring his arm.

name—and his own), basically because he cannot bear to be in the same room, or the same paragraph, as Jim Silberman. But you know, and I know, and he knows, and Jim knows, and others know, what a crucial part you played in the genesis and organization and editing of this book, even if it is not formally expressed (as it should be). I take the liberty of sending you a copy of the advance proofs, which you may very much recognize and feel as *your* book.

For various reasons, production and publication has been delayed. [. . .] Therefore the US publication is rescheduled for the first weeks of January. [. . .] I am feeling mildly annoyed with Summit about the whole thing. I expressed this, though not too strongly—because I *like* him—to Jim: I also said that if there had been someone like *you* around, this would not have happened, and that I was now concerned with the next (the Tourette) book, which would need exceptionally discerning and sensitive editorial guidance and help. [. . .] I asked him whether, if I worked with you, Summit would accommodate you and pay you appropriately. He said yes, of course, and that I should contact you. But, as I said at the start, I somehow couldn't, or didn't [. . .] because I wouldn't want to make any working between *us* contingent on any arrangement with Summit. [. . .]

What I am leading up to, I suppose, is [. . .] whether we might make some sort of *private* arrangement between us, my retaining you as—I can think of no term general enough!—but as some sort of literary/intellectual factotum, inciting, organizing, typing, editing, etc.: a Tourette book, in particular, but also other things: odd pieces, case-histories, essays, lectures, etc. I *need* someone to turn to—though quite WHAT would be involved, I don't know. Sometimes, as you know, I can write enormously in a very short time—and then I may fall into a dead period when I find it difficult to do *anything*. And, of course, you yourself are working full-time; you have [. . .] I know not what other commitments and demands on your own time. [. . .]

Anyhow, do, please, give me your thoughts on all this—how *you* feel, how *you* stand, and so forth.

And, again, more thanks than I can express for your crucial part in making *The Man Who* . . . possible.

Yours,
Oliver

To Samuel Sacks

SEPTEMBER 21, 1985
119 HORTON ST., CITY ISLAND, NY

Dear Pop,

First—we have exchanged telegrams—all my best wishes for the New Year. [. . .]

This is something of an "in-between" time for me—it always is, when I am *waiting* for a book, and I have not found it too easy to settle my mind on anything much. An unexpected interest this summer has been *deafness:* the problems of the congenitally deaf, and sign language. I will be doing an Essay-Review on this [. . .] for the *NY Review of Books*—but in so doing have aroused my own interest in the subject, and the unexpected light it casts on language and languagelessness—this makes it a "neurological" subject, like Aphasia. Last weekend, in my super new-car, a Toyota Cressida—which I love and which, for the first time in years, lets me feel safe, and gives me some pleasure, driving—last weekend, then, I drove up to Cape Cod, and then took the ferry over to Martha's Vineyard, an island (next to Nantucket) settled in the 1640's and where, in the remoter, more isolated (and more inbred) villages there was a singular hereditary deafness for 250 years (first noted in the 1690's, the last deaf disappeared in the 1950's). What is fascinating here is that *everyone learned sign language,* hearing and deaf alike, so there was none of the usual isolation, discrimination, etc. which can make the lot of the deaf so painful elsewhere. I found it a very moving and interesting experience.

My Chair* at Sarah Lawrence has an inaugural Lecture soon, and will then settle down to a series of Thursday evenings, 2–3 hours each, in which I will deal with various aspects of Human Nature—Memory, Language, Identity, etc., as these may be illuminated by Neurology. It will be my first experience with undergraduates—philosophy students, etc—I hope it goes well. Things will be recorded—perhaps it can provide a foundation or form for a book.

I have also now been formally invited to a "Regents" Professorship at the University of California, and will spend February in residence

* Really a one-term series of lecture-seminars called "On Being Alive."

on the Santa Cruz campus (this is fairly close to San Francisco)—it will get me out of the dreadful weather here, and into a State I have never ceased to love. [. . .]

I will be breaking the fast with [Madeline Gardner] and family. All of them, as well as the Weinbergers* up in Canada, join me in wishing you a Happy New Year, and, soon, a Happy Ninetieth as well!

All my love,
Oliver

IN THE FALL OF 1985, *The Man Who Mistook His Wife for a Hat* was published in the United Kingdom.

To Richard Gregory

DECEMBER 26, 1985
119 HORTON ST., CITY ISLAND, NY

Dear Richard,

I just received a copy of your review in *New Society*, and must write to you at once to express my gratitude and pleasure. I cannot write much more than this (or will be in trouble if I try), because I have a nasty 'flu with a highish temperature, and will fly into febrile tangents if I try to think (but perhaps I do that anyway).

There is *no-one* whose judgement in "our" realms I respect more than yours, and as your review of *Awakenings* (can it be 12½ years ago?) was crucial in allowing *me* to come to terms with it, so your review of *Hat* will assist me to come to terms with what may be a considerably more eccentric production. [. . .]

I am not sure how much I myself can answer the question of how much is actually observed and how much "interpretative creation," as you so nicely put it—they seem to merge into one another, hintings into guessings. The theorist and/or the dramatist in me may push towards some sort of completion or formulation, even leap to this, without my being quite aware of what is happening. When I showed

* Other cousins.

"Witty Ticcy Ray" the piece on him—perhaps in itself an impropriety or cruelty, like showing the mirror to Jimmie,* but I have never known how open to be in this delicate and difficult matter of writing about my own patients—he first said, severely: "You take a few liberties!" But when I took out my red pen, and said "What shall I remove?" he said "It's *essentially* true. Leave it as it is." I am somewhat haunted by all this. I was going to quote in my preface Molloy's words: "Perhaps I am inventing a little, perhaps embellishing, but on the whole that's the way it was"—but Colin forbade me! On the whole I have found it difficult enough to try to imagine how things were, or might have been, with so many of the patients: perhaps the essential question has always been: "How is it, how might it be, with such-and-such a person in such-and-such a situation?" and I cannot disentangle observation and interpretation in the attempt to reach some construction. But perhaps I should have sought to be more conscious, or more explicit, in laying bare my assumptions. The next book, I think, will be a series of "Essays"—thematic rather than narrative in approach—and an Essay may permit what a Tale does not.

I have felt for a long time that there was a certain parallel between us, your examining Illusions as I approach Diseases, both of us seeing in these "simplified" situations the possibility of delineating Mechanisms, or Strategies, "normally" invisible. I suppose this has always been the justification of "pathologizing," and the analogy so often made with experimenting.

Beyond this there lies a deeper parallel, in that we are both driven, by our experiments, or observations, to Philosophy—or, rather, to "philosophize." This is oddly rare, even suspect. Did not Rutherford† say to his new workers at the start of each academic year: "I'll have no discussion of *the Universe* in this laboratory!" You and I, without losing our scientific marbles (at least you don't lose yours, I may sometimes lose mine) feel *compelled* by our observations to think about Inner Worlds, Inner Universes, the constructing powers and mechanisms of Mind. [. . .]

I am very glad you spoke, in your review, of the *dignity* of the

* The amnesic patient described in "The Lost Mariner."

† Ernest Rutherford, a Nobel laureate in physics.

patients (even though often subject to the most terrible *in*dignities). I don't think of this much, but I suppose it is implicit, and forms, somehow, the *moral* centre of the doctor/patient relation and work; that provided one *respects* the patient, one can do or say almost anything (if one does not, one may become a Doctor Moreau or Mengele). I hope I respect them, even though I often laugh at or with them; it is curious how much laughing there can be in the most awful places and situations. People often say "How can you work in these frightful grim places, with these incurable diseases?" but there is dignity and humour all the way (otherwise it *would* be intolerable). [. . .]

I hope you are in good health and spirits, and effervescing as usual with new work and ideas. And I thank you more than I can say for your lovely review.

All my best wishes for the New Year,
Oliver

To F. Robert Rodman

JANUARY 26, 1986
119 HORTON ST., CITY ISLAND, NY

Dear Bob,

[. . .] I have felt, if not actually exhausted, *close to* exhaustion for the last 3–4 months; and if not actually ill, then *close to* illness (the very words "close to" are ominous, evocative: should one think of Health and Illness as *Lands*? Contiguous, adjacent, whatever? Like Gowers' "Borderlands of Epilepsy"?). I did get a horrible and persistent 'flu around Xmas—an unprecedented bronchitis/pneumonitis, fever around 102. Then England—rushing round, and a *very* distressing situation at home: my brother, floridly psychotic, physically attacking my old father—and everyone, apparently, in a state of laissez-faire, saying there was no call to *do* anything unless there was an "emergency." Then back here, to a mountain of piled-up work, correspondence, etc, and a mounting mountain of "publicity" etc. associated with *Hat* here—and promptly fell ill again, I thought with a Shigella, a beastly dysenteric thing with a high fever, but it subsided after five days. Maybe it was a relapse of the 'flu. So I am not just speaking metaphorically when I say "close to illness," even tho' the "illness" is (mercifully) trifling. The whole *organism* is exhausted—I have never felt so unfit, so stressed, so lacking basic physical resistance to trifles. I can only hope that I find California "relaxing"—I dare not hope that it would actually be "refreshing" or "regenerative"—and not an ordeal which I have to go through.

Clearly, to use the Californian word, I need some "Stress-Reduction"—I have never been so conscious of "Stress." Clearly there must be some good thinking and doing in the *economy* of life—I find (for example) that I am receiving nearly 5000 letters a year, many interesting, many with attached manuscripts, theses, books, theories, poems, etc. And I feel compelled to answer them all. (One of the reasons why I sounded low when we spoke was that I had been answering letters since five in the morning.) Paradoxically, with finding myself the world's correspondent, my *genuine* correspondence—as with a few choice spirits and minds like yourself—my *genuine* correspondence has taken a beating. I no longer have time, leisure, to write proper letters, since I am too "busy," every day, "dealing with correspondence" (to say nothing of a score or more phone calls). And sleeping, and eating, and exercise, is all awry—I eat grossly, sleep poorly, have had no exercise for months. Bad. Bad at an *animal* level. The *animal's* in bad shape.

But what worries me more is "spiritual" sickness—I mean neurosis (though not *just* neurosis). Shengold suggested, with some hesitancy, some months ago, that I should consider going *deeper* with him—there has been no "analysis" for years, but a (relatively superficial) "supportive" therapy. Obviously more than this, but still not deep. This has kept me alive, kept me going, kept me "functioning," in a fashion and perhaps made possible my brief *bouts* of creativity (I was going to write "ecstasy," but am not sure how much ecstasies are connected with creativity. I have known no ecstasy of mind since 1982, though for the 15 years before this I had several ecstasies a year, some a few minutes, some—sweet and mild—lasting weeks). He has hesitated, and I have hesitated, he feels hesitation is right—for I may not *tolerate* anything "deeper." Perhaps I only "function" at all through keeping the deep at a remove. He also observes that I don't *complain*, say, of sexual deprivation—though this is absolute, and has been, for at least seven years (tho' I complain frequently, and increasingly, of "loneliness" and "boredom"). But what *most* concerns me—for I would, not gladly but resignedly, suffer any instinctual or relational gratification if I could hold on to some intellectual or creative depths—what most concerns me is *the feeling of becoming superficial.*

This has, perhaps, come to a sort of climax with *Hat*. Everyone says "Ah, it is *brilliant*" (and so it is), but I say to myself, and others perhaps say, "But compared with *Awakenings* it is *shallow*." It lacks the passionate intensity, the passionate feeling for others' lives, that that showed. I don't know whether others feel this, but *I* certainly do. It seems to me that I no longer care as deeply for my patients as I did—so much so that on the rare occasions when I find myself caring, I have to fight back tears. This is partly because of a change *in* the patients I see—a third of my patients at Beth Abraham, for example, are in COMA. Not much relating with them, as a start. But there is also a change in *me*. I started to feel it (I think) in 1979—I see, as I leaf back through my notes, a perceptible decline in concern and passion starting then (it was in 1979, coincidentally, that I [. . .] perhaps closed my heart to any future or possible love-relationships). [. . .]

Perhaps it is also, specifically, something to do with *enterprises*—*Awakenings* was a grand enterprise, so too was my Tourette thing. (I am violently riled that I have not felt able to write about [Tourette's], because this *could* be a great book, on a great theme, while all the rest, even if "brilliant," is piddling and fiddling and trifling around.) I have not had any real *enterprises* since '76, and I am not hopeful that I will have any "enterprises" again (I could not help contrasting this feeling, wistfully, and rather enviously, with *your* feeling of bubbling with a thousand, or at least *lots* of things to do). I am not actually *doing* anything. I have nothing to do. I am drifting, and *Hat* was merely a random collection of odds-and-ends caught while drifting. Or is this a self-libel, a falling-back into the depressive position?

I was reading a biography of Reich* which the publishers sent me some months ago. With admiration, with pity, with all sorts of feelings. How *mad* his later enterprises became; and yet how they *energized* him and those who went with him. I found myself thinking: what is "better"—a *mad* enterprise, a *mad* energy, or none at all?

* Wilhelm Reich (1897–1957), physician and psychoanalyst.

But merely to ask such a question is itself depressive. Because these are not real choices: the real thing is *good* enterprise—a *good* enterprise, lots of them, full of *good* feeling, *good* energy. This is what I felt, almost flame-like, flaming in you today.

When I spilled some of these feelings over to Eric—I do not generally spill them: only occasionally, and to very old and special friends like you and Eric—he said tersely, "But you're having no sex." And I have a feeling that Shengold said something like this—but more delicately and deeply. Is the superficiality of my work, then, due to superficiality of relationships—to running away from whatever has deeper feeling and meaning? Is this perhaps spoken of, in a camouflaged way, when I describe the "superficialization" of various patients—tho' in their cases ("A Matter of Identity," "Yes, Father-Sister," etc) it is due to organic disease? I speak of the fascination and horror of "equalization"—and clearly I see this as some sort of allegory for myself. But it is a false one, because it excludes dynamic. What is going on in myself is distancing, remove.

I had difficulty responding to your book,* partly *because* of its so-manifest warmth and concern—both for your patients, and the predicaments and problems they exemplify; partly, specifically, because it made me yearn *as* a patient—"How lucky his patients are to have him!" I thought—and contrasted this with my feeling of unsatisfactoriness in my own therapy. Though it was just about this time that Shengold made his suggestion of "going deeper."

This suggestion at once bothers, frightens, pleases, and reassures me. I thought, first of all, when he spoke of "a change," that he was going to *dismiss* me from treatment: "I've done all I can for you—now manage on your own!" I was then rather moved that after 20 years (!!) he should still consider me "worth" extra work. But then I was frightened at this confirmation (as I saw it) of the depths of my own superficialization and neurosis. Then excited at the notion that perhaps something was *possible*—when I had come to consider boredom-depression-loneliness-cutoffness-meaningless the "norm" and necessity for me, from now on: a sort of grim (occasionally ironic) "holding-action" until death put an end. Then alarmed again

* A draft of *Keeping Hope Alive: On Becoming a Psychotherapist.*

at the notion of what "the depths" might bring up. Then shoulder-shrugging, with the thought "Nu? So what have I to lose?" But what one has to lose, of course, may be just that quasi-stable if fragile "functioning." So there is reason to hesitate, and, in any case, what with rushing here-and-there, London, California, Texas; with this nutty *akathisia,** NOTHING can be started. I must get back, settle down to dailiness/regular life in New York, *before* anything "deeper" is even physically possible, and this, in the nature of things, can't be before the end of March.

I don't know why I have written all this—and very likely I will not post it. It does not merit or demand a "reply." I am sorry to burden you with all this trash. But it is not so much to burden you as *dis*burden myself. And perhaps some of it is due to, an artefact of, exhaustion, and the factitious life-weariness which this (like a migraine) can sometimes bring.

I slightly fear leaving New York, leaving Shengold, as the corollary of my month in California. But perhaps Santa Cruz will turn out well. I will certainly *strongly* hope to see you out West.

All my love to you and Kathy,† the girls (women now), all your tribe.

Love,
Oliver

THE MAN WHO MISTOOK HIS WIFE FOR A HAT was published in the United States in January 1986, with a modest first printing of about fifteen thousand copies. It received more attention than anyone expected for a collection of neurological case histories, thanks in part to a review in the daily *New York Times,* and it slowly began to find an enthusiastic audience. Still, OS's publishers were surprised when it appeared on the *Times* bestseller list in March. It remained there for six months, and OS was increasingly deluged with correspondence from both fans and academics, as well as invitations to lecture.

* A neurological term for restlessness.

† Rodman had remarried.

To Donald Campbell[*]

CORRESPONDENT

APRIL 19, 1986
119 HORTON ST., CITY ISLAND, NY

Dear Sir,

Thank you for your charming, open and friendly letter. I am glad you found all sorts of "resonances" in yourself to some of my *Hat* patients. Certainly (although I hope I have avoided any gross projections of myself) I could not have written them had I not felt something of their own "deficits," "excesses," dispositions, *in* myself. They represent extreme examples—pushed out of balance, perhaps—of elements we *all* of us have. Coincidentally I seem to have many of the peculiar strengths and weaknesses (but perhaps one should not use value-words like these) you describe in yourself.

I do not know the "answer" to your particular conundrum—if, indeed, it requires one. (This too I share, at least in the allied art of "speaking," I *cannot* prepare, or remember, anything—but, by God, I can improvise, at least if "mood," atmosphere, etc. are favourable.) That you lack the "reportorial," but have the "improvisational," I think you should see as a *gift*—not a "conundrum." I think that one needs to recognize and *honour* one's deepest dispositions (which, in any case, one cannot change). One must be "oneself"—whatever this means!

With kind regards,
Oliver Sacks

* An English correspondent who wondered why, though he loved acting and was able to perform improvisations onstage, he was completely unable to memorize a script.

To Samuel Sacks

APRIL 21, 1986

119 HORTON ST., CITY ISLAND, NY

Dear Pop,

Sorry I have let so long elapse without a letter—I seem to have been constantly on-the-move, travelling, lecturing, getting distracted and exhausted.

There are certainly two sides to this business of—*being known,* and suddenly finding oneself an object of ceaseless attention and demand. I begin to long passionately for peace and quiet—I receive at least fifty or sixty letters and phone-calls a day, and, if anything, this number is *increasing*!

I WILL make up a parcel with lots of reviews, interviews, etc. but their number increases every day. I herewith enclose the latest best-seller list from the (New York) *Times*—as you will see, my book is now number 3! And a copy of an advertisement inserted in the *Times* last month, which will give you some idea of the widespread favourable opinions. But clearly, it is neither critical notice nor "marketing," neither reviews nor interviews, which finally *launch* a book, but a sort of gathering murmur, mouth-to-mouth, in cafes and street-corners, in every corner of this immense land, a sort of deep resonance in the hearts of quite ordinary women and men, which makes these clinical tales, seemingly remote, so familiar.* It is impossible to sell 70,000 copies (the number sold so far) without this deep and widespread emotional response—I get hundreds of letters (from North Dakota, Oklahoma, the most remote and far-out places) about "Jimmie" (The Lost Mariner), for example: he seems to have become a figure who means something to people, almost a friend, so I get dozens of letters asking about him, "How is Jimmie?", as if they knew him, and wanted to know him. It really is a remarkable phenomenon.

I find too (I am less happy about this) that countless people want to make plays, films, even operas and ballets, about my "characters"—thus there will be an *opera* about "Hat" (the title-piece) performed in

* This was certainly the case in 1986, before the internet and at a time when independent booksellers dominated the scene.

London in June;[*] there was already a sort of jazz-oratorio about "The Twins," performed in Edinburgh and Glasgow (without my knowledge or consent); people want to make a ballet about "The Disembodied Lady" !! and so on. These secondary things have some power to bother and embarrass me, and I may take steps to prevent their coming into being; unless, of course, it is a work of high art—like Pinter's. This too is very curious: I seem to have provided a stock of *images*—myths or fables for our time—and I thought I was merely writing simple case-histories! [. . .]

Erik (Korn) was just here from London, doing a bookfair here in New York. He is my oldest and dearest friend—and it is lovely to see him. Jonathan (Miller) is also here (and there), as mobile as myself, or more so, quick as quicksilver, I can't catch him. [. . .]

I think/hope I will come to England next month. [. . .] If I cannot make the Seder, I should make next month, and will try to get back to England for visits every few weeks or months. Certainly now I can *afford* to travel as I wish, and to take off time as I wish (except I feel anxious about leaving my patients, so it needs to be short, frequent travels, not extended ones).

My best Pesach wishes and love to you and Michael, and all the family in England.

Love,
Oliver

THOUGH HE CONTINUED HIS PRACTICE at the Little Sisters homes and elsewhere, seeing perhaps several hundred patients a year, OS had also become, in a way, neurologist to the entire world. Letters from readers and specialists continued to accumulate, and this became a rich extension of his medical practice, as people wrote to him about their own odd experiences. In early 1986 he received a letter from Robert Isaacson, a painter who had suddenly and puzzlingly lost all of his ability to see

* A chamber opera adaptation with music by Michael Nyman and libretto by Christopher Rawlence.

color. OS spent a good deal of the next few years investigating this with Isaacson, and eventually published an essay about him, "The Case of the Colorblind Painter."* This project marked the beginning of his engagement with a number of neuroscientists, and the growing use of new brain-imaging technologies such as MRI, fMRI, CT, and PET scans opened up whole new avenues for studying the brain in real time. People began talking of "neuroscience" as distinct from neurology.

One reviewer, John C. Marshall, an English neurologist, reviewed *The Man Who Mistook His Wife for a Hat* for *The New York Times Book Review* on March 2, 1986. In an otherwise positive review, he expressed a single reservation:

> *Dr. Sacks' admirable presentation of seemingly bizarre phenomena and difficult concepts is marred by only one weakness. It is stylistic, but nonetheless extremely annoying. He constantly plays naive about the neurological literature. He would have us believe that an experienced neurologist could fail to have read anything about many of the standard syndromes of behavioral neurology until he had himself seen a particularly pure case of the condition in question.*

In fact, OS protested, his "playing naive" was probably more a product of his spending far more time with each patient than an average physician would, thereby uncovering a more nuanced picture. Often he was able to find descriptions of the odd syndromes he was seeing in long-forgotten papers from the nineteenth century, which he regarded as more detailed and thoughtful than the current literature. While some of these older papers would have been available in medical libraries, many of them came to OS from rare-book dealers who specialized in science and medicine.

* OS referred to Isaacson by a pseudonym, "Jonathan I." This essay, which appeared in *The New York Review of Books* (the journal's first ever issue printed in color, to demonstrate what the painter could not see), was later revised and became the first chapter of *An Anthropologist on Mars.*

To Antonio Damasio

NEUROLOGIST

MAY 5, 1986

119 HORTON ST., CITY ISLAND, NY

Dear Dr. Damasio,

I have a bad (or good) habit: namely that I remain unacquainted with the neurological literature, bye and large, *until* a patient with a particular problem appears and excites my attention. *Then* (and only then) do I start to look for precedents and colleagues.

Thus although I think I saw, in passing, some of your previous articles in *Neurology* on central achromatopsia, prosopagnosia, etc. they did not sink in, because I had never then seen *a patient* with such problems. Now, having recently been consulted by a fascinating and gifted patient, a painter, with rather sudden onset of achromatopsia, possibly preceded by some alexia, I have started to cast around me: and, of course, come upon your own superb article in Mesulam's book—by far the clearest and *deepest* discussion of the subject I have seen (and, of course, the extraordinary brain-imaging illustrating your article).

I am wondering, first, whether it would be possible for you to send me copies of your articles in *Neurology* 31, 32, 33, 34 and any others you may not have listed in the bibliography. Second, whether you have, or would advocate, any particular tests or "protocol" for the examination of such a patient. I should add that (in my own rather crude testing) I have observed deficiencies in gray-scale discrimination as well as total color-blindness (with *no* color-anomia or disorders of color-association that I can detect). Low-contrast differentiations are impaired, high-contrast ones exaggerated. There are some hints of prosopagnosia,* if not in real life, then when given rather confusing and ambiguous pictures full of faces. He is an exceptionally intelligent man and is doubtless using *inference* far more than he realizes. Though greatly upset by his loss of color-vision, he is now doing some black-and-white painting, with a drama of contours, and modulations of hinted depth and movement, which are quite astonishing. Also, for the first time, some sculptures. After 3½ months, he

* Inability to recognize specific faces.

is beginning to accept he may never recover color-vision (tho' one hopes he may), but is turning with extraordinary creative intensity to the exploration of form, movement, depth. He has had CAT scans and NMR* read as "normal"—*is this consistent with his having genuine* (associational) *defects*? Clinically, so far as I can judge, he is genuine, and I note you conclude that "the exact location and range of the lesions that can cause achromatopsia is a matter of current research." Would a PET scan be useful? Have you other suggestions?

I am sorry now I did not talk with you more in San Francisco, though I greatly enjoyed meeting you there, and was very conscious of an exceptional cast and calibre of mind. Indeed, I think I said something impertinent like "How come someone as intelligent as yourself is slated to talk about brain-imaging?" not realizing, or forgetting, your fundamental contributions to perceptual neuropsychology.

But, of course, I had not then met this extraordinary patient.

I would be most grateful for any thoughts on the subject you may be able to share with me.

With kind regards,
Oliver Sacks

FRANCIS CRICK AND JAMES WATSON famously discovered the double helix structure of DNA in the 1950s. Crick later turned to investigating the nature of consciousness, focusing particularly on the neural correlates of perception. His 1994 book, *The Astonishing Hypothesis,* would make the case that human consciousness could be entirely attributed to its neural activity, and proposed that the most tractable way to study this would be through focusing on the primate visual system. This research, of course, overlapped considerably with OS's phenomenological investigations into patients with visual problems. At one dinner, OS sat next to Crick, and they discussed the colorblind painter and other visual-perception anomalies. Crick urged OS to send him more details.

* Nuclear magnetic resonance, an imaging technology related to MRI.

To Francis Crick

BIOLOGIST AND NEUROSCIENTIST

MAY 30, 1986

119 HORTON ST., CITY ISLAND, NY

Dear Dr. Crick,

It was an extraordinary pleasure and privilege meeting you—a little like sitting next to an intellectual nuclear reactor! I never had a feeling of such *incandescence*—wonderful to have such a bright, sharp, laser-like mind illuminate problems and predicaments I *see* (but whose significance and explanation is quite beyond me).

I *would* like (though I also fear!) to present some of these clinical observations on vision to people whose power to theorize so exceeds my own.

I take the liberty (with some trepidation) of sending you *Leg* (the UK edition), which contains an account of how I (partially) lost, then recovered, stereopsis (pp. 119–120), and a (perhaps-too-ornate) account of a visual migraine, with "mosaic" and "cinematic" vision (pp. 66–70). Also an article ("The Great Awakening"), which preceded my book, describing some similar, and other, phenomena in l-DOPA patients.

I was speaking to Dr. Damasio today, who told me he had seen you in Boston, and that you had mentioned my colorblind patient to him. I hope to *film* the patient (and some of his astonishing productions) this weekend (first for an NBC segment, then, hopefully, NOVA);* and hope too that Dr. Damasio will be able to go over him in Iowa. It is a great privilege to encounter patients such as this—remarkable (and *articulate*) experiments of Nature.

With my deepest respects, and best regards,
Oliver Sacks

OS DESCRIBES IN *ON THE MOVE* how he met Larry Perrier in a New York City gym in 1966. Larry was an unusual man who spent many hours each day walking long distances, often in

* Though the filming did occur, it was never broadcast.

flip-flops. When he visited New York, he often slept on OS's couch.

To Larry Perrier

ITINERANT FRIEND

OCTOBER 26, 1986
119 HORTON ST., CITY ISLAND, NY

Dear Larry, old friend,

I was so moved to get your letter—I don't know that you (we) ever wrote before.

Sorry I'm not answering sooner, but I have been finding life killing (rather than thrilling) with this tremendous barrage of attention, and hardly seem to have any time for myself (or my friends). The only *real* peace this year was a time in New Zealand, the sweetest and sanest place I've ever been (if my odd life has a "Part III" it may be there, or in Australia). The travels I've *had* to make are much less pleasant—but (lest I sound petulant) I have *also* been very warmed by the response to my book, and found a lovely welcoming wherever I've been, so it's an odd and complex mixture. I *do* meet more people—in a way now I'm a "celebrity" (awful word!)—but this also emphasizes a sort of incorrigible aloneness and apartness. At least I *think* it's incorrigible; though to my great surprise—for I had not had such feelings, I had thought my heart closed, for 20 or more years—I have felt a sweet tender feeling, a sort of mild "falling in love," with someone just this month.* I say nothing, I do nothing, I can see nothing coming of it, yet the sense of warmth (and *imagined* companionship) is like rain on my dry soil (I seem to be mixing my metaphors!). The feeling seemed to lend a special glow to the Fall colours when I went to Lake Jeff† last weekend, and to remind me (what one sometimes forgets) that life *can* (or could) be deeply happy.

* OS developed a romantic attachment to a neurological colleague, and although it was not reciprocated as such, the two remained good friends.

† OS frequently spent weekends at Lake Jefferson, in the Catskills, to swim and write, and Larry sometimes accompanied him.

Whatever the difficulties (which I don't minimize), I think your continuing relation with Blanche *crucial*—you were so evidently transformed by it when I last saw you. (Even Helen* remarks on this. She always asks about you, says, "He looked so good. He was a different man.") So I don't know whether, for either of us, it *is* "the same old story." Some of it is old—but there is surely something new.

And, as you say, it's the *Journey,* first and last (with all the arrivals and departures, meetings and partings, any journey involves).

I have some thoughts of getting another, larger house—perhaps (I have always loved it) that waterfront one, with the little cottage on its lawn, at the end of the street. If ever I can get it, the cottage is yours.

We have to *see* each other one of these days.

All my warmest best wishes,
Fondly,
Oliver

To Francis Crick

JANUARY 11, 1987
119 HORTON ST., CITY ISLAND, NY

Dear Dr. Crick,

When we talked at dinner in La Jolla several months ago, I mentioned the "kinematic vision"—seeing successive positions without intermediate movement—experienced by a number of my (migraine and chemically-excited) patients. You asked if this was ever a *permanent* condition; I could not answer.

I find now that it *can* be experienced on a permanent basis—this is beautifully described and discussed in a paper by Zihl et al. ("Selective disturbance of movement vision after bilateral brain damage," *Brain* 1983, pp. 313–340).† [. . .]

* Helen Jones, OS's housekeeper.

† Josef Zihl and his colleagues described the uncanny experiences of a woman whose brain, following a stroke, had become unable to construct motion—instead, she would see a collection of "stills," somewhat analogous to strobe photography.

I have now written up a longish report on my achromatopic patient: he is still totally color-blind a year after his accident. Only in the actual writing did I finally come to "see" various things—above all, how color is, indeed, a (cerebral-mental) *construct,* as movement is for Zihl et al.'s patient. I think I may have *resisted* this notion (for it is not at all "difficult" in an intellectual sense) due to some emotional cleaving to a naïve realism, a feeling that we are "given" color, form, the whole world—when, it is borne in on one in every way, one has to *construct* it (or perhaps "construe" it). I suppose this is a "Kantian" view (tho' muddled by his unfortunate apriorism).

I have *wanted* to read *Neurophilosophy,** which you spoke so highly of, but have found it difficult to get a copy—I *must* do so, because so much of my own work (whether I like it or not) seems to be a sort of clinical epistemology (or clinical "ontology" or "neurophilosophy").

I hope you received the copy of my book *A Leg to Stand On,* with its descriptions (for what they are worth) of some of my own (visual) problems (or illusions) at the time.

I hope you are in good health, and that I may have a chance to visit you when I am in La Jolla next month,

With kind regards,
Oliver Sacks

To the Editors, Social and Health Issues Review

JANUARY 26, 1987

119 HORTON ST., CITY ISLAND, NY

Dear Sirs,

[. . .] I am, of course, most moved by, and deeply grateful for, your selection of me as your Writer of the Year award.

I confess, I do not find it at all easy to *say* why I do what I do, or what indeed I "hope" to do. Because I do what I do spontaneously and "unthinkingly"—at least without *conscious* deliberation or intention.

* By the philosopher Patricia S. Churchland.

Perhaps *this* needs to be said—to be said in addition to, or in place of, the 300–500 words I have enclosed. [. . .]

My writings are neither "popular" nor "academic," not *designed* to be either; indeed, as I say, they are not "designed" in this sense at all, but come out, erupt, often very suddenly and unexpectedly, because of some inner need to articulate and communicate (*something*—I don't quite know what) which is deep inside me.

I feel a need to write this letter by way of confessing the inner truth, or psychological truth, about my "writing"—whereas the enclosed statement, tho' perhaps more suitable for publication, may be more conscious or superficial.

I have *always* found it difficult to say, or even think, why I do what I do!

Again, all my thanks,
Oliver Sacks

[Essay attached:]

I am sometimes asked whether I am a physician or a writer—I am *both,* always, I cannot separate them. I feel that my task, as a physician, not only includes the "usual" medical functions (diagnosis, recommendations, treatment, care), but extends beyond this—to an attempt to grasp, or "imagine," *the world of the patient,* both those alterations induced directly by altered neural functions, and the altered life-situation, the predicament, he in consequence finds himself in. I want to grasp his predicament—and I want the patient too, if he can, to grasp it, so that all his resources (physiological, mental, and moral) may be mobilized for the act of survival, leading the richest possible life under and despite his altered circumstances. I want to imagine his "world," *and also to convey it,* as best I can. Why? In order to enlarge the sympathetic imagination of those who may hear or read me; to make them feel that disability and patienthood are not remote or repellent, but integral parts of being-alive, being-human, adapting, and affirming. I want to assist—not by campaigning, but by careful description and analysis—a feeling of fellowship for the afflicted and ill, our sense of them as fellow-beings striving with adversity, as all of us do, in our different but kindred ways. I want finally, through these true tales and "parables" of affliction and survival, to illustrate human universals, *our own* situation, and (what I see as) the universal human condition.

To Elkhonon Goldberg

NEUROPSYCHOLOGIST*

APRIL 18, 1987
119 HORTON ST., CITY ISLAND, NY

Dear Nick,

It was a particular pleasure dining, talking, being with you last evening, and an illuminating one. [. . .]

I feel a need in myself to turn away (at least for a while) from these (now-too-easy) narratives, and do some "hard" thinking. I also find a wish and need, almost unprecedented, to work *with* others, friends/colleagues whose minds and energies I feel congenial and complementary to my own. I was never able to work *with* anyone before, nor had I any desire to. There has been a very good collaboration with my old friend and (ophthalmological) colleague Bob Wasserman for a year now, centering on our achromatopic patient, but then spreading to thinking and working together on other aspects of visual processing. This has stimulated us both, "pollinated" us both, produced (and will produce) good work of a sort which neither of us, perhaps, could or would have done alone. Bob is also more "scientific" than I am, can wrestle with physical and physiological theory in a way which I cannot (or cannot any more), and I respect his powers here immensely; but, conversely, or complementarily, he respects *my* powers of vivid and dramatic (and even, as you put it, "captivating") formulation and presentation. So we can make a good "team." And I think, after last evening, that you and I could also make a powerful "team" (I was going to say "troika," but that has three horses). Certainly, as a start,

* Goldberg had emigrated from the Soviet Union, where he had worked closely with A. R. Luria.

this matter of hemispheric adaptation* (and "allocation") is one which intrigues us both; is radically more interesting (and probable) than the conventional ideas of fixed and static roles; could be supported, at least suggested by an immense amount of clinical, experimental (and perhaps "personal") data; and *needs* to be presented in a way which commands attention. For better or worse (or both) I find now that I *can* command attention. This pleases me a bit, but it also frightens me, because it lays an added responsibility on me. I feel I must be careful what I say, lest I (unwittingly) mislead others—or myself. Bob Wasserman, a man of absolute integrity as well as fine creative intelligence, has been an exemplar and reminder of intellectual responsibility in our working-together, and I think you would be the same.

It may be, in some self-accusing part of my psyche, that "responsibility" gets equated with "respectability." [. . .] If others (some others) see me as a "pop" neurologist that is not my concern; but I must not see *myself* in that light; I must see myself as having some central seriousness and integrity; and one way of assisting this is to join forces, here and there, now and then, with others who *do* have the clearest seriousness and integrity, so that I can be recalled to my better self, and ascend with them to that elevation one needs, where the air is pure, the climb is fun, vast vistas unfold, and there is a ringing sense of reality. At such altitudes, which are the *proper* realms of the mind, there is a mixed sense of awe and humility, and transparency, and delight. [. . .]

Best,
Oliver

* Goldberg was particularly interested in understanding how the brain's left and right hemispheres seem to specialize in different realms but can adapt otherwise in response to injury.

To Susan Sontag

JANUARY 8, 1988
119 HORTON ST., CITY ISLAND, NY

Dear Susan,

I *meant* to have written to you a few days ago—I was away (Holland, England) over the holidays, but found the *New Yorker* with your *wonderful* piece in it on my return.* [. . .]

You *had,* of course, said something of the incident—you told me of it (very briefly) some months ago, and I could not help thinking, "How extraordinary. My God, she must *write* this!"—and now you have, with extraordinary beauty and feeling, and a wonderful sense of irony and affection for that long-ago self. But I suspect that it has been, pretty much, "a secret," for the most part (tho' it evidently needed to "return" and come up, and come out—so rightly—now)—and I see why it should be: and am slightly reminded of an occasion in *my* far-off, fervid, science-intoxicated boyhood (I too must have been 14 at the time) when (with Eric Korn) I sought audience with the great Julian Huxley† (he received us, our two precocities, with great courtesy and—I suspect—amusement, tho' an embarrassed amnesia prevents me remembering anything else). And there were earlier times—I must have been 10 or 11 (for he died in '45)—when I would hang around Hanover Terrace in Regents Park watching for a glimpse of my idolized H. G. Wells, and try and nerve myself, for hours, to knock on his door—I never did. But none of these experiences has the richness of yours, nor would say much—as yours says so much!—about such meetings and earlier selves. . . . And the continuity and chasm between "then" and "now"—Your one-but-last paragraph is the heart of the whole piece.

I can't help wondering, now you have released *this* "secret," whether there are not *other* scenes—and "secrets"—*needing* to be told. Your piece has the *tone* of the best autobiography.

* "Pilgrimage" was a memoir of Sontag's bookish youth; she wrote of her rapture upon reading *The Magic Mountain* and then, to her surprise, having tea with Thomas Mann.

† Julian Huxley (1887–1975) grew up in a scientific family (his grandfather Thomas Huxley was famously known as "Darwin's bulldog"). The younger Huxley became a zoologist and an influential popularizer of science. His brother Aldous was the author of *The Doors of Perception.*

I hope we can see each other soon—tho' I think we have some difficulty, tho' we are not (or are we still?) children, of dealing with *our* admirations now, for each other, so that a ghost of embarrassedness still makes our own intercourses a bit difficult. I feel this for me/you, as I do for Jerry Bruner, and as I did, most overwhelmingly of all, with Luria (correspondence with him was the most important of my life, but what I most desired—to see him—I could never make or allow myself to do).

I will be away, travelling, for several weeks, off and on—with a sort of playful travelling, a *license,* normally not "allowed," but made possible, "licit" now, because my photographer-travelling-companion (Lowell)* is *also* a *Touretter,* and our travels are *also* life-studies of Tourette's, which will be embedded in, and take the form of, a travel-narrative (?"Travels with Lowell"). I have pen in hand, he has camera—I have never had quite a relationship or a collaboration like this before, and what will emerge (hopefully!) will be something new, a sort of travelling clinical anthropology, as I go here and there, seeing Tourette's here and there, with Lowell whom I half see as a sort of Tourettic Tom Jones (or Tristram Shandy, perhaps, I don't know what I mean). I have "written up" the Alberta experience†—it was a not-wholly-coherent 25,000 words, which has been cohered (but perhaps also impoverished) now to 12,000; have all the "material" for the Holland-and-England section, and now want to go exploring and Touretting with him in Mexico and S. America, and maybe in a "primitive" culture (perhaps he has a "brother," a ticcing witchdoctor in New Guinea). So, we'll be on the road, I guess (from Jan 26) for 4–5 weeks.

I have (I suspect, thanks to you)‡ these City Arts things in San Francisco and Portland, tho' I don't know what to *do,* how to *present* self/work, and feel somewhat uncomfortable. These will be among the *last* such talks/performances that I do, because I feel they are "bad" for

* Lowell Handler.

† His travels with Handler to a Mennonite community in Alberta, Canada.

‡ Sontag was on the board of City Arts and Lectures and had encouraged her colleagues to invite OS.

me, emotionally and morally, and a (tempting!) distraction from the hard and solitary (but for me, ?for us) the *only* "real" thing—which is writing (whatever "writing" is—*e.g.* how complex a psychological act, and transcendence, you have made in writing "Pilgrimage").

All my love,
Oliver

*To Lars Erik Böttiger**

PHYSICIAN, WRITER

MARCH 7, 1988
119 HORTON ST., CITY ISLAND

Dear Dr. Böttiger,

Your previous letter did arrive, but I was away at the time, so not able to answer it. This has been, and is, a time of unusually intense travel, especially connected with my interest in Tourette's syndrome, and my desire, among other things, to see it in many different cultures and situations, all round the world. Tourettological "house calls," if you will, round the globe—or, perhaps, a sort of clinical anthropology. [. . .]

Apropos of your question: I do not know what started me as an author—nor am I sure that I see myself as one. On the other hand, I cannot recollect a time when I did not "write," when I did not have a need to cast or recast experience(s) into words. I seem to have had a pen in my hands since childhood (they are always *inky,* and have been so, now, for almost fifty years).

My "pre-clinical" writings, if I can so style them, especially took the form of travel narratives, landscapes, and "physiognomies"—portraits of people met on journeys (or, less specifically, on the "journey of life"). It seems to me that these narratives were the *precursors* of my clinical narratives; thus I think of a piece I wrote (when working/travelling as a trucker) which contained detailed (and, as it were, analytic) portraits or narratives of my two trucking companions—or, in

* Böttiger was the head of medicine at the Karolinska Institute in Stockholm.

effect, "case-histories" (perhaps cultural more than biological ones). These early writings are much in my mind now, for my Tourette-exploration is going hand-in-hand with travel, a sense of the patient (the "case") contexted in space and time, and the case-histories (if I can manage it) will be *embedded* in a travel journal, perhaps in somewhat the fashion of an anthropological journal. This way I will try to fuse (without damage to either) the two sorts of writing I love: the sense of a (biological) "case," and the sense of its human and cultural "context" (of course one has only to think of Sherlock Holmes's London or Wittgenstein's Wien to see how beautifully this can be done).

I hope I *will* be around to meet you.

With best regards,
Oliver Sacks

*To Tim Murphey**

CORRESPONDENT

MARCH 30, 1988
119 HORTON ST., CITY ISLAND, NY

Dear Mr. Murphey,

Many thanks for your rich letter, full of energy and ideas, and the papers you attach. In no special order: Yes, absolutely, the "other side" of body-image losses are all the *incorporations* (into body-image) of one's skis or one's scalpels—or, for that matter, the car or the plane one pilots. Nothing feels "natural," "comes naturally" *until* it is incorporated. How right you are about the feelings when one takes one's skis off!

In learning sports (or, for that matter, ballet-dancing, or pancake-tossing) one *has,* I think, to start with a "mimesis"—no abstract description or explication can be (to use our word again) incorporated as part of self.

I suppose *something* of the sort is operative in language

* Murphey wrote to OS with observations on the relationship between music and language, as well as the embodiment of learned movements.

acquisition—when people are said (for example) to "have an ear" for another tongue. My father had, has, a marvellous ear—and can (within minutes or hours) produce a sort of Swahili (or whatever) which *sounds* reasonable, and can communicate something, even tho' it is totally defective in vocabulary and grammar. My mother, more intellectual, but with less of an ear, would laboriously acquire the syntax and words—but be unable to open her mouth, or understand anything, in another language at all. My father, I should add, has a particularly good melodic "sense" and memory.

I am intrigued at your suggestion about "tricking" [Chomsky's "Language Acquisition Device"] into operation via music and song—you must tell me how the project goes. I do see (or rather hear) how many of my aphasic patients can "embed" (otherwise lost) language in recitatifs or songs. In a sense the "Man Who" was re-articulating all his actions by songs—"scoring" the world, his actions, as a musical script. And one sees again how Parkinsonians tho' unable to walk, may be able to dance; and though unable to talk, may be able to sing. I describe this power of music with them in many contexts in my book *Awakenings* (indeed I quote one Parkinsonian patient, herself a music teacher, who said she felt she had been "unmusicked" by Parkinsonism, and had to be "remusicked" to recover her own (kinetic) melody).

I too hope I can write many (at least some) more books. They may not be the most "scientific," but minute *descriptions* will always have value. Which is as well, because I am basically a describer.

With kind regards,
Oliver Sacks

To John Maguire

REVIEWER

MARCH 30, 1988

119 HORTON ST., CITY ISLAND, NY

Dear Mr. Maguire,

[. . .] Many thanks for sending me your reviews of *No Sense of Place**—it looks quite fascinating, in all sorts of ways.

I do not have a television set,† and have not thought too much about the nature of "authority." But I found myself at an extraordinary event, just two weeks ago—the uprising of the deaf students at Gallaudet,‡ their insistence on having a deaf President, and their saying in effect: "You hearing people stop legislating for us. We are not children. We are not incompetents. We will—thank you—decide for ourselves," in effect terminating the long, paradoxical reign of *paternalism* here. They objected, amongst other things, to some physicians, notably otologists, on the Board—feeling that they were seen by them merely as defective ears, not adapting people. The question of the "old-fashioned" doctor who says "I know what's best for you" may, indeed, be on the way out—I wrote about this, in a sense, in my book *A Leg to Stand On* (I was the patient here, and it is about "patienthood"). And yet a physician (or parent, or adult) *needs* authority. As someone who grew up in the pre-TV age, I see all sorts of social changes round me—I strongly suspect that the prevailing illiteracy is partly due to television—but had not really thought about other social changes. [. . .]

I like to think—your reviews make me wonder—about the *positive* power of TV: for example, to present, in the most sympathetic and

* Joshua Meyrowitz's *No Sense of Place: The Impact of Electronic Media on Social Behavior* (1985).

† OS was not a regular television watcher, but he acquired a television soon after this; he used it mostly to watch PBS or episodes of *Star Trek*.

‡ Gallaudet University, in Washington, D.C., was established in the nineteenth century for the education of deaf and hard-of-hearing people. Students there had objected to the naming of a hearing woman as president of the university. Their protest, Deaf Prez Now, became national news, and they succeeded in educating many hearing people about the importance of preserving their native language, ASL. A compromise was reached, and I. King Jordan—who had lost his hearing at the age of twenty-one and had some facility with ASL even though it was not his first language—was appointed president.

illuminating way, (say) people with diseases, or (what often seem) repugnant or stigmatizing or alienating conditions, in the hope of *reducing* the repugnance, etc. This, indeed—tho' not consciously—is one of the roles of my own writing, to enlarge the sympathetic imagination; and I carry a camera as well as a pen, because I think the image here is even more potent than the word.

Again, thanks
Oliver Sacks

To William Stokoe

AMERICAN SIGN LANGUAGE LINGUIST*

MAY 25, 1988
119 HORTON ST., CITY ISLAND, NY

Dear Bill,

What a brave, fine, heart-rending letter—I feel for you. I had no idea of the tragic situation you were having to face at home.

I don't know that I can tell you anything you have not already discovered for yourself—the patient, the family, are the only "experts" here. The good moments are very real—the EEG can suddenly change, become almost normal, in these moments; and though they are sometimes "spontaneous," "out of the blue"—if anything is—they more usually come when some personal association is touched (as with us all, perhaps). But it can be *so* striking when so much of the mind's formal powers are gone. A smile, a touch, a shared memory, a piece of music, smells, Spring (or something seemingly trivial, like a "Dunkin Donuts") can suddenly elicit one of these brief, sometimes tantalizing, but very real "good moments." But certainly your presence, and (redoubled) love and care is crucial.

But (if I may say this) you must not cut yourself out of life, out of *your* life, too much for her. You should not feel you have to be with

* OS got to know Stokoe's revolutionary work on the linguistics of ASL (American Sign Language) during a visit to Gallaudet. Stokoe had confided in him that his wife of many decades had advanced Alzheimer's.

her every moment of the day. It may indeed be that *someone* needs to be with her every moment of the day—but you must share the task with others (for her sake, as well as your own).

I would not say any of this if you did not ask me—and I may be speaking improperly. After all, I don't *know* your wife or yourself, and there are no general rules—there are only individuals. I hope that what I have said does not produce *dis*couragement—or offence. [. . .]

With warmest regards to you,
Oliver

IN 1988, OS PUBLISHED "The Divine Curse," an essay in *Life* magazine about a community of Mennonites in Alberta, Canada, with a very high prevalence of Tourette's syndrome. Orrin Palmer, a young doctor who also had Tourette's (and whom OS had met a number of times), wrote to express his reservations about the article.

To Orrin Palmer

PSYCHIATRIST

OCTOBER 12, 1988
119 HORTON ST., CITY ISLAND, NY

Dear Orrin,

Many thanks for your letter of the 9th. It was indeed, as you say, quite an angry letter, but I am glad you were able to write to me—for such anger is best expressed, if it can lead to anything constructive—though sorry, and surprised, you should have had such reactions. I will add that I have had upwards of a hundred letters about the LIFE article—many from people with Tourette's—and yours has been the *only* one with such negative reactions.

But, as you say in your postscript, it was not your intention to offend, but to open a discussion, and I shall take it in this spirit. I think your reactions are legitimate, as I think my own writing and viewpoint are legitimate. It will help to clarify both—for these illustrate (as I wrote in my earlier letter) the possible *ranges* of reactions to TS, and the range of *perspectives* in which it can be intelligibly seen.

But there is a need, first, to dissociate myself from certain things I did *not* say or write, but which were put in my mouth, as it were, by editors or journalists. I was angered myself by the proposed title of the piece, and threatened to withdraw it if they did not change it. I was informed that I could not withdraw it, and that it was the Editor's prerogative to give a piece any title he chose. There was nothing else I *could* do in the circumstances. [. . .]

Let me come to the meat of the matter. You take objection to my speaking of "prehuman" behaviours in TS (and interpret this as my seeing people with TS *as* prehuman). Here I must remind you of my own considered words in the matter, which you may find in *Awakenings* pp. 292–3 (enclosed).* I *did* indeed think that a number of my post-encephalitic patients, when excited by l-DOPA, showed phylogenetically-primitive ("prehuman") behaviours, probably due to excitatory lesions in the upper brainstem. I never saw such behaviours in any other patients—until I encountered some patients with severe TS (and this because such patients, like some post-encephalitic patients, may also have—even if not anatomically demonstrable—pathophysiological excitations at this level in the neuraxis): my interest in TS, as you know, originally stemmed from my observations of excited post-encephalitic patients, and my seeing—as von Economo had done in the acute epidemic—a sort of "tourettism" in *them:* I think P-E syndromes provide the best model we have for certain Tourettic syndromes. [. . .]

MY OWN perspective is indicated in these pages from *Awakenings,* where I speak of "phylogenetic depths which all of us carry in our own persons," and of such behaviours as showing us that "man's descent is indeed a billion years long," i.e. my observations and my perspective, here, is a *Darwinian* one—which is why I quote a relevant passage from Darwin. I think *all* of us carry our phylogenetic (as well as our ontogenetic) history within us—and that certain unusual conditions (viz. some forms of post-encephalitic and Tourettic disorder)

* This reference is to a footnote in the 1983 American edition of *Awakenings.* The same footnote (number 42) can be found on page 55 of the 1990 edition.

can, through a physiological excitation or disinhibition, reveal these. My excitement at observing this was a purely scientific one—and no different from my excitement at observing that "branchial myoclonus," so-called, which may be associated with excitatory lesions of the lower brainstem, causes rhythmic movements of the palate, stapedius, thyrohyoid muscles, etc, i.e. vestiges of the gill-arches: that it is, in effect, a persistent or vestigial gill-movement, something of immense phylogenetic antiquity, which shows us that something as primitive as gill-movement and its neural control is still part of the repertoire of the human nervous system (which shows us, if you will, "the fish in man"). Obviously I do not view patients with branchial myoclonus *as* fish! Any more than I see people with post-encephalitic or Tourettic syndromes as pre-human or non-human. I think that someone drawn to biological psychiatry, as you are, should be deeply interested in such biological possibilities, and am amazed that you could so misinterpret what are (and intended as) objective scientific observations as a sort of "racist" slur on people with Tourette's—the imputation that they are less than human. [. . .]

The second general theme in your letter has to do with what might be called "play"—the tendencies of people with TS (as I see it) to "play" of various kinds. I first wrote of this with regard to some post-encephalitic patients (see p. 289 from *Awakenings,* enclosed),* and then with regard to some Tourettic patients (I speak in Witty Ticcy Ray of "an odd elfin humour and a tendency to antic and outlandish kinds of play"). You take great exception to this—even though you must have observed it yourself in innumerable people with TS. It seems to me that this stems from your misinterpreting (my use of) the word "play," as you misinterpret (my use of) the term "primitive." I think there are (at least) *two* sorts of "play" here. One—the more usual sense—is play in its "creative" sense, the sense in which Winnicott writes about it in *Playing and Reality,* and in which he sees it as the one way of escaping or mastering or transcending "the bondage of the instincts." What is here involved is a response to psychological

* This reference is also from the 1983 American edition of *Awakenings.* A slightly revised version of the same footnote (number 28) can be found on page 26 of the 1990 edition.

or social *situations*—and such "situations" are all too common for people with Tourette's—"carrying off" such situations, situations of potential uncontrol and embarrassment, with wit and high humor. This is, if you will, a defensive or adaptative form of play. A second form has a clear physiological quality—I have seen similar forms in postencephalitic patients, and in some patients with cerebral disinhibitions ("*witzelsucht*,"* etc)—and this is "biological," not defensive or adaptive, and has a driven quality, half-compulsive, and half-playful. You must have seen countless examples of this—for example, Lowell, in some of his excited moods, opening provocative or (playfully) outrageous conversations with everyone in the street (or, in his own words, "playing the Tourettic fool"). Or the Tourette-twins, in Atlanta, yelling "Shark, shark!" on a crowded beach.

There is, of course, a great *range* of TS—(as I indicated in "Witty Ticcy Ray") from very simple forms (which are hardly more than myoclonus) to the most elaborate behavioural disorders. Bye and large, I think one can hardly even *consider* tourette's (if one wishes to go beyond a sort of idiotic DSM-III type listing) WITHOUT considering "play" of various sorts. And it is at this point, as Lowell himself indicates, that Tourette's can become interwoven with the highest (and, potentially, most creative) parts of the personality. *You seem to see* (my use of) *"play" as something "frivolous," "not serious," derisory; on the contrary, I regard play as the highest form of human activity.* It is in this sense that Winnicott speaks of it; in this sense that Huizinga writes of "Homo Ludens." Thus, when I speak of the "playful" in relation to Tourette's, I am speaking of adaptive and creative functions of a high order. [. . .]

I am not unacquainted with affliction—as you seem to imagine (in, for example, your image of the scoffing aristocrat). I have spent more than twenty years among patients with the most dire neurological afflictions. *Awakenings*, itself, is an essentially tragic work—a study of affliction, and a meditation upon it. And the sense of affliction, if it be crushing, hardly permits any room for humor or play—although it permits (and, indeed, demands) some sort of accommodation, otherwise life would be intolerable: I speak of such

* A compulsive need for inappropriate punning or joking, due to abnormalities in the frontal lobe.

accommodations—courage, resignation, the sense of "the will of God," etc.—in *Awakenings*—and Simone Weil ponders these matters in her profound and poignant essays on affliction and the love of God. It is crucial, if one has been afflicted, that one does not give way to bitterness and rancour, to cursing the world (or, like Job, cursing God); it is essential that one preserves one's self respect, and does not see oneself as reduced to indignity. I hope, for your own sake, that you can do this.

I understand this deeply—because I have worked with the most deeply-afflicted for years. I have a profound respect for Affliction, and the Afflicted—this is at the heart of my physicianhood and personhood—and, had you not been carried away, you would surely have seen this.

The third and last general point in your letter refers to what might be called "the medical model" of Tourette's—or, as you put it, "Tourette's as a medical condition"—your feeling that this is the *only* way of regarding it, and that I have transgressed it.

While I fully respect the medical model—and my own writings (*Migraine*, etc.) are examples of it—I think that one must go BEYOND it, especially when dealing with such a matter as Tourette's. The medical model, one might say, deals only with "pathology" and "therapy"—with Tourette's, for example, its "signs and symptoms," its presumed or actual biochemical substrate, and ways of treating this. Dr. Shapiro,* here, has performed a major service in delineating Tourette's as a medical disorder, and the "purely medical" approaches to it—Lowell says, jokingly, that he is "the world's most powerful dopamine-antagonist."

But, as you should be the first to know, Tourette's is not *just* "signs and symptoms." The concept of disease itself has to be expanded—Jelliffe and Mackenzie wrote of this with great insight in regard to encephalitic and postencephalitic pictures—to include not only the direct effects of the lesions or dysfunctions, but *all the adaptations and accommodations the organism may make,* at every level from the neurological—to the social. Properly defined, Medicine *is* the study of these adaptations: I indicate this in the preface to *Hat,* and in the

* Arthur K. Shapiro.

Mackenzie[*] quotation I use as epigraph: "The physician is concerned (unlike the naturalist) with a single organism, the human subject, striving to preserve its identity in adverse circumstances."

It is for this reason that I have expanded my presentations from case-histories or "pathographies," in the narrow sense, to human narratives or biographies, that I have presented what might be called *the human face of disease.* This human and humane attitude has been almost universally recognized in my orientation and writing—you will find an article on it, indeed, in last month's issue of the *Annals of Internal Medicine*[†]—and it is ironic, therefore, that you tax me with exactly the opposite, with an *in*humane (or, to use your word, "dehumanizing") attitude towards, and descriptions of, patients.

I would indeed, at this point, especially with something like Tourette's—but also with a condition like deafness, which I am deeply interested in (I take the liberty of enclosing a recent article on the Deaf, and am currently completing a book on the subject)—go even further. It has been felt by the Deaf that they have been subjected in the past to an exclusively "clinical" or "pathological" attention—seen as diseased, defective, otologically impaired, "oral failures," etc—and not in their own terms (as people with a language of their own—Sign; who form communities and cultures with their own centers and perspectives). In one sense (a purely "clinical" sense) a deaf person may feel himself handicapped; in another (a "cultural") sense he may feel himself different, a different being, living in a different mode of being, from the Hearing—but feel that mode as legitimate and complete. Thus the purely clinical or "pathological" attitude to the deaf must be supplemented by a different and more neutral attitude—one which one might call "anthropological." This sees the deaf on their terms, not "ours." Although one would hesitate to push the analogy too far with Tourette's, I think it necessary to see (at least) some people with Tourette's in terms other than the strictly pathological or medical. One of the profoundest and [most] open-minded observers of Tourette's—your colleague Dr. Ben van de Wetering, in Holland—has said that at times he cannot see TS just as a disease, but has to see it "as a mode of being." This may scarcely apply to some people with

* Ivy Mackenzie (1877–1959), a Scottish physician.

† "Toward a Romantic Science: The Work of Oliver Sacks" by Alan G. Wasserstein.

TS—it may scarcely apply to yourself—but it is a form of description that is legitimate, and (I think) demanded, in relation to some others, and to what might be called "a Tourettic life." *This is not antithetical to the medical model—it supplements it by an ontological, or anthropological model.* And it is one of the things which I shall try to use, if relevant, in my own studies of TS. Though, on the other hand, one needs all sorts of advances in neuroscience, one *also* needs, with Tourette's, such a "neuro-anthropology."

This is not at once obvious, or easy to take in, in those with a medical background and training; indeed it arouses an immediate resistance. I have had to struggle with this resistance myself. And you too, if you are to expand (as I hope you will) into the fullest intellectual and human understanding of Tourette's—you too will have to deal with this resistance.

I understand fully that such a broadening of perspective may be felt as quite threatening to you—as threatening your "identity" *both as patient and doctor.* It is simpler to see TS as "a purely medical" condition, requiring only "a purely medical approach." You have been *conditioned* to think this way—both as a patient and as a doctor. You are subject (how can you not be?) to crushing forces—both the forces of your own Tourette's, and your patienthood; and the forces of medical convention, and a narrow view of "doctorhood."

A purely medical way of looking at Tourette's promises (or seems to promise) not only an intellectual simplification—of something which may be bewilderingly complex, not only in itself, but in its interactions with the personality—*but a way of controlling it.* It is easier to say "You have—or I have—a purely medical condition, which can be controlled and treated, hopefully, in a purely medical way." And yet—this may not do justice to the very deep and complex reality. And I think a broader and deeper way of considering Tourette's, therefore, has to evolve.

You yourself—as a young man who both has Tourette's (and must have an extraordinary "inside" experience and knowledge both of its physiology and of ways of adapting to this) and is also a biological psychiatrist—are in a unique position to understand it, and to explore it, with a depth as yet never done. Having Tourette's, of course, can make this more difficult—one can be imprisoned (so to speak) in the perspective of the sufferer, as one can be imprisoned in the perspective of the "purely medical"—yet it can also assist one to a wider perspective. I am sorry you have misunderstood my own gropings towards a wider perspective—and hope they may be clarified by this letter; and I hope that you yourself will move beyond your present perspective into a wider and more spacious one. It will be a great struggle—intellectual and personal—and it may not alter your Tourette's as such; but it will reduce your *feeling* (at times intolerable, I am sure) of conflict and constraint, and give you one of spaciousness and freedom.

With all best wishes,
Oliver

12

Adaptations

1989–1995

DURING THE LATE 1980S, OS began reading the work of Gerald Edelman, who had won a Nobel Prize for his work in immunology and had then turned to the study of consciousness. Edelman's theory of neuronal group selection, or neural Darwinism, proposed that the brain was endlessly adaptable, with experience strengthening certain groups of neurons while other groups of neurons, less called upon by experience, might be rededicated for other purposes (there was a certain analogy here to the immune system and its ability to learn by exposure). OS was very excited by this vision, feeling that it proposed a plausible mechanism for clinical syndromes he had long observed—for instance, the way that body image could rapidly remap, as it did when the immobilization of his injured leg caused its complete disappearance from neural representation. Edelman expounded his theory in three major works: *Neural Darwinism* (1987), *Topobiology* (1988), and *The Remembered Present* (1989).

To Robert B. Silvers

APRIL 2, 1989 (9 P.M.)
119 HORTON ST., CITY ISLAND, NY

Dear Bob,

I find you phoned about an hour ago when I was out to dinner—I just tried to call you back, but you had left for the night, and (I guess) your trip to Europe. It was very thoughtful of you to call.

I just got back myself from a marvellous trip to Italy—not quite my

Italian Journey (though I saw the tree under which Goethe conceived his Metamorphosis, his *Urpflanze*, in Padua), but the sort which casts life in a different, fuller light than ever before (at least this is how I feel at the moment, tho' the euphoria is rapidly evaporating). There was a lovely symposium in Florence (from the Art of Memory to Neuroscience), organized by Pietro Corsi (whom you know—what a fine and brilliant man he is). [. . .] I went with Gerald Edelman and Israel*—Gerald opened the conference with a stunning presentation, and Israel and I closed it with our own presentations. I had met Edelman before, but never been tête-a-tête with him for three days on end. It was an overwhelming experience. He is a bit of a tyrant (Israel calls him "Mussolini"!), but a real genius, with a grip and focus which never let up for a moment, and an almost quixotically-huge mental vision. I had read *Neural Darwinism* and *Topobiology*—neither of them easy to read—but feel I didn't really understand him or his theories until I was exposed to them at point-blank range for 72 hours non-stop in Florence (and with Israel, who is a marvellous expositor, and who sees all sorts of implications and applications which the more theoretical Edelman himself may not at once see, with me as well). I am a "convert," I think. I feel Edelman may have provided the first, the *only* non-reductive theory of mind we have, the only one rooted in known (or probable) brain-anatomy. I have babbled vaguely about "a neurology of self" for years, but I think Edelman has provided the theory for such a neurology—a theory of individuation, of the historical development of the individual, of the uniqueness of mental disposition and contents, of biology as biography. I suspected this when I was reading about the visual evolution of the congenitally deaf (which seemed so puzzling on any "pre-programmed" Chomskian basis), and phoned Israel up at the time, saying "Doesn't this *sound* rather 'Edelmanian?'"), and I am sure of it, in a very much more general sense, now. So my visit to Italy was doubly memorable—for giving me such a sense of the Renaissance—the past—and, with Edelman, such a sense of the future (or what I would *like* to see as a possible future of neurology).

The manuscript of his third book *The Remembered Present: A*

* Israel Rosenfield, a mathematician, physician, and medical writer whose books include *The Invention of Memory: A New View of the Brain.*

Biological Theory of Consciousness has just been dropped at my house today—and I think it more radical, more wide-ranging, more pertinent to ourselves, our own mental life, than either of his previous books. [. . .] But it is *difficult*—partly because of the sheer complexity and density of the thought, partly because he is not the most transparent of writers. I think that such a book cannot be merely reviewed—but has to be expounded and explained, and all its relevances and resonances explicitly brought out. I think its publication will be a major intellectual event, and that it deserves this. And I think there is no-one better qualified (indeed no-one else at all qualified) to do this than Israel, who is closer to Edelman's thought than anyone else, but also himself imaginative, wide-ranging, critical, and in a position to bring out the clinical, literary, and varied other resonances and relevances. I think he may need another 10,000 word essay to do this, because there is so much that is new, and was not even hinted in the earlier work(s). And Edelman is *not* an assimilated or accepted part of the neuroscientific climate—he is widely misunderstood—or completely ignored (even Crick, who is a genius, and whom I greatly like and admire, seems completely to miss the essence and importance of his theory). [. . .]

I spoke in Italy of my Italian artist, Franco Magnani,* and his "autobiographical" memory, and his extraordinary paintings of a Pontito he has not seen in 30 years. It was fascinating visiting Pontito myself, going into the inside of another man's memories and visions and art, so to speak; finding how incredibly exact his memory was—but also how *transforming;* the art of the transformer, the dreamer, was everywhere apparent. I had a feeling of Luria's Mnemonist, and Freud, and Proust, all mixed up together, and illustrated by Franco—*his* quest to recapture (or, rather, reconstruct) the past. I also spoke of Stephen†—and *wanted,* when I went on to England, to initiate contact with him and his mother and teacher. Unfortunately, by a piece of bungling, not irremediable, but annoying, his mother had been upset, and this was not possible. I will have to make a special visit later to see them.

I certainly want to make proper studies of *both* of them. I am sorry

* Described in "The Landscape of His Dreams."

† Stephen Wiltshire, a savant artist, is described in "Prodigies."

you didn't get sent Stephen's pictures—what I would like, when you come back, if you have time, is to visit you with *both* Stephen's and Franco's pictures, and discuss these with you; as well as talking about Italy, Edelman, or anything else.

I hope you have a good time in Europe—and look forward to seeing you on your return.

Best,
Oliver

PS Thank you too for your kind words about *Seeing Voices.*[*] It seems, in some ways, more and more inadequate, as I learn and think more, and especially as I am now *learning* Sign—but I *have* to leave it at this point, with the thought that I might return to it, or the subject, in future works or editions. Perhaps *everything* one writes is "obsolete" (at least in relation to oneself) by the time it gets into print!

To Gerald Edelman

NEUROSCIENTIST

APRIL 16, 1989
119 HORTON ST., CITY ISLAND, NY

Dear Gerald,

[. . .] I found it an enormous pleasure and privilege being able to talk with you when we were in Italy—and if the Italian visit gave me a tremendous sense of the Renaissance, and the past, talking with you gave me an excited sense of a future (or a possible future), the sense—as you put it (*Consciousness,* p. 66) that one might come to enjoy "a *deep* view of how the brain functions."

When I first came to this country, as a young man, in 1960, I made a pilgrimage to Marvin Minsky,[†] and his then recently-opened Experimental Epistemology lab—I thought I might find the *deep*

* *Seeing Voices,* OS's fifth book, would be published in August 1989 and dedicated in part to Silvers, who had stimulated OS's interest in the deaf world by asking him to review several books in 1986.

† Minsky was a cognitive scientist at MIT interested in artificial intelligence.

view there. And when Hubel and Wiesel* (or, before that, Lettvin, and the MIT people) published, I got a shiver, and wondered if "the deep view" lay there. And then, I think, I fell into a sort of depression—an *epistemological* depression, if I may so put it—in which I ceased even to use the word "consciousness"; or, at least, to think of it as something for scientific enquiry (thus I declined to go to another conference in Italy a couple of years ago, *on* "Consciousness": foolish of me—but that was my state of mind). You speak of something akin to this as being, perhaps, widespread—with neuroscience not daring to tackle anything so "global," and philosophy (for the most part) as being unproductive or absurd. And, most specifically, of the inadequacy of "functionalist" approaches, the "computer" view, which has so dominated (even tyrannized) people for the last 30 years. Given all this, and a certain sense of impasse and despair (epistemologically), I found it exciting, in the extreme, to listen to you, and to get a *glimpse* of a quite other way of thinking.

In the preface to *Awakenings,* I quoted from Keynes' preface to his General Theory: "The ideas which are here expressed so laboriously are extremely simple and should be obvious. The difficulty lies, not in the new ideas, but in escaping from the old ones, which ramify, for those brought up as most of us have been, into every corner of our minds." And he speaks of the composition of his book as having been "a long struggle for escape" for him—a struggle which it must also be for any reader who hopes to understand it. I feel somewhat similarly about the ideas you express in *Neural Darwinism, Topobiology,* and now—with such rich resonances and implications—in your new *Consciousness* book.

You yourself, perhaps, have completed your escape—but you must not underestimate the struggle which some of your readers may have to go through; not because they are stupid, nor because your ideas are obscure (on the contrary, they *shine*); but because understanding entails a "conversion" of mind—or, at the least, an ability to detach oneself (if only for a while) from habitual modes of thought, in order to savour a quite new mode of thought. There will be sudden "Yes! I

* David Hubel, Torsten Wiesel, and Jerry Lettvin all elucidated feature detectors in the visual cortex and much else.

see it!" and "Yes, he's right!" and "Yes, that would explain x, y, and z, which I could never understand!"—and sudden backslidings, when the thought seems to elude one, and one cannot "see" it any more. This, at least, I have found in myself—and in some of my friends who are studying your works. [. . .]

This letter is an interim, an acknowledgement, something to say how much I am appreciating the contact with you and your work, how exciting I find it, and the (epistemological) hope that it gives me. And because I have been rudely silent since we met.

May I also, in the meantime, take the liberty of sending you these proofs of my new book*—you will find brief reference to brain "plasticity" (p. 102), and to Neural Darwinism (pp. 115–6). (Most of the book was written earlier, in a rather "Chomskian" frame of mind, and it was only at the proof-stage, in the galley-proofs, that the little Edelmanian section was put in. It is very inadequate, but it showed a direction. I wonder what *you* think about the issues—particularly, the rather novel sorts of perceptual categorization and learning, which the deaf may go through.)

I am sorry if this letter is now too long!

With all best wishes,
Oliver

P.S. I should say that one of the things which brought my dissatisfaction with some existing theorizing and thinking to a sort of crisis was encountering a patient with achromatopsia—reading Land, and meeting Zeki.† (I have not yet got on to Chapter IV where you discuss these specific extrastriate operations.)

With my colleague and friend Bob Wasserman, I wrote this up—and in the *NY Review of Books* (not exactly perhaps a place for neurology or neuroscience—tho' Israel and I are rapidly making it so!). I don't know whether you saw it. I take the liberty of enclosing a copy.

Subsequently, with Zeki, and Ralph Siegel at Rockefeller, we have

* *Seeing Voices.*

† Edwin Land, inventor of the Polaroid process, wrote about color constancy, among other visual topics; Semir Zeki, in London, was studying color processing in the human brain.

been exploring the man and his problems more—we had an exhibit together at the Soc. for Neuroscience in 11/88, and hope to have a more formal publication.

Israel (who met the patient on the weekend, as well as Bob, Ralph and myself) had an extraordinary notion for trying to "trick" or "induce" the brain to construct color some other way—an idea which could never occur to one thinking in terms of fixed functions and "centers." It would be spectacular, theoretically, as well as therapeutically, if anything like this could be accomplished.

You have met my young friend and colleague Ralph Siegel,* who is now completing his postdoctoral stint with Torsten Wiesel. He has been doing very beautiful work on movement-perception (and movement "construction") in monkeys; and is, I think, a young man of exceptional originality and intelligence, whose mind tends to move towards more general issues—and (I think) to such a "deep" view of brain function as is intimated in your own work. If this is not improper of me, I would like to say that I think very highly of Ralph—as well as finding him a stimulating and enjoyable companion, and an immensely hardworking and energetic collaborator and assistant—and would recommend him to you in the strongest terms.

SUSAN SCHALLER, an American Sign Language interpreter, first contacted OS in 1987 about a book she was working on, to be titled *A Man Without Words.* She described the plight of a young man she called Ildefonso, who was born deaf and grew up in rural Mexico. Never exposed to sign language in his childhood, Ildefonso was completely languageless as an adult, able to imitate or mime but not to truly communicate with others. He was isolated in much the same way that Helen Keller had been before she learned tactile sign. Painstakingly, Schaller set about teaching Ildefonso language, in the form of ASL (for him, the breakthrough word was "cat").

* Ralph Siegel, a neurophysiologist at Rutgers, was perhaps OS's closest intellectual companion in the late 1980s and the 1990s, guiding him through the work of Crick, Edelman, and others, over countless dinners and early morning bike rides.

In July 1989, Schaller accompanied OS to Gallaudet University in Washington, D.C., where they attended Deaf Way, a gathering of thousands of deaf people from all over the world.

To Susan Schaller

WRITER, SIGN LANGUAGE INTERPRETER

JULY 21, 1989 [NOT SENT]
119 HORTON ST., CITY ISLAND, NY

Dear Susan,

Thank you for your good letter—it was fun having you along (you not merely interpreted, you *intermediated* for us, as well as being your interesting self), and added to the enjoyment of the trip. I was delighted and inspired by—oh, everything in Washington, and want to write about it; but I have had a hard and unpleasant and exhausting week, and haven't been able to compose myself—or anything. But I think I can hope for some peace (and production) now.

One reason, perhaps, why I have not plunged into writing (other than sheer exhaustion) is that I have been pondering precisely the question you raise about the definitions of "politics" and "culture," etc. [. . .] The Revolt (Revolution) last year was about Oppression and Victims, Power, empowering, dispowering, etc. Whereas this year—the Festival—was about Celebration, about richness, and the enrichment of Self, Community, Culture. Were the two events, then, completely unrelated?

Clearly they were not—although (to some extent) they had different origins. The Festival was conceived in '87 (or before)—but it could not have attained its extraordinary size, and richness, and confidence, and vitality, had it not been for the great (political) victory and liberation of last year. And (I had been chewing on this thought even before your letter, but your letter helps to catalyse and clarify) this year's event was "political" also—but political in a broader and *quieter* way. As if the (political) aim of self- and cultural development (& autonomy) HAD to pursue two separate (yet inseparable) paths—the path of violent Action, and the path of a quiet and ongoing process (you put this much better in your letter).

Clearly I am not (and perhaps no-one can be) "apolitical" in this wider sense of "Politics." I am all for development and autonomy and creative joyous power—how can one not be? (unless one is an Oppressor or Soul-Murderer, as the Deaf must, sometimes, have seen the Hearing). Yet I am against Rhetoric and Polemic and Inflammation and Belligerence. I might almost add—against Aggression too, and this reflects a rather profound internal difficulty of my own. That in myself, due to a particular sort of victimization in childhood (the sort which Shengold calls "soul-murder"), much of one's aggressiveness is turned inwards—into masochism. So instead of fighting (the good fight) I—I break a leg. (My "accident" in '74, which almost killed me, occurred when I was struggling to write a "political" and polemical book about a terrible soul-murdering, Behaviour-Mod hospital-ward I worked in). I have something so damaged in me that I *cannot* speak out, but tend to submit (with rage, and inappropriate guilt, and then murderous self-destructiveness). *Thus I have to leave (this sort of) politics to others.*

I think I may have mentioned that once when my cousin Abba Eban was doing an academic year at Princeton I asked him how he found it: he made a gesture of boredom, sighed, looked and said: "I pine for the Arena." He is, indeed, as you know, magnificent in the Arena. He changed world-opinion, made Israel *possible* (to as great an extent as one person could) *by* speaking for it, fighting for it, so eloquently in the UN.

But I do *not* pine for the Arena. I am scared of the Arena—and have reason to be (because of my compulsion to run, sacrificially, on the swords of my enemies).

So *I* must confine myself to what (you well call) "the other half" of politics, the study of educational process, of culture, of art, of *becoming*. [. . .]

Kate and Allen* both liked you and enjoyed getting to know you, and I know would send their love if they knew I was writing to you.

Keep in touch,
Yrs,
Oliver

* Allen Furbeck, KE's husband.

To Susan Rutherford

DEAF STUDIES SCHOLAR

JULY 30, 1989

119 HORTON ST., CITY ISLAND, NY

Dear Susan,

[. . .] I thank you for all the generous things you have said about my book (I hope they're true!), and, as I say, am largely in agreement with your reservation—really, your single, large reservation. I *do* think there could (and should) have been more expansion and depth with regard to the areas of anthropology and culture—but I do (at least) touch on this, and feel that it is not wholly absent, or denied, and that there is not (as you later say) "a major hole" in the book.

Something happened—and something is happening: my book is (as its subtitle indicates) a "journey"—and a journey which is far from complete(d); perhaps more than anything else I have written I see it as provisional—as a book to be expanded, to evolve, in the future—and, hopefully, future editions—as I myself, as my journey, will do. For me, it is (above all) a journey from a largely "medical" perspective (though I would prefer to see it as physiological or biological) *towards* a cultural one. I do, clearly, have quite a strong resistance and reservation to even speaking of "Deaf culture"—you discuss such resistance, in general terms, in your *SLS** paper. The resistance (even incomprehension) may have strong personal (even pathological) roots in myself: thus although I am English, Jewish, etc. I do not see my "identity" in these terms—and to some extent resist them when applied to myself (at least in a categorizing way). Where medical case histories commonly start by speaking of a "Caucasian" or "Black" Male, I never use such terms. I do this (it is largely unconscious) in order to focus on the *human,* and to prevent attention being given to the "categorical" (but I think there may be some confusion, in my mind *between* the "cultural" and the "categorical"). I think too a little problem in my own childhood—being, in effect, abandoned by my parents, and placed for four years in the hands of strangers (this is partly a generational thing—it is what happened to hundreds of thousands of children in 1939 when there was a governmental order

* The journal *Sign Language Studies.*

to "evacuate" them)—led me to "animal" phantasies of one sort and another: specifically, of being brought up by animal parents (kindly wolves)—like Mowgli (in *The Jungle Book*—Kipling had a similar experience, with his parents, at the same age).

I maintained this "Wolf-Boy" phantasy (abetted by having the middle name of "Wolf") for fifty years—indeed it was only when I started to read about the Wild Boy of Aveyron, etc. (I may be exaggerating a little) that I became entirely grateful for having language, a culture, being part of the human race.

I have had odd difficulties, *generally,* with the notion of "culture"—both denying my own, and "not seeing" other people's. It has been, curiously, this journey with the Deaf which is starting to break down an almost lifelong confusion and resistance—and it, in turn, has been leading me to other languages and cultures: thus I never used to travel *except* in English-speaking countries—but this year I have been to Italy and Japan—and had a fascinated sense of the rich difference and uniqueness of *their* languages and cultures.

I am probably saying too much—but I feel a need to confess my own *peculiar* resistance to the notion of "culture" (and not merely Deaf culture). [. . .]

I need to repair this specific omission, and redress an imbalance generally—as you say: the most I can do at this point [. . .] is to write (if I can) a concise and powerful piece *about* Deaf Way (I was startled to find it completely ignored in the media—the media which all jumped on the Deaf Revolution) which (belatedly, but better late than not at all) will focus on (and give a proper expansion and depth to) the notion of Deaf culture, and which also—in text or footnote—will indicate your own contributions. [. . .] This way, I hope, I can make up to Deaf culture in general—and you and your work in particular—what is not mentioned, or mentioned adequately, in the book.

Again, all my apologies, all my thanks, all my best wishes—and my hope for future contact.

Oliver

To Jerome Bruner

COGNITIVE AND EDUCATIONAL PSYCHOLOGIST

AUGUST 23, 1989

119 HORTON ST., CITY ISLAND, NY

Dear Jerry,

I don't know where you are at the moment—I hope you and Carol are having a good summer away from this mad city, away in Ireland or wherever. For my sins I have been staying here (tho' I will nip to London next week for a few days).

Thank you enormously for your good words about *Seeing Voices.* They give *me* a feeling of confidence about a book I HAVE to feel so uncertain about (because I'm not a linguist, not deaf, not anything—an Outsider), and I think they will be very valuable in helping the book (about which I have a special, tender feeling—because I regard it as a book FOR the deaf, as well as about them).

The day after my birthday—I was so glad you could come, tho' feel that I fluttered nervously between too many people, and couldn't talk properly, or quietly, to anyone!—I went to Washington where there was an astonishing Deaf Way Festival—with nearly 10,000 deaf people from all over the world (and speaking more than eighty different sign-languages) came together for an unprecedented "Celebration." The concept of "Deaf Culture," which was never too clear (or plausible) for me, became clear—and rich. But what I found most startling was the way in which signers from all over the world would *improvise language* (in a way which has not been sufficiently analysed or understood). Finns, Japs, Venezuelans, deaf people from completely different *linguistic* areas, making contact and communication which no speakers from so far apart could have done. There has already, I understand (I did not know this when I wrote the book) been something of a PanEuropean deaf community emerging (evolving) in the past 20 years—with a degree of country-to-country mobility, and migration, and marriage, far beyond that of hearing people in comparable economic/professional positions—and now, clearly, there will be the beginnings of a global community. It will be odd if the sense of *the local*—in terms of the provincial, and frontiers, etc—become lost in the deaf! Perhaps there is some analogy to other supranational communities (like the Jews)—don't know.

I couldn't put any of this in the book—it was already set—but perhaps can write something for future editions (if any).

I hope you can come to a book-party (at NY Academy Sciences) on pub-date, Sept 12.

Best to you both,
Oliver

*To Peter Weir**

FILM DIRECTOR

AUGUST 26, 1989
119 HORTON ST., CITY ISLAND, NY

Dear Peter,

Delighted to get your letter—I should have delayed my own, because I saw *Dead Poets Society* almost immediately after I wrote, and thought it *magnificent*—one of your finest, surely (though your films are all so different that one can't use any sort of scale, they are all so completely realized and individual). I couldn't help wondering whether autobiographic elements, transformed, were at work in its inception—certainly, for me, the film's power came partly from its evoking intense personal reminiscences; but, as you know, it has had an astonishingly universal appeal here.

(In my ignorant way) I had not, in fact, seen Robin Williams *until* I saw him here—and having since seen him in *Good Morning, Vietnam* etc. I am extremely glad I saw him in your film first. I thought he gave a wonderful performance—performance *as,* incarnation *as,* the schoolteacher—that you had brought out marvellous "straight" acting; whereas in *Vietnam* I had the sense of him playing himself, of an incredible virtuoso exhibition of his wild, unique talents, but not wholly of "acting" as such. So I found D.P.S. extremely reassuring.

* OS met Peter Weir when Weir was considering directing the feature film of *Awakenings.* He turned down the project, but they remained friendly, and OS would sometimes visit Weir when he traveled to Australia. Meanwhile, Penny Marshall had taken over directing the film, and OS had recently met Robin Williams, who would play the doctor character in the film. Weir had directed Williams in *Dead Poets Society.*

Now I have met him, and found him (not only coruscatingly, quasi-tourettishly spontaneous and funny, but also) warm and empathetic—I was most moved by the patience and tenderness he showed when I took him to meet and listen to two of my ageing postencephalitic patients. I feel happier (tho' not really happy!) about the question of the "real me" and the fictional one. The script has improved a lot, though the general question of "reality" and "fiction" still worries me, as well as the particular one of the real and fictional "me." But at least I can be assured that Robin will infuse all his intelligence and warmth and humor and personality into the part, and bring it to life, give it a reality (not necessarily "mine") if anyone can. (When we went, with Robt De Niro, and Penny etc. to see a psychiatric ward in Bronx State, Robin *suddenly* exploded, in the car on the way back, with the most incredible, phantasmagoric replay of the entire scene, taking on different voices, different personae, with kaleidoscopic rapidity—it was a most amazing, even *neurologically* amazing, eruptive achievement; I don't know whether to call it "inspiration" or "Tourette's" but it certainly showed an amazing, explosive, phantasmagoric-comic unconscious; which one sees in genius—and also (sometimes) in Tourette's.)*

Robert De Niro seems, at first, a polar opposite—shy to an almost pathological degree (I felt this, since I sometimes have this frozen shyness myself, and want to melt into the nearest wall)—but he has been becoming easier and warmer each time we've met. I am going to London tomorrow, will meet him there, and show him the last-remaining group of post-encephalitics (just 9, out of the 20,000 originally brought to the Highlands Hospital in 1920). Certainly, if anyone can study them, incarnate them, become one, he can—tho' there's also a danger of clinical literalness, and the need for something which is also dramatically powerful, art. I am quite uncertain, and curious, as to how this sort of thing can be achieved (if it can). I will try to show Penny and the actors the clinical realities (of patient, of doctor, of hospital, etc—and then, I think, disengage myself, and watch from the sidelines; since filmmaking is as deep a mystery to me

* Because the postencephalitic patients had sometimes acquired tics not unlike those of Tourette's syndrome, OS had introduced Williams to a number of friends with Tourette's.

as book-making to them, and so completely different from the work of "literary" creation (or *is* it? Perhaps it is not—I can imagine *you* doing both, with equal ease and power).

I very much look forward to you coming to New York again, and to Thai dinners in the rain!

Affectionately,
Oliver

PS I herewith return your Sartre, with thanks.

To Catherine Barnett

POET AND WRITER

OCTOBER 25, 1989
119 HORTON ST., CITY ISLAND, NY

Dear Ms. Barnett,

Many thanks for sending me the current *A&A.*[*]

I read your article at once—it seemed to me brilliantly researched and (no less crucial!) beautifully *balanced;* I do congratulate you.

The issue here, in the most general terms, has to do with "the neurology of imagination," I suppose, and this is a subject I am increasingly fascinated by, and am writing a series of essays on—all of which arise from clinical explorations, mostly explorations of some of my own patients who happen to be artists—tho' it would be important to expand this to include artists in the public realm such as de Kooning. So I was particularly fascinated, and particularly sorry to have been too involved with other things (this has been a frantic time) to talk with you while you were researching his story.

I find myself both agreeing (on some points) and disagreeing (on others) with Gardner,[†] Damasio, etc. I have myself *seen* all sorts of skills (including artistic ones) preserved, or largely preserved, even

* Barnett had written an article for *Arts & Antiques* magazine about the artist Willem de Kooning, who by that point had developed some dementia, and the question of whether the condition had affected the quality of his paintings.

† Howard Gardner, known for his theory of multiple intelligences.

in the advanced stages of dementing diseases (such as Alzheimer's)—indeed I think this preservation, if not of artistic skill, at least of aesthetic and artistic *feeling*, fundamental—so much so that at hospitals where I work we have art therapists of various sorts (music therapists, drama therapists, visual art therapists, etc.) to "bring out" the still-preserved artistic potentials of our Alzheimer patients, and to show a path, or way of access, into these minds which may otherwise be so impenetrable to the usual verbal and conceptual channels. A colleague of mine specializes in this and has some remarkable paintings done by people with Alzheimer's so advanced that they have become incapable of *verbal* expression, and (manifestly) damaged in (many) cognitive and conceptual capacities. Such paintings are *not* mere mechanical facsimiles of previous work—but can show real feeling and freshness of thought. And this in patients who are not even artists, or who were, at most, "amateurs" before becoming ill.

With a master, and a genius, like de Kooning, I think it may be very difficult to dogmatize—for if "ordinary" people can show so much painting skill, and feeling, and spontaneity, despite advanced cerebral disease (I am assuming, at the moment, that this *is* what de Kooning has), tho' it is far from clear to me from such accounts as you give (one remembers how Pound, for example—or Swift, for the matter—retreated into a terminal silence, but a *willed* silence, not an incapacity for language)—if ordinary people can do unexpectedly fine painting in face of advanced disease, what may the potential of a de Kooning not be? (Similar thoughts arise, I think, with some of the late sculptures of Henry Moore, when *he* was into his eighties, and "not quite all there" in certain ways.)

What the neurological basis of such preservations may be I will not hazard—but obviously there must be one, if such preservations occur. I doubt if one can think in the simplistic terms of left and right brain or frontal vs. occipital etc. But even in face of *diffuse* brain damage, there can be sudden, remarkable, if transient "restitutions" of function—I describe this in *Awakenings* (xerox enclosed), and always find it astonishing. Importantly, the electroencephalogram (EEG), which may be very slow or disorganized in such patients, may become virtually normal in these (usually) brief periods of restitution.

What it comes to, finally, is that, whilst avoiding anything "mystical" (all thought, all consciousness, all art, so far as I am concerned, must have some biological basis), one must equally avoid all dogmatism—and perhaps hold off premature interpretations or speculations *until* one has deep and reliable observations. At the moment we cannot say, we do not know, what is going on in de Kooning's mind, or his latest productions—and (to quote a phrase of William James in another context) "they forbid a premature closing of our accounts with reality."

With all best wishes,
Yours sincerely,
Oliver Sacks

PS I don't know if you have correspondence columns. But if you do, you should (after checking with me) feel free to publish this letter in whole or part.

To Jane Gelfman

LITERARY AGENT, JOHN FARQUHARSON, LTD.

NOVEMBER 8, 1989
119 HORTON ST., CITY ISLAND, NY

Dear Jane,

[. . .] The film of *Awakenings* is moving ahead—they are shooting every day (sometimes for 16 hours a day!), and do so till the end of January. It is a *vast* enterprise, organized like an army or battleship—at least I keep being reminded of this: Captain Penny at the wheel, the great cameras swivelling like gun-turrets, the AD's like First and Second Mates ordering people around, with a strange lingo heard nowhere else ("Lock it up guys! Respect for the actors!!"), and the order "Roll!!!" resounding from deck to deck. It is very exciting seeing this strange half-surreal half-authentic thing made real—in particular I am fascinated by the actors' ways of conceiving and imagining, the acting *investigation,* so akin to, yet so different from, a scientific investigation.

I hope you can come on the set one day—filming is mostly in an old hospital in Brooklyn. [. . .]

Otherwise, I have been travelling and talking, to and fro, in relation to *Seeing Voices.*

I cannot judge if it is a "success" in commercial terms—it is selling very well, but not wildly, so I gather—but then I never conceived it, nor "success," in these terms. I believe I have written a good book, and a powerful one, and that it may cast an unexpected light and make people think. Certainly the Deaf themselves feel I have done something which a hearing person rarely does—enter their world, listen, begin to understand, try to represent. It has certainly demanded an odd shift of perspective.

I have been very exhausted these days—with the demands of book promo, *and* the film (where they seem to want me on the set all the while, and I have to try and disengage)—and eager to get back to my own real work, whether it be my so-long-incubated (and perhaps "doomed") Tourette book, or a series of studies (with artist-patients or artist-subjects of various sorts) on "the neurology of the imagination." [. . .]

Anyhow—thanks for everything—and hope to see you on the set!

Best,
Oliver

*To Jonathan Mueller**

NEUROLOGIST

DECEMBER 19, 1989
119 HORTON ST., CITY ISLAND, NY

Dear Jonathan,

[. . .] Altho' I am very honored and touched by your sentiments, and suggestion about an "Oliver Sacks Dept." (or whatever), I think this is not what I want myself. One can't make a comparison with the "Augustus S. Rose Dept"†—he *founded* it, he was in charge of it for 20 years, and then he retired, while still taking much interest in it—it is *properly* called the "ASR" Dept. Whereas I have had nothing to do

* Mueller, who worked in the neurology department at UCSF, had suggested that the university might create a center named for OS.

† At UCLA.

with the UCSF Dept, nor am I retired. I don't think it appropriate to have a dept. named after myself. (Auden was embarrassed when they named his little street in Hinterholtz "Audenstrasse," but then he *lived* there, was a part of the village.) [. . .]

I hope I can avoid some of the evil weather here. It is *very* subzero, and snowing again, and my wretched unploughed street is a lethal slope of ice. I remember all too well practically being under house arrest for a whole February, once, when this happened before.

I had an amusing little role in the movie Tuesday night—was Santa Claus, bellowing "HO HO HO MERRY CHRISTMAS . . . HELP THE NEEDY . . . THANK'EE sir, thank'ee . . . GAWD BLESS YOU SIR" etc. I could not put on a Jolly Green Giant accent, and (to Penny's annoyance) sounded more like the Monty Python show. Don't know if this little Hitchcockian cameo will survive into the final film!* They take a week off Xmas–New Year, and then they complete the shoot in January. Meanwhile, with Kate's help, I am trying quietly to get a 1990 *Awakenings* into shape, but find myself rather useless and exhausted all the while. I seem to have a very negative image of myself, Life, work—but am glad the Channel 13 portrait† presented such a positive (and engaging) one. It almost made me like myself for a few days, and feel my absurd life might have a grain of use or sense. But now that glow has gone away.

Anyhow—to you, a Merry Xmas (ho ho ho!) and a really new and deeply fulfilling New Year—

Love,
Oliver

* It was well below freezing that night when the crew filmed at Rockefeller Center, working into the early morning hours. As it happened, Werner Herzog (another director who had been interested in filming *Awakenings,* though it had already been optioned by then) was in town. He stopped by to observe the filming and was amused to see the crew of about two hundred people. "I would have done it with five people," he commented. (The entire scene was cut from the final version of the film.)

† *The MacNeil/Lehrer NewsHour,* on PBS, had produced a fifteen-minute profile of OS in conjunction with the film's release.

To Robin Williams

ACTOR

MARCH 10, 1990

119 HORTON ST., CITY ISLAND, NY

Dear Robin,

I was hoping very much I might see you in San Francisco, or perhaps visit your ranch—but the damn back which prevented my going to the wrap party has continued to plague me, and prevented my going to CA.

I find quite a sense of loss now the shooting is over—which surprises me, in a way, because I often cursed it when it was going on! But it was also very fascinating in a way, and a real delight—and honor—to work with people like you and Bob (etc.). And there was a real camaraderie on the set—it became a sort of a family, a sort of home to go to—which I quite miss now it is all over, and now I am back to my usual solitude (and writing).

I was very moved by getting your copy of the *script* (with all its evocative photos), and the beautiful leather-bound copy of Steve's* script, with your lovely inscription—thank you so much for both! I will certainly treasure them in my (chaotic) archives. [. . .]

Various Touretters are going to be on the Geraldo [Rivera] Show of March 23 (I may have the date slightly wrong)—I declined to be on it myself. I hope it is decent and dignified—I have not been wholly reassured by Shows I have seen before. I think there is a great need for a *real,* full, free, hour-long Show or Treatment or Presentation of Tourette's, uncontaminated by interruption, commercial pressures, or "bleeping"—such as you approached Cable TV about. I *very* much hope that it is possible for you to proceed with this. [. . .]

I do find—this has been the case for many, many years—that my heart goes out, for some reason, to all these Touretters, who suffer so much, and get so misunderstood and isolated, but who are nearly all intelligent and brave and open and funny—I have long wanted to write a book about them (and had planned to do so as far back as 1975!), about—and also *for* them. (My book about, and *for* the Deaf, comes from the same impulse—but the feeling towards Touretters is

* Steven Zaillian, the screenwriter.

stronger). And you, I suspect, may share a little of this feeling now, especially after meeting Shane,* with whom you instantly found affinity and made contact, and who is, in the deepest sense, a sort of brother.

So I hope some of these things will be possible. [. . .]

I am adding (part new material, part old) about 50,000 words to *Awakenings*—all in the form of Forewords, etc, Prefaces, and now (seven!) Appendices—the *text* (I hasten to add) remains the same. [. . .] Appendix 6—which I have yet to write [. . .] will be about the many (and *all* wonderful) dramatic representations of AW there have been: I will say the most of the film, for here (unlike the others) I could see in detail, and was sometimes myself even part of, *the process*—and it is the Process, the nature of the Dramatist's and Actor's Imagination and Presentation of Reality (too many Capitals!) that I want to give a glimpse of. [. . .] Certainly all of these, but above all the film and the filming, have been for me quite extraordinary experiences.

And *real* ones too: as real in their way as working with the original patients, and as real as the thought and dialogue, the "scientific" model-making and world-making which followed.

But you—as actors, as dramatists—are also making worlds; and though these are "illusions," they are also full of truth.

I have never known any actors before; nor have I been much of a theater- or film-goer; but I think these experiences have changed me (or will). [. . .]

My warmest best wishes to you and Marsha—and hopes to see you in New York, or California, before long.

Love,
Oliver

---

* Shane Fistell, an artist who has rather severe Tourette's syndrome.

To Jennifer F.

HEARING MOTHER OF A DEAF CHILD

MARCH 30, 1990

119 HORTON ST., CITY ISLAND, NY

Dear Mrs. F.,

Many thanks for your letter.

My phrase may have been a bit too sharp—and I certainly appreciate the (above-all) *complex* and delicate (even paradoxical) situation of parents who find that they have a deaf child.

My sole concern, like yours, is that there should be really good *first* language—given this, everything else can follow. Bilingualism, trilingualism, good intellectual development. The overwhelming danger is NOT achieving a good first language—or not achieving it before the age of (say) five. And this is all too common, tragically so, as you know, among the deaf, and a reason why the level of literacy may be so low among them.

I also think it very important—as you do—that the deaf child be at-home in the "dominant" culture of his own country, English, or whatever; so that he can belong to *it* no less than to his own (perhaps-more-limited) culture.

Where we differ has to do with the *means* of achieving this. It is *not necessary* to have spoken or signed English in order to acquire a good knowledge of English. One can become an excellent reader and writer of English *without* the use of spoken or signed English. One can acquire English equally well if one's first language is Sign (say BSL).* And Sign is much easier, even neurologically easier, for the deaf child to acquire—even though it may be a bit more difficult for his parents to acquire. A deaf child fluent in BSL can *easily* learn to read and write English as well. He becomes—bilingual: BSL and English can both be first languages (even though it is reading and writing, rather than speaking or signing, which gives him access to it). All fluent signers, in any case, will become fluent in signed English, or whatever, besides their own BSL (or whatever).

You do not make clear in your letter whether or not your child

* British Sign Language, which has different etymological roots from American Sign Language.

has much to do with other deaf people—and you write in terms of "owning" and "belonging" as if you felt he was in danger of being kidnapped by the Deaf Community. Your child does not "belong" to you (nor to anyone else)—but he will have the richest *relation* to you to the extent that he is able to grow up as a free and independent being, with a sense of *his own* language and community and culture, as well as yours. He can be, if you let him, a member and inheritor of BOTH worlds. Surely this, rather than any exclusive belonging, is the ideal solution.

But the problems are very profound, because they relate to questions of *identity,* and the way in which, perhaps, a (prelingually) deaf child needs to have *two* identities, not one. You love your child, he is your child, he is "yours"; and yet you are different, and he cannot be yours exclusively—he has also a need to develop in his own way. But this does not mean he gets lost to you, or taken away by another community.

With kind regards,
Oliver Sacks

ISABELLE RAPIN WAS A PEDIATRIC neurologist and one of OS's closest friends. OS would often join Rapin and her family on weekends at their rambling house on the banks of the Hudson in Athens, New York. At times he would stay in their barn for days on end, writing, with periodic breaks to swim in the river or take the children for rides on his motorcycle. Rapin was also one of his earliest guides to the worlds of autistic people. She introduced him to Jerry C., a young man whom OS wrote about as "José," the "Autist Artist" in *The Man Who Mistook His Wife for a Hat.*

To Isabelle Rapin

JUNE 14, 1990
GRAND HOTEL FIRENZE, FLORENCE, ITALY*

Dear Isabelle,

How nice seeing you again on Friday—we must have more evenings together: I wonder if we can thrash out autism the way we thrashed out deafness? (I think I must have more clinical experience, however; I was happy to renew contact with Jerry C. today, who is now doing VERY well in a residency in the Bronx. I will try to see him regularly, and perhaps sometime you might like to see him again too).

[. . .] I think, with Autism, I will (besides a piece on Stephen†) try to make some wise synthesis, or simply an appreciation, of the five different books on the subject I was sent last year, drawing attention to their different virtues, and being (as Schein‡ would say) sparing of criticism. I would certainly think Uta Frith§ may have grasped or unified the central cognitive deficits—but I am not persuaded the problems are *merely* cognitive (but it is even more certain that they are not merely, or at least primarily, emotional/dynamic/motivational). Uta, you remember, starts with an "exemplary" or prototypic description of autism—a "composite." You, on the other hand, always speak of the *range* of disorders and behaviours one may see. She has a need (at times excessive) to discern a central or "core" disorder—and maybe this sometimes causes her to be selective, to force things into a unity, or omit things which seem outside it.

Can one have a syndrome which is ALSO a range?¶ I think so—I think this is exactly what one has with Tourette's, where no two cases are the same. (And where, perhaps, one would not speak of

* The address is from the hotel stationery, but the letter was likely written and posted in New York.

† Wiltshire (see letter dated July 18, 1990).

‡ Jerome D. Schein, a writer and professor at Gallaudet.

§ A developmental psychologist who did pioneering research into autism and dyslexia.

¶ The term "autistic spectrum" had not yet been widely adopted. Even as late as the 1990s, autism was poorly understood by most, and not seen as part of a spectrum except by a very few specialists.

the same condition were there not innumerable intermediates and overlaps.)

As I write I am reminded of Wittgenstein's thoughts on how one defines a "game," whether or not there is an *essence* of game (gaming). And his deciding there is no such essence, "no one thing common to all games, but rather a complicated network of similarities overlapping and criss-crossing, sometimes overall similarities, sometimes similarities of detail," "family resemblances"—"games" form a family.

Perhaps it is like this with Tourette's—and autism.

I am rambling—and it is late. And I have to be in good shape and good voice tomorrow to make a recording of *Seeing Voices,* a shortened (but not greatly shortened) version of the book for audio-tape.

See you soon.

Love,
Oliver

STEPHEN WILTSHIRE, ANOTHER AUTISTIC ARTIST, was passionate about drawing buildings, the more complicated the better. By the age of seven or so, he had been recognized as a child prodigy, able to draw a building from memory after seeing it for just a few seconds. When OS met him in 1988, Stephen was thirteen. He had just published his second book of drawings and had come to New York to draw famous buildings there.

To Stephen Wiltshire

ARTIST

JULY 18, 1990
119 HORTON ST., CITY ISLAND, NY

Dear Stephen,

Your *wonderful* drawing of our old house in London (and of the room inside) just arrived. This was the house I was born in, the house I have known all my life—and your drawing (which was so exact, and so full of feeling) will bring it back to me, remind me of it, for the rest of my life.

I am very sorry I was not there when you were drawing it that day—especially as it was my birthday as well. But I was very glad to see you the previous weekend, and delighted we could spend a little time together. How is the *fish* your sister bought? That was a lovely little drawing of it you did in my notebook.

I couldn't see you when you were drawing our old house in London because I was already in *Rome,* in Italy (which is very different from *Venice,* where you had gone earlier). Rome is full of OLD buildings and ruins—some of them are more than 2000 (two *thousand*) years old! People have tried to imagine what Rome was once like, how the buildings must have looked in Ancient Rome long ago—I enclose two such "reconstructions" of how Rome might have looked (and if you ever come to Rome you can see the original models, which are wonderful and fill a whole room). I hope you may have a go at drawing the Colosseum—or to give it its proper (Latin) name, *Amphitheatrum Flavium Constantini Aetate,* this incredible round building, with statues in the arches, where fifty thousand people (as many as a football crowd!) could watch gladiators or horses. Or you could try and draw some of the old buildings in ancient Rome (*Roma Imperiale*).

I think your new book (*Floating Cities*) is going to be just beautiful—and (who knows?) you may do a fourth book, of ancient Rome and other cities (Athens, Jerusalem, etc.)—a book of cities as they once were, and can be reconstructed: a book of Ancient Cities.

I hope you, and your mother, and your sister, are all well—and I look forward to seeing you again when I next come to London.

I hope you are enjoying life—and your new school.

You were a wonderful companion when we all went to Russia together!

Best wishes,
Oliver

To Hugh S. Moorhead

PHILOSOPHER*

NOVEMBER 11, 1990
119 HORTON ST., CITY ISLAND, NY

Dear Professor Moorhead,

Many thanks for your charming letter—I am glad you are enjoying *Hat,* and happy to inscribe your copy. (I think, however, my best book is *Awakenings*—a new edition comes out next month—which is, perhaps, even more explicitly concerned with SURVIVAL, psychic survival, transcendence, in the face of lifelong, devastating, sometimes grotesque illness.)

Though I seem to write about disability and disease, I am always concerned with patients' ability, and will to survive—to make meaning, and dignity etc. out of their so-shattered lives. (And this is surely the moral theme of A. R. Luria's wonderful book *The Man with a Shattered World.*) In this sense I find my patients an inspiration.

When moods of defeat, despair, *accidie* and "So-what-ness" visit me (they are not infrequent!), I find a sense of hope and meaning in my patients, who do *not* give up despite devastating disease. If they who are so ill, so without the usual strengths and supports and hopes, if *they* can be affirmative—there must be something to affirm, and an inextinguishable power of affirmation within us.

I think "the meaning of life" is something we have to formulate for ourselves, we have to determine what has meaning for *us.* (This search is quite explicit even in retardates, as with Rebecca—see *Hat,* p. 185.)

It clearly has to do with *love*—what and whom and how one can love. I only feel alive, discern "meaning," when (in some sense) I am in love. The object of love need not necessarily be human—my own most passionate loves, for years, were in physics and chemistry, and these remain for me as something to go back to.

I do not think that love is "just an emotion," but that it is *constitutive* in our whole mental structure (and, therefore, in the development of our brains). (I was very delighted with a recent book *Love and its Place in Nature,* by Jonathan Lear.)

* Moorhead was chair of the philosophy department at Northeastern Illinois University and edited an anthology called *The Meaning of Life* (1988).

I envy those who are able to find meanings—above all, ultimate meanings—from cultural and religious structures. And, in this sense, to "believe" and "belong."

I woke this morning with an oddly-happy feeling (from a dream of which all but the "mood" was forgotten) with an intense sense that it was Sunday, that church bells were ringing, that it was time to go, with my family, and neighbours, to church. Thus my Sundays, indeed all my days, would be ordered, centered, consecrated, given meaning. Then, as I awoke fully, the feeling went away—it was not Sunday, I am Jewish, and I have not been to synagogue since being a child (and even then, it seems to me, with more wistfulness than conviction).

(My old father died, at almost 95, a few weeks ago—and going to synagogue, reading the Bible, was central in *his* life, and gave him comfort, orientation, "meaning," to the end. I have just returned from gathering-together his effects—hence, I imagine, the timing of my dream and mood.)

I do not find, for myself, that any steady sense of "meaning" can be provided by any cultural institution, or any religion, or any philosophy, or (what might be called) a dully "materialistic" Science. I *am* excited by a different vision of Science, which sees the emergence and making of *order* as the "center" of the universe. I find David Layzer's book *Cosmogenesis: The Growth of Order in the Universe* particularly cogent and pleasing.

I *know* that human beings (and perhaps all organisms, at least from frogs up), and their nervous systems, are built to generalize and learn from experience, tho' obviously only human beings need or devise "philosophies."

I have, especially in the past, been excited by philosophies—I felt, with an intense reading of Kant in 1979, that *everything* was illuminated by his philosophy. The feeling, however, did not outlast 1979—and I have had a certain sense of philosophical disillusion and "dryness" *since* then.

In conclusion, I do not (at least consciously) *have* a steady sense of life's meaning. I keep losing it, and having to re-achieve it, again and again. I can only re-achieve (or "remember") it when I am "inspired" by things or events or people, when I get a sense of the immense intricacy and mystery, but also the deep ordering positivity, of Nature and History.

I do not believe in, never have believed in, any "transcendental" spirit *above* Nature; but there is a spirit *in* Nature, a cosmogenic spirit, which commands my respect and love; and it is this, perhaps most deeply, which serves to "explain" life, give it "meaning."

But—being who I am (and not, say, an Einstein), I need constant *human* contact, and above all, contact with my patients, to re-affirm the sense that life indeed has meaning. Meaning is given to me, daily, in seeing patients, and in practising as a physician—and in trying to be (in Nietzsche's words) "a *philosophical* physician."

I have exhausted my pen—and doubtless your eyesight and patience—with this rambling, floundering letter; and, on top of this, I have not answered "the question"! But life, for me, *is* rambling and floundering, sometimes creatively, to some never-attained goal, but to hopefully some higher level of understanding and integration. This struggle *is,* for me, the motive and meaning of life.

You know, you *did* write to me (I think) some years ago, when you were researching your book. I may have failed to reply because I was feeling depressed, or nihilistic in regard to life's "meaning."

This reply is too confused and contradictory, I suspect, to pull anything out of it. But you are welcome to try!

With all best wishes to you—and thanks!

Oliver Sacks

To Lillian Tighe

LAST OF THE POSTENCEPHALITIC PATIENTS

NOVEMBER 21, 1990

299 WEST 12TH ST., NEW YORK*

Dear Lillian,

Sorry I have been out of touch—I have been travelling almost constantly. My mind, however, has been very much on *Awakenings*—and on *you*. It was marvellous (and brave!) of you to put up with us

* OS had established an office space in Manhattan, where he would sometimes spend the night, though he still lived mostly on City Island until the mid-1990s.

all—to allow all these visits to you in your place, and, even more, to come out to the set in Brooklyn. Everyone there admired your humor and spirit, and how strongly and gracefully you live, despite the long illness. You were really an inspiration in making the film—and I have said this in this new edition of *Awakenings*, especially in its final pages (pages 385–6). (You will find further references to yourself on pages 63n, 245n, 271n, 313, 315, and 369: and an acknowledgement on page xiv.)

I very much hope it is possible for you to come to a preview of the film, scheduled to be shown in a theater downtown here on Dec 6.

Again, all my thanks, all my best wishes—and hopes to see you again soon.

Oliver

To Stephen Jay Gould

EVOLUTIONARY BIOLOGIST, AUTHOR OF *THE MISMEASURE OF MAN*

DECEMBER 21, 1990
299 WEST 12TH ST., NEW YORK

Dear Professor Gould,

An early-morning malaise was instantly turned into delight on finding an inscribed copy of *Wonderful Life* from you, and your notes inside. I had often wanted to write to you—for I have admired you for many years, and try never to miss your articles in *Natural History*—but always felt too diffident: but now you have given me the opportunity!

Specifically, when preparing a new edition of *Awakenings* (which I herewith enclose) I found myself thinking hard about history, and about the "evolution" of illness, neurological syndromes, responses to l-DOPA, etc, and feeling more and more that a *theory* of contingency was needed. Three (?four) such approaches excited me: those of chaos (and?catastrophe) theory—which, in a preliminary form, got worked up into my Appendix #6; that of Edelman (which found its way into "Neurology & The Soul," my *NYRB* piece); and *your* (and

Eldredge's)* theory of punctuated equilibrium—though it seemed to me that you were moving (in *Wonderful Life*) to a still-more general theory of history and contingency. I originally had in mind *three* Appendices to *Awakenings*—a Chaos-Catastrophe one, a "weather" one, an "Edelman" one, and a "Gould" one—but through laziness, or stupidity, I did not or could not carry through this idea—and reference to you, instead of being an essay, or a reconsideration of the whole book, as it should have been, only emerged in a tiny footnote (on p. 230). At a deeper level I suppose I remain unsure as to how to relate your evolutionary-phylogenetic considerations with ontogenetic ones—hence my inability (at this time) to say anything more on the subject. On this, as on so many subjects, I would love to meet you, and talk with you.

My own first love was biology (and palaeontology)—I spent a great part of my adolescence in the (fossil invertebrate galleries of the) Natural History Museum, in London. (And I still go to the Botanic Garden almost every day, and to the Zoo every Monday.) The sense of diversity—of the wonder of innumerable *forms* of life—has always thrilled me beyond anything else; and I partly see my patients (some of them, at least) *as* "forms of life," and not just as "damaged" or "defective" or "abnormal," etc. My oldest friend is Erik Korn, whom you know; he and I, and Jonathan Miller (we were all at school together) had an absolute passion for taxonomy—Erik's special passion was for holothurians, Jonathan's for polychaetes, mine for cephalopods. How much greater this would have been had we not been constrained by the "old" view of the Burgess shale!† Erik was overwhelmed by *Wonderful Life*—he wanted to review it, I know (but may have been *too* overwhelmed, and I know it filled him with a sort of regret for the life in Zoology which he gave up for books—though, really, his life is still, equally, if through books, *in* Zoology).

I did not know that you had an autistic son—he sounds quite

---

* Niles Eldredge.

† The Burgess Shale fossil deposit in the Canadian Rockies preserved soft tissue and many organisms previously unknown. For decades after its discovery in 1909, it was assumed that these creatures must somehow fit into modern taxonomic groups. Gould proposed, in his 1989 book *Wonderful Life,* that Cambrian life had been much more diverse than previously believed, with many of the Burgess Shale fossils representing now-extinct lineages.

extraordinary. I am bewildered and dissatisfied by all the different approaches and theories etc. there are about autism—I had been going to review half-a-dozen books on the subject for the NYRB, but got cold feet as I found what a battleground the whole subject was, and an exacerbated sense of my own inexperience and inadequacy. I have spent a lot of time recently with Stephen Wiltshire, an enormously gifted autist-artist in England, and hope I can say or show something of his mind, his singularity, his "style," at a deeper level than anything I have written on the subject—tho' here I find myself prevaricating and postponing, because I cannot pretend I really *understand* what goes on inside him. I do not know if you have seen any of his drawings—his third book of drawings is being published (in England) next month—(and he is only just sixteen now!).

Again, my deepest thanks for your book and notes—and for *everything* you have written over the years. I do hope we can meet in the New Year,

With kindest regards,
Oliver Sacks

EARLY IN 1991, DUE TO budget cuts, New York State laid off or asked for the resignations of twelve hundred doctors, nurses, and other employees of psychiatric hospitals statewide, including one in the Bronx where OS had worked since 1966. Many were appalled by this further deterioration in the already minimal services available to those with serious mental illness. The following was OS's response to receiving an "employee horizontal reassignment form."

To New York State Office of Mental Health

JANUARY 22, 1991

I do *not* resign my position at a hospital where I have worked for 24 years. You will have to fire me! Nor am I prepared to appropriate anyone else's job.

Oliver Sacks, M.D.

AND IN A LETTER TO his English colleague Gerald Stern a few months later, OS wrote:

Being laid off at the State Hospital does not matter to *me* that much (though I will miss many patients, whom I had got to know well in my quarter-century there), but these lay-offs and cut-backs throughout the State, and the States, are going to have a disastrous effect on medical care in general here. As you say about the NHS, this is partly the inevitable collapse of romantic notions—but much else too.

To Nancy Best

CORRESPONDENT*

MAY 5, 1991
299 WEST 12TH ST., NEW YORK

Dear Ms. Best,

I was touched by your letter, its sensitivity and obvious concern. I am less sure how to reply to it, and not sure at all how you will like my reply. [. . .]

I am glad you liked the chapter on Organized Chaos, and this is exactly where I might organize a reply. It so happened that the very day I received your letter I heard that Ilya Prigogine† was lecturing in New York. His first lecture (damn it!) I missed, the second was a revelation. To see this great old man, so calm, with a piece of chalk in his hand, writing up differential equations and weaving them and his thoughts into poetry, bringing together a tremendous, luminous synthesis of the universe—*that* was a wonder. I found this, I found him, a great comfort, a comforter—I wanted to run out in the streets afterwards shouting "It's all OK. It's all alright! I have it on the highest

* In the revised 1990 edition of *Awakenings,* OS added several new appendices, including one called "Chaos and Awakenings," which dealt with the emerging fields of chaos theory and self-organizing systems in relation to the oscillations of parkinsonism. Nancy Best resonated to this concept of self-organization, but also wondered what OS thought about the theological implications of his work.

† Ilya Prigogine (1917–2003), a Nobel laureate, was a pioneer in the study of self-organizing systems.

authority!"—and the feeling of reassurance, of bliss almost, stayed through the weekend.

In particular, he presented (in theory, in outline) a picture of a completely self-organizing universe—self-organizing from the Big Bang, through galaxies and chemical compounds, through higher and higher levels of nervous activity, to the human brain and mind: all this in terms of unstable dynamical systems which, if they tend to chaos, also tend to self-organizing structures (what he calls "dissipative structures"). At one point he said "Nature is like a gigantic brain"—and that gave us all a shiver, a sense of awe.

It seems to me (but perhaps this is only becoming clear now, in our time and generation) that the origin and evolution of the universe (up to and including the human mind-brain) is entirely explicable in natural terms, specifically in terms of such self-organizing systems. Nature itself is seen as active, unlike the passive, clock-like Nature of Paley* (which needed a clockmaker). Or to use Nietzsche's metaphor: "The world is a work of art that is creating itself"—it paints itself, it stands in no need of a Painter. I myself do not feel anything *missing* in this picture, or any need to posit a creative Spirit. Nor, I think, have I ever felt that need—at least in terms of the creation of Nature and Man.

Having said this, I *do* recognize a need and quest for a sense of the timeless, of Eternity; for a sense of Peace; and transcendent wonder. I understand well what Auden meant when he used to refer to the world as "a Sacred Allegory." Or, at least, there are *times* when I know what all of these things mean—and then they vanish, and leave a sense of desolation. So, perhaps, there is hope for this soul yet!

Again, all my thanks for your fine letter,
With kind regards,
Oliver Sacks

* William Paley (1743–1805).

To Marsha Ivins

ASTRONAUT

JULY 10, 1991
299 WEST 12TH ST., NEW YORK

Dear Marsha Ivins,

It was a pleasure and a privilege meeting you at the Aeronautical and Space Museum last weekend, and hearing (tantalizingly briefly!) about some of your experiences in your twelve days in space. As a neurologist, and one especially interested in "body-image" and the way(s) in which one perceives one's own body, in different circumstances, I could not help feeling that you must have acquired a unique experience (and insight) into such things.

I have heard accounts of altered bodily perception, and states of disorientation, etc. after various forms of nerve damage, or damage to the vestibular system; and I have heard accounts of how different the body feels when diving (I used to scuba-dive myself); but I have never heard a detailed account of how one experiences one's body (and "world") in free-fall, as *you* must have done for 260 hours.*

I take the liberty of sending you one of my books—I don't know if you've ever read any of my stuff. The narrative called "The Disembodied Lady" relates to the experiences of a patient who had lost almost all proprioceptive sense—and hence felt "disconnected" in a fundamental sort of way. A former student of mine† has written a book about another such patient, in which he notes, at one point, that [. . .] the patient, "as he lay there flat on his bed, had the distinct and frightening sensation of floating . . . but this was not the relaxed feeling associated with swimming, but an almost unimaginable complete absence of feeling . . . He could feel nothing from the neck

* In a letter to Ivins of June 16, 1992, just before she was due to fly a new shuttle mission, OS added:

> I suppose it is partly the novelty of the perceptions themselves, but, equally, how the brain/mind can re-organize itself, in hours or days, so that you can manage without an "up" or "down," without a sense of gravity, with objects behaving in the most bizarre fashion; all of this you, as both sensitive human being and scientific observer, are in a unique position to feel and analyse.

† Jonathan Cole, OS's former student, wrote *Pride and a Daily Marathon,* about his patient Ian Waterman. Cole became a close friend of Ivin's as well, and together they set up some experiments with Waterman.

downwards . . . [and] he had no idea of where the various bits of his body were *without looking at them.*"

In the very brief time we talked (as we were being photographed!) together, you said, if I did not misunderstand you, that if *you* closed your eyes when in space, you lost all idea of where your body was, or how it was oriented, and would have to open them and *see* (and that then, wherever you saw your feet, that was "down"). You also said that you had kept something of a diary, or a journal, when you were aloft. I am wondering if it would be possible for me to see any of this, or to learn more from you about the changes in bodily perception (and sense of orientation) when you were in space, and the ways in which you (and others) adapted to these. After writing "The Disembodied Lady," I would like, I confess, to write "Lady in Space," with your help or collaboration, about experiences which must be fundamental, but (at this point) so strange to most of us—tho' in the next century, perhaps, they will be common knowledge.*

Do tell more!

With all my thanks, and best wishes,
Oliver Sacks

IN 1991 OS AND I TRAVELED to Washington, D.C., where he testified before a Senate subcommittee about the value of music therapy. There we met Mickey Hart, the Grateful Dead percussionist, who invited us to an upcoming show at Madison Square Garden. OS was seated onstage, behind the band, and my husband and I had the unlikely experience of seeing OS bopping away, looking strangely like Jerry Garcia's twin. Both sported generous gray beards and were dressed in large black T-shirts. Later that week, on September 17, 1991, having sampled one Grateful Dead concert, we arranged to take OS's patient "Gary," a blind, amnesic Deadhead, to another one of their shows. (OS wrote about this in "The Last Hippie.")

* Soon after this letter, OS's briefcase was stolen from a hotel, and with it all his notes and thoughts on this potential essay. He never felt able to reconstruct them.

To Frank Wilson

NEUROLOGIST, AUTHOR, AND MUSIC THERAPY ADVOCATE

SEPTEMBER 16, 1991
299 WEST 12TH ST., NEW YORK

Dear Frank,

Many thanks for your good letter—fascinating with your parkinsonian lady!—photos, Ted Rust's article, etc.

Very exciting about developments in DC—this MAY be the week for us, for music etc. etc. I went with Jonathan* to a Grateful Dead concert. I had never been to a Rock concert in my life before—and *loved* it!! It was so dynamogenic that it caused me to dance for two hours—now I have an effused knee and can only hop. Incredible to see the vast audience (18,000) in a sort of group rapture, *all* psychically-neurologically completely "entrained" by the music: as direct an example of music's power as I have ever seen—and an overwhelming one because of the numbers involved. There really was (as our ancestors would have said) a "neurogamy" involved.

I am intolerably busy, as always. Could someone make a *transcript* of my Rochester talk—it was quite slow, and did not contain too many words—and then I'll get to it. I sent a short piece (just my Washington statement) off to Rosalie Pratt, and I will be writing a much longer piece with Connie† and try to place it somewhere prominent.

All best wishes to you,
Oliver

* Jonathan Adelstein, a Senate staffer who had put together the music therapy hearing.

† Concetta Tomaino, a music therapist and cofounder of the Institute for Music and Neurologic Function.

To Rodrigo Delfino Nascimento

MEDICAL STUDENT

SEPTEMBER 16, 1991

299 WEST 12TH ST., NEW YORK

Dear Sir,

I was enchanted by your letter and language—and have even enlarged that portion of your letter where you express the hope that I may continue my work "until your cerebral cortex and your myocardial fibres allow (I beg God's mercy these will for still a long time)" and pinned it on my notice-board.

I was a bad student in medical school, cut most of my lectures, did not read my books—but (this was perhaps in a more leisurely era, in England, in the 1950's) I was permitted to spend as much time as I wanted with patients, listening, looking, pondering, reflecting. We did not have DSM-III then; nor were there all the tormenting exams one has now. My "models," I suppose, were my parents, both doctors—and the strong sense they had for *individuals* and their "stories." I grew up, for better or worse, in an atmosphere of medical "stories"—but stories which always had a human, no less than a clinical, side. I grew up feeling the need for *description*—that experience could not yield its full meaning without this. If I went to medical school now, I am sure I would be flung out.

What can I say to you? Stay with it, and (sooner or later) things will get different, and you will emerge from medical school into the *spaciousness* of real life.

Good luck!

Oliver Sacks

To Thom Gunn

APRIL 1, 1992
299 WEST 12TH ST.

Dear Thom,

I was away (Australia/NZ)—on my return found your new collection of poems, for which deep thanks.

A few I had read before—especially "Lament," which overwhelmed me—most were new (to me) and quite marvellous. You speak in so many voices now, you embrace more and more, and yet it is always your own voice, which I first heard (and loved) in *Fighting Terms* so long ago. I especially liked "Nasturtium"—I hope you may write more poems like this, celebrations of brave plants in vacant lots, ditches, crannies etc (you remember how "Hajdi Murad" came, or came back, to Tolstoy when he saw a thistle?)

There are times when the human, and even the animal world, seem past bearing, and then I turn (back) to plants—and *their* metaphors. I very much liked "Otter" and "Pitcher Plant" too. At other times of "going back" I return to the water, to swimming, for which I get an ever-greater passion (Australia + New Zealand were wonderful for this—as for their fauna + flora—it was a relief from "the human condition"). I was given a proof of a book I found mysterious and amazing—*Haunts of the Black Masseur: The Swimmer as Hero,* by Char. Sprawson (forthcoming from Cape). Do you know of this—or of him? (blurb enclosed). It seemed to me (like motorcycling perhaps?) to reveal a deep, acted-on, but not fully acknowledged, part of myself—maybe others will find this too—especially if they are (of the trans-cultural "brotherhood" of) swimmers.

I take the liberty of sending 2 recently published pieces of my own. I hope you too are enjoying (while sickness and madness ride around us) good health and a steady wind.

Love,
Oliver

To Jane Mathias

PSYCHOLOGIST*

JULY 14, 1992
299 WEST 12TH ST., NEW YORK

Dear Dr. Mathias,

I am not sure whether I can answer any of your questions, or should try. I live somewhat like a somnambulist, and the reasons for my actions and thoughts are often opaque to me—nor do I want to get caught in the unproductive circularity of self-reference.

The clinical life (like life itself) is full of chances—it seems to me that it was chance which brought me into contact with the postencephalitics of *Awakenings*, chance which brought me into contact with my first group of patients (*Migraine*), chance which determines who contacts me and whom I respond to, and perhaps it was chance which caused the accident I describe in my "Leg" book. The scientific life is much more systematic—one has a thesis, or a hypothesis, or a set of ideas, or a technique, and one explores in these terms. Whereas the clinician never quite knows who is going to walk through his door, phone him, whom he will encounter on the ward, or anything else.

So, perhaps, there is a life rich in chances, or vicissitudes. And the question then is: which does one *select*—or which select one—How does the particular concentration and focus come about?

The patient population may seem unpromising in the extreme—what a dead-end, I first thought, when I joined a migraine clinic. But then (as I indicate in my preface to *Migraine*) I found the subject, and the patients, opened out in a wonderful way—so it seemed to me (for a while) to be the most enchanting of subjects. *Every* subject promises enchantment, if approached as a microcosm of human nature, or Nature; every subject becomes a key, or a door, or a window (whatever metaphor you like to use), opens out more and more widely, and surprisingly—if one lets it.

Recently (for example) I was phoned by a man blind from infancy,

* Mathias had written to say that she admired OS's work and wanted to know more about his influences and teaching approach.

who has recently been given sight through surgery.* But he can't "see" too well—he (and his brain) never *learned* to see. This seemingly rather special and narrow situation opens into the most general (even philosophical) issues—as when Molyneux,† in the 1680's, wrote to Locke about such a patient, or (rather) trying to *imagine* such a patient.

My "teacher" in a way was A. R. Luria—especially the Luria of *Shattered World* and *Mnemonist,* and the one who as a young man was influenced by Pater's "Imaginary Portraits,"‡ and decided *he* wanted to paint "*Un*imagined" ("real") portraits"—portraits at the intersection of biology and biography. I cannot take more time on this letter—but in my CV there are a number of more general articles (precisely *on* some of the questions you ask), and I must leave you to explore yourself. But I would not waste too much time on *me*—rather, get on with, find, discover, create, *your own* work and path.

With best wishes,
Oliver Sacks

* This encounter, with the man OS would call "Virgil," led to the case history "To See and Not See."

† In 1688, the scientist William Molyneux wrote to the philosopher John Locke to pose what is sometimes called "Molyneux's problem": whether someone born blind who learned to distinguish a sphere from a cube only by touch could—if somehow given sight—translate this knowledge to sight alone.

‡ Walter Pater's collection of fictional character studies dealt with artists and philosophers, and was influential through the twentieth century.

To Vernon Mountcastle[*]

NEUROPHYSIOLOGIST

AUGUST 5, 1992

299 WEST 12TH ST., NEW YORK

Dear Vernon,

Thank you so much for your letter—I will, if I may, convey your appreciation of his book to Jonathan Cole.

I was moved, more than I can say, by your beautiful description of the joy(s) you have found in laboratory work—and people imagine that science, and scientists, are "dry"!

I was reminded of the way in which Bertrand Russell describes his joy at first reading Euclid (this he compares to first love), and how Chandrasekhar (in *Truth and Beauty*) describes *his* delight in science (tho' this, for him, means *theory* above all), where your ecstasies come from *discovery* (but perhaps, at this level, they become one).

Perhaps a man like yourself should not cease, never cease, laboratory work—did Sherrington cease, or visit his labs right through?

I found it painful, and poignant, to think of you ceasing lab work and hope/am sure you will find some way of resuming it (if this is your wish). Did you know Purdon Martin, the neurologist? *He* was devastated at his formal retirement but then, unasked, unsolicited, started a new chapter of his life, going to the Highlands Hospital in London, working with postencephalitics—beautiful, leisurely work, full of discoveries, which he pursued, in his own way, past the age of ninety—this last chapter of Purdon's life was the most astonishing of all.

I had a sort of pang on reading your letter, because *I* once dreamt of being a "real" scientist and discoverer myself, but (I sometimes feel) *betrayed* these aspirations. But the same life, the same gifts, are

* Mountcastle (1918–2015), a distinguished neurophysiologist, discovered the columnar organization of the cerebral cortex in the 1950s and later studied the neural correlates of perception. He wrote to OS to say that he had recently retired from laboratory work, adding, "I am still actively engaged in scholarly activity, but I miss laboratory work in a way that is difficult to describe. It has always been my heart's joy, and my own experience has always been that even the most trivial original discovery of one's own evokes a special kind of ecstasy—it is almost like falling in love for the first time, all over again!" OS so loved this quotation that a copy of it hung above his desk for many years.

not given to us all—and perhaps the clinical life, so incoherent and "impure" compared to the scientific one, has its own discoveries, and beholdings, and even ecstasies too, tho' perhaps these are always "complicated" by a sense of the suffering or affliction entailed—so there cannot be an *unclouded* joy.

But there is nonetheless a strong pleasure and fulfilment in trying to draw the many strands of a patient's life into a coherent clinical narrative or "case-history"—and perhaps (as I hope in my more sanguine moments) such case-histories also have their uses.

I take the liberty (after all this!) of sending you one of the *least* "medical" narratives I have written*—but I hope you may enjoy it, and get *some* sense of an odd phenomenon, and a man, and a life.

With my warmest regards, and looking forward to our meeting again,
Oliver

PS Friend Ralph (Siegel) asks me to send *his* regards too.

PPS Sorry about my writing—I should have typed this.

To Marina Molino Ronza

ART TEACHER

DECEMBER 7, 1992

299 WEST 12TH ST., NEW YORK

Dear Mrs. Marina Molino Ronza,

Many thanks for your letter (and, I think, your FAX last week), and for sending me a copy of your husband's article.

I am interested (and moved) that my own observations on patients (often with degenerative brain disease) can have resonances in an area which would seem so remote—namely, your own work with drawing and creative development in primary school children. I was very struck, many years ago, when I worked with children, to see how many of them had "breakdowns" in their 6th and 7th years when they

* "The Landscape of His Dreams."

first went to school. I was supposed to diagnose neuro- or psychopathology in them all, but finally I got so puzzled that I started visiting nearby schools, and there perceived what you center upon: that the natural spontaneity and freedom and creativity of the child can get appallingly inhibited or prohibited with the uncreative character of the school, and that it is precisely this violation (this near "soul-murder," at times) which causes the "breakdowns." So, by the same token, one desperately needs to have (*somehow* or other) the sort of "room," the creative freedom, the child needs—without, however, forfeiting the structure, and perhaps the formal instruction, he may also need—as has happened in all-too-many "free" schools. [. . .]

With kindest regards,
Oliver Sacks

To Robert Bogdan

PROFESSOR OF SOCIAL SCIENCE AND EDUCATION

MARCH 27, 1993
299 WEST 12TH ST., NEW YORK

Dear Dr. Bogdan,

I have been away (in Guam)—hence my belated reply to your fascinating and moving paper on "A 'Simple' Farmer accused of Murder."

(Driving into town this morning I heard of a documentary *Brother's Keeper*—I am not sure whether this is about the Munnsville episode, or some other: I had thought there *was*, already, a film about Delbert and his brothers).*

I have immediate (positive) resonances to what you write about—not least personally knowing three brothers in Sullivan Co, NY, also referred to (affectionately) as "the boys," and occasionally as "simple" or "hillybillies," but never in derogatory or stigmatizing or excluding terms—they are totally part of the community and always have been.

* *Brother's Keeper* explored the "Munnsville" incident, in which Delbert Ward, a farmer, was accused of killing one of his brothers but acquitted for lack of evidence. The police had forced him to sign a written confession that, because he was illiterate, he was unable to read.

I have known them myself for 25 years, and occasionally slept over in their ramshackle house. Having a personal relation to them myself I find it impossible to attach diagnostic or pathologizing terms to them like "mentally retarded," "schizoid," "deviant" etc (altho' I am sure all these—and other things—would be said if, like the Ward brothers, some special incident were to bring them up for "diagnosis").

In a less personal, but more "medical-anthropological" sense my experiences have been with high incidences of diseases in various "geographical isolates" [. . .] and, just last month, as I mentioned, I was in Guam, where there is a very high incidence of a neurological disorder (among the Chamorros) called "lytico-bodig." In the 1940's, in some villages, scarcely a household was unaffected, and it accounted for almost 50% of adult Chamorro deaths: it is now in decline, but there are still some hundreds of people affected.

Virtually all the reporting of lytico-bodig has been of a strictly medical (and usually) epidemiological sort—I do not know of any studies of the sort which need equally to be done—of how those affected (and their families/friends/neighbours/communities) *feel about* the disease, "perceive" it, to what extent exclusionary or stigmatizing forces are active (there used to be a leper colony on Guam, incidentally), and to what extent the sick remain totally part of the community. I would say that the latter is strongly predominant, especially in the smaller, most affected, most rural communities—they are never shunned, and almost never institutionalized. All the reasons for community acceptance you analysed—native birth, lifelong residence, extensive family, shared values, independence (tho' this last, obviously, changes as the disease advances)—are strong; there is no suggestion (I should add) that the disease is contagious. There *is* some suggestion that it may be caused by eating the seeds of a local plant—a cycad—and this suggestion itself has caused some local indignation (because "fadang" or "federico," as the Chamorros call it, has been a staple of their diet, and prized and seen almost as their "special" food for two centuries or more).

This, of course, is a very different situation from the Munnsville one. [. . .]

I am very ignorant in this sphere. Can you tell me of studies of communal reactions to (locally) high incidences of endemics of illness? There are a number of such "geographical isolates" around the world—or have been. One such is the endemic of Huntington's chorea in Venezuela; another (if it still exists) of a cerebral degeneration (an adolescent neuro-lipidosis) in a village of Huguenot stock, in Nova Scotia. These are hereditary endemics; then, of course, there are the affected and the survivors in tragedies like Minamata Bay (mercury poisoning)—and (I guess) Hiroshima.

Anyhow, thank you again for sending me your beautiful and careful study of Munnsville—and let us keep in touch.

With kind regards,
Oliver Sacks

OS CONTINUED TO WRESTLE WITH his piece about Stephen Wiltshire and his thoughts about autism in general, including Asperger's syndrome—a type of autism then almost unrecognized except by a few autism researchers, and virtually unknown to the public. He went to Colorado to meet Temple Grandin, thinking the encounter might end up as a footnote to his piece on Stephen. Instead, it sparked a lifelong relationship, and a new essay that would become the title chapter of *An Anthropologist on Mars.* He also visited schools for autistic children and corresponded with many of the leading researchers.

To Bernard Rimland

AUTISM RESEARCHER*

APRIL 21, 1994 [NOT SENT]
299 WEST 12TH ST., NEW YORK

Dear Bernard,

Thank you so much for your nice letter of the 1st. [. . .] This has been an oppressively busy time, not least because I keep writing and rewriting and revising and rerevising a long piece on Stephen Wiltshire (the

* Rimland was a physician whose own son was autistic. OS was so hesitant about being drawn into the controversies and passions rampant in the autism world that he decided not to send this letter at all. (He had discovered how easily people might quote or misquote him, in or out of context, in their own published work.)

autistic savant in London whom I am sure you know about). I have been finding it a *most* difficult piece to write—partly because I have known Stephen for many years (since '87), partly because I am *still* not sure what his "inner life," mental processes etc. are like—nor how he does his miraculous drawings (and, now, his music too). Perhaps also because the whole business of a savant prodigy going "public" raises delicate issues which one hesitates to discuss. However, the piece is in galleys now, and I hope it may appear in the *New Yorker* this spring.

I went up to Williamstown last weekend, to see Jessy Park and her parents—they have *such* a loving (but realistic) understanding of her, and her art. I wish *they* would write another book—it is 25 years and more, now, since Clara wrote *The Siege.** I went up with Isabelle Rapin, who is a very old friend of mine, and had wanted to meet the Parks since she read it in the Sixties. Isabelle also read *your* book when it came out, and was recommending it to me in the Sixties; but, at that time, I had no interest in autism, and scarcely knew of its existence. [. . .]

For myself, I esteem *all* of you, who have spent so many years observing, pondering, writing about autism, although I am not unaware of some of the internecine dissensions between A and B, B and C, etc. etc.

This was very much my feeling when I entered the Deafness world, and found so many brilliant people, with different perspectives and different angles, often in strong disagreement with one another. Equally with the Migraine world, the Tourette world, all the other worlds I have "visited" though I think Intellect, and Dissension, reach a special pitch in the Autism world!

I have, for better or worse, never felt myself a primary worker in any of these fields—certainly not in Deafness, or Autism, perhaps a little bit in some others—and my own impulses are for synthesis and reconciliation of views (reconciliation is not always possible with *people*!), and a general eclecticism and empiricism. I also have a strong sense of the biological variability of many disorders, and thus of the (possibility of) variability in their treatments. When Gowers

* Clara Claiborne Park's book *The Siege,* about her daughter, was an early classic on autism.

writes, in the 1890's, regarding the treatment of migraine, "the measures that do good in one case will fail in another, apparently quite similar," I think his statement still holds a century later. I suspect too that something similar may hold regarding Tourette's syndrome, and Autism—that *some* patients will do well with some (medicational) measures, whilst others will show no response to them.

I feel I may need to say all this partly because you have sent me copies of your own (and others') observations on [vitamin] B_6, magnesium etc, and some of your correspondence with Isabelle and others. I myself, as I have said, am not a primary worker here, and have had *no experience* of any of these biological interventions. My *suspicion* is that they may help some people with autism, but not others—but I have not reviewed the literature, nor do I feel competent to do so. It would obviously be very good all round if a broad clarification and consensus could be reached. I have no special feelings "for" or "against" any mode of treatment—whatever works, so far as I am concerned, works, and should be tried (even if the mechanism of its working is wholly unclear). [Certain] beans were a "folk-remedy" for parkinsonism for centuries—but only found to contain l-DOPA (and, indeed, to be the richest natural source of it) in the 1970's. Digitalis and colchicum were "unorthodox" remedies for centuries—I think we are still uncertain why colchicine works. [. . .]

So, after all this preamble, you will understand that I have a neutral position, or no position, or an entirely open-ended position, regarding most of the advocated treatments of autism. I also have to say that my own interest is less in treatment and especially in looking at phenomena, and a life, and perhaps at various theories—this was even true of migraine (tho' of course, when I saw a patient with migraine, I was prepared to offer the entire pharmacopeia, to say nothing of biofeedback, behavioural therapy, acupuncture and analysis). I do not *have* any patients with autism,* and my experience here is, in a way, very limited. I have probably seen no more than half a dozen people with autism *at any depth,* whereas I have seen thousands with Tourette's and migraine. [. . .]

* OS had known a number of patients with autism, including several he wrote about in *The Man Who Mistook His Wife for a Hat.* But he did not regard himself as an expert in this realm; nor did he offer any "therapies" for people with autism, apart from sympathetic attention.

I have rattled on, and must stop—but I felt a need to *define* this and that.

With all my thanks again, and best wishes,
Oliver

To William Swann

PSYCHOLOGIST

MAY 16, 1994
299 WEST 12TH ST., NEW YORK

Dear Bill,

Many thanks for your letter and two chapters, all enjoyed. I will reply relatively briefly, because I have a painful shoulder (indeed it will be operated on tomorrow), and typing is difficult. [. . .]

The "case" of Virgil is one of seven case-histories I am now putting together for a book,* and (it seems to me) *all* of them, in their way, describe or intimate something similar—the transformation of identity forced upon individuals by sudden, profound sensory or neural insults or inputs. In the case of Virgil (as of so many such patients) [. . .] the change is very difficult, tho' a few, apparently, can change and adapt very well.† [. . .] Similar considerations, in a way, arise with enterprises to augment or restore hearing in the deaf—especially, now, with the development of cochlear implants. You may know Harlan Lane's latest book here (*The Mask of Benevolence,* Knopf—originally, and perhaps better, entitled *Let Me Be Deaf*).

Another patient, an artist who became suddenly totally colorblind in his sixties (a head injury), whose case I originally published in the NYRB in '87 (but have now extensively rewritten) *was* able to "redefine" himself or adapt in a most creative way, to harbour the birth of a new Achromatopic sensibility and identity and creativity,

* The book, *An Anthropologist on Mars,* would be published in 1995.

† Virgil, who had been functionally blind since early childhood, had, as an adult, been offered cataract surgery, which would "give" him sight. What his physicians had not considered was how difficult it would be for his middled-aged brain to make any sense of this new input.

after the color-normal one had been extinguished in him. But there was a terrible (and dangerous) interim between these, when [. . .] he wandered between worlds—"one dead . . . one not yet ready to be born," or whatever Matthew Arnold says in "Dover Beach."* And yet, "adapted" though he is, in a way, he cannot help looking back, wistfully, to time when his vision was full.

Regarding the *Awakenings* patients, [. . .] with someone like "Rose R." the problem was far more radical, because her whole *identity* belonged to another period. There was none of the continuous evolution (for better or worse!) the rest of us have, and she had to "adapt," or redefine herself, quite radically, with the medication, astonishingly. *She* failed, which is not surprising, but many of the others succeeded to some extent. [. . .]

There are so many examples of this sort of thing—indeed I think "medicine" or "the clinical" must itself be re-defined, so that it anticipates, discusses, and studies all such changes, and thinks in terms of "identities" rather than "norms."

Now I have selfishly ranted on about *my* patients and attitudes etc. and not properly "critiqued" yours—but, as I said, I am relatively ignorant of the *social* "shackles" you talk about so well (tho' very conscious of the *neural* shackles instead). Both, obviously, are absolutely crucial.

Thank you so much for sharing your work with me,
Oliver

IN THE SUMMER OF 1994, OS revisited Guam to examine the curious neurodegenerative disease endemic there, which so many scientists had tried and failed to elucidate over a span of decades. (In various ways similar to parkinsonism, to dementia, and even to ALS, it was regarded as a possible "Rosetta Stone" for neurological diseases in general.) He was accompanied by a

* This thought is Matthew Arnold's but from his poem "Stanzas from the Grande Chartreuse":

Wandering between two worlds, one dead,
The other powerless to be born.

BBC film crew that was making a series of six films called *The Mind Traveller,* based on OS's work. They also visited the tiny, isolated atoll of Pingelap—OS had heard about a community of people there who were congenitally completely colorblind, or achromatopic (a neat contrast to the "colorblind painter," whose achromatopsia was acquired late in life). Eventually, these travels in the South Pacific would lead to a book, *The Island of the Colorblind,* published in 1996.

To Marcus and Gay Sacks

SEPTEMBER 9, 1994
299 WEST 12TH ST., NEW YORK

Dear Marcus & Gay,

I have only just got back from my travels—to a huge backlog of mail, of everything; and, of course, your full letter of the 16th.

The experience—of Pingelap, Ponape, Guam, Rota etc.—was enormously rich, and totally exhausting. It rained constantly, the temperature was always over 90 (often over 100), and the working days, all of them, were 16 hours long—so I had almost no time for myself, and would have had even less for anybody else. I think we did some grand filming in the Islands—there has never been documentation of the achromatopes of Pingelap/Ponape—and I *hope* I can recapture some of the experience in writing too, tho' what with the filming, the exhaustion, the diarrhoea [. . .] and the rain, I did not feel up to making many notes at the time. I hope the essential things are somewhere in my memory, and will come back when I try to tell the story.

My shoulder—done in part arthroscopically, and part conventionally (acromioplasty, removal end of clavicle, repair of rotator cuff)—now feels very good, and I was able to do hard overarm swimming of a sort which had become almost impossible before. But, as you say, this is probably one of the many prices of my foolish weightlifting in youth. When I was in hospital for the second ruptured quad, in '84, a former weightlifting friend of mine wandered by (with two hip replacements at the age of fifty)—and we agreed what fools we had been.

You seem to have been having spectacular weather in Sydney—I wish *I* could catch that sort of weather when I was there, instead of the hot and sticky Februaries I seem to land in (and it will be that way, I guess, if/when I come to Canberra for the wedding* in March). I hope the drought is not too serious—I remember how frighteningly brown Tasmania looked, Gay, when we flew there some years ago. [. . .]

Unlike you, I still use my (now 25 year) old IBM—and get it repaired when it needs it. I loathe the new typewriters—I like the pressure of the old keys, the way they bang, and stick—most pleasing. I remain among the few computer-illiterate people left—but Kate converts my writing/typing into beautiful word-processing (so I am spoilt). I also remain a two-finger typer—which was disabling when the shoulder was out.

I had a memory (or is it a pseudo-memory) that Ma once showed me a model (or original) of our grandfather's safety-lamp† in the Science Museum in South Ken. Do you have a similar memory? I asked the Science museum about it, and they said they had no record but they were kind enough to send me the details of his patents. I enclose a copy of these for your amusement. [. . .]

So I will write again soon, but I did want to answer your own letter while it was fresh—and, of course, to wish you all the best of New Years.

Do let me know wedding details as soon as you can, and *if* I can I will come over for it in March. [. . .]

All my love,
Oliver

* Of Marcus and Gay's daughter, Carla.

† OS tells the story of the Landau lamp in *Uncle Tungsten.*

To Antonio Damasio

OCTOBER 13, 1994
299 WEST 12TH ST., NEW YORK

Dear Antonio,

It was a delight being able to spend time with you and Hanna in such congenial circumstances—I was very taken with the College, the "wholesomeness" of the mid-West (as you put it),* and, of course, the incredible calibre of the Conference itself—an astonishing combination, as well as the relaxed way we could all chat informally into the early hours in the guest-house. What with this, and completing your book,† I feel I *know* you a little better now!

I don't have the book with me as I type [. . .] but I admired your discussion of "somatic markers," and your general reflections on Embodiment immensely. I feel that this is what I have (roughly and inchoately) thought myself for years, but you give it an explicitness, a fullness, and a musing on possible neurobiological and neuro-evolutionary bases which is stunning. You have really written a book to be very proud of, and one which will surely play a very significant role in orienting and guiding thought/research in the future.

The description and discussion of your frontal lobe patients—their complete loss of "secondary" emotions, with complete preservation of "primary" ones—was particularly fascinating (and I could not help wondering how much this might apply to some autistic people too). Luria, who plumbed the neuropsychology of such patients, would have been particularly enthralled by your observations/thoughts here—do you have separate articles on this?

Had I read the second half of your book when I was in Minnesota I would not have said, impudently, that the "self" wasn't being adequately considered. The heart of your book, I suppose, is that (in a way which is different from Gerry's,‡ but parallel, or perhaps convergent) you bring the "self" (as a continually-renewed construction—not

* During these years, Damasio and his wife, Hanna Damasio, a neurophysiologist, were at the University of Iowa.

† Damasio had just published *Descartes' Error: Emotion, Reason, and the Human Brain.*

‡ Gerald Edelman.

a thing), and emotion, and the body (not just the brain) where it belongs, in the very center of neurology, and with such a wealth of clinical observation and other evidence as nobody now can gainsay.

What would William James say now?

My warmest regards to you and Hanna . . . and hopes to visit,

Oliver

To Brian Friel

PLAYWRIGHT

OCTOBER 13, 1994
299 WEST 12TH ST., NEW YORK

Dear Mr. Friel,

I have been reading *Molly Sweeney* with mixed feelings—admiration at its fineness (I think you are very gifted), and some disquiet at the unacknowledged way in which you have appropriated so much from my own piece "To See and Not See."

It is indeed a very creative use of my piece, as *A Kind of Alaska* was a very creative use of *Awakenings;* but where Harold Pinter sent me the manuscript, and noted (on its publication, and in its performances) that it had been "inspired" by *Awakenings*, you yourself have been somewhat less straightforward.

I (and my Agents) will await your reply.

Yours,
Oliver Sacks

To Brian Friel

OCTOBER 18, 1994
299 WEST 12TH ST., NEW YORK

Dear Mr. Friel,

Thank you so much for your letter—I am delighted to have some personal contact with you, and only wish there had been some before your play was published.

I did not know that you had cataracts in both eyes—my sympathy—and, of course, I well understand why this should have stimulated, and given a poignant personal interest to, your interest in seeing, blindness, and restoration of vision.

I have admired your own work, and was especially fascinated to read your latest play, *Molly Sweeney.* I think it is a quite remarkable play, which brings out all the subtleties of experience and feeling as only a juxtaposition of stream-of-consciousness soliloquies could. I appreciate the formal similarities between it and *Faith Healer,* and the many resonances between the two plays. I perceive too how you have turned to seeing/not seeing as a metaphor—I too, though confined to the realm of clinical fact, am very conscious of the metaphorical potential of many of the cases I explore; and this, no doubt, is one of the reasons why my own case histories have had such special resonances for dramatists and others. This, finally, is why *Awakenings* was the inspiration of Harold Pinter's *A Kind of Alaska* (as he was the first to acknowledge); and, I suspect, why you gave up your original idea of using [Richard] Gregory's S.B. for your central character, and turned instead so massively to "Virgil."

Though you say that Molly is entirely fictional (I leave other characters out for the moment), and has no real antecedents except Grace Hardy, her personal and clinical history, as you give it in the play, is a virtual duplicate of Virgil's. Molly, like Virgil, is a massage therapist; she is blind from early life with a retinitis and cataracts; she has first one cataract, then the other, removed in middle life, and following this is thrown into a state of agnosia and confusion; this is followed by episodes of impaired gnosis, then blindsight; and finally a strange reversion to blindness.

The extreme particularity and circumstantiality of this history differentiates it wholly from Gregory's S.B., or Valvo's cases, or of any other cases in the world literature. Specifically, no other neurologist has spoken of "agnosia" in this context, and no one else has described agnosic episodes—or blindsight. These differences arise because the physiology in these cases is *not* identical. Virtually all the patients described in the world literature have problems with the *media* of the eye—the lens or cornea—not with the retina as well, as is the case with Virgil. It was the grossly impaired retinal function, a receptive inadequacy, which led to episodes of agnosic blurriness and blindsight. The other blind patients in the world literature, with normal retinal function, have not suffered from either agnosic episodes or blindsight.

This extra retinal pathology and its consequences are fortuitous, peculiar to my patient, and irrelevant to an allegorical story of restoration of sight. Nonetheless, *Molly Sweeney* follows Virgil's history with extreme literalness. Whatever other qualities you have endowed Molly with, her clinical history is a virtual duplicate of Virgil's.

Beyond this identity of clinical history, there are frequent close similarities (and sometimes identities) in wording; thus on even a quick examination I perceived the following:

Friel: There were scars of old disease, too. But . . . no current active disease process. So that . . . her vision, however impaired, ought to be stable for the rest of her life. (MS, p. 27)

Sacks: Examination, I was told, suggested the scars and residues of old disease but no current or active disease process; and this being so, Virgil's vision, such as it was, could be stable for the rest of his life. (NYer, p. 62)

Friel: She could distinguish light and dark; she could see the direction from which light came; she could detect the shadow of Frank's hand moving in front of her face. (17)

Sacks: Virgil could still see light and dark, and the direction from which light came, and the shadow of a hand moving in front of his eyes. (59) [. . .]

Friel: But Molly's world isn't perceived instantly, comprehensively. She composes a world from a sequence of impressions. . . . What is this object? These are ears. This is a furry body. Those are paws. That is a long tail. Ah, a cat! (35–36)

Sacks: He would pick up details incessantly . . . but would not be able to synthesize them, to form a complex impression at a glance. This was one reason the cat, visually, was so puzzling; he could see a paw, the nose, the tail, an ear, but could not see all of them together, see the cat as a whole. (64)

I could cite many other examples.

Finally, whatever the metaphoric and other differences, there is considerable importation of what I might call the moral atmosphere of my piece—the over-enthusiastic spouse, the over-ambitious doctor, and the relatively passive blind patient who becomes their agenda.

Thus to my mind (and to others) your borrowings are both deep and extensive, even though they may indeed have been "altogether unconscious." I am very well aware of the power of such unconscious absorptions—the creative state seeks for its own truth only, and is not concerned at the time with the distinction of "inner" and "outer," who said what first, where things come from. It is only later that one may perceive what has influenced one, what one has used. It is then, recovering consciousness, that one recognizes influences and sources, and pays them their due acknowledgement.

Early in 1982 Harold Pinter sent me a manuscript of *Alaska* with a charming letter. When I commented on some of the similarities to *Awakenings*, there was an initial response (from his agent) to the effect that "There are many postencephalitics. There are many 'awakenings.' Mr. Pinter himself has read all the sources. *Awakenings* was only one of many." But then Pinter himself, with great courtesy and delicacy, cut through all this, and said, "Of course *Alaska* was inspired by *Awakenings*!" This acknowledgement was put up front in the programs, playbills and publications of his play (see attached photocopy), and he agreed to pay me 20% of the net receipts from the play. I feel that it did no harm to Pinter, nor to Peter Brook,* to indicate their indebtedness to a distinguished clinical source—quite the reverse: it moved their material from the realm of the purely imaginary or metaphorical, and gave it the density of the actual, the factual—and the same, surely, applies here.

Molly Sweeney is at least as close to my case history as *Alaska* was

* Peter Brook had directed *L'homme qui . . .*, a stage production based on *The Man Who Mistook His Wife for a Hat.*

to *Awakenings*. Thus I do not think it would be fair of you to expect me to be satisfied (in the light of everything I have shown above) with a mere listing as one among many sources. I really feel that some sort of exclusive acknowledgement is called for, something wholly separate in tone and place from the acknowledgement you may make to other sources which constitute a mere background to the subject. I would hope that some mutually acceptable form can be arrived at, perhaps something akin to the acknowledgement and short history Pinter put in *Alaska* (xerox enclosed), or something as brief as "Inspired by Oliver Sacks's case history 'To See and Not See,' and the long, strange tradition of such histories."

I feel it is imperative to resolve this swiftly, so that it may be included in programs and playbills for the forthcoming London production of your play, as well as in any future printings and publications it may have. My own new book (*An Anthropologist on Mars*), I should add, will come out in January, and contains an enlarged version of "To See and Not See," along with a bibliography in which I speak of your play (along with earlier literary treatments of the same theme—by Wilkie Collins, Gide, Patrick Doherty, etc.), and its striking resonances with my own piece.

I continue to look forward to an early and amicable resolution of these matters—and, indeed, to a day when, unencumbered by such things, we will be able to share a meal, and some thoughts, together.

Sincerely,
Oliver Sacks

To Brian Friel

OCTOBER 19, 1994 (4 A.M.)
299 WEST 12TH ST., NEW YORK

Dear Mr. Friel,

Let me say by way of an early-morning postscript what I should have said, more strongly, at the start of my letter. I think *Molly Sweeney* is a very powerful play, a beautiful work of art, unmistakably and uniquely your own. And though I was constrained to speak of "borrowings," this in no sense diminishes (for me or for anybody) your own originality, for you have *refracted* whatever you have used through your own creative powers. "Borrowing" implies no disrespect for your work or originality—on the contrary, I feel *honored* to have provided any observations or thoughts which might have inspired, and been used, so creatively, by an imagination such as yours. I should have made this more explicit in yesterday's letter. I see you, more than ever, as a major and original artist.

Again, yours sincerely
Oliver Sacks

To Brian Friel

OCTOBER 20, 1994
299 WEST 12TH ST., NEW YORK

Dear Mr. Friel,

I was delighted to receive your friendly FAX, and to learn of your willingness to make an acknowledgement of the form discussed.

I have, first and foremost, regarded this whole business as a matter of courtesy—and now that the courtesy of an acknowledgement has been granted, I have no interest whatever in pursuing any share of Royalties or legal action (as your Agents seem to fear).

We have a gentleman's agreement, there is good faith—and this, as far as I am concerned, can end the matter.

I would enjoy meeting you, of course, when I am in London, or you in New York.

With kind regards,
Oliver Sacks

To John Bennet

NONFICTION EDITOR, *THE NEW YORKER*

NOVEMBER 7, 1994

299 WEST 12TH ST., NEW YORK

Dear John,

I have just got some proofs,[*] and hasten to send one off to you—for you, after all, have been a chief midwife to the pieces (5 of them, anyhow).

The four which the *New Yorker* published have, as you see, been somewhat revised, at times enlarged—and (of course!) footnoted. As for Stephen ("Prodigies") I remain, perhaps we all remain, uncertain about this. [. . .]

Ren[†] will have told you, I imagine, about the business with Brian Friel, and the closeness of his play, sometimes, to "To See and Not See." But this has been amicably sorted out, and Friel has provided an acknowledgement and this, first and last, was what was called for. I suspect his play *Molly Sweeney,* just opening in London, may be coming to New York about the same time as the book comes out—and this may lead to further resonances (good ones, I hope!).

It has been a long time since we were in touch, and I have done a lot of (sometimes exotic) travelling in the meantime. The visit to Pingelap (see enclosed map), "the Isle of the Colorblind," was quite amazing, and I am trying now to reconstruct and write this *as* a piece. I went with a BBC film crew, and this had advantages and drawbacks: on the one hand, they were able to arrange all sorts of things—e.g. the achromatopic night-fishing etc, which I could not have done, and the power to *document* a unique situation at every level from the physiological to the clinical to the social-anthropological but what with cameras looming, for twelve hours a day, I did not make the notes, do the *writing on the spot* I usually do, nor have the sense of intimate personal-clinical contacts I usually have. I can (and must) describe some *individuals,* but I must also picture a *community,* and an island, and life as it is for them on this tiny coral atoll. One of our party was a charming and learned Norwegian neuroscientist, Knut Nordby,

* Of *An Anthropologist on Mars.*

† Lawrence Weschler.

himself an achromatope (like his brother and sister—the incidence of this* in the general population is extremely low, about 1 in 50,000; whereas, on the island, a quarter of the population are carriers)—and it was extraordinary to see the instant community of experience, sensibility, language etc between him and the islanders; his excitement at his finding his (retinal) brothers-and-sisters, so to speak, on a remote Pacific atoll and theirs at also finding one of themselves, a white brother, a European, from afar—one obviously esteemed and distinguished, able to travel independently all over the world (where they are sometimes seen as "disabled" and "second-class citizens" on the island, and he, of course, was manifestly neither). [. . .] As yet I am not quite sure how to *begin.* I was in Guam, then—but you know all about the Guam disease; and then in South Africa, seeing, besides the incredible political transformation (and its uncertainties), a wonderful range of Botanical Gardens and Ferns, so *this* too, as a subject, stays at the back of my mind (and often comes to the front too).

I must not weary you with too long a letter—and had better stop.

I hope you enjoy *Mars*—and that we will be in touch again soon,

Best!
Oliver

To Joanne Cohen†

A FRIEND WITH TOURETTE'S SYNDROME

JANUARY 6, 1995
299 WEST 12TH ST., NEW YORK

Dear Joanne,

Thank you for your letters.

The second one first—is easy to answer, because it contains news

* I.e., total color blindness, or achromatopsia.

† Cohen wrote in response to "A Surgeon's Life," a *New Yorker* profile in which OS described Mort Doran, who, despite his severe Tourette's, was able to fly his own plane as well as perform his duties as a surgeon. (OS used a pseudonym for Doran, "Carl Bennett," in honor of his *New Yorker* editor, John Bennet—but Doran later revealed his real name.)

of your success in school, and how much you've been enjoying it. [. . .] For myself, I think I'm well (or as well as one ever is in the seventh decade), the shoulder-surgery worked beautifully, and I'm back to long swims. I am excited—but also nervy—about the new book coming out soon.

Your originally-suppressed letter of Sept is not so easy to answer—but I am glad you now have enough faith in me, and yourself, to know I can accommodate it. Your major point (and you are not the first to make it, nor am I wholly unconscious of it) is the notion that I may get so "fascinated" by Tourette's that I fail to recognize the all-too-real suffering and disability it may cause—and the isolation and stigma and demoralization which may go with it too. I do recognize this split, or potential split, in myself—indeed I confessed it (back in 1970) in the preface to my *Migraine*—where I spoke of being "delighted" at the complexity of some case-histories, and of the sense that every patient with classical migraine "opened out, as it were, into an entire encyclopedia of neurology." But then (I continue in the next paragraph) "I was recalled from my neurological preoccupation by the suffering of my patients and their appeals for help." A life with Tourette's is infinitely more complex than a life with migraine—Tourette's itself is infinitely more complex than migraine, not least through psychological (and moral) dimensions peculiar to it. [. . .] I do not think I am unconscious of the suffering, disability etc, so much as wanting also to present some of the more "positive" aspects of the disorder, or of the inner resources which it may call up (these include courage and humour, and a tolerance of others' intolerances, which you yourself have to a very high degree). I may err in overemphasizing the "heroic" aspects, and underemphasizing the "victim" ones (and again your story—from the hoarse shouting in the night to the paternal intolerance to the university discrimination [. . .] etc. etc.—certainly brings out these tragic aspects too). But I hope I am learning a better balance (and *you,* amongst others, must teach me here—not least by violent and honest letters such as your now-unsuppressed one). Certainly if I told *your* life story it would be different from Doran's—less "upbeat," more disquieting, tho' (finally) not less inspiring. [. . .]

Love,
Oliver

To Robert B. Daroff

EDITOR IN CHIEF, *NEUROLOGY*

APRIL 14, 1995
299 WEST 12TH ST., NEW YORK

Dear Bob,

Thank you so much for the preprint of Chatterjee and Southwood's paper, which I found quite fascinating (you know my tastes!).* Their observations, investigations and discussion are very careful indeed, and as good, I think, as anything in the literature. And yet I do not remain entirely convinced, because I am not sure how much some of the questions actually demand "imagery" (as distinct from "visual knowledge"—or, rather, knowledge obtained in the first place through vision, and then perhaps retainable despite the absence of perception *and* imagery). [. . .]

Some of this uncertainty of interpretation was present, I think, with Jonathan I., the colorblind painter, with his devastated V4's.† It was extremely clear that he did not need to *image* color(s), but this did not impair his ability to "*know*" them. He "knew" the exact green of van Gogh's billiard table (and could even give its reference number on a Pantone chart) because he had studied it carefully before, and indeed reproduced it when in art school, before his cortical colorblindness. It was *established* knowledge. He became uncertain of some colors, seemingly, more than three years after his injury—they had become indifferent to him, were no longer part of his visual thinking; but he would never have lost the knowledge that, say, a cucumber was green. The image yes—the "knowledge" no.

It seems to me that this question of "image" vs. "knowledge," and the difficulty of interpreting some of the responses, might be more fully discussed by the authors.

These are first impressions and thoughts—I will read the paper

* "Cortical Blindness and Visual Imagery," published in *Neurology* later that year.

† V4 is a section of the visual cortex partly devoted to processing color.

again (it is admirably readable), and if I have further thoughts take the liberty of writing again.

With all my thanks, and best regards,
Oliver

PS It is the *differences* between the 3 patients that I find especially fascinating.

To Daniel Dennett

PHILOSOPHER, COGNITIVE SCIENTIST

APRIL 20, 1995
299 WEST 12TH ST., NEW YORK

Dear Dan,

I just read your beautiful article "Darwin's Dangerous Idea" in the current *Sciences,* and immediately felt the impulse to write you—and send you a book (I guess my own, perhaps very limited theme of adaption and reconstruction etc. in clinical contexts might be seen as a "Darwinian" one too). Reading your piece I am (again!) hit by how well you write, how well you think—and (no less important) how sensitively you feel—I think we are all lucky to have in you such a *tender* intellectual searchlight, and one which can illuminate so many different realms (tho' perhaps, as you imply, they are all the same realm). Certainly darwinism has been a "universal acid" for me, which *had* to extend itself—downstream—prebiotically—and upstream ("psychically") as soon as I had grasped it. And it seems incredible, *now,* on the level of reason, that any intelligent person could think otherwise—I was startled at Paul Davies' words as you quoted them (because it seemed to me, when I met him some years back, that he was not only intellectually but *emotionally* satisfied by the concept of an evolutionary universe, self-organization going the whole way up). Tho' I remember him adding, uncertainly, that such a view might be a "substitute" for a religious one, which should have been a warning; for clearly, emotionally, it is not a substitute, which may be why he now, apparently, allows "God" to re-enter. I think, for myself, the emotional craving for a "higher" principle is not too strong—tho' obviously (as you bring out with your very moving remembrance of "Tell Me Why") there has been a time with all of us when that "Why" was so strong, and there has to be a sadness in outgrowing this.

I have also just found *The Third Culture*[*] in today's mail—and look forward to seeing what all of you are saying. What an exciting time this is!—I think it is chiefly just this sense of a "third culture" which, for me, wards off various physical (and metaphysical) depressions, and keeps alive a sense of romance (and I got furious with Brian Appleyard[†] for so misunderstanding this).

With all good wishes,
Oliver

---

* A book by John Brockman in which Dennett and two dozen other philosophers and scientists wrote about recent developments in science. The "third" culture refers to a melding of C. P. Snow's two cultures: science and the humanities.

† Bryan Appleyard, a journalist, had recently published *Understanding the Present: Science and the Soul of Modern Man,* which declared that science could not explain consciousness or provide value and meaning to spiritual life.

13

Syzygy

1995–2003

To Kay Redfield Jamison

PSYCHOLOGIST

JUNE 19, 1995
299 WEST 12TH ST., NEW YORK

Dear Dr. Jamison,

You must think me very churlish not to have responded to your letter of February 6, and the inscribed copy of *Touched With Fire.* But these just got into my hands TODAY, because my fucking College (Einstein) lost or hoarded the last six months' mail, and just sent it on—a cartload of it—TODAY.

By coincidence I read a galley of your new book* just YESTERDAY, and was meaning to write to *you* to say how fine I thought it—it gripped me from the first page—and I could not stop to the end. I think it a very *brave,* as well as a brilliant and often very beautiful book. Perhaps *because* it was a personal account it affected me more deeply than *Touched By Fire* [*sic*], tho' that was certainly an extraordinary piece of research. I made a (too-) brief reference to it in *Mars* (p. 165)—which I hope, however, will not be taken as sardonic.

As you bring out, continually, and powerfully, in your new book, one's attitudes, interests, judgements, researches, *have* to be conditioned, to some extent, by one's own experience. I sometimes wondered whether I was manic-depressive myself—at least for a period

* *An Unquiet Mind,* Jamison's memoir of her own bipolar illness. In an earlier book, *Touched with Fire,* she had explored the relationships between manic depression and creativity.

of about fifteen years from the age of 32 onwards, when I would have periods of intense intellectual-emotional excitement, with exaltation, and sometimes ecstasy, during which thoughts would rush through me—far more than I could use (at such times I would sleep minimally, and fill a notebook in a matter of hours). (My first book *Migraine* was written in such an excited-exalted state in a nine-day period in which I scarcely went to bed.) And then there were the troughs or depressions—tho' I am not sure that this is the right word. The spiritual word "arid" seems more appropriate, for I would feel chiefly *dry*, dull, empty, rather than *sad* inside. These excited periods always went with creativity—but (it seemed to me) they were called into being *by* stimuli to creativity, *by* the excitement of creativity itself—not the other way round. They were never seasonal or cyclical. The notion that there might be an autonomous mood-disorder, that the determinants of my life might not be "inspirations," but, so to speak, subcortical in origin, distressed me and also seemed at odds with my experience; and THIS may be a reason why I did not find myself fully in sympathy with the theme/thesis of *Touched with Fire*—tho' it may be that I over-simplified this in my mind.

I never had lithium or anything else. In the past fourteen years I have moved into a more "equable" state, which perhaps lacks intense affects (in either direction), but also the sort of "inspiration" I lived for and loved. Nonetheless, I have continued to work and write after a fashion (I take the liberty of sending you a copy of my latest book, *Mars*) altho' feeling that something magical has gone from my life. Whether this is "Wordsworthian" or biological I do not know, but it gave me a special sympathy with the later parts of your new book, and your own (sometimes) mixed feelings regarding the effects of lithium. I met Robert Lowell once (in '74), and he expressed some similar feelings—I mention this, briefly, on p. 274 of *Mars*.

My reservations are very minor compared to my admiration of your work.

With best wishes,
Oliver Sacks

To Paul Theroux

WRITER, AUTHOR OF *THE MOSQUITO COAST*

AUGUST 15, 1995
299 WEST 12TH ST., NEW YORK

Dear Paul,

That was a quite marvellous weekend; I enjoyed myself so much, and felt privileged to enter a little into your own memory-palace (both literal and metaphorical).

I return "Memory & Creation"* (I xeroxed it). How nice that we converge on Bartlett† (from different directions, as it were)—but I was a little startled (tho' after the weekend, less so) at the weight you give to memory-systems and "mnemonics" (clearly you *are* surrounded by "reminders" of all sorts, as Freud was by his statuettes). It may be that we are rather differently constituted this way: I *never* had "a manageable sub-Funes system" for converting anything into images—on the other hand, I seemed to have quite a good memory in my own way (I used to drive Jonathan‡ and others mad by reciting Poe short stories, and longer ones, verbatim. They were not consciously "memorized," or translated into images, they just seemed to sink in after a reading or two; and my favourite reading at one point was the 3000-page *Handbook of Chemistry & Physics* (29th edn.), which I knew virtually by heart. Kate, knowing of my love for this, got me a copy of the new (75th edn.) for my birthday—but it was the 29th which I could page through, and "read," in my mind). On the other hand, my house, unlike yours, is almost devoid of (memory-) objects, and perhaps very impoverished this way.

Jonathan's house (have you been there?) is incredibly rich in the Memory-Palace sense (as well as in the 50,000 book library, scholar's sense), and Jonathan's memory is very amazing: your mentioning how you recalled the names of your schoolmates long ago reminds me of how *he*, a few years ago, seeing a picture of St. Paul's School in

* Theroux's 1991 essay was published in *The Massachusetts Review* and later included in his book *Fresh Air Fiend.*

† Frederic Bartlett, author of an influential 1932 book, *Remembering.*

‡ Miller.

the 1940's recognized and named all six hundred people[*] in the picture (I don't think *I* could recognize a single one). On the other hand, I probably bear the "stories" of tens of thousands of patients in my mind—everyone I have seen, as a physician, for almost forty years. I also used to be good at remembering (or reconstructing) labyrinthine proofs and arguments. I guess there are all sorts of memories.

I absolutely agree that writing itself is a sort of memory (device), and that as one writes (whatever one writes "about") one learns, one burrows more and more deeply into oneself; gains, regains, oneself and, at the same time, transcends it. I was interested at the extreme importance you give to "fiction" as providing not so much an escape from life, as enrichments and multiplications of it—and yet I think that phantasy and imagination are exercised as much, altho' differently, in science. (Edelman sometimes calls science "the exercise of imagination in the pursuit of reality" or something of the sort.) I am babbling. [. . .]

I must stop and return to Liveing![†]

Again, my thanks to you both for a wonderful weekend,
Oliver

PS I enclose some oddments written at different times, and slightly hyperbolic & embarrassing to me now, about the importance of Liveing for me in the past. Now I must write something more reasoned, less personal.

---

* Though Miller indeed had a remarkable memory, this is almost certainly exaggerated for dramatic effect.

† OS was writing an introduction to a facsimile edition of Edward Liveing's nineteenth-century work on migraine, which had inspired him to write his own *Migraine.*

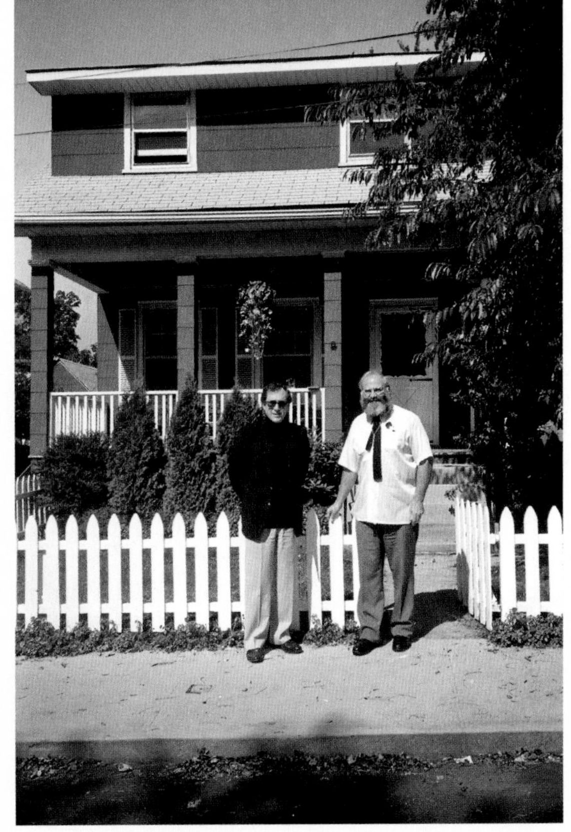

ABOVE: Publishers and writers at the tenth anniversary of Picador Books in 1982. *Left to right*: Tim Binding (editor), Michael Herr, Salman Rushdie, Bruce Chatwin, Clive James, Adam Mars Jones, Mike Petty (editor), Russell Hoban, Hugo Williams, Tom Maschler (publisher), and OS. Seated are Sonny Mehta (publisher) and Emma Tennant.

RIGHT: OS with an unidentified visitor in front of his house on City Island, late 1980s.

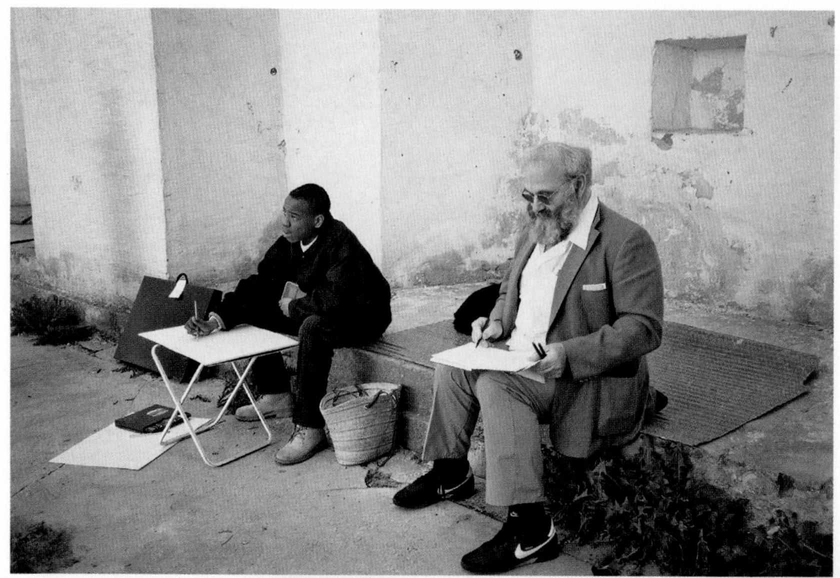

ABOVE: Drawing side by side with Stephen Wiltshire, savant artist, in Moscow, 1990.

RIGHT: Marsha Ivins, a NASA astronaut and veteran of five space flights.

BELOW: A few hours after returning to Earth following her 1994 shuttle mission, Ivins and OS discuss her sense of proprioception and orientation in microgravity.

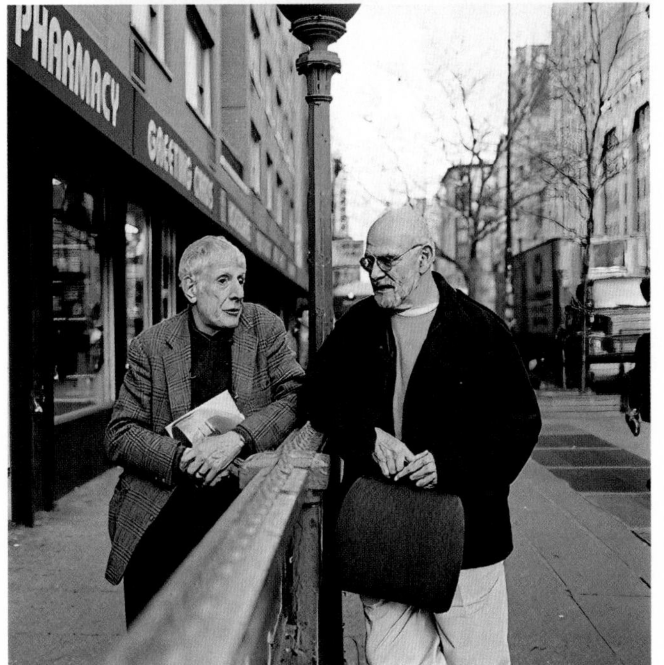

ABOVE: OS, holding his back cushion, briefcase, and umbrella, near Jonathan Miller's home in London, ca. 1990s.

LEFT: With Jonathan Miller at a Fourteenth Street subway entrance, New York City, ca. 1990s.

Cousins.
ABOVE, OS with Abba Eban in 1993; BELOW, with Robert John Aumann in 2013.

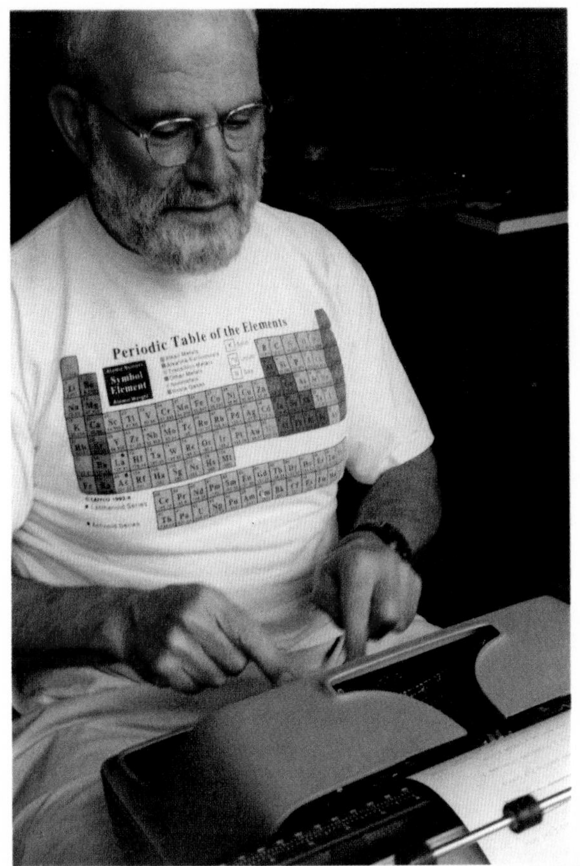

LEFT: Typing in his office, wearing a favorite periodic table shirt, ca. 1999.

BELOW: With patients at Beth Abraham Hospital, ca. 2009.

ABOVE: With Bill Hayes at Jonathan Miller's house in London, 2015.

LEFT: Dan Frank, OS's editor at Knopf for twenty years.

BELOW: OS and Jonathan Miller in London, 2015.

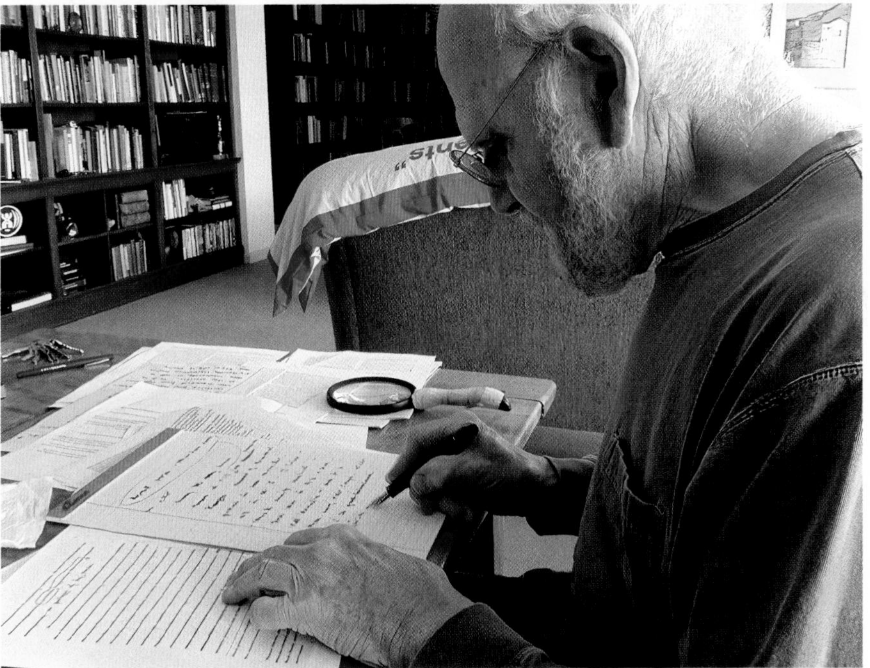

TOP: With Kate Edgar, visiting Peru in 2006.

BOTTOM: Revising an essay, 2015.

O. W. SACKS, M. D.
11 CENTRAL PARKWAY
MOUNT VERNON, N. Y. 10552

(914) 668-3796

Nov 13/82

Dear Bob,

A letter - why not? The rumble of train wheels, the passing scenery, the sense of a journey - I am on my way to Boston. The helter-skelter to a wayside station (for I have intercepted The Great Train en route): just catching it - in a very Eric-like, not Freud-like, way; the feeling of a jolly company aboard; the fob-watched conductor - Father, Shepherd - full of train-history, train-lore; what a real feeling of Departure & Travel this gives - how unlike the excruciating boredom of flying - the uniformity of all airports and planes, and the vacuous uniformity (at 35000 ft) of a featureless blue sky -

11/15, Back in NY (and its mood), after a delightful liberated weekend, not since [illegible] ... Eric sends his warmest regards. His California tour - hence also my visit - last week of February - Till then, then ...

The first page of a 1982 letter to Bob Rodman.

To Dan Frank

EDITOR, ALFRED A. KNOPF

AUGUST 21, 1995
299 WEST 12TH ST., NEW YORK

Dear Dan,

I am so glad we met last week, and could talk over some of the questions of the islands experience—and, hopefully, book—to and fro. We are also getting to know each other a little—this is not easy, because we are both (I think) shy.

With Kate on holiday, I had a little "holiday" myself—not to Labrador (as I had intended), but to the XIXth Century, where I could reconstruct an atmosphere which calmed me, protected me from the (dis)stresses of the present, and gave me an odd sense of anchor and "home." I completed quite a long essay on my "ancestor" Edward Liveing—and Kate, who is now back, is going over this now. It is in somewhat the same *mood,* I suspect, as my Humphry Davy essay, though the connections (Liveing's book, approach, directly stimulated mine) are much closer.

With this on its way now, I turned to my (April '93) Guam MS, last night—and was so bored I lost consciousness after about twenty pages! I have just completed it this morning. The tone of the entire thing is wrong, it seems to me—certainly, at least, for the first 50 pages. As soon as I start talking about some of the patients, the tone changes, and they, at least, come alive. What is missing (I suspect you feel the same) is a sense of *the island* itself, in its islandness, its geography, its plants (I went fern-collecting with a botanist in the jungle there), its reefs (I went snorkelling [. . .]), the physical and human landscape of *the place* which needs to be always there, as background, to [. . .] the house-calls etc.

I think I need to reconceive, redo, the whole thing—putting it in the form of a personal narrative (like Pohnpei, a journey following the one to Pingelap in '94) but drawing on the earlier stuff for "data" of all sorts—from the patient-portraits to the "medical" stuff about lyticobodig and its "mystery" and investigation etc (tho' *this* must somehow be embedded in narrative, perhaps in small portions, and not as a great medical-historical "lump"). Then, I think, after Guam, our flight to (little) Rota, with its cycad forests, medicine men, indigenous nature & culture more intact—and meeting old Bill Peck, who, in his own long life and person, can bring *all* the island themes together. I only HOPE I can do this! [. . .]

Thanks!

To Abba Eban

COUSIN, ISRAELI STATESMAN

NOVEMBER 7, 1995
2 HORATIO ST., NEW YORK*

Dear Aubrey,

I had meant to write to you last week to say how *awed* I was by the CD-ROM of *Heritage*†—I had never, in fact, seen a CD-ROM before (and this, clearly, was marvellously and responsibly done, although the doing, the adding to, can go on forever), and it brought all sorts of images to mind: an electronic Talmud, an infinite encyclopedia (a Borgesian dream), or one of those Pascalian spheres whose centre is everywhere and whose periphery nowhere. I can see that such an instrument will revolutionize education (or could) and history, and your *Heritage,* in particular, will gain a prodigious new dimension of reference from this.‡

I had meant to write this, and write it now, but now the hideous events in Israel have overtaken us, and I have been glued, like

---

* In September 1995, OS moved his Manhattan office to a larger apartment in the West Village, where he could stay over more frequently, avoiding a daily commute to his house on City Island.

† *Heritage,* a nine-part public television series on Jewish history and culture, was narrated by Eban and accompanied by a book version.

‡ Eban replied to this note a week later from the Hotel Delmonico, where he often stayed when in New York.

"Dear Oliver," he wrote,

> I, like you, am awed by the evidence of the technological revolution. I only fear that the proliferation of options may exceed our capacity to absorb them, whereupon we shall find ourselves on a treadmill chasing satisfactions only because they are available. There are Greek myths, ending in tragedy, about the terrors of excessive satiety and they will not be myths much longer.

everyone else, to the television for days.* I know you had known Rabin for many years, and played a considerable part in his life, and the personal loss to you must be very real. So much has been said, but what you have said (on Charlie Rose etc.) has always been full of emotion but untouched by sentimentality, and (like Kissinger's words) gives the measure of the man most faithfully. You, of course, were already clear thirty years ago as to what needed to be done, and what he was trying to do now. [. . .]

One doesn't know what to say after events like these, but I felt I had to write.

My love to you all,
Oliver

To Merlin Donald

COGNITIVE NEUROSCIENTIST, AUTHOR OF *ORIGINS OF THE MODERN MIND*

NOVEMBER 17, 1995
2 HORATIO ST., NEW YORK

Dear Merlin,

Good, as always, to hear from you.

Yes, I stay in touch with Susan Schaller, and I am sure she'd be delighted to hear from you. [. . .]

When Susan started writing about "alinguals," nobody believed her; they said no such creatures existed, and she had a hard time getting due recognition. But now many people have taken up the theme, including Ursula Bellugi† et al. at the Salk.‡ (There is an interesting piece about Nicaraguan alinguals and their (communal) creation of language, in the latest *Scientific American*.) When I visited a

* Prime Minister Yitzhak Rabin was assassinated on November 4, 1995.

† Bellugi was a cognitive neuroscientist who was a pioneer in, among other things, linguistic studies of ASL.

‡ The Salk Institute for Biological Studies in La Jolla, home to Bellugi and other researchers, including Francis Crick.

deaf-blind summer camp near Seattle this summer—itself astounding for its revelation of tactile language, tactile intelligence, the potential richness of a world based on the tactile, etc.—I encountered one fine, dignified Phillipine man of about 60 with no language, but the most eloquent, concrete "home-sign" and mime; he "told" us a beautiful story of how he had seen a bird swoop to catch a fish in mid-air. But what the bird was, where the incident occurred, when, etc., could not be conveyed. And yet there was a sort of *mythic* generalization. I think it would be hugely valuable for someone like you to explore the mind of such a man . . . for you could do so in a way which neither Susan, nor Ursula, nor anyone else *could* do. So I am delighted your interests have veered in this direction.

Let's keep in touch.

Best,
Oliver

To Stephen Jay Gould

SEPTEMBER 5, 1996
2 HORATIO ST., NEW YORK

Dear Stephen,

Sunday was fun, with the five of us—tho' I have to remember that rum is not a soft drink, and stick to Diet Coke!

Here is the final proof of *Island*—it goes off to the printers on Monday, and (I guess) the first UK copies will come out by the end of the month. I feel *infinitely* less anxious about the book since you "vetted" it with such care (tho' I occasionally left the original locutions intact—e.g. calling the "Lost World" creatures Jurassic, even tho' you reminded me that some were Cretaceous; but I thought "Mesozoic" too much of a word at this point). Since some things slipped, inexplicably, between Galley and Page proofs—such as the last word of BOOK I!—I shudder to think what may happen between now and the printed book, and pray there will be no inverted pictures as in your one (tho' I hasten to add that only a minority of readers will notice the inversion in yours, and no-one will blame you for it, and you should not obsess on it—because you have written a magnificent and far-ranging book, and *this* is what people will respond to, and not a printer's error. I sense I am speaking to myself as well as to you!).

I was excited to hear that you had written the central part of a long-incubated history of evolutionary theory—for this *is* the most central part of yourself, and will be the deepest self-expression and fulfillment when you complete it. But having done the central portion, you can now take "time off" and help *Full House* on its way, without feeling a painful sense of interruption. (Not that this sort of travelling for books is exactly "time off"—it is as exhausting as anything I know.)

I envy you this central interest, and identity; I often think that, as a clinician (and perhaps a dilettante) I myself just pick up odd things which come my way, and play with them, though they then seem to get charged with unexpected significances, or to allow views I had never expected. But I have never perhaps had a sufficient passion for a General Theory, or Theory generally (an incoherent passion for Darwin is as close as I come).

Let's see each other soon again—and again, more thanks than I can say for your great help—and your friendship.

My love to you both,
Oliver

*To Jane Goodall**

PRIMATOLOGIST

SEPTEMBER 28, 1996
2 HORATIO ST., NEW YORK

Dear Jane (if I may),

Some of what you describe sounds like visual agnosia—in particular "prosopagnosia." One does not have to have a lesion or a disease

* Goodall had written to OS that she had only very recently discovered that "what I have always considered a uniquely embarrassing/irritating mental failure is not peculiar to me!" She went on to describe her own complete inability to remember and recognize individual faces.

(like "the man who") to have this—there is a huge variation in visual-gnostic abilities, and there are some (sometimes very gifted) people who lack visual gnosis almost completely, on a congenital basis. I enclose the story here of a "Dr. S"—a "gifted" psychoanalyst.* When I read it, I thought I was reading about *myself,* because my defects in this area are very severe too. I have constant embarrassment due to my failure to recognize people—Kate, my assistant, has to explain that it is "nothing personal," and that I may fail to recognize *her* after 10 years of daily contact—and I am also constantly getting lost due to my difficulty in recognizing places (or left/right) etc. In writing about Dr. P., "The Man Who," I felt I was partly writing about myself. You will find some interesting self-descriptions of such difficulties in facial recognition in Bertrand Russell's autobiography—and his "strategy" (making a mental, verbal *list* of "salient characteristics") for dealing with this. He himself ascribes this to "coldness" or "social indifference," but it sounds more neurological to me.

In other ways visual memory and recognition can be very acute. As a fern- and cycad-lover, I instantly recognize *them,* as, I am sure, you instantly recognize your (non-human) primate friends and subjects.†

On the language-problem I cannot comment. (I am hopelessly monoglot myself.) Your grand work does not, however, seem to have suffered in the least (and perhaps this is true of mine also)! (despite all these "problems").

with best wishes,
Oliver

* OS had reviewed, for *Neurology,* a 1992 book: *Mental Lives: Case Studies in Cognition,* edited by Ruth Campbell. This included the story of "Dr. S."

† Goodall replied that it was no easier for her to recognize individual chimps, and that her cinematographer also had the same difficulties. She went on to say that both of them also got lost easily. (OS would later write, in his article "Face-Blind," that agnosia for faces and places are often linked.)

To Lewis Frumkes

WRITER AND HUMORIST[*]

FEBRUARY 20, 1997
[2 HORATIO ST., N.Y.]

Dear Lewis,

One of my favourite words is APOCOPE—I use it (for example) in "A Surgeon's Life" ("the end of the word omitted by a tactful apocope," *Anthropologist on Mars.* Vintage p. 94).

I love its *sound,* its explosiveness (as do some of my Tourettic friends—for whom it becomes a 4-syllable verbal tic which can be impacted or imploded into a tenth of a second) and the fact that it compresses 4 vowels and 4 syllables into a mere 7 letters. I also like SYZYGY (for some of the same reasons and because it has 3 distinct meanings).

Oliver Sacks

PS I also like "ZAROB" and "UBE."

To Jared Diamond

GEOGRAPHER, AUTHOR OF *GUNS, GERMS, AND STEEL*

APRIL 14, 1997
2 HORATIO ST., NEW YORK

Dear Dr. Diamond,

I have been meaning to write to you since your review of *Island,*[†] and reading your splendid piece in the May *Discover,* I can refrain no longer.

I am in strong agreement with all you say there—and well remember my own passionate indignation at the indignity of Sagan's rejec-

* Frumkes was working on a book called *Favorite Words of Famous People.*

† Diamond reviewed *The Island of the Colorblind* for *The New York Review of Books.*

tion.* I was deeply moved by Stephen (Jay Gould)'s appreciation of him, a few weeks ago, in *Science.*

What you say about younger scientists being afraid to address themselves to the public is painfully true—my own mentor, A. R. Luria, was sixty-five when he published *The Mind of a Mnemonist* (although the work—and relationship—it tells of occurred thirty or more years earlier). And I have a vivid memory of the reaction of the publisher's medical "reader," when I submitted my own first book (*Migraine*) for publication—I was 34 then. "It is too easy to read," he said. "This may arouse suspicion. Professionalize it!"

It may be that in the sphere of clinical medicine and neuropsychology etc. there need not be the absolute gulf between "technical" and "popular" language there has to be, say, in physics or genetics or molecular biology; I (like to) think that Luria's late books, and (hopefully!) my own, can address themselves *equally* to both scientific and general audiences; but the balance is a difficult (and delicate) one to maintain.

I am most appreciative of the care and generosity with which you reviewed *Island*—which is certainly among the oddest of my books. (I was not unaware of the risks in venturing beyond my "usual" clinical realm—into botany, colonial history, etc—and using a style or format which was sometimes impressionistic; but the experiences themselves were both so wide-ranging, and so brief, that I felt I could not honestly attempt anything more comprehensive or systematic.) You were certainly right in indicating that more should have been said about "founder effects" etc. on Pingelap, and about "prions"—and I will, at the least, have footnotes on these in a coming edition. (Indeed, by a series of coincidences, your review of *Island,* being sent the proofs of Richard Rhodes' new book,† and having a patient with Creutzfeldt-Jakob have all converged to make me think intensely

* Carl Sagan (1934–1996), a planetary scientist at Harvard, was denied tenure despite, or possibly because of, his huge following as a popular author and television personality.

† *Deadly Feasts,* a book about mad cow disease and other prion diseases, including Creutzfeldt-Jakob disease. Creutzfeldt is a very rare degenerative disease leading to personality changes, dementia, and death. Like kuru and other prion diseases, it is spread by contact with infected tissue.

about prions etc., and to write a little about them in my own review of Rhodes' book.)

What is "proper" or "permissible" in scientific writing has been argued, as you know, since the founding of the Royal Society—Sir Thomas Browne (whom I adore!) was rejected by the RS, despite his obvious eminence & suitability in other ways, because *his* writing was felt to be too personal and "fine"—a sort of precursor to Sagan's rejection.

I very much admire *your* work—and *your* new book—and write now to express my thanks and appreciation—

Oliver Sacks

To Austen Riggs

ZOOLOGIST*

APRIL 14, 1997
2 HORATIO ST., NEW YORK

Dear Dr. Riggs,

Many thanks for your letter. The question(s) you raise are not easy to answer, not least because of the unclear nature of so much dream-experience (in all of us), and not knowing how much it is "transcribed" (e.g. into sensory images, perceptions etc.) in the telling. One may "know," or "be given to understand" that someone in a dream has red hair—but does one actually *see* it as red?

For myself (I have normal color vision) I am unable to say, in the vast majority of my dreams, whether they were in colour or black-and-white—they do not seem clearly in *either* mode. But occasionally (for example) I dream that I am making a color print, using the old dye-transfer method (which I used to do many years ago)—I "see" the yellow, the magenta, the cyan diapositives in my dream, and align

* Riggs, a zoologist in Austin, Texas, wrote in response to "The Case of the Colorblind Painter" to observe that his own dreams were devoid of color, even though he had normal color vision.

them carefully so the picture bursts into full color. This, for me, is one sort of color dream. In another sort, color may have a special significance *e.g.* the blue of a particular cycad species. Perhaps one only "constructs" color in dreams to the extent that it is "needed," or carries some special significance or use. A different sort of color-dream (for me, and for many others) has to do with some organic disturbance—a migraine, a drug-effect, a fever, a delirium; such dreams seem to me far less conceptual or symbolic, and far more sensuous (and scattered) than "normal" dreams—and rich, thick, almost lithographic colors, and iridescences, and iterative geometric patterns etc. seem characteristic in these. One may have these in a semi-waking state, and see the colored images or patterns projected on the ceiling.

I think the question—does one dream in color, do one's dreams contain colored elements, has some similarities to questions like: do one's dreams contain *movement*? Do they contain depth—specifically, stereopsis? (I sometimes dream I am looking thru a stereoscope—stereo-photography, like color-photography, was once a passion of mine—but otherwise I have no idea whether I dream stereoptically.) Movement is nearly always mentioned in dreams, tho' I recognize (in myself) a particular category of dream which much more resembles motionless dioramas—I describe some of these in my new book (*The Island of the Colorblind*). Curiously, before I started on the actual writing of this, I kept having repetitive "black-and-white" dreams—dreams of the island, the vegetation, sunsets etc. which looked a little like old movies, and made me think (in my dream) that I had become an achromatope myself.

Typewriter ribbon finished. Will try to write legibly next page!

I wondered if these dreams—in which black-and-whiteness *per se* was so important an element—represented a witty prompting from my unconscious that it was time to write about the Island.

With the colorblind painter, damage to (the V4 areas of) the prestriate cortex knocked out *all* forms of "color-construction"—not just perception, but (chromatic) imagery, memory *and* dreaming. People with damage to the primary visual cortex (V1), cortically blind, seem to lose all visual dreaming.

To return to your questions—I would suspect that if color is not memorable in your dreams, it is because it is rarely a significant element in them. What we need, among other things, are sophisticated forms of brain-imaging (capable of differentiating activity in different areas of the visual cortex) of people *as* they dream. There might be many surprises—including (who knows?) a surprise for you: I can imagine that V4 might be very active in yr dreams, that you might in actuality dream in color, but have no memory of this on waking since the color lacks significance.

Enough!

With kind regards,
Oliver Sacks

To Jim Poyser

JOURNALIST

JUNE 30, 1997
2 HORATIO ST., NEW YORK

Dear Jim,

Many thanks for your FAX—I remember, most gratefully, your fine profile of me.[*] [. . .]

I am not sure that I can begin to answer your question(s) properly, for I regard risk as an essential part of life—at least, of a life which seeks adventure, novelty, uncertainty, surprise, or at least permits these; and a life without these is no life at all. Again, risk has to accompany every wish, expectation, hope, desire; every promise, every trust; and, for that matter, every hypothesis, every theory; to say nothing of every relation(ship). I cannot imagine a life without risk(s). And if one regards life (following von Neumann)[†] as a game—(or, following Pascal, as a wager)—then the *defining* idea is Risk.

Your first and third questions relate to (my) work. I am not especially at risk (I think) from infectious disease—tho' I contracted chickenpox in 1968, just fourteen days after I had examined a patient with shingles. I have to have tuberculin tests now yearly—every

* Poyser had written about OS for *Nuvo*, an alternative newsweekly in Indianapolis.

† John von Neumann (1903–1957), a mathematician and computer scientist, was also interested in game theory.

health worker must. I work predominantly with an elderly population, in whom pneumonias, infected decubiti (bedsores) etc. are all too common—but my own immunity, immune system, seems fairly robust, and I do not seem any the worse for being with them. I have had almost no contact with patients with AIDS, altho' I have often thought that I would like to work more closely with such patients.

There are other sorts of risk which are much more in my mind. For example, in 1967, when l-DOPA became available, both its promise, and its risk(s), for my post-encephalitic patients, seemed enormous to me. It might, I thought, "awaken" them—but awaken them to a world no longer their own. It might, I thought, act as a stimulant—but it might stimulate manias, tics, deliria, nightmares, uncontrols which had been present before they became parkinsonian, and which might re-erupt, if they were treated, once again. (I speak of these doubts in the 1990 edition of *Awakenings.*) I have constantly to ask whether a treatment may do harm—and harm may result from the best of intentions. In "To See and Not See" I ponder whether "the gift of sight" did not pose a major risk to the patient, who had been so well adapted, for half a century, to being blind.

There seems to me major risks in having power, or assuming responsibility, both of which are inseparable from the life of a physician. I try, now, not to promise, even implicitly promise, anything—other than listening and attention and honesty. I say to every patient "I don't know if I can help, but I will try to understand and to discuss every issue carefully with you." There is risk in the (inevitable) idealization(s) which accompany all doctor-patient relationships: I will try to bring these to consciousness, to discuss these—as "transference neuroses"—with some patients; and to work these through, quietly, with others.

My larger projects, my books, I regard as *major* investments, adventures, and (so) risks. One puts all one has—one's heart, one's reputation, one's whole life—into these, without any certainty as to how they will "come out," or be received. There is always great anxiety about separation, letting go—letting this intimate piece of oneself into the world. I regard the first six months (or so—sometimes it is much longer) after the publication of a book as a time of real risk—one is dangerously vulnerable, annihilable by bad reviews. But somehow, thank God, one gets through this—perhaps because of some still-deeper confidence or security, at a level deeper than the obvious vulnerability. That is, if one truly feels one has done one's best, and that it is a good piece of work.

You asked for a couple of sentences, and this is becoming a confession!

I think I might be perceived, in some ways, as a risk-taking person (why else should I be covered with scars?)—traipsing alone up a mountain (*A Leg to Stand On*), flying in a small plane with a Touretter (*Anthropologist*), and all sorts of things—such as you mention—in *Island.* Some of the risks might be seen as reckless, some as courageous, most as part of the job-in-hand. I did not regard sakau* as a risk, it is not considered risky—whereas (many years ago) I *did* risk my life with drugs. With swimming (and, since *Island,* scuba-ing) on reefs, including night-dives, and dives below a hundred feet, the idea of risk (if I take reasonable care, and have a responsible "buddy") is not in my mind. I am not fearful, in general, of other forms of life—if I let them alone, I feel, they will do the same for me. I have a certain fatalism when things are out of my hands (e.g. plane rides, being struck by lightning etc.). On the other hand, the daily commute to the office from my home in the Bronx, on bad roads, in bad weather, with millions of frequently incompetent drivers, seems to me increasingly unacceptable as a risk, and one which I am now bringing to an end.

I must also bring this overlong (and perhaps risky!) letter to an end—please feel free to ignore it, and write in your own way!

My best—
Oliver

* Sakau, or kava, is a mildly psychoactive drink that OS sampled when he was in the South Pacific, as he wrote about in *The Island of the Colorblind.*

To Jonathan Cole

SEPTEMBER 20, 1997
2 HORATIO ST., NEW YORK

Dear Jonathan,

Thank you so much for your fine, warm letter—(I think) I have been a little depressed lately (for unclear reason or reasons), and it did me good to read it. [. . .]

For myself, I have found it a bit hard to settle back to New York and neurology after the Amazon*—and am still there, in a sense, vicariously, through reading. I am pondering Ageing, somewhere, in the back of my mind—and, as a start, have begun to make house-calls, to visit the aged and infirm populations around Beth Abraham, with the visiting nurse.† A single half-day of this, in which I saw so much: not just illness, but poverty, loneliness, community, courage etc.—what real life can be like in the Bronx—has made me think how much Medicine itself misses, is impoverished, by *not* making house-calls, and just seeing patients in hospitals or offices. Of course this is not the case in England (and Canada and Australia)—but house-calls have been extinct in the States, virtually, for 50 years or more. (I was, in part, incited to start on home visits by reading a fascinating paper published in Scotland in the 1950's, "Confusion in the Home.")

But also (and in part because I don't want publishers, agents etc. breathing down my neck, and saying "How's the Ageing book coming? What have you to show us" etc.; I want to pursue the research quietly, privately, in my own way) I am wondering about collecting various pieces published here and there over the years, and turning *these* into a book. There are several categories—biographic essays (as of Humphry Davy & Freud), neurological pieces, and personal memoirs of one sort and another ("Water Babies" etc.), and I am not sure whether to mix these or keep them separate.

* OS had recently returned from a trip to Brazil, during which he explored the Amazon River and the Pantanal (where he fell in love with the gentle, giant rodents called capybaras).

† OS had long believed in the importance of seeing patients in their homes, but in this case he had begun visiting very elderly people who lived independently, with the idea of writing a book about healthy aging. (This book was never written, but some of his thoughts on the topic were included in a short essay, "The Aging Brain.")

At the moment, I am revising—rather, reconceiving rather radically—an (unpublished) essay on Music & the Brain[*] I wrote about three years ago. It is a subject which takes one in the most unexpected directions.

Again, thanks—and my love & best wishes to you all,
Oliver

To Alison Samuel

EDITOR

OCTOBER 23, 1997
2 HORATIO ST., NEW YORK

Dear Ms. Samuel,

Thank you for the proof of Elizabeth Mapstone's book.[†]

It is clearly well-researched, and -organized, and -written, but I am the wrong person to comment on it, because I never felt there *was* any significant difference between men and women in Science, the Arts, Argument, or anything else; so I find the book—and all similar books—incomprehensible (I am prepared to believe that this is a strange gender-agnosia, or gender-denial, on my part; indeed I think the word "gender" should be confined to grammar!).

Can I suggest that you send the proof (if you have not already done so) to my colleague and friend Deborah Tannen?

Yours sincerely,
Oliver Sacks

* A version of this essay was published as a foreword to a book edited by OS's good friend Concetta Tomaino (see Selected Bibliography) and eventually evolved into a book, *Musicophilia*.

† *War of Words: Women and Men Arguing.*

To Freeman Dyson

PHYSICIST, AUTHOR

NOVEMBER 30, 1997
2 HORATIO ST., NEW YORK

Dear Freeman,

[. . .] For the last two or three weeks I have been thinking and writing about the passion I had for Science—in particular Chemistry, and most especially the Periodic Table—between the ages of 10 and 13. I was doing Classics then at school—competently but indifferently, but spent every minute I could in the Science Museum and Library, and doing experiments in a little lab of my own. I had some encouragement from a favourite uncle (he manufactured "Tungstalite" lightbulbs, and I would go to his little factory in Holborn and see the heavy tungsten powder compressed and sintered and drawn into wire)—I called him "Uncle Tungsten"—but most of the passion was autonomous and so exothermic that it needed no stimulation beyond books and museums, (and, later, a couple of friends who shared it).

But it was not just excitement—it was peace, it was security, it was friendliness, it was a haven. I thought of the Periodic Table as a sort of magical map, or world, or garden, with qualities of the magic garden (in "The Door in the Wall") Wells writes of, and "the landscape of his dreams . . . the fauna and flora of his own secret planet" G. K. Chesterton speaks of—and was sure, from the start, that it was a deep representation of reality, and not simply a useful device. When I was 13, I got my school certificate, and moved over to the Science side at school. And with this (tho' there were doubtless other factors at work) my passion for Chemistry, it seems to me, promptly ended, was "killed." Perhaps I should have stayed with Classics, as the High Master suggested?

Forgive this gratuitous reminiscence—but picking up your grand essay "To Teach or Not to Teach" today I was intensely glad and grateful you had written it, not least because it allowed me to feel that my own (seemingly paradoxical) experience was neither unique nor perverse. I don't think I was able to recapture the delight and playfulness and passion of those early days until I finally got out of school—which for a physician, a neurologist, can last 20 years.

This mood of reminiscence was prompted, in part, by a

chemist-friend, Roald Hoffmann, sending me a little bar of tungsten (like the bars of W my uncle gave me) [. . .] and re-opening what used to be my favourite book (combining, as it did, the intellectual excitement, the *fun*, of Chemistry and the Periodic Table, a morning freshness and an infinity of footnotes in all directions)—Mendeleev's *Principles.* [. . .]

I hope I can visit (all of) you soon.

My best,
Oliver

To Peter Swales

FREUD SCHOLAR

DECEMBER 30, 1997
2 HORATIO ST., NEW YORK

Dear Peter,

How very nice to hear from you again (and at this season which evokes memories of a goose and a Christmas-pudding we shared in Mott St. one year).

I enjoyed "Freud, Filthy Lucre, & Undue Influence" (as I enjoy all your writings)—for its wit and its vivacity (which are always captivating!), but, no less, for its detailed documentation and ~~synthesis~~ argument (I was going to write "synthesis," as you see, and you mustn't take "argument" amiss: after all, Darwin wrote that *Origin* was essentially one sustained argument from beginning to end).

Many years ago (indeed the last time I met him, in Feb '73, was when I was in a passionate Leibniz-phase) I said to Auden that I could hardly believe what I had read about Leibniz' "meanness." Surely such a man, such a philosopher, I said, must be "above" all material concerns and pettiness, of an immense generosity in *every* way, etc. Wystan smiled at my naivete, and said that *he,* as a start, would have no more difficulty believing Leibniz to be a skinflint than Newton a murderer (which he was, in effect, when Master of the Mint). I don't quite know why I make this tangent—I guess it is to say that whilst I am prepared to believe the worst about the Master's money-hunger, exploitations, influencing unduly, sexual peccadilloes etc (and, of course, "liberties" in some of his case-histories)—yet this does not make him less interesting to me. (Nor more interesting.) Whereas I am *deeply*, radically, disquieted (say) by Heidegger's pro-Nazi attitudes etc, and feel these *do* often infiltrate/contaminate his thought—so that this *cannot* be dissociated from what one knows about him. Deep waters.

Though (as you know) I have myself been in analysis (or whatever) for more than 30 years—it will be thirty-two this January—I rarely think about Freud, and see him neither as a "hero" or a "villain"—though certainly as remarkable and important. *How* remarkable, *how* important, it is impossible to say; nor to make ready comparisons—how does one compare apples and oranges and plums and pears?

Perhaps now, additionally, I shy away from the subject—I had a feeling that I was retreating into palaeobotany in the last year or two (tho' if you get to read it, you may be amused by my little biographic footnote on Marie Stopes,* whose movement from cycad sexuality to human sexuality proceeded, poetically, through her thwarted love for a Japanese botanist—the one to describe, for the first time, the motile spermatozoids of gymnosperms)—and now I am (perhaps) beating a further retreat, away from the organic altogether, to my first love (*in*organic chemistry), which I had a grand passion for in the years just before adolescence—so my next writing (book, piece, whatever) will be an attempt to remember/reconstruct the life, feelings, explorations of an odd, perhaps gifted chemistry-mad boy of eleven, in the last years of the War, in London.

Given this interest at the moment I could not help noticing the (envious) reference to Auer von Welsbach in Freud's "top-hat" dream.† Von W was a great and dedicated chemist of the rare-earth elements—the separator of Mosander's "didymium" into two separate elements, and the discoverer, in effect, of element 71, which he named "Cassiopeium" [. . .] (tho' the credit of this was taken from him by

* Though Marie Stopes (1880–1958) became famous in Britain for her advocacy of birth control and women's rights, she was also a paleobotanist.

† In "The Dream Mechanism," Freud describes his own dream of holding a glass top hat in his lap. This reminds him of Carl Auer von Welsbach's huge commercial success with a new type of gas mantle widely used for street lighting until the age of electric lights.

slick Urbain, a French chemist, who buttered up the Parisian Academy of Sciences by calling it "lutetium").* But v. W was *also* (without ceasing to be a very important chemist) an entrepreneur passionate for fame and fortune (which he was to win with his gas-mantles). The at-first civilized dispute between him and Urbain† became exceedingly ugly (and somewhat partisan and nationalistic—the French touting "their" chemist, and element; the Austrians "theirs"), and lost *both* of them a Nobel prize. Both behaved very "ignobly" (in a moral and personal sense) at least in their later squabbles, but both were great chemists who deserved Nobels. Perhaps I am again (tho' this was not in my mind at the beginning of the paragraph) searching for some sort of "extenuation" here, as with Freud—that if someone behaves ignobly or even monstrously at times it has no *necessary* bearing on their work (etc.).

Thank you so much for the charming postcard of the Vienna Palm House and your friendly message. I spent more time in the Schönbrunn Palmenhaus than in Freud's house, and it (along with other botanical gardens in Austria and Germany) inspired a footnote—one of twenty *new* footnotes—which will appear in the paperback of *Island* next month (I enclose a copy).

All my best wishes to you, Peter, for 1998!

Oliver

To Richard Gregory

JULY 20, 1998
2 HORATIO ST., NEW YORK

Dear Richard,

I was delighted (as always, but especially now) to get your letter—tho' very sorry you could not join us on the boat on Saturday.‡

* Lutetium was so called after Lutèce, the Roman name for Paris.

† About priority in the discovery of element 71.

‡ To celebrate OS's sixty-fifth birthday, on July 9, 1998, we organized a boat party, circling Manhattan. The cake was decorated as the periodic table of elements.

Whatever internal feelings or knowledge one may have of oneself, there is another sense of one's self, and one's life, which can *only* come from one's friends (students, colleagues, family, etc)—you have to see them, perhaps have them (or as many as you can) all gathered together at a celebration or party, you have to *see them* all together to know who you are. There were, I don't know, a hundred, nearly 200, people on the boat, who would have seemed, to a passer-by, another boat, bizarrely unrelated, heterogenous. What, who, could have possibly brought this unintelligible collection together? And it is *you*—yourself. You brought them together: their radii intersect in you, and yours in them. I am putting it badly—but it is a sweet and extraordinary experience; and I know you too had it at that grand and lovely party *you* had in the Exploratorium for your 70th.

A 65th I have a clear memory of is Auden's.* I was 38 then, and thought Gosh, how grand, how old, what could 65 *be* like? And now I'm there, here, I don't feel grand, or old, tho' (I like to think) some sort of "integration" has occurred. I see this in many of my contemporaries and friends—in Jonathan† (who will have *his* 65th next July), in whom so many seemingly autonomous or divergent (even conflicting) streams of mind seem now to have coalesced into a single broad river; tho' he, we, will continue to move in unexpected directions, to leap, to surprise even ourselves, to our dying day. [. . .]

God knows whether any of our "Works" will last—I am much more confident of yours than mine.

You sound chock-a-block full of your habitual, marvellous, inextinguishable vitality—so many, many, many happy returns to you for yr 75th on the 24th.

All my best wishes, and love,
Oliver

PS Sorry for scrawl!

* In the margin, OS added: "Is 65 the lowest sum of 2 squares in 2 ways? 8^2 and $1^2/7^2$ and 4^2? Is this why one celebrates it (vs. 55)?"

† Miller.

To Rachel Herz

OLFACTORY RESEARCHER

JANUARY 9, 1999
2 HORATIO ST., NEW YORK

Dear Dr. Herz,

Thank you so much for your letter (of November 6) and the varied and fascinating reprints you enclosed—I have been away much of these last two months, and am sorry to be so slow in acknowledging these.

I am not sure that I can tell you too much more about the medical student I wrote about in *Hat,** but (it seems to me) I may have observed similar "hyperosmias" in certain categories of patient, namely those with Tourette's syndrome, with post-encephalitic syndromes, and with Asperger's syndrome or early infantile autism. I have never written at any length about these, though have made occasional short references to them (as in the enclosed xerox pages from *Anthropologist*). Whether one is seeing a genuine "hyperosmia," or whether it is (just) a heightened *attention* to smell, I am not sure. It tends to go (in all of the above) with striking sniffing behaviours—a sniffing and snuffling almost as conspicuous as looking and listening, in the rest of us. One must wonder (in a very different way!) about the heightened, or educated, olfactory sensibilities of those who work professionally with perfumes, wines etc (do you know Patrick Suskind's novel *Perfume*?).

I was intrigued by your review (with Trygg Engen) of Odor Memory—especially those parts dealing with its apparent sparing (or relative sparing) in patients with Korsakov's syndrome. I have also had this impression, though have made no systematic studies—and I refer to it briefly in another one of the *Anthropologist* subjects, Greg—xerox enclosed.

I wish I could write more, but my mind is on a very different subject at the moment—namely boyhood (my own), and an early passion—chemistry. This, for me, is an intensely sensuous no less than an intellectual passion: I cannot see lilacs in May without being reminded of the color of bivalent vanadium, nor (I am sure) will I

* In *On the Move,* OS revealed that this "medical student" was in fact himself, under the influence of amphetamines.

ever forget the smell of hydrogen selenide. I have only smelt it once in my life—when I made some, more than fifty years ago, and stunk out our London house—but (I suspect) the memory is ineradicable (even if I can't voluntarily evoke it). Though when I smell rotten radishes (not too often!) it brings back the *realm,* the general category, the "environs," so to speak, of H_2Se. A great deal of chemical expertise—and clinical expertise, for that matter—is (one suspects) dependent on these very accurate, very complex, and almost indelible memories of smell. Though I can't *voluntarily* evoke smells (as I can visual scenes, lines of poetry, tunes etc), there are often *involuntary* evocations or associations of smell. Thus when I recently visited our old house in London (which has been in others' hands, and completely redone, since my father's death in 1990), going into what used to be the dining-room immediately set me sniffing and smelling the odour of "Alicante" wine—the Sabbath wines used to be kept in the old sideboard there. Whether any trace of their scent could still have been lingering, or have impregnated the walls and floor, or whether it was a sort of hallucination, I don't know. It was similar when I went into the room which *used* to be my parents' "surgery"—I instantly seemed to smell iodoform, methylated spirit, "trilene" (a chloroform-like anesthetic which my mother kept in her obstetric bag), and other smells, a whole tapestry or symphony of them—either lingering there, or as quasi-hallucinatory associations. And the *emotional* power of these was quite overwhelming—I felt a child again, and the past as present, and then wept to think I was old and the past gone.

I was recently at U. Cincinnati, where my host was Steven Kleene. He too was kind enough to send me a bunch of reprints, assembled by Heather Duncan, the director of their Taste and Smell Center. I imagine you are in close touch with them. No-one *used* to pay much attention to olfaction once—and now it seems to be a "hot" subject, as it should be.

Again, many thanks for your letter and reprints—we must stay in communication.

Yours sincerely,
Oliver Sacks

To Susan Sontag

FEBRUARY 14, 1999
2 HORATIO ST., NEW YORK

Dear Susan,

I know (a little) about what a harrowing time you have been having,* and was not sure whether (or how) to write.

Wendy (Lesser) tells me you are feeling somewhat better now—and I wonder if you are up to a visit (but I don't want to intrude at a painful and private time).

I think of you, often, and hope things are better.

My love,
Oliver

To Roald Hoffmann

CHEMIST, NOBEL LAUREATE

APRIL 29, 1999
2 HORATIO ST., NEW YORK

Dear Roald,

I just got back from Australia yesterday—cemented bonds with family and friends, came to feel, more and more, Australia as a "second home." I met your cousins (twice), went swimming with them too. [. . .]

I enjoyed a visit—several visits—to the Australian Museum, especially to the wonderful mineral collection (the Chapman collection) they have there now; and a charming mineralogist/vulcanologist (Lin Sutherland) spent an afternoon showing me "behind the scenes," showing me, amongst other treasures, a rather favorite mineral—stolzite; raspite too—the tungsten analogue(s) of our almost-edible wulfenite, and with the same range of colors as this. I longed to get some—but it is (now) very rare, most of the (? unique) Broken Hill

* At this time, Sontag was battling uterine cancer. She survived this, as she had survived breast cancer earlier, but died in 2004 of leukemia.

deposits having been long since exhausted. Can one get *synthetic* (crystals of) lead tungstate? There were other, palaeontological wonders too—slabs of rock from an entire Devonian billabong, with thousands of fish-fossils most exquisitely preserved, and the *opalized* jaws of a Mesozoic Australian mammal (figured on a cover of *Nature* back in '85 or so). The *enthusiasm* in the Museum was very contagious.

There was a *meteorological* event of (literally) stunning proportions in my first week—a HAILSTORM of gigantic hailstones, many of fist-size. (The epicentre, as it happens, was right over us in Woollahra.) An immense, incredible *roar* at the start (I thought a jet was crashing into the house), then this tremendous crashing as icy rocks were hurled at the house (we lost four skylights, tiles, windows, car windscreens—the total damage in Sydney in excess of a billion $). I put some of these monster hailstones in the freezer, but thought they would not survive the journey back (one, fractured on impact, showed beautiful concentric rings of transparent and opaque ice, the visible evidence of its building-up as it fell). I thought of Thor throwing thunderbolts, or a shower of meteorites—it is amazing that so few injuries were sustained (tho' a neighbour, venturing out to protect his car, had his hand broken by the impact of a hailstone on it, and I heard of one man killed while in his car, by an almost kg-stone which went through the car-roof. [. . .]

My best,
Oliver

To Stephen J. Gould and Rhonda Shearer

JULY 22, 1999
2 HORATIO ST., NEW YORK

Dear Stephen and Rhonda,

It was lovely seeing you both on the 9th*—I had been feeling a bit low lately, and the party, and the people who came, cheered me up

* The Goulds had hosted a small celebration for OS's sixty-sixth birthday, for which guests had been invited to dress as a particular element.

immensely (and, I think, have got me back on a more positive track). Auden always used to say one *must* celebrate, celebrate birthdays, whatever, however one is feeling—and I have found this proves right, has worked out, again and again.

I would/should have written earlier, but I have been savoring your books—*Animal Magnetism,*[*] and *Rocks of Ages.* I am rather a slow reader, especially when I am also writing, and *Rocks of Ages* has set me to ruminating a bit (the more so as I have also been reading Jane Goodall's autobiography (subtitled "A Spiritual Journey") and seeing how complementary the two "magisteria" have been for/in *her.* I also had dinner a few days ago with a cousin of mine from Jerusalem[†]—a great mathematician, and also deeply Orthodox. When (cautiously!) I brought up Creationists with him, he said, with a smile, that there was one at the table—and I drew back, feeling I was on delicate ground. He is also a Pythagorean (and a Cantorian and Brouwerian) who feels that numbers exist, are real and eternal, as opposed to the transitory abstractions which are ourselves. (He is in a most painful state now, his wife died a few months ago—but beyond this there are states-of-mind in him I cannot fathom, I find unintelligible.)

I think this cousin of mine (who is a grandson of the firstborn, as I am the fourth son of the 16th born) is probably rather similar to my mother's father (his great-grandfather), who was also a mathematician and mystically-religiously inclined. This conjunction of passion for maths or the physical sciences with deep religious feeling was present in most of my mother's brothers, though diluted, perhaps, in my mother herself. My parents were "orthodox" in practice, but I am not sure what they "believed," nor that "belief" was ever discussed. What "religious" feeling I may have had myself as a child got rudely broken when I was evacuated at the age of six, and effectively cut off from family & community. When I see Giotto paintings I ache (perhaps we all do), but I have been (as long as conscious memory goes) a sort of atheist (curious, sometimes wistful, often indifferent, never militant), and this feeling deepens (or becomes indurated) with the

* It is unclear what book OS is referring to. Gould did not use "animal magnetism" as a title for any of his books or articles, though he discussed Anton Mesmer and animal magnetism in his essay "The Chain of Reason versus the Chain of Thumbs," collected in *Bully for Brontosaurus.*

† Robert John Aumann.

years (Jonathan Miller also calls himself "an old Jewish atheist"). I have every "respect" (at least in outward ways) for those who believe; I enjoy working with the Little Sisters of the Poor, and in an orthodox Jewish Hospital, as I have done for thirty years; but, in general, avoid talking on the subject. (I regard conversation on religion, sex or politics as *dangerous* here, except when one is with close friends—what an unholy brew of them blew up with the impeachment proceedings!) But having said this, I am still not sure that I in fact *allow* two "magisteria"—at least in areas where the natural or scientific apply—"anthropic principles" strike me as either ludicrous or tautologous; "creation science" (of course) I am wholly with you on (I did tell you, did I not, that Alex Ritchie was beaten up and mauled by creationists in Australia, when he simply asked a question about Noah's Ark at a meeting?); and talk of "parapsychology" and astrology and ghosts and spirits infuriates me, with their implication of "another," as-it-were parallel world. But when I read poetry, or listen to Mozart, or see selfless acts, I do, of course feel a "higher" domain (but one which Nature reaches up to, not separate in nature). Anyhow, I shouldn't be babbling in this way.

I found the excerpts from Darwin's letter(s) to Asa Gray enormously moving—I think I must read it/them in its/their entirety. I was much interested by your chapter on William Jennings Bryan, whom I had (ignorantly) regarded previously as a mere buffoon and a bigot; you illuminate his life and motives, so he becomes at least *intelligible* (intellectually and emotionally)—and it helps (at least psychologically) to *understand* "the other side." (I have not read Polkinghorne—do I have the name right?—the quantum physicist in England who is also a Reverend, and has written on the "parallels" between quantum theory and religion.) I love the way you organize *Rocks of Ages* in "parables"—Doubting Thomas etc. etc.—it seems a perfect way to do it. And how fairly, and irenically, you present the two realms—was Irenaeus actually irenic? I have known you as a battler, in fine polemic form (a hint of this comes through, when you talk here about creationism), but it is very good to see you in irenic form too. I think you have written a noble book in *Rocks* (as well as being erudite, charming, personable, as in all you write). Thank you so much for giving (and beautifully inscribing) me a copy—and for coming. I am around in the summer—let's see each other again.

My love to you both,
Oliver

To Howard Gardner

EDUCATIONAL PSYCHOLOGIST

AUGUST 2, 1999
2 HORATIO ST., NEW YORK

Dear Howard,

I have just received a proof of your new book,* which I shall devour with eagerness. I opened it, straightaway, to your discussion of "The Naturalist Intelligence," and am very happy to see it securely admitted to "Number Eight" among the Intelligences. (I have just been reading a proof of Jane Goodall's autobiography, and one sees how early and clearly this presented itself in *her*—though also how clearly a "moral" and a "spiritual" intelligence emerged in her too.)

As soon as I started reading, all sorts of (perhaps silly) questions came to my mind—questions you perhaps address in the book. But, precipitately and prematurely, I think I shall mention some of them.

Writing a sort of "scientific autobiography" of my own (a title which a Planck can use, but hardly myself!), I have especially been thinking back to my first passion, which was chemistry; and the nature of the "Chemical" intelligence, with its capacity (and disposition) to perceive, remember, order, thousands of sensory appearances and phenomena. It is very nicely described by Liebig in *his* Autobiography (see enclosed, from Brock biography of Liebig). I think this *sort* of intelligence is clearly akin to a biophilia, a naturalistic one—and of course many naturalists at this time (or perhaps a little earlier) were at once chemists, mineralogists,

* *Intelligence Reframed: Multiple Intelligences for the 21st Century,* an updated version of Gardner's hugely influential theory of multiple intelligences—linguistic, spatial, intrapersonal, etc.—put forth in his 1983 book *Frames of Mind.*

botanists (and often physicians)—like Berzelius, for example. In my boyhood there was a tendency to classify young "nerds" by whether they were prone to dissect animals, to play with chemistry sets, or to take radios apart. Now playing with chemistry sets, or having labs of one's own, is no longer possible—see my attached Op-Ed piece;* and since chemistry itself has changed so much (one no longer does bench-analysis, one no longer uses one's senses, one just pops the specimen into the mass-spectograph, and out pops a list of instrument-readings), it seems to me that this sort of Liebigian-Davyan-Faradayan "chemical intelligence" may now be all but extinct.

The whole question of "obsolete" or "extinct" intelligences, as well as potential ones, is hinted at, I suspect, throughout your new book: "new" ones associated with the rise of computers etc; "outmoded" ones—the chemical is an example. Also certain sorts of physical/clinical. My father, like all doctors of his time, was a virtuoso in percussing and ausculting the chest, he seemed to "see" the minutest abnormalities through his fingers and ears. Now radiology has displaced this skill, this "intelligence"—and perhaps Medicine, like Chemistry, is reducing itself to instrument-readings, moving away from its primal biological intelligence. (I wonder how one might classify "clinical intelligence"—it has obvious affinities to "naturalist" intelligence, but then again.)

I enclose the opening page of Max Planck's *Scientific Autobiography.* One sees here something very different from the "sensory" intelligence Liebig speaks of—a primary passion for Theory, for Law. I am not sure whether "mathematico-logical" is quite the right term for this—whether the old dichotomy of practical and theoretical intelligences should not be called upon too. Or the (Pascalian) dichotomy of "analytic/intuitive"—Keynes seems to bring this up comparing the economist's intelligence with the theoretical physicist's. [. . .]

And again—I saw Theodore Bikel in an astounding performance yesterday—what of the actor's gifts, the actor's "intelligence": is this just a sum(mation) of verbal, kinesthetic, interpersonal etc. or is there something uniquely "histrionic"? (It is often present from a very early age.) Are there "born" actors, like "born" musicians? So many questions.

* "Hard Times for Curious Minds."

Basically I just wanted to say that merely *opening* your book set my mind going at 1000 mph, and I am sure this will be so, once again, for a host of readers. So, all my congratulations!

With best wishes,
Oliver

AS THE NEW MILLENNIUM APPROACHED, a number of magazine editors asked OS for prognostications about the future, and occasionally he replied.

To Charles A. Riley II

EDITOR, *WE* MAGAZINE*

AUGUST 21, 1999
2 HORATIO ST., NEW YORK

Dear Dr. Riley,

Some question! What may the next 1000 years bring to the lives and images of people with disabilities?

It could bring unimaginable technical and medical advances—"neuroprotective" drugs to halt, slow, reverse or prevent all sorts of neurodegenerative diseases, like Parkinson's disease, Alzheimer's disease, etc. Nerve-growth factor etc—to promote healing and regeneration in the nervous system. Implants of foetal nerve-tissue—or (most probably, less problematically) of stem-cells, or embryonic cells, grown in tissue-culture—which could develop and replace nerve cells which have been damaged. A vast realm of gene-therapy, now in its infancy, will open up.

We will see surgery (our present micro-surgery is only a start) of an incredible minuteness—a sort of "nano-surgery," able to address individual cells, and perhaps individual organelles or molecules within the cells.

"Bionics," clearly, has almost unlimited potentials, and not just at

* *We* magazine, though short-lived, was dedicated to people with disabilities and their families and friends.

the level of bio-engineering new limbs and prostheses of exquisite sophistication, but providing new or replacement "modules" for the brain. It even seems to me possible that a sort of "electronic telepathy" could be provided for those who are "locked in" (from strokes or other mishaps) and unable to communicate in any of the usual ways.

But equally—and no less crucial—must be a change in societal and cultural attitudes to the disabled, a change in their "image" (and, of course, their own self-image). Not just an absence of discrimination and stigmatization; not just a sympathetic understanding and awareness; but a vivid appreciation and respect for the disabled as *individual people,* individuals whose perspectives are different from our own, and who may create lives and selves from a quite different center, and of the unique roles the disabled may play by virtue of this. Thus, there are (potentially) special forms of sensibility, of imagination, of language and culture, which one may see among the Deaf, and in some people with Tourette's syndrome—forms which need to be recognized and encouraged. Such a *positive* view of the disabled makes the very word "disability" quite inadequate, even misleading—a distinction made very clear by the Deaf community itself, which speaks of "Deafness" (with a capital D) as a special form of existential and linguistic and cultural individuality, as opposed to "deafness" (with a small d), which means "hearing-impaired" or "-disabled." This, of course, is a special case—but *all* people who are disabled, whatever the disability, have, have to have, "compensations" of various sorts, powers, possibilities, increased (and, so to speak, demanded) by their disabilities. So it is not just a question of "awareness" or "compassion," but a perception of the *positive* powers and roles of the disabled: this has already led to major changes of attitudes in the last 20 years, and one hopes these will deepen in the next millennium.

All this, of course, depends on our own physical and moral survival as a species—our not blowing ourselves up, or destroying each other, or destroying our so-vulnerable planet; but becoming more decent, and more civilized—wiser—as well as more "advanced," in the future.

Yours sincerely,
Oliver Sacks

To Peter Singer

PHILOSOPHER, AUTHOR OF *PRACTICAL ETHICS* AND *ANIMAL LIBERATION*

SEPTEMBER 9, 2000
2 HORATIO ST., NEW YORK

Dear Dr. Singer,

When Daniel Halpern* sent me a proof of your forthcoming *Writings on an Ethical Life,* I excused myself from providing a comment on the grounds that the issues you raise were "too deep and complex" for me to venture a comment without a careful perusal of the book, and that I was too involved in writing a new book of my own to give it the time it needed. All this was true, but not perhaps the whole truth—I have to confess that I had an incorrect idea of your viewpoints, because (to my shame) I had never read any of your works, tho' I *had* heard various misrepresentations of them. Now, having opened your new book, I have had to read straight through it (I have not yet quite finished), and I am so impressed by what I read that I feel I must write to you. I now strongly agree that this accessible "essential" Singer is an extremely good idea, and that its publication is very important.

I think I was introduced early (perhaps too early) to some of the considerations you raise, because my parents were physicians, and my mother, in particular, an obstetrician. When I was ten or so she told me how she (and the Matron or Nursing Sister in a Nursing Home) might drown anencephalic infants, or infants with severe spina bifida, at birth—"like a kitten"—and how if the head was too big to pass through the pelvis it might be crushed by a cranioclast. She sometimes brought these deformed or killed fetuses back to the house, and wanted me to dissect them. Thus (for me, as a ten year old) the obvious commonsense of what she did was infused or associated with a kind of horror. I do not know whether these drownings were first discussed with the mother or not.

As a medical student (more than forty years ago, in London) I was the "extern" on the Medical Unit when a friend of mine, a young man, a fellow-swimmer, was admitted with odd symptoms which turned

* Singer's editor at Ecco Press.

out to be due to acute leukemia. Medications did not slow his disease, and he soon started to have unbearable pain from the development of leukemic deposits in the spine. When he realized that nothing could be done, and that only extreme and increasing pain lay ahead he said he wanted to die, and one day I found him climbing over the railing on the balcony outside the ward. I pulled him back and prevented him throwing himself over. Subsequently his pain became so great that nothing—not even heroin—could hold it, and he screamed constantly, day and night, in agony, unless he was given a general anaesthetic. He cursed me for having prevented his suicide.

As a physician, a neurologist, working for the most part with gravely ill patients, or patients with long-standing (or lifelong) neurological disease (or damage), I often face—or avert my face from—some of the ethical considerations you raise. Perhaps I should say, rather, that I often keep my thoughts to myself. Twenty-five years ago or so I saw in our hospital a young woman who had suffered a devastating herpes encephalitis, and had been in a coma (with "decerebrate rigidity" and a profoundly abnormal EEG) for five years. She was never visited by her family, who wished she would die; but youth, and "good care" (tube-feeding, toileting, intermittent use of antibiotics, etc) kept her alive. I suggested, in my note on her, that she should be allowed to die, since she was incorrigibly and profoundly brain damaged, with no consciousness or prospect of consciousness, and because her life offered nothing positive to herself or anyone else. My then chief, the Director of the Hospital, was horrified by my suggestion, and said that I "belonged in Auschwitz."* When I protested that there was all the difference in the world between a truly hopeless patient such as this, and my *Awakenings* patients, or other patients in whom I was often the first to perceive or evoke the potentials for consciousness and improvement, he was unmoved. The young woman "brain-dead" from herpes encephalitis is *still* alive—she has been maintained now, in her coma, for thirty-two years.

I am not in private practice, and to some extent, therefore, I can or may avoid certain responsibilities, or at least direct actions, while

* See letter to Harvey Shapiro, November 10, 1977.

feeling free, as a "consultant," to discuss these with others. In a way which is either cowardly or "diplomatic" I refrain from stating my thoughts in public, even the (mildly) public form of hospital notes. I don't want to be called a Nazi again, nor outrage the sensibilities of those whom I work with (one of the chronic-disease hospitals, the "Homes," I work in is orthodox Jewish, and one is Catholic, run by nuns). But when I see patients who have suffered profound and irreversible brain-damage—whether from something like herpes encephalitis or a stroke, brain-injury, or the terminal stages of Alzheimer's disease (no-one is more active than myself in finding ways of approaching, or making life more worthwhile for patients with only *mild* or *moderate* dementias)—I will make a point of discussing their now purely biological (rather than biographic) lives with their relatives and their immediate caretakers, at least opening up the possibilities of discussion (and perhaps action) on a subject which seems so taboo, but is also so forced upon one now in the present state of Medicine. I wonder now, after reading some of your extremely clear (and candid and courageous and consistent) writings whether I should not "come out" myself on some of the issues you raise. But this is something I must think on, and these are my (as yet) private thoughts.

I have not had much philosophical background (though I did go to some of A. J. Ayer's conferences in the 1950's, when I was a medical student in London), but (as you say) "moral philosophy" at that time had little to do with the real world. I read Moore's *Principia Ethica*, and I read (immaturely) a little Bentham and Mill. Later, I read some Bonhoeffer, about "cheap" and "costly" Grace, and, in particular, his final essay in *Ethics* on "What is Meant by Telling the Truth?" This, if I understand what he was saying, was an exploration of how to tell it to the right people, in the right way, at the right time, which might have seemed a prescription for equivocation—but his own truth-telling, of course, landed him on a Nazi gallows. Reading your *Writings on an Ethical Life* brings home to me that I have not *read* any such writings for the last thirty years—and the beauty and importance of such writings, at least in hands such as yours. How (or indeed whether) I can/will be *affected* by such writings—relatively late in my life (I am in my sixty eighth year now)—I do not know; but I am very glad that Dan sent me the proof, and that (in a sense) I have made your acquaintance.

And I think it will be tofu, not tuna, for lunch!

With my best wishes to you, and for the new book,
Oliver Sacks

PS I think I have to differ from you (though I am not sure of the consequences of this, in ethical terms) regarding the uniqueness or otherwise of *Homo sap.* I think that there is not just a difference of degree, but a difference of kind, between human beings and the great apes (even though they have the beginnings of self-consciousness, and can acquire a considerable lexicon—I am not sure if it can be called a language—of signs). Language (for what it is worth) seems to be species-specific—thus a microcephalic "idiot," with a brain smaller than a dog, may automatically acquire a vocabulary such as the brightest ape, with intensive teaching, cannot. This may make it difficult to make direct comparisons, as on a scale, between human infants, retarded children etc. and other species. Having worked with severely retarded children (and adults), I would hesitate to characterize them as "non-persons." I would very much *like* to have worked with chimps and other apes—I am (as it happens) having breakfast with Jane Goodall tomorrow. Perhaps one needs a special (moral) discourse for human beings, and one which does not stem from "speciesism" or bias.

While a biographical consciousness may be (more or less) confined to human beings,* I think that biological consciousness, and the capacity to feel pain and suffer, developed, must have developed, quite early in evolution. Having had dogs and cats (though I am not especially an "animal-lover") I cannot doubt their capacity to feel—and my horror at the sufferings of cattle led to slaughter (I describe this in *An Anthropologist on Mars*) was certainly a factor disposing me to avoid meat. I am now, with uncertain feelings, chiefly a fish eater—I persuade myself that fish feel less—though I avoid eating squid, octopus, cuttlefish (which I suspect to have almost-mammalian intelligence and emotions). I am tending, gradually, to vegetarianism.

* Within a few years, with the burgeoning research on animal consciousness, OS, like many others, would begin to revise his opinion on this point.

I am confronted, almost daily, with dilemmas about patients. I (have to) regard human beings as "persons" so long as there is any vestige of consciousness left, and I feel the feeble and intermittent consciousness of the profoundly retarded or demented as essentially human even though it may lack foresight, retrospect, the biographical sense of a life. I believe that this is a valid observation, and not a reflection of self-indulgent speciesism.

To Lewis Wolpert

DEVELOPMENTAL BIOLOGIST

FEBRUARY 25, 2001
2 HORATIO ST., NEW YORK

Dear Lewis,

Many thanks for your letter and your (*Telegraph,* Jan 24) article on Belief. [. . .]

On the whole I very much like the thrust of your piece, which (I take it) will be the thrust, or one of the thrusts, of your book. I do think that there is, with us, a built-in need to tell stories, make narratives, make narrative *sense* of the world (physical, social etc.); narrative thinking develops very early, long before "paradigmatic" thinking (I very much like Jerry Bruner's book on these two forms of thought)—and one sees it hugely developed, for example, in people with Williams syndrome (along with their verbal and social precocity generally), along with profound defects in their comprehension of the simplest logical operations (especially spatial ones). What has especially fascinated me is the almost frenetic (and obviously, to the person, vital) story-telling which I describe, for example, in "A Matter of Identity" (*Hat*). And if it cannot be a rational story, it will be a fantastic one.

The need for causality, or an idea of causality, also seems to be very deep or fundamental in human nature—but I would not agree with you that, in its most elementary forms, it requires language. And indeed you yourself note, later in the piece, that "infants in the first year of life perceive the causal structure of simple collision events" etc. It does seem almost species-specific—one is amazed at the "elementary" mistakes which even the great apes make. By the same token, violations (or seeming violations) of causality are very shocking—I think here of the intellectual shock of quantum mechanics, and Einstein's obstinacy in the face of a seemingly statistical universe.

The question of scientific beliefs seems to me a complex one—for while they *should* be provisional, and subject to revision, they *may* be charged with an almost religious or mystical force, an emotional conviction, and an emotional investment, which makes them rather like religious beliefs, and as hard to give up. I think here of Eddington*—and his numerical-cosmogonic System—and his intellectual intransigence and aggressiveness when faced with contrary data and concepts (notably by Chandrasekhar).

I think (but will not dilate on this, because I know it is one of your "things"!!) that you go too far in ranking psychoanalysis with astrology, telepathy, homeopathy etc. I fully agree with you about the *latter* (Freud himself spoke about "the black mud of the occult" here), but a certain amount of psychoanalysis, at least (tho' by no means all), seems to me to allow of experimental or experiential confirmation (or refutation), and to have the requisite qualities of scientific hypothesis. The fact that many people *treat* it as a religion, or a cult, is neither here nor there. They are indeed an embarrassment to genuinely-striving psychoanalysts.

E. M. Forster, as you know, starts one of his essays with "I do not believe in Belief," and this is a sentiment to which I resonate, because I have never been sympathetic to religious or political or ideological convictions—and have been bewildered when I saw some of my friends carried along by these. On the other hand I do believe (but very quietly) in the methods of science, and if my strongest need is need for order, I will not accept any order other than that generated by a slow and patient science.

I have written too much, but you asked me to write.

I'm not up to thinking of mystical or hypnotic experiences in *this* letter—maybe another one. I hope you have had some personal experience of both—it helps! As it helped Williams James.

Will you also write about belief (if this is the right word) and

* Arthur Eddington (1882–1944), mathematician and physicist.

Art—the profound sense of "truth" or "verisimilitude" which a great painting or poem can induce—or will this stray too far from an already dangerously-large theme?

Do you know Maynard Keynes' memoir "My early beliefs"?

I have just (after 3½ years!) finished my own "memoir" of boyhood, and science etc—which is partly (I suppose) about the evolution and turmoil of my own beliefs as a boy, and certainly my need for causality and order.

I hope you are well—you certainly sound productive!

With warmest regards,
Oliver

To Frank Kermode

JULY 16, 2001
2 HORATIO ST., NEW YORK

Dear Frank,

I have finally completed (or abandoned) *Uncle Tungsten*—and have taken the liberty of asking Picador to send you a proof (it comes out in November).

I struggled with it for three years or more—often feeling that the attempt to weave together a personal and a scientific-historical narrative was an impossible or doomed one—but then, for better or worse, it all seemed to come together.

One takes leave of such things—as you, above all, know!—with very mixed feelings, relief and regret combined: "resurrecting" parents, early experiences, loves, hopes, passions, disappointments, and then leaving them (having imprinted them with one's transformations and distortions etc) once again. I think I feel a little more "accommodated," a little calmer, since writing *Uncle T*—as if doing so put some of my life in perspective *for me;* but whether the book will have resonances for others is not something I can judge for myself.

I have *no* impulse to continue, to add a Part II etc, and have turned with relief to an entirely different writing project—writing about a trip to Oaxaca last year, which though primarily botanical, led me in all sorts of other directions too, all the wonder (and sometimes horror) of meso-America. I read Prescott* as a boy, and it all came flooding back to me. Though my own interest, especially, was in the *stuffs* the New World gave us—chocolate, cochineal, rubber, tobacco, etc etc, and the extraordinary agricultural knowledge they had. But it was great fun to be with the American Fern Society—to see the enthusiasm (and energy, and erudition) of these often marvellous amateurs—and to be reminded of the importance of amateurs in science—at least the "field" sciences, like geology and botany.

I continue to see long, grand reviews by you in the NYRB on all sorts of topics—and I was deeply interested in your thoughts on memory etc. in *Index.*† I wish I had your ability to write these somehow compact yet complete pieces on such a range of topics, but I feel, more and more, that I have to write *books*—at least as long as I can, and I am afraid of getting "diverted" from this.

I hope you are well, and that I can see you, perhaps, when I come to England for the publication of *Uncle T* in November/December.

With kindest regards,
Oliver

ON SEPTEMBER 11, 2001, the World Trade Center towers were destroyed in a terrorist attack, killing nearly three thousand people. The rubble continued to burn for weeks, filling Lower Manhattan, including Horatio Street, with toxic smoke. OS was in Ithaca, New York, when the attack occurred; he returned to the city a few days later.

* William H. Prescott was an eminent historian, whose *History of the Conquest of Mexico* (1843) and *History of the Conquest of Peru* (1847) remain widely read.

† Kermode's article "Palaces of Memory" in the journal *Index on Censorship.*

To Barnet Koven

A YOUNG FRIEND

OCTOBER 15, 2001
2 HORATIO ST., NEW YORK

Dear Barney,

I was delighted to get your letter (of Sept 15)—and yes, thank you, I (and everyone I know) is OK, though many friends have lost friends, colleagues, neighbours etc. in this awful business. I was away—teaching at Cornell, in Ithaca—at the time, and both glad and sorry not to have been right here when it happened. Some people who lived close by and were evacuated from their houses, stayed over in my apartment for a few days.

I went to Ground Zero last night—the pyre is still fuming, and one needs to wear a gas-mask—one cannot realize the scale (and horror) of the thing from photographs alone. Nor the comradeship of the Firemen, Policemen, Sanitation Workers, Salvation Army, Red Cross etc. who are all working (sometimes 16 hours a day) at the site—I spent the whole night talking with them.

I am glad that you, at least, are building, while others are destroying. A solar-powered fan and dog-house (what sort of dog do you have?). Do you know how the sun itself is powered? While you were in British Columbia I was also at high latitudes, on the other side—in Iceland and Greenland, and saw orcas, whales, seals—and one polar bear. It was so beautiful I felt I wanted to spend the rest of my life there.

Glad you're enjoying school now—and what a lovely picture of your sister and yourself. I should have replied earlier, but I delayed because I wanted to send you a copy of my book, and the copies have only just come in. So here, with my love, is *Uncle Tungsten,* which is (partly) a portrait of my life and family etc. when I was about your age (and a little older).

Let's see each other again soon!

Oliver

To T. J. Hymson[*]

CORRESPONDENT

NOVEMBER 23, 2001

2 HORATIO ST., NEW YORK

Dear Mr. Hymson,

Thank you for your intriguing letter.

I am not surprised that you experience the sort of visual patterns and elaborate scenes that you describe because I too am subject to somewhat similar ones. I refer to this (on p. 39 of the paperback of my *Island* book) when I write of "a strange visual excitement such as I am sometimes prone to, especially at the start of a migraine—endlessly moving vistas of breadfruit trees and bananas on the darkened ceiling," and then, later, with the superadded effect of "sakau" (kava) "floating over coral-heads. Lips of giant clams, perseverating, filling whole visual field. Suddenly a blue blaze. Luminous blobs fall from it. I *hear* the falling blobs distinctly" etc (ibid pp 89–90). Not mentioned here is how, walking back with an ophthalmologist friend in the inky darkness, I spoke [. . .] of the incessant, intricate patterns weaving before me—and how he, to *my* surprise, said that *he* did NOT see anything when he closed his eyes, or was in pitch darkness. So that something I had taken as a sort of universal was, apparently, quite unusual.

Although I mention migraine, and drugs, in this context, and they can certainly heighten the imagery, I think I have it, to some degree, all the time—tho' perhaps not as vividly, or elaborately, as you do. There seems to be a certain *sequence* in such visions, going from simple geometric patterns to complex "scenes"—and such sequences are seen in a variety of conditions (migraine, drugs, sensory deprivation—or just "naturally"). [. . .] Some people are far more prone to such "visions"—there must be a physiological basis for this.

Further: Such "visions" seem to be produced less (as well as being

* Hymson described his ability to visualize, with closed eyes, finely detailed "movies." He wrote, "I've been aware of these movies since I was a child, and spent many hours happily sitting with my eyes closed watching the show play on. It was not until my teens that I realized other people didn't have them too. I can remember hearing people talk about being bored, and not really understanding. 'Why don't they just close their eyes and watch a movie if they're bored?' "

less visible) when the eyes are open, and the brain receives a "normal" visual input. It is almost as if, in the (relative) absence of such input, the brain starts generating its own activity, unrestrainedly, and a very special form of this may now be happening with your superadded retinitis pigmentosa (I am sorry to hear about this). Such imagery with impaired vision was described by Charles Bonnet, and is sometimes termed Charles Bonnet syndrome—and here one is tempted to draw analogies with phantom limbs, and with the tinnitus (and sometimes musical hallucinations) which may occur with diminished hearing, especially "nerve deafness."

All this emphasizes that the brain is not a passive thing, waiting to be stimulated by the senses, but incessantly active, incessantly generative (even "creative") in its own right—the simpler patterns arising from spontaneous activity in the complex cyto-architecture of the visual cortex (and reflecting this), but the more complex images determined by the experiences, sensibilities, imagination etc. of the individual. The absolute marvel of the brain is the lesson of all of this—but whether this helps one to accept, or adapt, or "appreciate," these unasked-for visions is quite another matter!

With best wishes,
Oliver Sacks

To Kay Redfield Jamison

DECEMBER 22, 2001
2 HORATIO ST., NEW YORK

Dear Kay,

I am reading your latest book* with—I don't know what words to use—a sort of appalled fascination (for I have lost many friends and others to suicide, and the lure and compulsion is one I have known on and off for most of my own life), and admiration at your bringing together so much, from personal accounts to medical statistics to historical descriptions etc. with such power and grace. Admiration too

* *Night Falls Fast: Understanding Suicide.*

at your bravery, from your autobiographic Preface to your steadily outfacing the Monster for what must have been years of research and preparation. And all this at a time of especial personal distress—I hope this is a reasonable time (so far as any time can be in the circumstances) for your husband.*

I enclose a copy of *Uncle Tungsten,* though I feel it may have a sort of levity which may not accord well with the gravity of your book—but you will find (what someone called) "a noisy subtext" under the lightness of the text. I don't know whether I should have written of my brother's psychosis—the more so as he is alive, but tragically depressed and delusional (he has had so little of a life, and was so blessedly gifted to start with); I often wondered if I was going to follow him into schizophrenia, or a bipolar illness, but I seem to have settled for "mere" neurosis. But I am very glad I was able to complete *Uncle T,* and I have some sense of connection, and reconciliation, now which I never had before. At a lighter level it has been great fun to write, and to re-live the early, almost Edenic delight in science—and I hope this communicates itself, and perhaps has resonances for you as well.

My warmest good wishes to you both,
Love,
Oliver

PS I had only read PART I of your book when I wrote the above; as I have continued to read I have been steadily more impressed—and shaken. The "case-histories"—of Meriwether Lewis, the young airman John Wilson et al—appall me. Chapters 8 and 9 equally—they should be compulsory reading everywhere. I think of friends—among them a notable number of physicians—who have made suicide attempts, and several who have achieved their aim. I wonder, specifically, about some of the patients I have been seeing at the NYU Clinic where I now work. And, very specifically, about people with severe Tourette's, who besides the impulsiveness which so often goes with TS, are apt to have all their emotions *amplified.* To the list on pp 272–3 I would add those with severe Tourette's, some of whom achieve particularly

* Jamison's husband, Richard Wyatt, a neuropsychiatrist, died in 2002 after a prolonged illness.

bizarre and terrible self-destructions—I think of one (adolescent boy) who climbed into a drainpipe from which he could not extricate himself—tho' this was not perhaps "suicide" in the usual sense. His death sent a wave of horror through the Tourette community; and Touretters can be dangerously suggestible.

I think that you show here a combination of realism and understanding—"toughness" and "tenderness," in James' sense—I have never, or only very rarely, seen before. And your writing soars at such times, becomes as fiercely and compassionately eloquent as in your autobiography. Perhaps only someone who, like yourself, has actually been through it, *and* is a physician and a scientist (and a close reader of poetry and personal accounts and biographies etc) *could* write like this.

I feel almost afraid to read more—but cannot stop!

The (1994) Szasz case seems to me especially shocking—I did not know about it.

I am very glad (I am not organizing thoughts here, just darting about, as you see) that you have good words to say for ECT*—I had one patient (she was also parkinsonian) whom drugs had not touched, who said to me, "Forget your 'principles,' your 'distaste' for ECT, for God's sake, and *treat me.* Shock saved my life twenty years ago, and I need it now." (She got it, and it did save her life.) So Himmelhoch's words, which you quote on p. 251,† seem to me exactly right.

I am especially glad to see your passionate defence of psychotherapy, therapeutic contact, as often crucial in addition to medication—and your quotation of Morag Coates's words.‡

There is a very personal feeling here—I feel my own psychiatrist (to whom I have been going since 1966) has repeatedly saved my life. I have rarely been explicitly suicidal, but rather parasuicidal (a term

* Electroconvulsive therapy.

† Jamison quoted Jonathan Himmelhoch, a psychiatrist: "The narcissistic ponderings of psychiatrists whose political agenda supersedes their clinical experience must not be allowed to keep patients who are suffering the most severe form of pain from relief."

‡ Coates had written:

> It is not only the doctors who perform hazardous operations or give life-saving drugs in obvious emergencies who hold the scales at times between life and death. To sit quietly in a consulting room and talk to someone would not appear to the general public as a heroic or dramatic thing to do. In medicine there are many different ways of saving lives.

which does not seem to be used much now, and which I do not find in your Index) with drug-overdoses, reckless and self-destructive behaviours of all sorts—none of my friends expected me to make it to 40, much less 50. I am more grateful than I can ever express at having made it, thus far—one has always to say "thus far"—to 68, somewhat scarred and battered (it is true, but, for the most part, alive and even creative). Those who do not know me well see me, imagine me, as successful, self-assured, stable, the demons exorcised—but I know better. They may rail at my "dependence" on a psychiatrist—but I know the thoughts, the behaviours, the very real dangers, which still threaten or lure me in his absence(s).

I didn't intend to say this—but you, above all, will understand.

Now I have to complete your book!

Oliver

To Karl D. Stephan

ELECTRONIC ENGINEER

JANUARY 7, 2002

2 HORATIO ST., NEW YORK

Dear Dr. Stephan,

Thank you very much for writing to me—one of the pleasures of writing *Uncle T* has been the great range of personal letters and "resonances" which have been sent to me, some from Londoners, some from evacuees etc.—but many from people like yourself (like *us*) who had "a love affair" with science or technology at a young age—the age which Faraday would address in his Christmas Lectures (designed for "a Juvenile Auditory," which meant boys and girls of twelve). I was, as you say, extremely lucky, privileged, to have a large, curious, encouraging family—and to have the *freedom* (along with certain wise constraints!) to explore and follow my bent at a young age. This may, paradoxically, include the freedom of *not being taught* (in a formal, formulaic way, at school)—this is an issue which Freeman Dyson addresses in a wonderful, partly autobiographic essay "To Teach or Not to Teach."

I end *Uncle T.* on a somewhat puzzled note—I was not sure, nor am I clear now, as to what "happened" at fourteen or fifteen, to break the almost trance-like love-affair with chemistry. I would like to say that it was followed by, or mutated into, an equally passionate "affair" with cells, genes, biological systems, organisms, but I don't know whether this was the case. I think it was only years later (after the endless suppression of enthusiasm, playfulness etc in school, medical school, residency) that I recovered that youthful enthusiasm—tho' now in the so different realm of human behaviour and its vicissitudes. I think I was probably "bright," even very bright, but "brightness" is not enough to carry one beyond a "wunderkind" stage unless there is some deep, ongoing development and involvement; and it was this, I think, which miscarried when I was 14–15, and which I was only able to recapture, years later, in a clinical context, as a physician.

Despite a phantasy (or two), I do not think I would have made an adequate chemist. I do not think I had enough feeling for *structural* chemistry (and I am defective in powers of three-dimensional visualization). I suspect I could not have made a good research chemist (and later attempts to become a research physiologist did not fare well). It may be that, as a physician, receiving letters, questions, requests, from all directions, and free to follow this and that as I wish, that I have found my own desultory, agile, unsystematic identity and "role"—as a "fox" interested in many things, and not a "hedgehog" capable of pursuing one big thing.

I sometimes say something of the sort when I am asked to speak at Commencements—that one may misjudge one's own proclivities and talents, and find all sorts of ambitions and aspirations disappointed; but that, on the other hand, one may *discover* all sorts of powers which one had no idea of. This has certainly been the case with me. Having said this, I do have a wistfulness (occasionally envy) of those who *do* seem to know at an early age what they want to do, and what they are capable of doing, and who then (often despite great obstructions) forge ahead and do it—it is this which seems to me heroic. And maybe *this* is genius.

This is not a very coherent answer to your searching questions!

With best wishes,
Oliver Sacks

To Kalman Cohen[*]

MATHEMATICIAN, COLLEGE FRIEND

JANUARY 8, 2002
2 HORATIO ST., NEW YORK

Dear Kal,

Delighted to hear from you, and to learn something of your life in the 40-odd years since we met. It sounds like a very good life, a satisfying one, in work and love—though I cannot imagine anyone as intellectually driven as yourself being "retired" (and suspect you will use "retirement" to become more productive than ever). Last night, for no particular reason, I picked a biography of C. S. Peirce[†] from my bookshelves, and started reading about his passion for logic—I suspect that a memory of the young you, the passionate mathematical logician, was behind my choice.

By another coincidence—or the one which put the young you in my mind—I have an archivist here who has been going over the myriad boxes of papers, letters, diaries and what-not which I have accumulated over the last half-century; he has just got to "C", and yesterday came up with three letters of yours—I take the liberty of enclosing xeroxes of these. (No idea whether you kept copies of your letters—I, at least then, did not keep copies of mine, and wonder how I replied to yours.) You will be specially taken, I think, with your letter of 3/22/56 when you speak of your meeting, and becoming engaged to, your future wife! I was intrigued to re-read, in your 1955 letter, that you were working on a chess *learning* machine (did you envisage, then, that one day such a machine would beat a grandmaster?)—in that same letter you ask "Why don't you try to come to the United States, Ollie?"—as, five years later, via Canada, I did. I am touched by your memories of Seders at 37,[‡] and (in retrospect) by your "tremendous faith" in my intelligence and ability—a faith I was very far from sharing myself. I see, looking back through those old letters, how tormented I was by my "research"—it did, indeed, end badly, or inconclusively—and perhaps this did something to show

* Cohen was one of OS's closest friends during their time as Oxford undergraduates.

† The philosopher Charles Sanders Peirce.

‡ OS's childhood home at 37 Mapesbury Street in London.

me that if I had strengths, they were not in pure scientific research, or in anything academic—although it was only years afterwards, seeing my own patients, and finding myself outside Academia that I started to find my own "voice," and to write the peculiar narratives/essays which I have been doing these last 35 years. I think I had to leave what I saw as the stifling, or rigidly-stratified atmosphere of academic Medicine in England to do this. And the States, for me (as Australia for my older brother) seemed to promise "space," freedom, interstices in which I could live and work, in a way which England did not.

But I think the real impulse to come to America had much earlier origins than this—and dates back to a time soon after WW II—as I suggest in a couple of pages (pp. 278–280) of *Uncle Tungsten,* which I enclose. Or still earlier.

Am I an American now? Don't know—though I felt like a New Yorker, with my fellow-New Yorkers, after 9/11. Am I English—I still have a UK passport. I must be some sort of amalgam, as all of us are. But I was oddly pleased, a couple of years ago, when Queen's* made me an Honorary Fellow, and I have enjoyed visits, and staying there since—which is how I got your address, and wondered if I might see you there again.

I do come to the Chapel Hill area every so often, and it would indeed be a special experience to see you again—and, of course, to meet Joan too—and other family of yours!

Affectionately,
Oliver

---

* The Queen's College at Oxford University, OS's alma mater.

To Daniel Pinchbeck

WRITER

OCTOBER 20, 2002
2 HORATIO ST., NEW YORK

Dear Mr. Pinchbeck,

Many thanks for your letter and book*—I am glad our friend Ted Mooney suggested you send it to me.

I read it with a mixture of pleasure and disquiet—pleasure at your adventurousness and honesty, and at the excellence and charm of your writing—the man on p. 201, for example, with "an appealingly attenuated goat face"—but disquiet, increasingly, as the book went on, at your abandoning a neutral position, and becoming not just an acute and empathetic observer (or observer/participator), but—a *believer.* The line seems to me a very important one: there is no doubt, for example, of the depth of understanding and imaginative sympathy with which Gershom Scholem explores Kabbalism, but there is never the tone of convert or evangelist (any more than there is the tone of a sceptic). Thus I was fascinated by your accounts of iboga in Gabon, your sensitivity to culturally-mediated drug experiences, and your criticism of self-observations (from Weir Mitchell to Aldous Huxley, etc) which, however acute in themselves, are not in (or sensitive to the role of) a traditional and sacred context. But disconcerted at the way you yourself "go overboard," and are more and more disposed to belief in a supersensible, supernatural world (and to a correlative *dis*respect, if I can so express myself, for the natural world). Thus when you speak (p. 265) of the visual organization of the DPT† experience as "far beyond anything that the synaptical wiring of my brain could create," I want to say, "What do you know about the potentials of your own synaptic wiring? Who are you to set limits to it?" I want to contrast this *cavalier* approach to the physical with Spinoza's attitude. "No one," he writes in the *Ethics* "has hitherto laid down the limits of the body . . . no one has as yet been taught by experience what the body can accomplish solely by the laws of

* *Breaking Open the Head: A Psychedelic Journey into the Heart of Contemporary Shamanism.*

† Dipropyltryptamine, a psychedelic.

nature . . . the body can, by the sole laws of its nature, do many things which the mind wonders at." And this could include *all* your DMT/DPT experiences, and all the other experiences (yours and others) which you relate in your book. That they can be conjured up by the brain/mind is no disrespect to them, and (in a sense) irrelevant to their interpretation or "meaning."

For myself I cannot conceive (nor have ever been able to, nor wished to) a "real" world, that is a Primary World, other than the one we live in (your "consensual" world). I am, of course, (or hope I am) sensitive to the inner worlds, the Secondary Worlds, of poetry and art and dreams and myths, and of my patients and friends, and of people I meet; but I cannot imagine taking them for *another* Primary World. But you, apparently, feel free to posit or believe in another Primary World. This comes out very clearly on pp. 221–2 where your friend Robert speaks of having been visited by mushroom beings: "I asked myself whether I believed Robert's stories, and I decided that I did. I don't mean that I believed he believed them; I mean I believed the things he described had actually taken place." This goes beyond imaginative sympathy; it is crossing the line into (identification and) belief. And if you allow a belief in mushroom-spirits, you open the door to everything—Archons, Elves, what you will—a whole world, or worlds, of phantastic beings which a non-believer, a "rationalist," would find physically, biologically and ontologically impossible (nor can recourse to novel physical concepts like superstrings or supersymmetry allow them). I have to ask myself: do you *really* believe (in) all of this? And, if so, what should we understand by "belief"? [. . .]

I should say that I find the *subject* of hallucinogens, psychedelia, whatever you wish to call it, deeply interesting, and not in a merely academic way. I have tried many things myself—tho' not DMT!—and I did so in the hope of recapturing that sense of wonder which I had felt so strongly as a boy, but which seemed to desert me when I was about fifteen. I think I had some drug-visions which "inspired" me—as when, "stoned," in 1967, I saw the neurological heavens open before me, and migraine shining like a beautiful constellation. I felt then that I should become "an astronomer of the inward," and this led me to write my first book (*Migraine*) and others. This vision had the reality, or irreality, or super-reality, of a dream; but I did not believe, when I came down, that there was *actually* such a heaven.

On another occasion (jumbled by artane, nasty stuff) I had a long (philosophical) conversation with a fly on the wall—the fly had a tiny but very clear, rather Bertrand Russell–like voice—but I do not now *believe* in articulate flies. I can appreciate how a brilliant chemist (like Kary Mullis) might find his chemical insights sharpened by LSD (as Kekulé, after pondering for years on the structure of benzene, had a symbolic dream of snakes in a ring, and waking realized that benzene had a closed ring). I think there are real insights and illuminations which can come from drugs (as from dreams, and all "primary process" thinking), but which then need to be assessed, interpreted, carefully, by one's critical mind when one comes down. And, more to the subject of your book, I can imagine (to some extent, and your book has increased the extent) how a culture and its mythology and sacred symbols and rites and religion may be deeply and crucially dependent on ("sacramental") drugs. *This* seems to me fine and deeply interesting in your book—whereas I find your *own* "psychedelic journey" sometimes suggestive of "shamanistic regression" (as you put it)—or something more.

But it is not my function to sit in judgement (and on the very rare occasions when I write reviews they are always in the mode of appreciation). I don't think that I am the right person to review a book like yours—altho' it would be interesting to compare it with many of the other books you cite (you have read very widely!), with John Horgan's forthcoming (or perhaps just published) *Rational Mysticism,* etc. etc.

We all have Journeys, and mine is not yours. I have, I hope, some intellectual curiosity and playfulness, but very little in the way of metaphysical need or hunger. I do not crave Belief. (Perhaps this has something to do with our different bringings-up and backgrounds.) But I can admire your adventurousness and boldness and sincerity, and am glad you wrote to me and sent me your book,

My best wishes to you and to it,
Oliver Sacks

PS: A tiny factual correction: speaking of ergotism you say "St. Vitus' Dance"—this should be "St. Anthony's Fire."

To Dave Draper

BODYBUILDER

DECEMBER 11, 2002
2 HORATIO ST., NEW YORK

Dear Dave,

How very unexpected—and nice!—to get your two inscribed books, and to see (forty-odd years later) how good you look, and how well you write.

I leafed through *Brother Iron, Sister Steel* straightaway, and felt very nostalgic at your descriptions of "The Dungeon" as it was in the early Sixties. I was more of a power-lifter myself, but I remember Hugo Labra very well, Zabo, and some of the others you mention. I vividly remember the enormous, unwieldly (and dangerous!) dumbbells—but the biggest of all you do not mention and that was the 375 lb dumbbell which I once saw Charlie (Chuck) Ahrens side-press in the yard of his own gym—he and Steve Merjanian, both with 60+" chests, completely filled a VW beetle. I have vivid memories of the lifting platform at POP—it was here that I challenged Dave Ashman to a front-squat contest, and beat him (I did 575 to his 560)—but he was a wonderful Olympic lifter, and I couldn't snatch to save my life. Then there was Sidney's down on the beach at Santa Monica—it was here that I met Dave Sheppard, Jim Hamilton and others, who all became good friends; then, up in Venice, there was old Mac Batchelor, with the strongest hands in the world.

I read (with mixed feelings) your excellent sections on Injuries, Aging, etc. I managed to get complete quad tendon ruptures on *both* legs (in '74 and '84, first left, then right)—an unusual injury—and couldn't help wondering if I had overdone it years earlier with squats (I had a routine of fives—5 sets of 5 reps with 555 lbs every 5th day)—but I had good surgery, and the old legs are still powerful. Have also had rotator cuff surgery—should have done the rotator cuff exercises, but in the old days one never thought of them.

I was intrigued to see how young you were when you started handling iron, and dreaming of the-body-to-come. I had a relatively late start—in my twenties (23) and had been inspired especially by photos of Reg Park (whom I later met at the Empire Games in '58). But I was equally (or more) drawn to Olympic and power-lifting—especially after seeing Bennie Helfgott at the Maccabi gym in London (he had two Olympic medals*—and the most beautiful technique: he never pushed himself unduly, knew his own body, and never suffered the repetitive injuries which so many lifters—Dave Sheppard, Shemansky etc. had. He is still in great shape in his early seventies, and I see him whenever he comes to New York). My other training buddy was Ken McDonald (now back in Australia). But (to my lasting regret) Ken introduced me to that most lethal of lifts—the stiff-legged deadlift—when I was pretty much of a novice. I did 525 lbs (I had only been lifting for six months or so), but three days later got an excruciating backache—and continued to get these, at intervals, for the next forty years (now, oddly, in the late sixties, they seem to have let up).

My chief activity now is swimming, which I adore—I try to swim 2000–2500 yards each morning—but I also do some (relatively) light weight-training at a gym twice a week. I have boiled myself down from the 270+ I used to be in "Dungeon" days to about 200, and try to keep in fair shape. So I guess that I too am a veteran of the Iron Game, in a small way. And, mercifully, I have had no real health problems (touch wood!) outside some injuries, and the slow changes that go with aging.

I take the liberty of enclosing a copy of my (most personal!) book *Uncle Tungsten* (ever thought of having weights made of tungsten? it's 2½ times denser than steel—a 16 lb tungsten shot is no bigger than a tennis ball). I think my fondness for metals, especially very heavy metals, goes with the fondness for weights. Enough said. Lovely to hear from you, and my very best regards to you (and anyone else who may remember me from those far-off days—I think George Butler would be one of them).

Oliver

* Helfgott was captain of the British Olympic weight-lifting team at the 1956 and 1960 games, though the medals OS refers to here were from other weight-lifting meets.

To Paul J. Kiell

PSYCHIATRIST, ATHLETE, AND AUTHOR

DECEMBER 12, 2002

2 HORATIO ST., NEW YORK

Dear Dr. Kiell,

Many thanks for your article(s), which I enjoyed, and which I think especially valuable as a counter-actor of the notion that there is something wrong, "unJewish," in physical activity and prowess. It is a common enough misapprehension, not least among Jews themselves.

For me (besides parental models—my mother had been a champion jumper at school, as my father was a swimmer) a very important contact, in my early twenties, was with the Maccabi Club in London—(I don't know whether the Maccabi had/have branches here, even whether they still exist). It was there that I met Bennie Helfgott and Laurie Levine [. . .] and many other Jewish athletes. We had a good power-lifting team (I was their mid-heavyweight), and enjoyed contests with other teams all over London.

I have two "ultraorthodox" cousins, but both of these have a passion for nature, hiking, climbing etc. One of them—a mathematician-Talmudist who lives in Jerusalem*—still skis and climbs high mountains in his early seventies: he describes the amazement of Tibetans at seeing someone with a long white beard as active as a mountain-goat at altitudes of 20,000 feet or more. They see no contradiction between their orthodoxy and their love of nature and physical activity.

I think the stereotype of the sickly, intellectually precocious, but physically feeble Jew needs to be contested—and, among its other excellencies, your article does just this.

Thanks again!

With best wishes,
Oliver Sacks

* Robert John Aumann.

To Peter Edwards

CHEMIST

DECEMBER 30, 2002
2 HORATIO ST., NEW YORK

Dear Peter,

It has been a longish time since we corresponded (still longer since we saw each other), but this time of year is a time for sending and receiving things. [. . .]

I wonder if you have been picking up anything special in the superconducting line with these extraordinary silver fluorides—I still have vivid memories of magnetic levitation with liquid nitrogen in your lab in Birmingham. [. . .]

Now *I* have a question. Kate got me some gallium for Christmas, which I keep in a smooth hemispherical glass (it makes a gorgeous mirror). But when it melts there is a strange illusion of transparency—as if one were seeing a transparent colorless fluid moving around *above* the background of silvery, unmelted metal (Roald saw this too). I wondered if this was an *edge* phenomenon, it seems to have a concave meniscus (unlike Hg's convex one), but I am (probably very stupidly) perplexed. From other angles, it seems that the liquid has a *skin* on it, with fine lines, textured, like the skin on milk.

It seems to supercool, so that once melted it stays molten for hours in quite a cool room. But then when it finally solidifies, it seems to do so in shallow quadrangles layered in strata, or odd zigzags like the fortifications of a medieval fort. I am not sure whether these are patterns of crystallization, or stratification patterns as one has with drying mud.

I am sure this is all childish stuff, but you must be patient with me, and explain what I am seeing. You might even have fun playing with gallium yourself.

(I sometimes dream of the 200 tons of ultrapure liquid gallium in a solar neutrino detector under the Caucasus. My love of metals and my love of swimming come together in the phantasy of swimming in, or *on,* this great pool of liquid gallium. There was an attempted *heist* a few years ago—thieves came with a truck and a siphon; but it was foiled at the last moment.)

Anyhow, write to me, if you have time, tell me about gallium, about silver fluorides (if this is not secret), and, in general, how you are and what you are doing,

My warmest good wishes for the New Year,
Oliver

To Harriett James

SPEECH PATHOLOGIST

FEBRUARY 25, 2003
2 HORATIO ST., NEW YORK

Dear Ms. James,

Many thanks for your letter and Tito's book,[*] which I find quite extraordinary.

It is unfortunate, as you say, that facilitated communication, so called, was such a mix of the genuine and the influenced.[†] I like the way Tito himself says "I need mother or my class teacher to sit by my side not as facilitators, but as my *environment*" (this made me think of Winnicott's phrase "the facilitating environment," and it opens up a far broader and deeper picture of what goes on; and one needs someone as rare and insightful as Tito to bring this out). Certainly you are right to seize on this quality of *presence* as crucial (I do my own best work when my editor, and perhaps mother-figure, Kate, is in the next room—this sort of facilitating environment is needed by all of us).

I personally am in no position to explore these issues myself further, but they *need* to be explored, and at the highest and deepest level. I wish I knew who to suggest—it needs to be a (perhaps Winnicottian) psychiatrist. If I have any brainwaves I will let you know, but

* *The Mind Tree* by Tito Mukhopadhyay, a young man with autism.

† Facilitated communication, where a "facilitator" helps nonverbal people write or type, remains controversial. Some critics feel that the writing that results is unconsciously influenced by the facilitator.

for the moment I can only endorse what you say, and hope that the genuine aspects of facilitation can be extricated from the cloud which has surrounded facilitated communication, and explored.

Sincerely,
Oliver Sacks

To [first name unknown] Godwin

CORRESPONDENT

MARCH 20, 2003
2 HORATIO ST., NEW YORK

Dear Dr. Godwin

Many thanks for your fine and full letter—I never quite know what the post will bring me! [. . .]

With regard to the (vexed question of) "repressed memory"—this is not something I have personally studied, nor one on which I would care to say anything publically. I *did* write about it, in a sense, in my case-history "Murder" (in *The Man Who*). [. . .]

There is no doubt, psychologically, neurologically, about our enormous suggestability and impressionability (both for good and evil), and our inability, frequently, to distinguish primary experiences, primary memories, from apparent (or secondary) memories which have arisen from description or suggestion. I had to acknowledge this in myself, with regard to a (very vivid!) "memory" described in my autobiography *Uncle Tungsten*—of an incendiary bomb landing in our back garden (see p. 23 in *UT*). Another bomb memory—of a large, unexploded bomb—was confirmed by my (older) brother Michael, but regarding the incendiary bombs he said "You never saw it." I was stunned, and said "But I can see it *now,* clearly, in my mind's eye. Why?" "Because our brother David wrote us a letter," Michael said, "a very vivid letter. And you were enthralled by it." Yet learning that I never saw the incendiary bombs with my own eyes, but had only a *constructed* memory or image based on a description, did nothing to lessen its vividness. It still seems real—though it is not. Such "secondary" memories are common, or universal—and they may be, or become, indistinguishable—phenomenologically, and perhaps physiologically—from primary ones.

But this is somewhat different from "repressed" memories, for these are infused with a violent affect, unconscious conflict, trauma etc, which ordinary secondary memories (often quite bland and neutral) do not necessarily have. The whole business of repression, and repressed memories, has been a minefield of ambiguities, assertions, counter-assertions, retractions etc. since Freud first believed that many of his patients had been sexually abused in childhood (as they told him), but then came to dismiss such memories as phantasies. (No doubt he went too far both ways.) There *are,* alas, cases of sexual (and other) abuses of children sometimes—but with nowhere near the frequency which "memory" (and an atmosphere of accusation) might suggest. A veritable social hysteria seems to surround this subject—as well as the personal hysteria which sometimes generates such memories ("Hysterics suffer from reminiscence," Freud remarked).

I am saying too much—and too little. Perhaps I should simply reiterate the fact that our minds are enormously suggestible and influencible (especially where such charged dynamics are involved), and that false (or secondary) memories may seem very real (and may perhaps come to have even the neurological underpinnings of genuine, primary memories). The whole subject is in need of the subtlest exploration.

I hope these thoughts—all very unsystematic—may be of interest to you.

With kind regards,
Oliver Sacks

14

Snapshots

2003–2006

To Christof Koch

NEUROSCIENTIST

MAY 8, 2003

2 HORATIO ST., NEW YORK

Dear Christof,

Many thanks for your charming letter—I am glad *Uncle Tungsten* had such resonance for you. I know Peter Atkins personally, and *love* his *Periodic Kingdom;* it certainly conveys the *enchantment* of the Kingdom better than anything I have ever seen, and (I suspect) it was in my mind when I wrote (p. 28) about a magic "kingdom" or "garden" of numbers (maybe an essay of Freeman Dyson's called "Ramanujan's Garden" was also in my mind—it is impossible to trace all the antecedents of one's thoughts). It is very sweet of you to think of sending me the Atkins book. I am sorry that neither "[The Case of] Anna H." nor "The Mind's Eye" reached you, and I enclose these.

I thank you for the summary paper you wrote with Francis—and, as it happens, I have read it, and several times, with great attention and admiration. One reason for doing this is that I have especially been thinking of motion perception, and its disorders, and its possible mechanisms, myself, as well as some of the broader problems of consciousness. (I think I got scared off the whole subject back in '92, and took refuge in chemistry and botany for a decade; but now I am back.) I take the liberty of enclosing this* too, and will send a copy (as well as an *Uncle Tungsten*) to Francis as well, especially as it draws

* An untitled essay manuscript, eventually called "In the River of Consciousness."

on and quotes your paper with him. It is only a manuscript, rough and provisional, though I hope it may become an article for the *NY Review of Books.* I was at Cold Spring Harbor a few days ago, and was told there that Francis is not too well, though fully and wonderfully himself.

With kindest regards,
Oliver

To Francis Crick

MAY 10, 2003
2 HORATIO ST., NEW YORK

Dear Francis,

I hear news of you from many sources—from Ralph Siegel, especially, from friends at Cold Spring Harbor, and from Christof Koch whom I met again recently at CalTech, and have been corresponding with—and you have especially been in my mind since I read your paper with him in February *Nature Neuroscience.*

I was very fascinated by this, especially your "snapshot" proposal, because I have long been interested in motion-perception and its disorders, and have often wondered whether one uses "frames" or "snapshots" in real-life, as well as in the cinema.* I had already started writing about this when I received a copy of your paper with Christof—and now (though I don't usually send out manuscripts which are still being revised) I take the liberty of sending you my "consciousness" piece, because it draws heavily on your own thoughts towards the end.

I also enclose a published piece ("The Case of Anna H.") about a patient of mine with a visual alexia etc—and (finally!) a copy of my book about boyhood and love of science etc, *Uncle Tungsten.* I hope this may give pleasure, and not be a burden to you. I have heard

* The hypothesis Crick and Koch were working on proposed that a visual scene consisted of discrete "frames" or "snapshots," which were endowed with continuity by the brain rather than by external reality.

that you have been in poor health lately, but Christof tells me you are undaunted, and still reading mightily. I often think of you with great admiration and affection, and send you my warmest good wishes,

Oliver

To Gerald Edelman

MAY 25, 2003
2 HORATIO ST., NEW YORK

Dear Gerry,

I had meant to write to you, thanking you for your recent paper (with Tononi),* which I found immensely fascinating—and stimulating. Indeed it provoked a (perhaps unwise!) addition, or change of direction, to something I was writing—this was basically part of a mass of writings about "Time," especially, time, motion, continuity etc—thoughts provoked by recent biographies of Muybridge and Marey, a boyhood interest in ciné-, stereo- & color-photography/perception, the extraordinary accelerations and retardations and standstills of movements *and* thought-flow in some of my "awakenings" patients etc which had been occupying me throughout April, when I had a lot of time on my hands, having had an injured right shoulder re-operated, and hence an excuse for deferring most of my usual activities, and being able to type with one finger of my left hand almost non-stop for four weeks—all this (forgive my syntax, this now-labyrinthine sentence)—and then I got *your* article, and ventured onto the charged and "dangerous" realm of the "basis" of consciousness; and added to what I had written a sort of "coda" on some of your thoughts (and Francis's). As I say, I had meant to send a letter, *this* letter, or something like it, but then I had suddenly to leave for Australia—my brother there having become suddenly ill with cancer of the pancreas—so the letter wasn't written, and Kate, meanwhile, sent off my (*very* rough, and perhaps needing-to-be-radically-revised) manuscript. [. . .]

I was intrigued, while in Sydney with my brother and his family,

* Giulio Tononi.

to contrast the behaviour of his 3-month-old grand-daughter with an 8-month-old child of a neighbour's—to see how the younger one had almost no use/control/image/consciousness of her body—or so it seemed to me—while the 8-month-old child was full of skilled and purposive movements of all sorts. I feel I need to observe babies closely (and, maybe to read Piaget again too). I mostly see the *decline* of consciousness in my injured or elderly patients, so it was lovely seeing something of its dawn instead. Our birthdays—my 70th, you have a year or two on me—come up in July; if you are in NYC, I hope you might come to a little party I am having on the 9th. And I hope you—and Francis—were not miffed by my phrase about "grand old men" when, of course, *intellectually,* one is just warming up, just beginning to see how things really are. I certainly felt an almost-adolescent energy and intoxication with my burst of writing last month, and I imagine you get such bursts all the while. But Francis, I hear, is not in good health now—I have not had contact with him for several years. [. . .]

I take the liberty of enclosing (or did Kate send it?) a piece on visual imagery,* which was supposed to come out in the *New Yorker* in April, but got bumped because of Iraq; oh, and a little letter (from the 15 May *Nature*) on elephant motion, a tiny offshoot of my "Time" writings in April.

I do hope that Maxine is now restored to health, and I send you both my warmest best wishes,

Oliver

* "The Mind's Eye."

To Christof Koch

JUNE 18, 2003
2 HORATIO ST., NEW YORK

Dear Christof,

[. . .] First, you are very welcome, more than welcome, to use the *Leg* quote* in your book; I am delighted that (finally!) someone thinks the phenomenon worth notice (I am not sure that I have much capacity for abstract thought, but I think I am an accurate *describer*). Besides *Leg, Awakenings* and *Migraine,* I make a (very brief) reference to the phenomenon in *Uncle Tungsten* (p. 143), where I mention experiencing this (as well as other visual disturbances in migraine attacks) from an early age—and where I mention that such experiences, along with a passion for color-, stereo-, and cine-photography as a boy, were important in tilting me towards the later choice of neurology, and an interest in visual disorders in particular. I also make a (brief) reference to cinematographic vision (and some other visual disturbances) in my *Island of the Colorblind* (pp 87–90), after drinking sakau (kava), altho' I suspect (and hint) that I had already been having "a strange visual excitement such as I am especially prone to, especially at the start of a migraine." In general, my own experiences of cinematographic vision have nearly always been in the context of migraines (and often associated with scintillating scotoma etc. tho not invariably or necessarily so)—altho' I have had the experience once or twice from LSD, and, as mentioned, with the sakau. [. . .] Sometimes (as in the passage you want to quote) it has been associated with a geometrized, "mosaic" vision—but again this association is not invariable or necessary.

The acutely disordered context of cinematic vision makes it difficult to get reliable estimates of its rate—but my own experiences,

* Koch quoted a passage about what OS calls "cinematographic" or "cinematic" vision:

> I asked [Nurse Sulu] to look at the picture, talk, gesture, make faces, anything, so long as she moved. And now, to my mixed delight and disquiet, I realized that time was fractured, no less than space, for I did not see her movements as continuous, but, instead, as a succession of "stills," a succession of different configurations and positions, but without any movement in-between, like the flickering of a film run too slow. She seemed to be transfixed in this odd mosaic-cinematic state, which was essentially shattered, incoherent, atomized. (*A Leg to Stand On,* chapter 2.)

and those related to me by patients, suggest a *relatively* "slow" flicker, or succession, in the 6–12 hz range. The very word "flicker" seems to suggest something faster—on the verge of "fusion"—whereas it can definitely be a slower and more leisurely succession (and often with persistence and overlap of the separate images, but—again—not necessarily or invariably so). In some attacks, with a scintillating scotoma in addition, I have felt there might be a synchronization or resonance between the two phenomena. I tried hard, when I was working in a migraine clinic, and later seeing my "awakenings" patients, to get EEG's *during* attacks (and once thought I might have caught one, with a concurrent slowing of alpha rhythm)—but the attacks proved to be too sporadic and unelecitable (tho' migraines, especially visual migraines, can sometimes be provoked by strobe illumination, like photic seizures etc, and I experimented with this a little). I wondered about transcranial magnetic stimulation when I was writing and thinking about Isaacson, the painter with central achromatopsia—I wondered if I could be made to experience what *he* experienced—but at that time (late 1980's) there was some talk about elicitation of seizures or permanent damage from TMS, so I didn't press the matter. [. . .] I cannot answer the question about relative movement—I think I was probably rather immobile, partly because I was enthralled, partly because I was immobilized and in bed with the leg and its cast. I don't *think* there was any relation between the leg syndrome and the migraine—though when I wrote to A. R. Luria about the leg, he wondered whether such syndromes were commoner on the non-dominant side. I have not discovered—but I have not systematically searched for—other descriptions of cinematic vision in the clinical literature. [. . .]

If I have another attack of cinematographic vision—I have not had one since 1994—I will try to make some better observations, in particular as to the effects of head-movement!

I have not yet heard whether the *NY Rev. Books* will publish the piece,* and, if so, in what form—tho' the Editor has told me that he "likes it," though also that it may be "too long," unless abridged. So I await further word. I have some impulse, I confess, to *add* to the piece, to deal with "the stream of consciousness" since the piece started off,

* OS's essay "In the River of Consciousness," about theories of visual consciousness.

initially, as a sort of meditation on "time," and one especially stimulated by the Wm James quote. Now looking through Chapter XV of the *Principles* ("The Perception of Time"), I find myself fascinated by the James Mill passage (pp 605–606)—which makes me think of some of my amnesiac (Korsakovian) patients. James himself is so rich, almost every sentence is pregnant in unexpected ways.

Meanwhile I can gaze, through my kitchen window, at a large exhaust fan (as I must presume it to be—tho' at first I thought it was a sort of anemometer) which glitters in odd ways when the sun hits it, and shows me all the phenomena of the continuous illumination "wagon-wheel" effect. At the same time I find a renewed interest in high-speed cinematographs for recording and analysing animal movement—and this led me to write a little Letter to the Editor on elephants' gaits* (this was in *Nature* last month), and another on insect flight (in yesterday's *Science Times*).† There are all these odd spin-offs to the central question(s) of Continuity vs. Discreteness, or how they are related.

I will write a (shorter!) letter separately to Francis—this one is, in a sense, to you both.

My thanks for your interest, and best wishes,
Oliver

To Francis Crick

JUNE 28, 2003
2 HORATIO ST., NEW YORK

Dear Francis,

I was delighted to get your letter (of June 5), and am sorry for this somewhat belated reply—altho' I wrote a long letter to Christof (which I am also posting today) addressing some of the questions which both of you raised.

* *Nature* titled this "Early Work on Elephant Gait Not to Be Forgotten."

† "Credit to a Pioneer," OS's letter to the editor of the *New York Times*' Science section, was about the work of an early physiologist, Étienne-Jules Marey.

I am glad you liked *Uncle Tungsten,* and that you had some personal "resonances" to it; I wonder how much "continuity" you see between that teenager at Mill Hill who talked about the "Bohr atom" and the Periodic Table—your "Double Helix" self, and your present self. (When Freud wrote a letter to Karl Abraham in 1924 he said, "It is making severe demands on the unity of the personality to try and make me identify myself with the author of the paper on the spinal ganglia of the Petromyzon. Nevertheless I must be he, and I think I was happier about the discovery than about others since"!!) I suppose we all have several, overlapping, mutually-penetrating lives.

I wrote to Zihl myself—I had met him, briefly, in '95—wondering about [his patient's] *experiences,* and in particular whether she had akinetopsia in her visual imagery or dreaming, but I could not get any more information about this. I get the impression that her "standstills" lasted for several seconds each, in contrast to the several-a-second "frames" of cinematographic vision (and the hours-long standstills of my ?"zombified" postencephalitic patients). Christof raised other questions about "cinematographic" vision, which I tried to answer (not too successfully!) in my letter to him.

I shifted ground, without quite meaning to (or realizing it) in my "Perceptual Moments" (or "Snapshots")* piece from the original plan, which was to go from motion-perception, the flow of visual awareness, to the (Jamesian) "flow of consciousness" and what constitutes "continuity" here (this may be an enormous subject, because, as far as I can see, James devotes a good deal of his vol I† to exactly this question). The "movie" analogy is obviously only good to a very limited extent. But I somehow departed from clinical/personal description (which is my strong point) to trying to say something about the *basis* of consciousness, the mind/brain problem etc—and this is not my own thinking—and can be a real minefield! I remember, when we first met, how you sat me down next to you at the dinner table, and had me tell you clinical stories, descriptions etc, and how these instantly generated hypotheses in your mind (it was similar with the color-blind painter, and the marvellous letter you sent me about him). I sometimes fear I am not very good at abstract thinking, and

* Working titles for an essay that would be called "In the River of Consciousness."

† William James's *The Principles of Psychology* (1890).

that my strength lies in description—and I am very sure of cinematographic vision as a *phenomenon.* It would be fascinating, as (you and) Christof suggested in his letter, if it could be *elicited* by TMS* to the motion-complex (or higher).

Sooner or later, I guess, we are bound to find ways of *showing* the activities of small sets of neurons rising and staying above threshold, and the correlation of this with (a) consciousness. We seem to be lacking in the middle, "meso" scale between the grossness of brain-imagery, EEG, TMS etc, and (multiple) single-neurone recordings. I have written a little about this in a later draft of the piece. But, at present, it is in limbo, and I still have no word about its publication (or otherwise) in the *N.Y. Rev. Books.* On the other hand, my piece on visual imagery in the blind(ed), "The Mind's Eye," will be in the (double) issue of the *New Yorker* which comes out on July 7.

My good friend Ralph (Siegel) tells me he will soon be in La Jolla again, and he can convey other thoughts, and my greetings, to you. I hope I can visit you myself when I am next there,

With all my thanks, and warmest good wishes,

To Deepak Chopra

ALTERNATIVE MEDICINE ADVOCATE

[AUGUST 2003]
2 HORATIO ST., NEW YORK

Dear Deepak Chopra,

There has been a surprisingly large "resonance" to my "Mind's Eye" piece, and I have been startled at the number of letters received, and their variety; but none of them startled me as much as yours! I am delighted, first, that the piece had resonances *for you,* and that you know some of my (miscellaneous) other writings—which include a couple of essays on consciousness written about ten years ago (and published in the *NY Rev. of Books*). I then, frankly, got so appalled at the size of the problem (and also the veritable spate of books

* Transcranial magnetic stimulation.

on the subject) that I retreated, in a manner of speaking, to earlier interests—botany and chemistry. But in the past few months I have been attracted back to mind/brain problems, the nature of consciousness etc, have again written on the subject (only a rough draft of a piece, and not yet published, or perhaps even publishable), and I am taking off over the weekend to La Jolla, where I will spend time with Crick and Koch, and with Gerald Edelman, whom I have been corresponding with.

I can, of course, only go a little way with them, as perhaps I can only go a little way with you, because I feel I am essentially a describer, a storyteller, a phenomenologist (or, to put things reductively, a clinician), and I never dare to go or think too far beyond clinical (or personal) experience. And yet, of course, one has to. I am very divided as to whether we will ever resolve the brain/mind "dilemma," as you put it (which is different in *character* from the relatively straightforward wave-particle "duality"), or whether (as Colin McGinn and other "mysterians" suggest) the problem is actually beyond human power to resolve, or even conceive properly—due to our "cognitive closure." Having said this I also have a feeling that there is no essential problem at all, but a single phenomenon which (so to speak) looks both ways, outward to physiology, chemistry, physics etc, and inward to introspection and subjectivity. But what *sort* of phenomenon could do this?

I have read the two very thought-filled papers by Herms Romijn—many thanks for sending these to me. He certainly builds up a wonderful picture of the dendritic tree as having "an infinite spectrum of possibilities to rapidly integrate in dynamic fashion the flow of information that continually arrives at its surface . . . by recording the information into a profile of highly ordered, electric/magnetic fields," with thousands of rapidly-changing "sinks" and "sources." He brings out the quantitative complexity of even a single dendritic arbour quite beautifully. And then, of course, there are 10^{10} of these—perhaps arranged (as Crick & Koch envisage) into "coalitions" of (perhaps) 10–100,000 neurones each; and all of these variously interacting, integrated, orchestrated ("bound") together.

No doubt there are enormously complex, continually shifting (electric/magnetic) fields—and yet I am not persuaded that it is "fields" of this type which are the crucial phenomenon—the phenomenon which, Janus-like, looks both ways. (I should say that I have been very attracted by some sorts of "field" theory in the past—particularly in relation to gestalt psychology, patterns perception, etc. etc.)

I confess to being out of sympathy with some of the additional notions which Romijn explores, viz. "regression" ("former lives"), a collective unconscious, archetypes etc; and the cosmic dimension of this ("the submanifest order of being" etc.) is quite beyond me—tho' I think Bohm was a genius, and (as Freeman Dyson likes to say) that Nature's imagination is far richer than ours . . . and far stranger. But, to come back to the grounds of my own thinking, my concern is always with the affected (and acting) *Individual,* the individuality of experience, of illness or injury, of all life's impacts, of adaptation, of strengths and weaknesses, of will & action, and I have little capacity or desire to go beyond this.

In general, I am not a theorist—and my books and writings, such as they are, are designed to provide *examples* of one sort and another, which others, more theoretically-minded, can perhaps use. I remember when I first met Crick, fifteen years ago or so, he took me by the shoulders, sat me down, and said "Tell me stories!" and for hours I told him (clinical) stories of different sorts, while he generated dozens of hypotheses—when I told him the story of my colorblind painter, he made suggestions which would have led to a whole century of research!

But I hesitate to theorize about consciousness even in a circumscribed body-mind sort of way, and could not possibly venture to anything beyond this—models which (as you put it) might go "a little deeper into the notion of resonance as a fundamental activity of a non-local ground of Being." This I have to leave to you—and perhaps a successor to your gifted friend, the remarkable but now-deceased Herms Romijn. [. . .]

With kind regards,
Oliver Sacks

To Bill Borden

PSYCHOLOGIST

NOVEMBER 17, 2003
2 HORATIO ST., NEW YORK

Dear Bill,

What a lovely, thoughtful letter—and what a gorgeous packet of good things from Berlin, Chemnitz, Buenos Aires (which I will mull over, and return to you when I see you at the NY Mineralogical Club next month—I have no idea what I will say, or present; or no idea *yet*).

Your travels with Allen* sound quite idyllic, in their combination of Museums, Botanical Gardens, Concerts (plus a conference or two).

For myself, I went with Kate to Stockholm last month—had a lecture or two there, and an honorary degree—but we were also able to go to Köping, the little town where Scheele† had his apothecary-shop and lab (the older part of the town has a very sweet, 18th-century, Scheele-like feeling about it—one could easily imagine him there); and to Ytterby (where, as you say, Y, Yb, Tb, Er‡ as well as Sc, Ho, Tm etc. were found, and a new scandium mineral, thortveitite (I have misspelt it) quite recently. And, very recently, tho' microscopically, a new scandium phosphate, the Sc-analogue of xenotime); the old mine is closed, but there are lumps of quartz, and some darker, heavier lumps (gadolinite, I liked to imagine) scattered around; and one can take a boat-ride through the archipelago to a little fortress-island which has a (tiny) mining/mineral museum, and photos of the active mining days.

And then we went to Tallinn, where there is a wonderfully-preserved medieval inner town; a quaint, crowded, idiosyncratic Natural History Museum; and (this I have never seen before or elsewhere) an amazing permanent exhibit of Lichens and Mosses (a huge range of these is endemic to Estonia). Thence to Prague—music

* Borden's husband, Allen Heinemann.

† Carl Wilhelm Scheele, an eighteenth-century chemist and pharmacist, discovered oxygen and identified many other elements, including tungsten, hydrogen, and chlorine.

‡ Yttrium, ytterbium, terbium, and erbium, as well as four other elements, were all discovered in a mine in the village of Ytterby, so naturally OS wanted to make a pilgrimage there.

everywhere!—and so back here. Kate joins me in sending our warmest regards to you both,

Thanks again!

Oliver

To Michael Herrick

EXECUTIVE DIRECTOR, WINDHORSE*

DECEMBER 22, 2003
2 HORATIO ST., NEW YORK

Dear Michael,

More thanks than I can say, and to all of you, for letting us visit you, talk to some of your clients, and get a good idea of what an extraordinary venture Windhorse is—I could not have got this sense adequately without actually visiting it.

I have been transcribing some of my scribbled notes, and thinking about the five people I saw: Jeremy, Aimee, Greg, Chaya, and David, and how utterly different and (in their so-different ways) intriguing they all were. Seeing people in their homes on Saturday was quite as crucial (perhaps more so) than seeing them in the office—I have always felt that "house-calls" were the most valuable. [. . .] I think it was very important that, after my years in the depressing (and often horrible) wards of a State Hospital, I was able to see a deeply hopeful, deeply different, enterprise like Windhorse, where there is no sense of an inexorably advancing organic disease (tho' the organic is no more denied than the precipitants and contexts of illness), but of a condition which, while it may be lifelong (in terms of an underlying vulnerability or propensity) is one which can be understood, fought, worked with, controlled, essentially by human means (tho' a *little* drug-assistance may be needed sometimes), and in a way which will allow a full and happy life. Jeremy put it very well when I asked him

* Windhorse Integrative Mental Health, a program in Northampton, Massachusetts, provides support to people with severe mental illness, using an approach based in part on the philosophy of Edward Podvoll, one of its cofounders. Podvoll was an intern at Mount Zion when OS was there, and went on to direct the Contemplative Psychotherapy graduate program at Naropa University.

whether he had wholly "recovered," whether he felt it was now "all over." He said, "Having had a schizophrenic break and recovered, you have to spend the rest of your life managing it, learning to recognize its intimations and ruses, the dark and cunning powers still in the mind. It is my belief that they are still there. 40% of my mind is schizophrenic, but there is 60% which watches and fights back."

David too spoke of "perpetual vigilance," a life, in some ways, spent on "the edge." The business of *motives* is very complex, for although, clearly, most of your people want to "function," live fully in the world, have relationships, work, be imaginatively free and creative etc. yet they have also seen (and, to varying degrees, explored) very deep and dangerous, but also rich and extraordinary, realms of the mind which the "ordinary" person knows nothing about—and to this extent they feel "special," and might not want it otherwise. (Tho' few would go as far as my brother, who speaks of the generality of people as "rottenly normal"—but he has had (on the whole) a tormentedly limited life, and is angry and resentful. I cannot help wondering how different this might have been if *he* had the advantage of a Windhorse fifty-five years ago.)

Again, all my/our thanks for your openness and generosity. I very much hope that I can visit Windhorse again (I see it, in part, as Ed's* legacy to the world), perhaps in the Spring. And do talk with Kate about the thought of a talk, or something, which might assist the place, help draw attention and funds to it.

With best wishes,
Oliver

OS ALMOST INVARIABLY REPLIED TO correspondents who were incarcerated, as well as to those who were either very old or very young. Emma Roth, a sixth-grade student, wrote to say that she had read three of OS's books and asked whether she could interview him for a science article she was writing. She offered to come at any time that would be convenient for him. "My mom," she noted, "said I could miss school for this and I

* Podvoll (see footnote at the start of this letter).

am SURE my school would understand." During the interview (Emma brought her pet white rat along), OS showed her his mineral collection; she wrote him again afterward, enclosing a selenite egg as a thank you.

To Emma Roth

MIDDLE SCHOOL STUDENT

DECEMBER 29, 2003
2 HORATIO ST., NEW YORK

Dear Emma,

What a grand surprise! Your charming, warm, funny letter—with its Thurber-like drawings in every margin—and the fabulous egg of *selenite* (along with its chemical composition and molecular weight).

No, I do not have any selenite, and I will cherish the lovely egg you sent—it has an extraordinary, eerie, moon-like glow (hence the name, I suppose), indeed it seems to have a light of its own. It immediately made me think of a short story by H. G. Wells ("The Crystal Egg"), which I must have read when I was about your age—I enclose a copy of this, hope you may enjoy it too. (In one of his early books, *The First Men in the Moon,* the insect-like inhabitants of the Moon are called Selenites).

I visited Stuyvesant High School a few months ago—or perhaps it was longer, because so many of the people there had memories of 9/11, the terrible events which occurred only a little way from the school—and was tremendously impressed by the "feeling" of the school, and its outstanding intellectual quality (my friend Roald Hoffmann, to whom I dedicate *Uncle Tungsten,* and who is a Nobel-prize winner in Chemistry, went to Stuyvesant, and speaks of it with great warmth). So, I think you should go there! Glad you liked "The Mind's Eye" in the *New Yorker* (I have another piece—"The River of Consciousness"—in the current *NY Review of Books*—I enclose this too). And schizophrenia, since you mention it, is a realm I am starting to explore—a landscape as strange and primitive (sometimes frightening, sometimes beautiful) as Yellowstone Park.

My great love, after minerals, was Animals too—you saw all the [models of] squids and cuttlefish etc around my place—and I have been wondering, recently, if some of them not only move, but perceive, and *live,* at different rates (I enclose a passage, an imaginative ride, from William James' *Principles of Psychology*—I suspect Wells read this as well, because in some of his stories *people* get strangely accelerated or retarded). Sorry to hear about your poor rat—I hope you get another one (but there is a real sense of grief when a pet dear to one dies). I wonder if rats think, or react, more quickly than we do? I am still wondering about [getting] one myself.

I hope *you* had a lovely Hanukhah (no idea how to spell it—Chanukah, perhaps?), and wish you a very happy New Year (and keep writing! I will look forward to your letters).

Oliver

NICK YOUNES, ANOTHER CORRESPONDENT, wrote to OS in 2001; his letter was so learned that OS assumed he must be an elderly chemistry professor. Younes explained that he was actually a high school student with an interest in philosophy, chemistry, and much else. They began to correspond regularly and soon became friends.

To Nick Younes

FEBRUARY 21, 2004

2 HORATIO ST., NEW YORK

Dear Nick,

I have just got back from Australia—my brother suddenly went into near-coma, and I rushed to what I thought would be his deathbed, or funeral—but, against all expectations, he then recovered, cerebrally, became completely himself, so that we could have a lot of brotherly (but, of course, farewell) conversation. He is pitifully weak and emaciated, and cannot, I imagine, last much longer. [. . .]

Although I had been told that my brother was "semi-comatose" I found his state more singular than this: he made (at least initiated) no movement, and (at first) was without speech, and his eyes were open but showed no following; then he would move and speak, but only in response. When I arrived in Sydney last Saturday night, he stuck out his hand in recognition, but otherwise seemed to be in a "zombie-like" state, with no spontaneity, no emotion, not personally there (or so I felt). And then, on Tuesday, against all odds, all expectation, he recovered full consciousness, and "self"—all his spontaneity, and humor and powers of reflection—and being himself a curious physician looked at me earnestly, and said "Well what do *you* make of it all?" His recovery from (what is usually an irreversible state of) "akinetic mutism" (usually associated with massive midline or medial lesions in mid- or fore-brain) filled me equally with gratitude and delight (at being able to hug him, talk with him, which I never thought I would do again) *and* a lot of neurological wonder, in particular a sense of how inadequate it is to speak just of "consciousness," and how there are many forms of it, all with their own neural substrates—so that, on Tuesday morning, there was a sudden resurrection or switching-on of "self," "ego," "higher-order consciousness," "meta-consciousness" (whatever term one uses), a sudden return to himself. He had some amnesia for the episode—and immediately asked "How long was I away?" [. . .]

Kate also put on top of my mail the package just received from Max* in England—viz a kilo of shining, incredibly dense (5–10 Gm) nuggets of iridium—(Max calls them "Smarties"). I had them by my bedside last night, and whenever I woke marvelled anew at the weight of this little plastic bottle, only a third full, which weighs so much—I would (of course) much prefer a single shining piece—and if Max can get the right sort of induction-furnace I will send them back to be melted into a single blob (a blob would be fine with me, it doesn't

* Max Whitby, a chemist and naturalist who worked in close partnership with Theodore (Theo) Gray, the author of a popular series of books about elements, molecules, and other subjects.

have to be a particular shape). I also got a lovely slightly bluish "drop" of osmium (weighing 10.3 Gm)—what a *beautiful* metal (I was just telling Daniel Brush* about it). And, at the other end, quite a large ovoid (67.5 Gm, but it only has an SG of 1.7) of beryllium. I hope Theo will be able to make some cylinders of various metals to "match" the tungsten cylinder you saw—I have to contact him. I think I will also get a kilo of depleted U (unless this imperils my Resident Alien status, I musn't imperil my Green Card in these paranoid days) and a largeish chunk of lithium (I think it is in argon, though I wish it could be in Perspex or plastic-wrapped so I could handle it). [. . .]

My best,
Oliver

To Kevin Cahill

PHYSICIAN

JUNE 1, 2004
2 HORATIO ST., NEW YORK

Dear Kevin,

[. . .] Thank you for the very full notes on the patient you are referring to me—we have indicated to him that I would be happy to see him, and that he should set up a time at my Tuesday clinic at NYU. [. . .]

Now something quite different! When I was in England last week I chanced to see a CNN section devoted to Lariam.† This came up because there have been further incidents of soldiers returning from Iraq in very disturbed states, sometimes going on murderous rampages, ending with their own (actual or attempted) suicides. Many (or all) of these soldiers have been on Lariam. Some high US official—Secretary of this-or-that—said that he doubted if there was any connection here, & that he knew of no studies regarding significant (neuropsychiatric) side-effects of Lariam, but the Army would "look into it." There was then an interview with a researcher, who

* An artist and jeweler.

† A drug widely used to prevent malaria.

noted that there had been careful, double-blind studies of Lariam in 2001 and 2003 which showed a 30% or higher incidence of "neuropsychiatric complications," and that for this reason the Canadian and Australian forces absolutely avoided its use unless it was strictly necessary because of chloroquine-resistant "malignant" (falciparum) malaria. She added that whatever malaria there was in Iraq was *not* "malignant," and that therefore there had been no good reason to use Lariam there in the first place. She expressed surprise at the Secretary's not apparently knowing of these things.

I was intrigued at all this, because—as you will recollect—I myself had very bizarre almost hallucination-like *dreams* on Lariam—they went on for weeks and were unlike anything I had ever experienced, and you said, when I asked you, that you thought the incidence of striking (and sometimes serious) neuropsychiatric "side-effects" of Lariam was far higher than the 1% the PDR allowed—and closer, perhaps to 30 or 40% (this was right back in 1995).

Wondering how the Press would deal with this CNN interview, I scoured the papers (I had no US papers available to me in London), and found a Lariam item in the *Guardian*—an item going in the other direction, for it described the grave illnesses (and sometimes deaths) of several English people from cerebral malaria, because they had *refused* to try Lariam (having heard exaggerated or hyped-up reports of this). Both the US soldiers and the English civilians would have been very much better off with *balanced knowledge* of the drug, the indications for taking it, the need for careful monitoring, etc. etc. I felt like writing a Letter to the Editor (? of the *NY Times*), but I wonder if you, as an expert in these matters, might write an Op-Ed. Do give me your thoughts.

Warmest regards,
Oliver

To Christof Koch

JULY 21, 2004
2 HORATIO ST., NEW YORK

Dear Christof,

Thank you so much for your patience and care with repeated check-up calls from the *New Yorker*—indeed, your help and thoughts throughout the incubation of this piece.* I have been irritated, but also admire, the fact-checkers' pertinacity and vigilance (they have phoned *everybody*—Libet, Ralph, Noyes & Kletti, etc.—they would have phoned Wm James if he were still alive . . .). I enclose what (other than any overlooked typos or other errors) will be the published version of the piece. [. . .]

When I was asked to identify a mass of old letters yesterday (they are all addressed to "Bob," and the question was *which* Bob—actually seven different Bobs, I found, were involved), I found this letter (xerox enclosed) to an ophthalmological Bob† (the one who worked on the achromatopic painter with me) about discontinuities of visual perception—so you see I was "hot" on this subject twenty years before it re-emerged (for me), writing to you and Francis, etc. Indeed I remember calling Richard Gregory excitedly, around 1974 or so, because my overhead fan seemed to be doing odd things in the even morning light. I seem to have a great capacity to "forget"—but then, years later, things re-emerge. I wonder if it is the same with you?

Is Francis holding on?‡ I hope, at least, he is not suffering too much—there is a limit to the powers of transcendence, even in a Francis (or a Hume), if pain, nausea, etc. become too great.

I normally go to Australia in August, but with my brother's death there (in March—I had a visit with him in February) I am not going to go again next month. I have no idea *what* I will be doing—but if my thoughts turn West (Ralph is enjoying himself in La Jolla)§ I will

* "Speed."

† Wasserman.

‡ Crick had been diagnosed with colon cancer.

§ Ralph Siegel, a neurophysiologist, spent summers in La Jolla, close to Crick and many other neuroscientists at the Salk Institute, the Neurosciences Institute, and UCSD.

let you know. Again, all my thanks, and warmest regards to you (and, of course, Francis).

Oliver

To Christof Koch

JULY 29, 2004
2 HORATIO ST., NEW YORK

Dear Christof,

I just heard, last night, from Ralph, of Francis's death.

I have some inkling of how close you and Francis were—and send you my deepest sympathy. I think the two of you had a unique relationship, at once intellectual and emotional, which brought a special light and creativity into both of your lives, and which must have been a joy and an illumination to you both. Even I, who only knew Francis a little, feel somewhat orphaned by his death, and for you this feeling, I suspect, may be overwhelming.

One knew he was critically, terminally ill, yet the absoluteness of death is no less shocking for this.

I am especially glad that I was able to re-encounter Francis last year, to correspond with him and you, to write and think in relation to his/your ideas, and, last August, to meet him again.

That sunny lunch, with you and Ralph, and Francis and Odile presiding, and the conversation going in all sorts of unexpected ways, and Francis's piercing mind and (mischievous!) imagination leaping out in all directions, and his grace and courage dealing with his illness—that lunch will stay in my mind forever.

I wish I could be with you, and Ralph, and Odile (et al.) now, but am confined (by sciatica) to staying home.

Again, my deepest sympathy and warmest regards,
Oliver

IN JANUARY 1996, OS ATTENDED the launch of a space shuttle mission, flight STS-72, at the invitation of his astronaut friend Marsha Ivins. Ivins was not flying on that mission, but she had been assigned as ground support for the astronauts' families, and she introduced OS to Sue Barry, a professor of neurobiology at Mount Holyoke. (Sue's husband, Dan Barry, flew many NASA shuttle missions, including STS-72.) OS noticed that Sue had a squint—that is, one of her eyes seemed to look inward—and he asked whether she had stereovision. She replied no, that she did not have good convergence and thus had no binocular depth perception (though she could infer depth from other clues). Now, nine years later, she wrote to OS to tell him that she had, through optometric vision therapy exercises, in fact achieved binocularity.

To Susan R. Barry

PROFESSOR OF NEUROBIOLOGY

JANUARY 3, 2005
2 HORATIO ST., NEW YORK

Dear Mrs. Barry,

I have vivid memories of that night, the eve of STS-72, and have received Christmas/New Year cards from the two of you (or *all* of you) over the years, but have not, I'm afraid, been anything of a correspondent.

But your letter of the 29th fills me with amazement—and admiration, at your welcoming your "new world" of visual space with such openness and wonder, even if it meant your developing a fear of heights in Kauai—and at your describing it with such care, and lyricism and accuracy.

Amazement, because it has been "accepted" for years (but clearly Dr. Ruggiero had evidence and thoughts to the contrary) that if binocular vision was not achieved by a "critical age" (supposedly of some months), then stereopsis would never occur. Talking to Jerry Bruner, the psychologist, who was born with congenital cataracts which were not operated on until he was eighteen months old, seemed to confirm this. On one occasion he told me how, lacking natural lenses, with their slight yellowish tint, he could see some way into what I would call the "ultraviolet." I asked him, breathlessly, what this was like. He answered "I can no more *tell* you than you could tell me what stereopsis is like." Bertrand Russell contrasts "knowledge by description" with "knowledge by acquaintance," and you give wonderful descriptions of how utterly they differ, of how the greatest formal or secondary knowledge can never approach actual experience.

I need to think carefully about what you describe, and perhaps discuss it, if I may, with a friend in visual physiology. I think your experience & account ought to be published, in some form or another, in view of the physiological or psychophysiological revision it seems to call for; and, at a more personal level, the hope it may give for those who have long "accepted," at one level or another, that they are condemned to live in a "flat" world. I also think that the sheer exuberance you convey, at a sort of visual rebirth, is the sort of thing which can remind us that stereopsis (like all our perceptual powers) is a miracle and privilege, and not to be taken for granted. If one has (say) stereopsis all the while, one may indeed take it for granted; but if, as with you, one lacked it, and was then "given" it—then it comes as a wonder and revelation. This too needs to be brought out.

For one reason or another, I have never taken *my own* stereopsis for granted, but have found it an acute or recurrent source of pleasure/wonder for much of my life. This led me, as a boy, to experiment with stereo-photography, hyper-stereoscopes, pseudo-scopes, etc (I am still, in my eighth decade, a member of the New York and also the International Stereoscopic Society). And it caused me to pay special attention to an odd experience, in 1974, when (due to visual restriction, or rather spatial restriction) I was to discover how my own stereoscopy had been "collapsed," and how it re-expanded over the course of an hour or so, when I was replaced in a large space (I enclose a copy of the relevant pages from *A Leg to Stand On*).

Jan 4. I have taken the liberty of discussing what you describe with two colleagues of mine (Bob Wasserman, an ophthalmologist, and Ralph Siegel, who works in visual physiology), and they were as intrigued as I was, and raised a number of questions. One such, raised by Dr. Wasserman, and related to your mentioning that your eyes converged a few inches from your face, was whether you were readily able to thread a needle—which he thinks would be very difficult without stereopsis. [. . .] Another question was whether the *vertical* misalignment, which Dr. Ruggiero picked up (and corrected, with a prism) had been present from the start, or whether it developed later in your life. And whether there is still what Dr. Wasserman calls some "micro-strabismus," even though this might not be symptomatic. Other questions relate to problems with motion-perception (at least perceiving when *you* are in motion).

As you perhaps know I have written about people who have no perception or idea of *color,* and would sometimes ask them how they conceived of color, and (if it were possible to "give" them the capacity to see it) what this might mean to them. The question is a tantalizing one, because there is no known way of "giving" an achromatope color and, additionally, some of them say that they think the sudden addition of color—which never having been perceived before, and so having no associations or "meaning"—might be very confusing to them. But it is clear that the addition of "depth" or "space" to your visual world has been (almost) wholly positive.

So many questions! Since you have favored me with your story, and ask my thoughts, I think (over and above anything I can say here) that I would like to *visit* you, and perhaps to do so in company with my old friends and colleagues, Bob and Ralph, who could explore and check aspects of your visual perception which I myself could not do. (The three of us formed a "team" when seeing the colorblind painter [as well as] the middle-aged man "Virgil," who was given vision after being virtually blind from birth. I wrote about both of these in my *Anthropologist on Mars* book.)

You asked for my response to your story, and perhaps this is too much of a response! (I am reminded of how, as soon as I heard Virgil's story, I wanted to fly down to Atlanta to see him.)

But give me your thoughts on whether such a visit, to meet you, and to explore various aspects of visual perception with you, would be agreeable.*

If it would be, we can work details later.

* OS did write an essay about Barry, titled "Stereo Sue." His interest and encouragement also led her to write a book about her experience, *Fixing My Gaze.*

Again, thank you so much for sharing your experiences and thoughts with me, and—really—opening a new realm.

My warmest good wishes to you and Dan,
Oliver Sacks

To Robert B. Silvers

JANUARY 11, 2005
2 HORATIO ST., NEW YORK, NY

Dear Bob,

It is a while since we have seen each other or corresponded—I hope you are well. 2004 was a sad and painful year with the deaths of many friends, from Thom Gunn's in May to Susan Sontag's last month (I had known both Thom & Susan since 1962).

It also saw the last illness and death of Francis Crick, whom I had known and corresponded with since 1986. I was very fond of Francis as well, and as I looked through our correspondence after his death I started to wonder about writing a "remembrance" of him. There have been many obituaries and recollections by his fellow neuroscientists, etc, but I think ours was an unusual relation because I was a clinician, and he was hungry to know how people actually experienced various disorders of perception and consciousness, etc.

In particular he was greatly interested in the Colorblind Painter, whose case (by Robert Wasserman and myself) you published in the Review in '87; and equally interested, fifteen years later, in "Perceptual Moments" and "The River of Consciousness"* which you published in January '04. (He also remarked to me, in one letter, how he had followed my articles on the Deaf which you published in '86 and '88.)†

* Working titles for "In the River of Consciousness."

† "Mysteries of the Deaf" and "The Revolution of the Deaf."

So, for many reasons, as soon as I had finished writing "Remembering Crick" I felt I should send it to you![*]

My best,
Oliver

To Sister Marie Geraldine Therese

LITTLE SISTERS OF THE POOR

APRIL 16, 2005
2 HORATIO ST., NEW YORK

Dear Sister,

I have often thought of you—you were the first Little Sister I ever met—and often asked about you, and now I get your wonderful letter and enclosures which reach across the thirty-odd years since we worked together at Notre Dame, the "old" Home on 183rd St.

I felt sad, bereaved almost, when you left the Home for—for Pawtucket—and imagined that you too would feel uprooted; but your letter brought home to me (which I have now seen for myself) how *all* the Little Sisters are your sisters, and *all* the Homes are Home, and how moving from one to the other is an essential part of your Service and Life. I heard then that you had moved—to Alabama (was it Alabama? Curiously I have just been in Alabama myself), and then lost touch with you. And now it is lovely to hear from you again, to read about your Jubilee, and that you are (apart from a little arthritis) full of health and activity—scarcely surprising with your mother living to 106 (you must have great genes!). My father lived to 94, practised Medicine almost to his last day, though finding it increasingly difficult to walk because of arthritis in the knees (but this did not affect his *swimming*—and he remained a grand swimmer even in his nineties). It is similar with me—I hobble a bit walking, but once in the

[*] This essay, "Remembering Francis Crick," was published in *The New York Review of Books*, as well as in *The Threepenny Review*, and it appeared, in somewhat different form, in *On the Move*.

water take off like a porpoise. I was, I guess, forty, give or take a bit, when I met you in 1973 (I forget which month it was), and am now advancing towards my 72nd birthday. I loved seeing all the pictures you sent, from when you were a young Sister to your Jubilee pictures now—and I enclose a photo of myself (I think it was taken five or six years ago—I am whiter now).

I was tickled that you recognized yourself, as the "nursing sister," in "The Lost Mariner." [. . .]

I was very fond of brilliant Sister Geraldine, and had a lot of contact with her in the late Seventies, when she moved, as I did, between the three Homes in New York (I started at Queen of Peace in January '76, and at Holy Family soon after). So tragic that she developed a brain tumor so young—I rarely go to Wakes, but I went to hers (she was in the Philadelphia Home then, I think). Very fond too of Sister Josephine, with her great kindness (she always made me Christmas puddings full of richness and suet—none of your instant, ready-made ones); she had a lovely dry sense of humor, and great courage—as she needed to when she found herself in the long Purgatory of parkinsonism. [. . .]

I still go to Queen of Peace, every fortnight or so—this had been a constant for almost thirty years. I was very sad when Holy Family Home closed, and sad too when Notre Dame (which seemed so comfortable and homey, and so close to the Arthur Avenue district where some of the Residents could go) gave way to the relatively isolated new Home on Baychester Avenue. This, in turn, closed; but now there will be a lovely new Home in the Bronx, built by and for the Little Sisters, and not taken over like the Baychester one. I hope to resume work there this year.

I am babbling on—but your letter has aroused so many memories and feelings in me.

Suddenly, as I write, a memory of Agnes G., in one of her manic moods, chasing Sister Geraldine and myself up three flights of stairs at Notre Dame. I am not sure of the year now (perhaps it was in the late Seventies, after you had left). Agnes, if you remember her, lived to the age of 99. There were several residents at Notre Dame who could remember the great Blizzard of 1888. And in 1976, if I recollect right, at the Bicentenary, there was a Sister (as well as some Residents) who had reached *their* 100th. And, as memories float up, I remember another centenarian resident (was it Mrs. C. ?), demented, and often agitated, but who could be calmed if given a doll, which she would then hold to her breast and nurse like a baby—a situation which caused some dilemmas for us all (for this strange "mothering" seemed so "inappropriate," and yet it calmed and organized her as nothing else could).

Enough, enough—I hope we can keep in touch, and (most especially) that some day I can visit you in Indianapolis, pay my respects, and remember old times.

With admiration, affection, and warmest good wishes,
Oliver Sacks

To Sharon Stone

DIRECTOR OF RESEARCH AND EXTERNAL PROGRAMS,
PARKINSON'S DISEASE FOUNDATION

MAY 26, 2005
2 HORATIO ST., NEW YORK

Dear Ms. Stone,

Thank you for your letter regarding the therapeutic value of creativity in Parkinson's disease.

This is a huge, intriguing, and complex subject—a very important one too—but I am not sure how to organize thoughts on the subject, what one should emphasize, where to begin.

My father was a resident ("houseman") to Henry Head, the famous neurologist, in 1916. Head, at that time, already had severe Parkinson's disease—and (my father told me) would festinate the length of the neurology ward at the London Hospital, and have to be caught by one of his own patients. But he was at the height of his powers and creativity at this time (his *Studies in Neurology* came out in 1920, his great *Aphasia* volumes six years later). My father would always say that he thought Head's work, his passion to understand and communicate, kept him going in face of a parkinsonism as would have brought a less creative man to a halt.

When I was a resident in neurology (at UCLA) in 1963, I saw an artist with very severe akinetic parkinsonism—hardly treatable at this time, for l-DOPA was still some years in the future. When he told me that he could still paint large canvases I was incredulous—but he invited me to his ranch to see for myself. He was brought in his wheelchair to a blank canvas, and a paintbrush put in his hand—he was completely motionless, and I could not imagine how he would proceed; but, suddenly, with a wide sweeping movement, he made his first brush-stroke, and from then on completed a large and exuberant painting without the least sign of his parkinsonism or akinesia. When he was finished he sank back in his chair and became almost completely motionless again. I was not sure whether to call this "kinesia paradoxa," or to see it as the therapeutic power of art in action. Subsequent observations showed me it was the latter.

When I started working with severely parkinsonian post-encephalitic patients in 1966—patients, for the most part, with overwhelming akinesia—I observed, almost from the start, the profound power of *music* with them. Patients virtually unable to speak or move could be "released" by the flow of music, and sing or dance freely. (I give many examples in my book *Awakenings*, and there are beautiful examples to be seen in the documentary of *Awakenings*.)

Here, it seemed to me, it was especially the rhythmic and propulsive quality of music which was so powerful—but this was only part of the matter, for rhythm alone was not as effective. It was clear that *melody* played a great part too, providing anticipation, "motor shape," and "images of the future" (as Bernstein puts it), so enabling stuck parkinsonian patients to think and move *ahead*, to think and move in long cadences, "phrases," instead of the jammed or stuttering tiny steps they otherwise made. Perhaps it is not a coincidence that Henry Head, himself so aware here of what he was theorizing about, spoke of "kinetic melody" as that which parkinsonian patients lacked and needed, when they were stuck in "kinetic stutter."

On one occasion I brought the poet W. H. Auden to see my [post-encephalitic] patients, and the power of music in particular. He immediately cited an aphorism of Novalis: "Every disease is a musical problem, every cure a musical solution." Certainly this seems to be so of parkinsonism, though the power of music ceases when the music ceases. But patients can carry music *with* them (in a Walkman or iPod), and sometimes they are able to move relatively normally if they can *imagine* an accompanying music.

Should one speak here of the power of music as being "creative"? I think one can, and must; and I think of the words of E. M. Forster in this connection: "The Arts are not drugs. They are not guaranteed to act when taken. Something as mysterious and capricious as the creative impulse has to be released before they can act."

And speaking of action, and Acting, I have seen parkinsonian patients profoundly disabled for most of the time, activated and able to act in a theatrical setting; and when acting a part able to speak and move in a way which would be impossible at other times. Here again, as with music (and painting) the therapeutic power is temporary—it can deliver the patient for a while, but when the action, the music, the painting, is over, there is a return to the parkinsonism.

But knowing that they can be "liberated," if only for a while, and respond to music, or perform it, or act, or paint, and in so doing reclaim, for a while, their healthy selves, their powers, is itself profoundly encouraging and therapeutic for patients with parkinsonism.

Moreover I suspect—as my father suspected with Henry Head—that the ability to turn to creative activity may slow the advance of the disease; and, even if it does not do this, will enable patients to resist its power, and the tendency to passivity, to fight it (and sometimes conquer it) for years on end.

These, then, are a few thoughts provoked by your letter! [. . .]

With kind regards,
Oliver Sacks

To Görel Kristina Näslund

PSYCHOLOGIST

MAY 26, 2005
2 HORATIO ST., NEW YORK

Dear Dr. Näslund,

Thank you for your letter. I will answer as best I can, sometimes (perhaps) giving you more information than you need, sometimes less!

I have always been exceptionally bad at recognizing faces—but not merely faces. I am bad at recognizing people. I am bad at recognizing places. I have left/right confusion. I have no sense of direction. [. . .] My eldest brother had the same (agnosic) problems as I do, though my parents, and my two other brothers, seemingly did not.

The "topagnosia"* (I mention this first, to dismiss it, since you do not ask about it) can be quite disabling. I constantly get lost, unless I follow fixed routes, and have detailed verbal directions. I do not consistently recognize my own street or the building I live in, let alone other people's streets and buildings. I have repeated *illusions* of recognition—déjà vu and jamais vu, both false.

Other forms of visual recognition are good—I am an amateur botanist, recognize thousands of plants (more easily, perhaps, than people!). I tend to remember not only page numbers of things which have interested me in books, but the exact appearance of the pages—the fact that the sentence which interests me is a third of the way down on the left, etc. [. . .]

Now your questions about (my) prosopagnosia:

Yes, it is a handicap, and an embarrassment—not so much growing up, as in my adulthood, and perhaps even more now I am ageing (I am 71). My assistant (Kate Edgar, with whom you have communicated) often says to people in advance: "Don't be upset if he doesn't recognize you. He may not recognize *me*—and I have worked with him for twenty years. Introduce yourself by your name." [. . .] But there are always uncontrollable (and sometimes comic) situations—"cutting" old friends at parties, or in the street; introducing husbands and wives to each other; greeting strangers, etc as old friends.

* OS is using this term as a shorthand for topographical agnosia. ("Topagnosia" generally means an inability to identify which part of one's body is being touched.)

I have significant problems in *self-recognition.* On several occasions I have almost collided with a large, bearded man, started to apologize, and realized that I have run into a mirror. The opposite has also occurred. Once, when having a coffee outside a restaurant, I turned to my reflection in the window, and started to smooth my beard, then realized that my "reflection" was doing nothing of the sort, but staring at me in wonder—this was a bearded man inside the restaurant, whom I had mistaken for my own reflection. So the author of *The Man Who . . .* has some personal experience of what he writes about!

[. . .] I try to pay conscious attention to build, clothing, unusual facial features—a long nose, shaggy eyebrows, moles, mustaches, etc) to compensate for the lack of automatic perception. I should add that I am highly sensitive to *movement,* to gestures, postures, what one might call "motor style," in striking contrast to some other forms of visual perception.

On one occasion I was questioned by the police regarding a young man who had (perhaps) committed a theft. They asked me if I knew him. I said, "Yes, I see him almost every week." They asked me to describe him. Trying to remember, I described him as short, tall, dark-haired, blond, thin, fat, etc. The police said that I had indeed given them a mass of contradictions. They said that if I indeed knew the man, how did I recognize him? "By his walk," I answered, and instantly imitated this.

You will see from this that it is not merely facial recognition I have difficulty with. Much larger categories may be unperceived, so unremembered.

Bertrand Russell, in his Autobiography, speaks of his difficulties recognizing people, and how, if he felt it especially important to recognize them, he would make a mental list, itemize "salient features" in his mind, which he would then try to match with subsequent appearances of the person (I am not sure how successful this was). [. . .]

It seems to me very odd that the subject is "totally unknown" in Sweden, but I am sure that when you publish your book (have you written articles on the subject?) people will see it, or confess it, everywhere. Someone has to draw attention to it. I am reminded of my experiences with Tourette's syndrome. The day after seeing a patient with it (in 1971), I saw three people in the streets of New York with it, the day after this another two, etc. I realized that I must have been seeing, but not seeing, TS all my life. It supposedly had an incidence of 1 in a million. I suspected from my street experience that it might be a 1000 times commoner. This, in fact, has turned out to be the case. Indeed its incidence approaches ½%, but it was "unknown" before 1970.

With kind regards, and best wishes for your book!
Oliver Sacks

To Bill Borden

NOVEMBER 17, 2005
2 HORATIO ST., NEW YORK

Dear Bill,

What a wonderful, rich letter—but you *mustn't* send me all these fabulous fossils, it is over-generous in the extreme. I have not unpacked them yet, but I will do so *at* the next AFS* meeting, so that everyone can share in the joy of opening—a palaeobotanic Xmas treat for all.

Glad you liked "Recalled to Life"†—I have known Patricia since '91, but it was only last year, almost exactly a year ago, that we went for our walk together, and I thought what a remarkable person she was, and determined, if it was agreeable, to write about her. It was—and she insisted that I use her real name, as did her daughters Dana and Lari. I feel honored that you will include it in your course. Continued accounts? I guess so. I continue to see patients and—don't know what to call them, collaborators, subjects. I do have another piece with the *New Yorker*—no idea when they'll print it—about a gifted woman, strabismic and stereoblind, who gained stereoscopy at the age of 50, and is enchanted by the new depth and richness of her world. [. . .] I think I probably mentioned her to you before. Otherwise, I have been writing clinical narratives/essays, my hybrid form, on patients with musical disorders, or musicians who have come to visit me—and I

* American Fern Society.

† An essay about a woman with aphasia.

think they will come together to make a little book on Music and the Brain. I begin to see the shape it might be. [. . .]

Kate and I did have a little European tour in September—through Belgium and Holland to Italy. We visited Geel,* which you must know about—its "hospitality" (much too faint a word!) to the mad for seven hundred years or more. It was very extraordinary seeing obviously and sometimes deeply schizophrenic people walk freely in the streets, talk, shop, be fully recognized as persons by the townspeople (who serve as foster-parents, or provide foster-homes, as they have done since 1280—so close that the people they take in really become part of the family, with close emotional bonds going *both* ways. I want (sometime) to write about community-care of the mad/schizophrenic, and I have never seen such a place as Geel.† It could/should be a *model* for the rest of the world—this place where someone can have the full *experience* of schizophrenia (positive and negative), free to *be* schizophrenic, while having a "normal" and often emotionally and socially rich life. I wonder how things would have been for my poor brother had a place like Geel existed in England. We did not see much art in Italy, but we had dinner with the amazing Rita Levi-Montalcini, who (at 96) is a work of art, a miracle, in herself (do you know her autobiography *In Praise of Imperfection*?).

Yes, I saw Studs's‡ book in manuscript (and blurbed it)—I love it, and *he,* tho' physically frail, is wonderfully alive in feeling and wit at 93—but you probably see more of him than I do (he was here in New York, gave interviews all day without a hint of fatigue, when his new book came out). Robbin§ and John Betts¶ (to say nothing of Kate & myself) miss you . . . COME SOON.

Love from us to you both,
Oliver

PS Thanks for the zeolite *stamps* and mineral postcards.

---

* A small town near Antwerp.

† OS published his essay "The Lost Virtues of the Asylum" four years later.

‡ Studs Terkel, the legendary radio journalist, who lived, as Borden does, in Chicago.

§ Robbin Moran, a pteridologist at the New York Botanical Garden, who shared OS's and Borden's passion for mineralogy.

¶ A mineral aficionado and dealer.

To Gayle Tanaka

CORRESPONDENT

NOVEMBER 23, 2005
2 HORATIO ST., NEW YORK

Dear Gayle Tanaka,

I was intrigued by your letter—thank you so much for writing to me.

That your mother, despite being verbally limited, can listen to a song for the first time and sing (it) back (with) the words is fascinating—and, amongst other things, shows the preservation of *musical* memory despite her increasing difficulties in other ways. And equally fascinating, and very moving (and very useful!) is her *appreciation* of music, and the way it can keep open your relationship & communication with her. (I am currently writing about such matters, and others, in a book about Music and the Brain.)

The *talking*—to the photos, the stuffed animals, the birds, the Japanese prints etc—raises a dilemma (whether to preserve "her hold on reality," or to encourage her "conversation," however inappropriate), as you say. You do not say anything about the *sort* of things she says to the photographs, animals, etc (tho' earlier you speak of her being limited to a set number of phrases and words). You will have to work this out for yourself; this may involve a re-definition of "reality" for her—a re-definition which one might call "regressive," but which might also be adaptive for her, not only in encouraging "a form of communication" but in surrounding her with a host of make-belief "companions"—but these should be *additional* to real people, not instead of them. I am reminded (tho' much is different) of some of my *Awakenings* patients, before they awakened, before they could move and talk (in reality), having hallucinatory visitors and companions, who (in a sense) kept them sane, allowed a sort of "social life" when none was possible in reality. The hallucinations were all given up when they found they *could* talk, move etc once "awakened." I also think of a VERY demented old lady I once saw, no longer capable of *any* speech or interaction with others, apparently, who would "mother" a stuffed doll if given one, croon to it, rock it, even offer it the breast. Her caretakers were uncertain whether to disallow this "unappropriate" behaviour, or to allow it when it manifestly gave the old lady so much pleasure, and also served to calm and focus her (when she did not have her "baby" she was agitated and chaotic). Neither of these situations are that close to your mother's—but I cannot prevent them from coming to mind.

I cannot advise you, but I hope you will tell me what you do—and how your mother does as time goes on,

With best wishes,
Oliver Sacks

To Tobias Picker

COMPOSER

[DECEMBER 8, 2005]*
2 HORATIO ST., NEW YORK

Dear Tobias,

I couldn't collect my thoughts too well immediately after the Opera, and with the acclaiming hubbub all around you—it must have been a marvellous moment!—but I really think you have written a masterpiece—a huge weaving together of brilliant orchestration, and [. . .] of your ability to glide seamlessly from atonalism through a whole range of style, including amazing percussion effects. Actually (as with *Encantadas*) I could not think of the Music, the Drama and the Staging separately—they came together as a marvellous whole.

It is easy to believe that it has demanded ten years of work, of your life—but, of course, you have also been doing so much else at the same time.

I cannot help wondering what will come next—though, at this point, with the huge achievement of *An American Tragedy* under your belt, you should be taking a long holiday, and not thinking of what comes next—tho' (as a sort of creator myself) I know that one

* This letter was dated November 8, 2005. But Picker's opera *An American Tragedy,* based on the Dreiser novel, premiered at the Metropolitan Opera in New York on December 2, so presumably the letter was written on December 8.

is *always* thinking, consciously or unconsciously, of new ideas, projects, etc. In particular, of course, I wonder if you have had any further thoughts about *Awakenings.**

I am currently writing a series of essays—there will be a dozen or so—for what I think will be a book on Music and the Brain—and I wonder whether, sometime, you might come and chat with me about *your* musical memory, imagery, thinking, modes of composition, etc. I think (if my memory doesn't deceive me) that you once spoke of having a "congenital disorder," and I thought you meant Tourette's. No, you said, it was your musicality, the precocious, often overwhelming musicality which had dominated so much of your life from childhood on—sometimes preventing you doing other things you would have liked to. Freeman Dyson made a similar comment to me about *his* "congenital disorder," in his case, an enormous mathematical propensity, which dominated his life and work until he was about fifty (since then he has been able to do many of the other things he always wanted to, and has written dozens of enchanting and profound essays on a dozen different topics).

Perhaps there is something very special about music and mathematics here. For myself, not having any particular overwhelming gift (but only an overall curiosity) I cannot quite imagine what it must be/must have been like for Freeman and you.

Anyhow, I would love a chat one day.

My congratulations again on a great work, a real magnum opus—and my best regards to you both,

Oliver

---

* OS and Picker began talking about a possible opera based on *Awakenings.* The opera was eventually written after OS's death, with a libretto by Picker's husband and creative partner, Aryeh Lev Stollman. It had its world premiere at the Opera Theatre of St. Louis in 2022.

To Stephen Whitzman

ACTOR

JANUARY 23, 2006

2 HORATIO ST., NEW YORK

Dear Mr. Whitzman,

You raise a deep issue—thank you for your fascinating letter.

Soon after *Hat* was published I started to distance myself from the notion (Penfield's notion) that there was, or could be, any actual replay of experience in the sort of seizures and stimulations he described. And this distancing was widespread among neurologists. My friend Israel Rosenfield published a book a year or two later entitled *The Invention of Memory* (I wrote a foreword to it), and by that time I was persuaded that all "autobiographic" memory ("episodic" memory, to use the technical term) was "constructed" in the act of remembering (and re-constructed, and re-re-constructed with further rememberings and re-rememberings). Freud, I think, came to think similarly, on the basis of psychoanalytic practice, that whatever archaic traces there might be, these were re-worked, "re-transcribed" again and again. And yet, and yet—this said, one cannot help wondering whether *some* memories, and perhaps some *sorts* of memory, are preserved in their original form. Traumatic memories, for instance, which may be repeated with hallucination-like vividness, detail and fixity. And, as you hint, perhaps memories of intense sensations (tho' it is interesting that *pain,* usually, can't be remembered, tho' if it is a specific pain, it is instantly recognized if it recurs). You might read Frederic Bartlett's great (1932) book *Remembering:* he is at pains always to use the verb, always to speak of remembering as an activity, an active process, and never to speak of "memory" as if it were reified, a thing. This said, I do have the impression that there are some kinds of memory which are NOT "Bartlettian," not "constructed"—and this category (roughly) corresponds with what psychologists call "procedural" memory (the memory of HOW—how to do things, skills, performances—rather than WHAT). I am currently writing about one profoundly amnesic musician, who has lost virtually all autobiographic and episodic memory, but retains all his powers of musical performance, etc.

All these things (among others) are revolving through my mind at the present time, and I cannot come to any clear resolution. I enclose a couple of short pieces on memory which I wrote—but you give me YOUR thoughts now!

With best wishes,
Oliver Sacks

PS Some of my later thoughts on memory are contained in my *Anthropologist on Mars* book, especially the chapter "The Landscape of his Dreams."

But I have never really addressed or thought closely about what you call "sensory memory"—it's a tricky area.

15

Visions

2006–2015

To Anna and Joseph Horovitz

COUSINS

JANUARY 28, 2006
2 HORATIO ST., NEW YORK

Dear Anna and Joe,

It is a while since we have written, but you have been in my mind for various reasons, not least that I have been writing about Music and the Brain—have actually written *fourteen* essay/narratives now and begin to think that these might be organized into a little book; and, of course, I associated you, *all* the Horovitzes, with Art and Music.*

A second reason is not so good. I am afraid the New Year brought me an unexpected problem, a sudden disturbance of vision in my right eye, which turned out to be due to a melanoma behind the retina. Fortunately these (very rare) ocular melanomas carry a much better prognosis than the more usual sort, and it looks as if this is contained entirely within the eye. I have had intense radiation to the eye, and the chances are that this will burn the tumor out completely. It may do so, however, at the cost of some vision—difficult to predict. Now, a couple of weeks later, the swelling etc is going down, and vision is improving. But it has been difficult to read ordinary print (my left eye has never been of the best), so I have been turning, more and more, to *listening*—"talking books," and especially music—and in particular, Joe, have been enjoying the CD you gave me with

* Joseph Horovitz (1926–2022), the husband of OS's favorite cousin, Anna, was a well-known British composer and conductor.

your Trumpet Concerto, your Oboe Concerto, Jubilee Serenade and Sinfonietta—what a huge range of styles and moods and modes you bring together so masterfully. And all of these, if I understand the label, are relatively early pieces. I think I have often been too busy reading to listen to music as I should, and perhaps the current difficulties reading have an up-side too.

I have hesitated WHETHER, WHAT, WHEN, HOW to "tell" people what I have got, but it "came out" with the Sackses [. . .], and I suppose it should be known (at least not be a secret) among the Landaus,* if any of them are interested. I would not like it to come to *your* ears, for example, without warning, or presented as more alarming than it is.

Cousin Robert John (Aumann) happened to be in New York this month, visited me in hospital, and told a funny story. His son was moving house, and the movers wondered why he was constantly on the cell phone.

"I shan't tell you," [the son] said. Their curiosity grew. He refused again, saying, "You won't believe me."

Their curiosity went beyond all bounds. "Tell us!" they said. "We *will* believe you. We *promise* to believe you."

"Alright," he said. "My father got the Nobel Prize."

"We don't believe you," they said.

I have no clear English/European plans at the moment, but am sure I will come sometime this year; and, of course, I always love to see you.

Kate too sends her best—*she* has had back surgery, but is recuperating well.

Love,
Oliver

* That is, his mother's side of the family.

To Rita Levi-Montalcini

NEUROBIOLOGIST, NOBEL LAUREATE

FEBRUARY 14, 2006

2 HORATIO ST., NEW YORK

Dear Rita,

I was delighted to get your letter (of January 20), and the extraordinary chapters from your book *Without Oil, Against the Wind* (I am not clear as to whether this *has* been published in English yet—it should be, it must be, because it is among your most beautiful writing).

I was deeply affected by your portrait of Max Delbrück,* which is so alive, so poetic, so tender, so rich in insight, so deeply personal, and at the same time only a fellow-scientist, or perhaps only *you,* could have written it. It is really a masterpiece of loving yet a clear-sighted biography (and Delbrück would surely have felt this himself).

Bravery, stoicism, humor, calmness (or "raging against the dying of the light")—but especially noble reactions in the face of death and mortality—seems to me as high a subject for a book as any, for there can be few situations which test us more. Fermi died nobly, I have read, his unique self to the last, whereas von Neumann, at much the same time, died abjectly, superstitiously—though I think no-one can be taken to task for the way they die.

I enclose (I hope it can be enlarged for your eyes) the little masterpiece which Hume† wrote, in a single day, when he realized (p. 278) that his illness was "mortal and incurable." His calm in face of death infuriated Boswell and Johnson, who quivered with fears of dying (and visions of hell-fire). They expected him to cave in when he was in extremis, but he retained his calm, noble spirit to the last (this was in my mind when I last saw Crick, and he too died a sort of philosopher's death, thinking, never complaining, to the very last hours of his life). [. . .]

I now find myself with more than a sympathetic understanding of impaired vision, because in the last days of last year I had a sudden though partial loss of vision in my right eye—I thought it was

* Delbrück (1906–1981) was a pioneering molecular biologist.

† OS frequently quoted the philosopher David Hume, and in 2015 would write a short essay ("My Own Life") about Hume's thoughts on death.

a hemorrhage, or a retinal detachment, but it turned out to be an ocular melanoma, unfortunately very close to the fovea. It has been treated now with intense radiation—the eyeball was partly detached, and a plaque of radio-iodine attached to the sclera—it is reattached, and now the tumor will (hopefully) be killed, tho' perhaps at the cost of some post-radiation necrosis of the retinal cells too. Unfortunately my left eye was never nearly as good an eye as my right, but perhaps it can be "tuned up" a little to compensate for it. I certainly had a sudden sense of mortality when the word "melanoma" was pronounced, but now I am assured that ocular melanomas carry a much better prognosis—but, as with you, magnified print, or a magnifying glass, may now have to become a part of life. But meanwhile I am back to writing on my music book, though it continually expands beneath me, and I do not know when it will see the light of day (spring of 2007, perhaps).

I greatly enjoy and value our correspondence, and look forward to hearing from you again.

With warmest good wishes for the New Year,
Oliver

To Nick Younes

MAY 20, 2006
2 HORATIO ST., NEW YORK

Dear Nick,

Very good to get your fine long letter, your essay on "Kant & Poetry" (Kant and—so much else, really) and on *Phedre.* I don't know that I can respond adequately (whatever this means) to any of them, but I'll let thoughts come, for what they're worth. They will mostly be associative and tangential. [. . .]

Yr letter, and your thoughts on Kant, your gropings towards both "truth" and freedom, residing in Science/Reason/Nature, but free to take off into the worlds of poetry, art, aesthetics, ethics (I do not see the word "religion"), the *seriousness* of your probings and gropings give me a sense of my own levity, or irresponsibility, or laziness, or perhaps (as you suggest) "maturity." But they also bring back the Sturm & Drang, the intense philosophical/epistemological turmoil, which seized me in *my* twenty-second year. I had taken my degree at Oxford, seemed committed to a scientific/medical life, but felt that *none* of the questions that really mattered had been touched—and, unlike you, I was rather a loner, did not, as you apparently, have a group of passionately-seriously-minded fellow-students to talk with. Jonathan Miller, at Cambridge, did have such a group (the Apostles); and so did Maynard Keynes—do you know (if not, I will send it to you) his essay called "My Early Beliefs"? Anyhow, I would go for long runs on the towpath by the Cherwell every night, jump in, swim, run again, for hours and hours. And then I would write furiously, questingly, self-questioningly, in my Journal, sometimes till dawn. Meaning, meaning (I never think about "meaning" now). I did not have the philosophical sophistication and careful reading of texts that you have, but I certainly turned to philosophy—at least to certain philosophers—if not for "answers," at least for help formulating questions. Coleridge saw all philosophy ("philosophizing," as Wittgenstein would prefer to say) as "footnotes to Plato and Aristotle," all men as "Platonists or Aristotelians"—though part of the unrest and "antinomy" with you, me too, was being *both*. But my tastes were much more for Spinoza and Hume. Hume relates how when philosophical inquietudes and melancholies got too much for him, he would go out, dine with his friends, play backgammon, make merry—feel human again. I, not so socially inclined, would retreat (or at least moor myself in) the body, in physical activity, in a (perhaps desperate) running and swimming—running to? Running from? Now I am far too arthritic to run (but I can still swim, and *do*, sometimes, try to "swim away" intractable thoughts). I was not really up to Kant then (nor have I ever been, really)—partly because of the *severity* of his style. Not that he isn't good for great images and metaphors occasionally ("the crooked timber of humanity," etc)—and, as is so central, clearly, for you—he does define the reality, the possibility, of freedom, and thus of poetry, etc. etc. [. . .]

I have also, just this week—a lot of reading, come to think of it, my voracity not diminished by having to use a magnifying glass!—read a very hard-hitting (not to say militant) book by Sam Harris called *The End of Faith*. (Curiously, even before getting your essay) I had been calling it a Kantian "critique" of Faith; and a much more civil book by E. O. Wilson called *The Creation,* which takes the form of an extended letter to a Pastor. "Dear Pastor," he says (I am paraphrasing), "We grew up in the same little Southern town. We were altar-boys in the same evangelical church. But then—then I changed. But although we have fundamentally different world-views—yours religious, Biblical, mine secular, Darwinian—perhaps we can agree on some values, and some courses of action; that the diversity and richness of life is not only sublime but intensely important, intensely worth cherishing and conserving, otherwise it, and we, and living creation here will go to hell." So he is reaching across.

I fear (as I warned you) that I have probably just "taken off" in all sorts of self-indulgent directions. I should have said, at the start, that I am *very* impressed by your philosophizing. [. . .] Your current reaching out now, I think, is much more than "a stage," and may indeed represent, for you, a radical transformation in perspectives and action. [. . .]

Enough. It occurs to me, though you do not mention it, that your birthday is in May—and last year (if I have not lost count) was scandium.* So here, to celebrate 22, a sturdy screw of titanium from your old friend.

Oliver

PS love to all the family too.

PS Hope to see you in the summer. I am going to Peru (with Kate and family) in the second half of June, but otherwise have no special plans for the summer. My eye seems to have settled down, though its acuity and color-perception are somewhat compromised (I see fields of *white* dandelions with it, species I have never seen before), and I may need surgery for my bloody back.

PPS Forgive all the typos, I can't see what I'm doing.

* OS enjoyed sending his friends birthday presents, little pieces of elements whose atomic number corresponded with their age. Thus, when Younes turned twenty-one, OS sent him a piece of scandium, and for twenty-two, titanium.

To Emilio Presedo

PHYSICIAN

JULY 1, 2006

2 HORATIO ST., NEW YORK

Dear Emilio,

I have just got back from Peru, where I spent the last two weeks, extraordinary weeks in which I realized some of the dreams of sixty or more years ago: seeing The Lost City of the Incas, Machu Picchu; the mysterious Nazca lines—giant geometrical figures, animals, zoomorphs, sometimes kilometers in length, on the tops of mountains, and only discernible from the air—but no-one was flying when they were made nearly three thousand years ago; and Lake Titicaca, at 4000 meters the highest (navigable) lake in the world, where I had a delicious but decidedly *cold* swim (9° C). I have brought back various things with me—including, alas! amoebiasis—which will, hopefully, resolve with treatment in a few days. But as yet I feel quite depleted and weak, low potassium, etc. [. . .]

I will also add that while in Peru, I chewed *coca* leaves, and drank a fair amount of (delicious!) coca tea (mate de Coca). I did not note much in the way of *systemic* effects, but I had a striking enhancement of involuntary visual imagery whenever I closed my eyes, or in the dark. Nothing resembling teichopsia,* but all sorts of kaleidoscopically-changing geometrical patterns—lattices, carpet patterns, moire patterns etc, which would sometimes seem to enlarge and "zoom" closer (macroptically), sometimes the opposite, and also to rotate, as if observed from different angles. Also—and this I almost never get with a migraine, which for me, at least, has primarily rather low-level geometrical patterns ("Geometrical Specters") of the sort one gets with stimulation of primary visual cortex—also countless "scenes," faces, images, landscapes—sometimes beautiful, sometimes grotesque—rapidly-changing (at perhaps five times a second), and so difficult to remember (and far too quick to make notes on). [. . .] It seemed quite different from the imagery of hypnagogic—or [hypno] pompic states, which (for me, at least) always have a strongly dramatic

* A sort of shimmering or scintillating vision sometimes seen as part of a migraine aura.

and thematic quality, more akin to dreams. Also somewhat different from (though in other ways more similar to) the rapid play of images, patterns, colors, etc. one can experience with mescaline or LSD. With these I have much more impression of "the psyche" being involved—whereas with the coca I had more the feeling of a sort of unselective excitation or stimulation of visual cortex, and at levels of association cortex, color coding, etc, as well as primary. I am very struck by the *speed* of the perceptual changes—I do not know if I sent you my article called "Speed," and so I enclose this as well. One feels there should be clear-cut neural correlates for this sort of visual "delirium," but our current technologies (fMRI, PET, MEG, EEG) are not adequate to record neural phenomena of this order—hopefully they will be, in a few years. [. . .]

Enough for this letter—I must go and rest my eyes now, and also my poor amoeba-ridden body.

With best wishes,
Oliver

To David Lee

BOTANIST

JULY 29, 2006
2 HORATIO ST., NEW YORK

Dear David,

Thank you very much for your letter (of 7/20) and "Chlorophilia."*

I cannot give you any systematic reactions (this may be as well), partly because my mind is very much on another, and not even visual, subject: viz, music. But here are a few wandering thoughts.

PERSONALLY, as you might judge from my fondness for ferns, mosses, cycads etc. I too am attracted to green, feel it as older, more elemental (in plants) than other colors. I have just been in Peru, partly drawn by the great Inca and pre-Inca ruins etc, but also by rain- and cloud-forest, tropical vegetation. Visually I find it the most

* Lee had enclosed the final chapter of his forthcoming book on color in plants.

satisfying in the world (I have photos of cool, green ferny glades all over my apartment), tho' I am, unfortunately, very intolerant of tropical temperatures. And I have just been up-State, especially attracted to one road near Cooperstown which is one long arbour, a corridor of green. Alienness, as you say, is sometimes associated, in science-fiction, with other colors, e.g. the *red* vegetation of Mars in *War of the Worlds.* (I think this was in my mind when I lashed out in *Oaxaca Journal* against the Echeverias planted in front of the Cathedral, and spoke of an "uncanny, red-earthed Martian landscape.")*

Ed Wilson's "biophilia," as I understand it/him, is not just a love for, a being primordially drawn towards, living creatures, but also to living *landscapes,* landscapes of a sort our ancestors must have known (tho' as you bring out, whether 5 or 50 million years ago is an open question). Is this "hard-wired," you wonder—or a cultural acquisition? (With "Nurture through Nature," of course, one has to think of genes being switched on in certain circumstances—perhaps we get switched on, genetically, with our first sight of green.)

In general, in the plant world, one associates green with *health,* though when I was on Lake Titicaca I saw large areas, the most polluted, covered by an almost fluorescently-brilliant green algal slime—almost too green, it spelt danger! An unrelated association, image, comes to mind: the beautiful, moss-covered trees in the Southern island of New Zealand, which I adored.

We *are* hardwired to distinguish between the animate and the inanimate—though this is usually associated with the perception of *motion*—of autonomous, animal motion, as opposed to/distinct from movement of/by physical forces. But the recognition of plants is primordial too—I once saw some deeply demented patients, with advanced Alzheimer's disease, unable to use (or recognize) knives & forks, cutlery, etc. planting a garden—and never putting in plants upside down. Their recognition of plants had long outlasted their recognition of human cultural artifacts.

On healing, it is crucial to have windows, views of nature, landscapes, and gardens. I do not know whether green is a crucial feature here, I imagine it is, as part-and-parcel of the ensemble which

* OS may have meant to type "plantscape" here, as in the passage from *Oaxaca Journal* he is referring to.

we recognize as plant life, "Nature." (Your section on horticultural therapy etc is especially good and detailed.)

But I do not think we have anything corresponding to a phytochrome.* A counter-thought (or maybe it's not counter) comes to mind—the deep pleasure in their island landscape, their recognition of everything, by the totally colorblind people in Pingelap (see my *Island of the Colorblind* pp. 32–33).

Writing about the "universality" of music—all cultures, all ages, etc.—I find a certain embarrassment in thinking of the *deaf*, who cannot hear, and in a sense cannot imagine, music.† What then of the totally colorblind—the achromatopes—who cannot perceive, or imagine, green or any other color(s)? Perhaps they have some equivalent of "green." But the vast majority of us are NOT colorblind or deaf, so I think one may generalize without being too held up here.

Going back to "therapeutic landscapes," gardens etc. I think of my own experience as a patient (see *A Leg to Stand On*, paperback, p. 131): "A peculiar delight suffused the garden outside my window. There had been no real outside before, no daylight . . . no grass, no trees, no sense of life. Like a man parched, I gazed thirstily, yearningly, at the green quadrangle, only now realizing how cut off from life I had been, in my sterile, windowless, artificial cubicle. No picture sufficed. I had to *see* . . . the garden, . . . the real world . . . outside."

Where you speak of the destruction of natural landscapes, I think of visiting Sao Paulo, back in '96, and seeing not a single tree on the enormous drive from the airport to the city. Partly for this reason, I found it the ugliest, most frightening city I had ever seen (there is, as you probably know, a wonderful garden, an oasis of green *in* the city, but in general, and in Lima too, there *was not enough green*).

I see how you have brought together all sorts of strands—natural, ecological, physiological, cultural, personal, mythical, aesthetic, therapeutic etc. etc. in this, your final chapter.

It has elicited *associations* in me—but no coherent "critique." So I

* A type of photoreceptor found in plants, bacteria, and fungi.

† Not entirely true, as OS would no doubt agree. Some deaf people enjoy dancing—though they may not be able to hear melody, they can feel vibration. Other deaf people, like Evelyn Glennie, are percussionists.

look forward the more to seeing your *entire* manuscript, and not just the summation or synthesis of its final chapter.

With all good wishes,
Oliver

OS KEPT ALMOST ALL OF the letters he received, variously organized by topic or correspondent. Though he answered many immediately, others he might keep for years on end, until he had a relevant reply. As he began to work on *Musicophilia,* he consulted his files and resumed epistolary conversations begun years before, such as this one with Andrea Bandel, an Italian violinist.

To Andrea Bandel

VIOLINIST

NOVEMBER 13, 2006
2 HORATIO ST., NEW YORK

Dear Signor Andrea Bandel,

You wrote to me nine years ago about the problems affecting your left hand while playing the violin, and how these had eluded diagnosis, been unimproved by various interventions, and were threatening your career. I suggested then that you speak to my colleague Frank Wilson, who was especially interested in the problems of performing musicians—but I heard no more.

I am writing, basically, to ask how you are now—and whether you are aware of the important advances in understanding and treating such disorders which have emerged in the last few years.

The disorder now has a name—neurologists call it "task-specific focal dystonia"—and there have been various recommended treatments: in Europe various forms of retraining are often used, but here in the States the situation has been very hopefully changed by the introduction of botox (botulinum toxin), which, given in the right doses and the right way, can make the dystonia much milder, and (at least in some people) permit them to return to their musical careers.

You have probably heard of the great pianist Leon Fleisher; he was disabled, in his right hand, for nearly forty years by a dystonia—he could only play with the left hand, but now, since receiving (periodic) botox treatments he has recovered (almost) complete use of the right hand, and is again giving concerts with both hands.

I do not know how things are in Italy, but here in the States there is now great attention to these (dystonic) problems as they affect some performing musicians, and their treatment.

Do, if you receive this letter, let me know your current situation.

I will add that I have been writing a book about Music and the Brain and have a section on focal dystonia which ends with Leon Fleisher's story. But I have introduced the whole problem by quoting from your eloquent, detailed and brave letter of 16 July, 1997, which has always remained in my mind. I hope this is agreeable—I am happy to use either your own name, or a pseudonym, whichever you wish.

I send you my very best wishes, this message of hope, and look forward to hearing from you,

Oliver Sacks

*To Wendy**

COLLEGE STUDENT

NOVEMBER 14, 2006

2 HORATIO ST., NEW YORK

Dear Wendy,

I should have written before—I very much enjoyed meeting you, the gleaning together (how nice, how right, to meet in an *activity*, a shared activity), and talking together—but I have been totally occupied and

* During this period, OS visited a number of residential therapeutic communities for people with severe mental conditions like schizophrenic or bipolar disorder. Wendy (her last name is omitted for privacy) was a very engaging young woman whom he met at one such place. She had written to ask: "Why, if symptoms of untreated mental illness are so maladaptive (and too often socially stigmatized), has it survived

preoccupied with my book, and couldn't think of anything else. Now it is done!—was sent off to the publishers yesterday—I can start to catch up on everyone and everything else.

Having said this, I am not quite sure how I *can* answer your long, moving letter in any adequate way. I don't know, as a start, how much one can talk about genetic/evolutionary "mistakes," for while there is natural selection there is no "purpose" (and perhaps no "progress") in Nature. More to the point, things are very complex, and there can be unexpected spin-offs in both directions, all directions. Sickle-cell anemia is a fatal disease, but sickle-cell *trait* (to cite a famous example) confers resistance to malaria. (I want to avoid a dramatic way of putting this, like "Thousands die that millions may live"—for there is no "that" in Nature.) (Congenital) deafness was seen by Alexander Graham Bell in very negative terms—he felt that the deaf should not marry or be allowed to have children, lest they propagate an "inferior variety—of the human race," but many deaf people enjoy rich, full, creative and happy lives, often *enriched* by the fact that they are deaf, and that there is a vital language and heritage and culture of the Deaf, and they would not wish to be otherwise.

But you know all these arguments, and quote some of them in your letter. Also Kay Redfield Jamison's notion that significant numbers of creative artists have bipolar cycles, and may do their best work when hypomanic (not manic). The other side, of course, is the devastating grimness of (psychotic) depression, the urge to suicide etc (which Jamison has also written about very powerfully—and must know in herself).

What seems to me less stressed, and most in need of stressing, is that you are *an individual*—unique—with gifts and genes which no-one else in the world exactly duplicates—and that means you have a true place and role in evolution, and in the present. That you have bipolar disorder, in a sense *are* bipolar, does not begin to encompass the whole of you—it is a *what,* while you are a *you.* You have to hold to this sense of a personhood ("personality" is not quite the word, it has got too Hollywoodized)—Coleridge talked about "personeity,"

---

the evolutionary cutting-room floor, so to speak? Am I just a mistake that somehow survived evolution's ax?"

which is deeper than any "condition" you have, and perhaps these (relatively) gentle years at Gould Farm will allow you to realize this (realize it, in both senses—understand and actualize). You have much to hope and to live for. So, my best to you & keep in touch.

Kate too sends her warm regards—

Oliver

To V. S. Ramachandran

NEUROSCIENTIST

JUNE 26, 2007

2 HORATIO ST., NEW YORK

Dear Rama,

I hope when we meet in August, that we can also discuss another matter that has engaged your attention—and which now, through force of circumstance engages mine, namely some of the forms of "filling in" which occur with a (monocular) scotoma.

The melanoma in my right eye has been persistent (persistence rather than "regrowth" is the soothing word my surgeon uses), and has required further lasering. And because it now envelops the fovea, this is obliterating central vision. I have lost the top half of this: when I look at people [with the damaged right eye only], for example, they are only visible up to about mid-chest. I will probably have further lasering next month, which may obliterate this remaining part of central vision, and then I will only have some peripheral vision, a sort of crescent from about 2 to 7 on a clockface. [. . .]

I have been puzzled at the different appearances of the scotoma—it tends, initially, to present as an intensely black, irregular "ink-blot," roughly corresponding to the irregular [. . .] shape of the melanoma [. . .]. But if, say, I am looking at a colored surface—any color—the scotoma starts to change, to "match" its brightness and color so that, after a few seconds, it becomes almost invisible—so much so that I would have difficulty believing that it was there, and may have to intrude a hand or finger and see by its sudden "amputation" that the scotoma, tho' now color- and brightness-matched, is still present.

There is also—but this is slower, and may be less successful—a matching of texture. A fine texture, as of a wallpaper and some rugs, gets well-matched—especially a fine liney or pebbly texture. The textural filling-in tends to start at the periphery of the scotoma, like ice crystallizing at the edge of a pond in winter, and then moves in to the center. If I look at a page of print, a sort of "pseudoprint" will appear in the scotoma in fifteen seconds or so—I say "pseudoprint" because I cannot actually read it (curiously too it may sometimes be smaller, have a different "font-size," or be topographically squashed). But with a gross chequer-board pattern, with a much coarser spatial frequency, the simulation or illusion (whatever word one should use) is much less successful. (I cannot help comparing myself to your flounders, or Roger Hanlon's cuttlefish, camouflaging themselves against different surfaces; I cannot avoid the feeling of *statistical* evaluations being made—especially with random, "pebbly" surfaces—perhaps distinct from interpolations or extrapolations with more predictable patterns.) With the letters there is often a "ghostly" or "smokey" or "semi-transparent" quality to the illusion, whereas the fine pebbly patterns are scarcely distinguishable from perceptual reality.

Another phenomenon, which seems to be quite different in quality (and presumably mechanism) from this slow, limited, gradual "filling-in" is an extraordinary heightening and perseveration (if this is the right word) of visual imagery—again forcing itself on my attention because my visual imagery (other than hypnopompic/gogic) has usually been so feeble.

Thus, walking yesterday with Ralph along 8th Avenue, I saw two men in white shirts walking towards us. Closing my eyes for a few seconds (I dared not do so for longer!), I could "see" them with almost undiminished intensity, *continuing to walk towards us*. So this was not an "after-image" in the usual sense—e.g. a brief continuation, positive or negative, such as one might have with a lighted filament, an illuminated figure, etc. It was like, and I do not know whether to use terms like "perseveration" or "computational," but what they were doing, walking towards us, vividly persisted. It was similar with traffic, including traffic moving in both directions. This sort of imagery tended to fade, or diffuse, or lose its sharpness, after a few seconds—but the brilliance and verisimilitude was so striking that sometimes I wondered if I had actually closed my eyes. [. . .] I also find myself thinking in terms of a perceptual "cinema," with the movie going on even though there is now no visual input. [. . .]

I regret that my retina (and my brain) should be the seat of all this—but since it *is* occurring to me, I feel I should try and explore it a bit. [. . .] I hope we can discuss some of these phenomena—I think you probably have extensive knowledge (and all sorts of thoughts) here, and I would be grateful if you could send me your (early) paper(s) with Richard Gregory, and anything you may have published on the subject.

I hope you are well, have a grand return-visit to India . . . and to see you in August.

Best,
Oliver

To Orrin Devinsky

NEUROLOGIST*

JULY 11, 2007
2 HORATIO ST., NEW YORK

Dear Orrin,

[. . .] I have received a very full narrative of B's life from her sister, P, who accompanies her today (along with Tabitha, who works at Randall House and has known B there for the last fifteen years).† [. . .] I have [past records of her consultations and tests], none of which show any progressive pathology.

The overall history is as follows: B was grossly premature, weighing only a little more than 2 lbs. at birth. Her first weeks were spent in an incubator, and she became blind (from retrolental fibroplasia) in consequence of the high oxygen levels to which she was exposed. When she was two and a half, she showed remarkable musical

* From the late 1990s onward, over many sushi dinners and long bike rides, Devinsky and OS formed a deep personal and professional relationship. Devinsky shared OS's interests in antiquarian books and the history of neurology, as well as much else. They consulted together on quite a few patients.

† Names have been abbreviated and certain details changed for privacy.

gifts—"savant" talents. She had shown behavioral problems prior to this time: would scream, hurl objects, bang her head, could not be calmed, *except when music was played* (she would explode within seconds of a record finishing, and her mother had to get a special phonograph which moved from one record to another with scarcely any delay). Thus (in an account from her mother), "one day I was at the piano. I was playing some nursery songs and she was on my lap. She pushed my left hand away and she filled in the left hand. She didn't want any help."

Although lessons were attempted (her sister writes), not much formal progress was made due to her behavioral difficulties. Her piano playing was not merely repetitive, it could be creative. She was able to play songs she hadn't heard played in a particular style in the style of other artists she had heard, to transpose pieces into different keys, and also to create her own songs if asked. She could play one song with one hand and another with the other. Of course, having perfect pitch, she could name any note played.

In an existence often sadly restricted by her limited intelligence (estimated IQ around 70) and behavioral difficulties—impulsiveness, violence, self-destructiveness (and, later, seizures), B's musical ability, and her prowess at the piano provided an oasis, an island of stability and pleasure, in her life. She enjoyed playing for herself, she enjoyed playing for others (she sometimes gave recitals), and she enjoyed playing with others—especially with her sisters: P, who played the guitar, and her younger sister J, who played the flute, though they were not as gifted as she was.

She also showed other savant talents at a very early age, being able to name the day of any date in the past (or next) two centuries, and possessing an apparently limitless memory for dates and what happened on them. This too was a source of pride, and sometimes amusement, for her—and, along with music, gave her a special place, a special role, in Randall House ("a Home to 15 men and women who face the dual challenges of visual impairment and mental retardation or developmental disability").

Despite her modest intelligence, B learned Braille, and became fond of reading both books and newspapers. When I asked her what she most liked, she said biographies. I asked if she had read a biography of Helen Keller. "Not a biography," she said. "Her *auto*biography."

(This made me wonder whether she had greater intellectual capacities than her IQ tests might have suggested.) She is right handed, and reads Braille with her right index finger. If this is cut or hurt, she will continue to use this finger if she can, for she cannot read with any other finger. (I mentioned here that the reading finger of Braille readers had an especially large representation in the brain—B seemed to understand this, at least the general import of this finger being special, and nodded.)

There seemed no special physical problems outside B's blindness and hearing impairments (this came partly from having had mastoiditis as a child, and partly from a cholesteatoma she developed as an adult). Her left ear was the functional one, and could be brought up to par with a hearing aid. B enjoyed the garden at Randall House, enjoyed outings (with someone) in the City, and enjoyed visits from and to her relatives, who were devoted to her, as her two sisters and elderly mother (who is in her ninetieth year) remain.

Her closest relationship, however, was with her father—he seemed to understand her, and to be able to joke with her, to bring out her lighter side, better than anyone else. And it was when he died [three years ago] that B descended into a tormented and obsessed state which she has been in ever since. At the same time two fellow residents of Randall House, whom she was friends with, and had known for decades, also died. And this triple bereavement pushed her overboard, or so it seemed.

Prior to this time the chief problem (outside her "behavior problems") were seizures—petit mal (which had first occurred in infancy) and then grand mal attacks (from about the age of five). Various drugs were used, and it was soon found that mysoline (primidone) seemed to provide optimal seizure control with a minimum of adverse effects. In 1987 she was taken off this, as a trial, and immediately had a return of seizures. In one grand mal attack, she fell down an entire flight of stairs and received severe head and other injuries.

She is now on other medications in addition to mysoline—seroquil, depakote, and trileptal, a complex situation which I will leave to [you] to advise on. She has been seizure free, clinically, for several months, although there is some spiking etc. in EEG.

In the past two years postural and motor problems have developed associated with a severe (and at times painful) scoliosis—this was very marked today. [. . .]

Following her father's death, and other deaths, at the end of 2006 B's state showed a striking and unprecedented deterioration. She became suspicious, agitated, frightened. Fearful that her sisters or mother would die, or had died, she kept making phone calls to "check" on this. Feared the cat would die. Feared other residents at the Home would die. Feared that she herself would die. These fears assumed delusional intensity and "rationalization." When she had a D and C (for minor problems) in 2005, she told her sister that without this she would have died. She fears she may have a grand mal seizure, an unstoppable series of grand mal seizures (as perhaps happened, with drug withdrawal, in 1987), though she is, in fact, seizure free at this time. She lives in mortal fear of a sudden heart attack—her father died of a heart attack in 2004.

She thinks that another resident at the home, Lisa, follows her around (she felt that Lisa had made her way to [my office] today). Lisa's special way of tormenting her is to play music nonstop in her room—nonstop, B clarified, meant 24 hours a day, for 365 (or, if called for, 366) days a year. I asked what music she played. B says that it is just two pieces, endlessly repeated: "The Heart Attack Song" and "The Grand Mal Song." She has occasionally hummed these; both, her sister says, are "atonal and disturbing," with no resemblance to classical music or the popular songs B used to play and enjoy. These hallucinatory songs are fraught with horror and threat. They seem addressed to her: they are "telling" her that grand mal seizures, or a heart attack, are imminent, may come at any time.

In addition to these songs, which she feels are "aimed" at her, B has been subjected, in the past 2–3 years, to a range of invariably negative suggestions and commands from other people. She will not allow that these are "voices"—indeed got angry when I used the term. She says they are "people," and that this is what they said (though, when pressed, she cannot say exactly when or where or how they said such things to her). These ostensibly well-meaning (but in reality malicious) "people"—physicians, others in the Home, perhaps fellow residents—have intimated to her that she is a "major epileptic," and that as a "major epileptic" she cannot, and must not, play the piano.

They have also forbidden her to exercise her calendar or calculating talents. They have forbidden her to take exercise or have physiotherapy.

In response to them B has ceased to play the piano, her great solace, her one real pleasure and talent, in an always narrow and dependent life; and they have forbidden everything else which would give her any pleasure or any relief. They are also preventing her from taking the needed physical care she requires for her worsening scoliosis and pain.

It is partly in response to them that she has ceased to use her hearing aids—she feels these may provide a portal for voices and "bad" music; but taking them out also reduces her possibilities of real interaction with others, and she was always, despite her tantrums and frustrations, a sociable creature who enjoyed chatting and hearing about other people's lives. [. . .]

When, after formal neurological examination, I took B to the piano, she became very anxious, and refused to play anything. [. . .] She kept voicing the fear that "something"—a heart attack? A grand mal seizure? might happen if she played anything. She said, "It's too risky," but she was also tempted, and wondered if she dared "take a chance." [. . .] When I asked her about favorite composers, she immediately said, "Bach, Beethoven, Brahms." Any more B's, I enquired? She added Bizet and Berlioz (but not Bellini or Britten). She then asked me what I liked, and I said Chopin. She asked if I played any of his études. (Her fingers were moving when she asked me, as if they itched to play an étude themselves.)

Later when I spoke, by mistake, into her right ear, which is deaf, and then said that I had right/left confusion, and had to feel which side my heart was on, I elaborated this, and told her how, as a medical student, I was asked to examine a patient and got frantic because I did not hear his heart on his left side. And how he saved me by whispering, "Listen on the right side, Doc—I've got dextrocardia." B laughed at this, showing her "light" side, her nice sense of humor. This was, in fact, the only "easy" moment during the time we met.

In clinical terms, one would have to say that this unfortunate patient has a psychotic depression, with great anxiety, phobias, nihilistic delusions, musical (and perhaps vocal) hallucinations, etc. A psychosis of a particularly vicious and malignant kind, one which makes life almost unbearable for her. Dr. D. wondered whether this psychiatric symptomology, while originating from loss of a beloved figure, indeed the simultaneous loss of three important figures, might "ultimately have an organic basis." A 5-day video EEG [. . .] showed no correlation of psychotic muttering or hallucinations etc. with epileptic activity (this does not wholly exclude a more general connection).

With regard to the musical hallucinations, these are clearly "psychotized," and yet one has to wonder whether they might have an organic kernel in her deafness and auditory deafferentation—the more so as B has now renounced or refused the use of hearing aids. (But she will not easily be persuaded to use them again, and might take any pressure to do so amiss.)

There have been trials of antipsychotic agents—these have not been effective so far, and may have produced some dyskinetic effects. [. . .]

I cannot help wondering whether shock therapy might be useful here. I have memories of one or two patients with severe obsessive-psychotic depressions who were recalcitrant to all medication but greatly helped by shock therapy. But this is just a thought—I must leave it to my colleagues to weigh the matter.

I do hope this poor tormented woman, so undone by loss and bereavement, can be helped to return to a happier life.

Oliver Sacks MD, FRCP*

To Georg Klein

MICROBIOLOGIST

SEPTEMBER 12, 2007
2 HORATIO ST., NEW YORK

Dear Georg,

I was delighted to get your letter. On your question, first, about Bartok (et al.) having Asperger's: I suspect that there is very little solid

* OS sometimes used the honorary title Fellowship of the Royal College of Physicians (recently bestowed), especially when writing to other neurologists in a professional context.

evidence in any of these cases (the evidence, of course, would have to be historical and anecdotal, because none of them could be "tested"), and that this posthumous diagnosing comes partly from people *with* Asperger's who are hungry for eminent exemplars, role models, etc. (It is similar with regard to Tourette's syndrome. Mozart, for example, is considered to have Tourette's, maybe autism too.) [. . .]

On Seymour Benzer,* I found your notes riveting, and kept pulling out my marker to highlight particular things. I am agog to see your essay on him—will it be translated into English (or, rather, *when* will it be?). I am happy to think I may have introduced the two of you at that conference in Berkeley. [. . .] In regard to your specific question, I came away from an afternoon in Benzer's lab feeling that insects were my "brothers"—that huge insights in (the genetics and molecular biology of *human*) behaviour could come from our little brothers, with their "100,000 transistor" brains (I also got this feeling, in relation to the "psychology" of insects when I visited Ralph Greenspan, at Neurosciences). And I get it from a lovely little Robert Frost poem—do you know it? called "A Considerable Speck." The conservation of genes (not only throughout the animal kingdom, but *all* the kingdoms of life) seems to me one of the wonderful and startling (and yet, when one broods on it, *right* and inevitable) realizations of these last decades. I have especially thought of this in relation to conservation of a handful of visual genes, determining eyespots in protozoa, chloroplasts in plants, eyes of fascinatingly-different but "parallel" construction in seven or eight animal phyla. I have some of the same feeling with regard to Kandel's wonderful studies of *Aplysia* (which you mention in your Benzer talk).

Speaking of Kandel, I have a new position at Columbia—a chair in clinical neurology & psychiatry, as well as being a "Columbia artist"† (whatever this will mean). I have become close to Kandel, hope to see a lot more of him (we even saw a patient together), and of course Richard Axel is there as well. How this will work out, I don't know—I have been a maverick, a nomad, a freelance, whatever, for the last

* Benzer (1921–2007) was a molecular biologist who studied behavioral genetics, often in fruit flies, genus *Drosophila*.

† Columbia University created this title for OS, to give him a joint appointment at the medical school and their undergraduate campus.

forty years, and entering Academia in my 75th year seems a little daunting. But I think it could be a very rich and exciting new chapter of life—and one needs such feelings as one faces the losses and "lessnesses" which go with age.

With regard to my eye—since you ask about it—the situation is complex, but on the whole reassuring. Areas of melanoma have persisted despite radiation, but can be extinguished (it seems) by being lasered, cauterized, though this, of course, destroys the areas of retina overlying them. Unfortunately the tumor is close to enveloping the fovea, and so with lasering I have now lost a good deal of my central vision in this eye (and may lose it all). But other than the loss of stereoscopy this seems to matter remarkably little in functional terms—I can still drive, bicycle, walk, swim, read, write etc (and could even if the eye had to be enucleated). I have tried to suggest a "deal" with the melanoma: "You can have my eye, but leave the rest of me alone"—and can only hope that it will abide by this.

This has not stopped me working—and I have a new book (*Musicophilia: Tales of Music and the Brain*) coming out next month—I will send you a copy. [. . .] I hope that you—both of you—continue in good health and spirits . . . and SWIMMING DAILY.

Kate joins me in sending our love,
Oliver

To Henry Nicholls

SCIENCE WRITER

SEPTEMBER 24, 2007
2 HORATIO ST., NEW YORK

Dear Dr. Nicholls,

Many thanks for your letter—I am glad you will be writing about Chemistry sets,* tho' I fear it may be a lost cause.

1) I think that a personal *experience* of chemical reactions is crucial

* Nicholl's article "The Chemistry Set Generation" was published in *Chemistry World.*

to get young children interested. The first experience might be (say) fireworks, on Guy Fawkes night, or seeing one's mother remove a red-wine stain with salt, or bleaching—but then *one has to experiment for oneself.* I do not think that there can be any adequate substitute for having a chemistry set (or a little chemistry lab), and doing experiments oneself—thinking them out, taking responsibility for them—and (occasionally) facing risks too. Failing an *active* experience like this, the next best are *demonstrations* by a teacher or lecturer (as Humphry Davy used to do, and as are, perhaps, still done—are they?—in the Royal Institution, and in schools everywhere). Films (like the lovely—and very funny—little RSC film on putting bits of alkali metals, Li to Cs, in water) have their uses too; but none of these can be a full substitute for planning, doing, reflecting on, *real* experiments of one's own.

2) Not in America, where there is a sort of Nursery atmosphere, and a hysteria about risks, insurance, etc. Maybe in England. It may be (this overlaps with the first point) that a *virtual* chemistry set could yield some excitements, develop skills, and, conceivably, a "nano-chemistry" set, where one is using milligrams or micrograms and not grams. [. . .]

But if chemistry sets (or something equivalent) cannot come back, a certain realm of childhood, youthful fun, joy, thought, responsibility, etc may be lost for ever.

I may, of course, just be an old fogey consumed with nostalgia.

With best wishes—and do send me your piece!—
Oliver Sacks

PS Forgive typing.

To Neil Shubin

PALEONTOLOGIST, AUTHOR OF *YOUR INNER FISH*

OCTOBER 31, 2007
2 HORATIO ST., NEW YORK

Dear Dr. Shubin,

I was thrilled when I read of the discovery of *Tiktaalik** a couple of years ago, and now (reading the proof copy of *Your Inner Fish,* which Dan Frank sent to me) I find it even more fascinating—and moving—as I read the full story, and how it fits into your life and work. Indeed I found *every* chapter of the book riveting—and am sending off a "blurb," tho' I fear it may not do justice to one of the most exciting books I have read this year. (Another, on the botanical side, was *The Emerald Planet.*)

My mother was a surgeon and comparative anatomist (her teacher was Wood Jones), and drummed into me, and into her students, that our own anatomy is unintelligible without a knowledge of the origins, and "precursors"—and how it becomes infinitely fascinating *with* such knowledge. I think the medical students in Chicago are very lucky to have you as a teacher, illuminating this for them.

Tho' my own life is a clinical one now, I have never left my early interest in palaeontology—and a few years ago, on a visit to Australia, I went with Alex Ritchie on a "fossicking" expedition to the Devonian billabong of Canowindra—he, of course, was hoping for such a discovery as you and your colleagues were to make in Greenland—

I was intrigued by your remarks on hiccups (pp 186–189), and thought I would find reference here to what neurologists (perhaps mistakenly) call "branchial myoclonus." (I once wrote a sentence or two about this in my *Island* book—see enclosed.)

Anyhow congratulations on conveying so vividly (and engagingly) a marvellous scientific adventure-story.

With kind regards,
Oliver Sacks

* Fossil evidence of this long-extinct fish was discovered on Ellesmere Island by Shubin and his colleagues in 2004. Tiktaalik has various features (such as the primordial beginnings of four limbs and wrist joints) that make it a missing link in the transition from fish to land dwellers.

To Orrin Devinsky

DECEMBER 6, 2007
2 HORATIO ST., NEW YORK

Dear Orrin,

I very much enjoyed my quiet Chanukah dinner with you all—thanks so much for letting me join you (I think Deborah does sprouts particularly deliciously!). The train back was uneventful—I fell into one of those strange half sleeps, both dreaming, and yet (at least intermittently) aware of everything—a sort of "doubled" consciousness.

Tho' Deborah said it was sometimes too much, your girls' adolescence is fascinating to an outsider—they seem to me, each time I see them, more and more individuated as young adults. I guess that's what adolescence *is*.

Their showing me the dancing figures with *your* face, on the computer, and seeing all the images of *my* face, played a part in an odd dream I had this morning—when I seemed to be in a sort of gaudy gallery, with "pop-out" versions of my face all round me (one would pop out of the wall, then disappear, and then another, and another). And then I, apparently introducing "the Sacks show," as I understood it to be, announced *you:* "that epilepsy Knight, Fleischman of the 5:30" (at this point, my alarm went off and terminated the dream). So—you are a "knight" of epilepsy, a valiant warrior battling the powers of ictal darkness and confusion. The "5:30" was the train we took, *your* train . . . but why "Fleischman"? My associations are with Flashman and Flash Gordon . . . and I had a phonecall yesterday from De Niro, in which he spoke of seeing Leon Fleisher at the Kennedy Center.

I am being too prolix, but it was a *fun* dream, and intriguing in its bringing together (and transforming) some of (what Freud called) "the residues of the day."

I enclose the two letters [from prospective patients] I meant to give you last night—don't know whether they are of any interest to you. Thank you for your two consults. J.B. sounds fascinating, and articulate, and colorful—even if she does have a little dementia. It sounds as if she may have a combination of peripheral and central factors, as with Z. I would like to see her. [. . .]

I hope my scrawl is not too difficult to read!

Thanks,
Oliver

PS: I hope we can set up my bike on a stand, and I think I have (after your demonstration) to yield to an iPod, and perhaps combine it with cycling indoors.

To Bill Hayes

WRITER, AUTHOR OF *THE ANATOMIST*

JANUARY 17, 2008
2 HORATIO ST., NEW YORK

Dear Mr. Hayes,

Thank you very much for your note.

I *did* read *The Anatomist* in proof, with great enjoyment and interest (my mother was an anatomist, and the first review I ever wrote, when I was eighteen, was a sort of essay on a just-published new edition of *Gray's Anatomy*). I am very sorry I did not provide a "blurb"—I meant to, but it was a distracted time with interviews etc relating to my own new book,* and I failed to do it, to do all sorts of needed things at the time.

But I am delighted by the advance praise the book *did* get, and the lovely review in last Sunday's *NY Times*—all richly deserved. And I would be delighted to meet you when you are in NY next month.

My best,
Oliver

PS: I enjoyed yr. previous books too.

* *Musicophilia.*

PPS: And I gave a proof of *The Anatomist* to my eye-surgeon, who has contributed 2 chapters on ocular anatomy etc. for the latest edition of Gray.

A FEW WEEKS LATER, on a trip to New York with a view to finding a job and moving there, Hayes met OS for a meal. Once back in San Francisco, Hayes sent OS drafts of a few essays he was working on. OS's response to these, below, was never sent—perhaps he felt it was too personal since they had only recently met.

To Bill Hayes

FEBRUARY 29, 2008 [NOT SENT]

2 HORATIO ST., NEW YORK

Dear Bill(y),

(I always find the informal variants more difficult to use—everyone called Stephen Jay Gould "Steve," I called him "Stephen.")

I was very moved by your letter and the two pieces—the "gym" one more than the London one (but this may just be that I read it first, and had got saturated with sympathetic grief myself, and could not take any more).* Yes, writing in your own way, as time and occasion prompt, and without any hurry, or planning, or system, I think that your book may turn into a collection of such pieces, a sort of album of grief and loss, as they thrust up suddenly, in different ways and places.

I have not read Bowlby's volumes for a long time, but they may well be the deepest study of these themes. (I seem too to recollect a book by Geoffrey Gorer?? called *Death, Grief and Mourning* . . . or perhaps I'm mistaken.) When Darwin lost his mother at the age of eight, he did not mourn, none of the family mourned, no mention of loss or expression of grief was "permitted." And years later, when

* The "gym" piece explored Hayes's grief after his longtime partner's sudden death.

a close friend suffered a grievous loss, Darwin wrote a remarkable letter of sympathy, adding that he, however, had never known any (such) losses himself. On the other hand, the death of his favorite child darkened the rest of his life (and, besides much else, destroyed what vestiges of religious faith he still had).

I have spent much of today with a friend who has also suffered a grievous loss (tho' not as grievous as yours)—the departure of his wife, who suddenly decided to leave him after fifteen years of marriage. She took off, finally, yesterday, and Peter's voice, when I spoke to him then, was sepulchral. But I was able, today, to persuade him to come to a mineral show, with friends; to have a swim; and to go to a reunion party (of our swim/triathlete group, now coachless).[*] So, he was distracted for eight hours—which was exactly what was needed. What I am saying, I guess, is that I hope *you* permit yourself distraction—dinners with friends (yes, you do speak of this), going to movies, whatever—distractions *as well as* consolations.

At some level, I suppose, I cannot really feel what you (or Peter) feel—for I have never had a relationship of fifteen or seventeen years; or, for that matter, any "relationships" at all. This, alas!, is a "forbidden" area for me—although I am entirely sympathetic to ~~(indeed wistful and perhaps envious about)~~[†] other people's relationships. This used to be a grief for me, but now solitude is so habitual that I scarcely mind it any more—and I do have, if not "relationships," many friends to warm my life.

Before I forget I should thank you for the lovely photos as well, though my association is a neural rather than a vascular one—I think of how Nabokov compared trees in winter to the nervous systems of giants. Formally, I suppose, there is little to choose between vascular & neural arborizations.

I enjoyed meeting you too—look forward to seeing you again.

All my best,
Oliver

* OS had joined this group in his sixties, and spent time with them each winter in Curaçao, doing open-water swimming. Their coach, Doug Stern, had recently died.

† Lightly crossed out in the original.

HAYES DECIDED TO MOVE TO New York a year later, and over time their relationship grew into a partnership that would last the rest of OS's life.

To Christopher Payne

PHOTOGRAPHER

SEPTEMBER 3, 2008

2 HORATIO ST., NEW YORK

Dear Chris,

I have just read your own really beautiful Introduction to *Asylum**—indeed it is so good that I am now even more uncertain as to whether *I* have anything useful to say, tho' I was especially struck by your paragraph about speaking to former employees—and how "through recollections filled with pride, humor and a sense of nostalgia, the empty buildings and vacant grounds came alive" (also, in the same para, how they felt the end of "patient labor" was so damaging to patients).

I have also read, in the last few days—was it you who recommended these?—*The Discovery of the Asylum* (Rothman), *The Architecture of Madness* (Yanni), and (most movingly for me) *The Lives They Left Behind: Suitcases from a State Hospital Attic* (Penney & Stastny).

But this read and said—what do *I* have to say?

I did spend 25 years working at Bronx State Hospital (1966–1990)—tho' this, unlike the great State Hospitals you photographed, was built only in 1963—and designed to be a short-stay hospital not a residential or custodial one (but, of course, it had "chronic" wards, where patients from Pilgrim and elsewhere were transferred). I did spend a year on one (hideous) ward—but for the most part worked in the Clinic, and did not have that much to do with the Wards. It was clear, however, that they varied greatly in character—morally, in all respects: some were full of life and interaction and community

* *Asylum: Inside the Closed World of State Mental Hospitals* is Payne's photographic history of the monumental but now abandoned and decaying state hospitals that had once been a source of civic pride throughout the United States.

and good care and doctoring—others (like the hideous Ward 23) the opposite.

My more intensive and intimate experiences, however, have been with Institutions other than State Hospitals—but with obvious analogies to these (and equally obvious differences), e.g. Beth Abraham Hospital in the Bronx (originally "Beth Abraham Home for the Incurable," possibly even "BA Home for the Crippled and Dying")—where I worked from 1966 to 2006. It was here, in 1969, that "Awakenings" occurred. BAH has not been pulled down, but its character has changed a good deal—it is much less a Home for patients with chronic/congenital neurological disease, as it was originally, and as it was for my first thirty years there, for now there has been a "de-institutionalization" of such patients (partly for budgetary reasons)—and the place is largely filled now with drug-addicts, and people for short-term rehabilitation after head injuries etc. I was VERY struck by the adverse effect, in BAH, of closing the Sheltered Workshop (because of "patient labor" laws)—for this had a special, happy, productive, social spirit, where patients could forget they were "patients," and just become fellow-workers, and companions. I was very struck too by the closing of "the old wards," the 32-bedded wards, which (amazingly) combined a sense of community with one of inviolable privacy (whether patients chose to close the curtains about their beds). When these were dissolved, in 1967, and replaced by 1-, 2-bedded rooms, much community was lost, and a new devastating sense of isolation and loneliness entered.

I have also worked with the Little Sisters of the Poor for more than thirty years—their three Homes in New York for the Elderly and the Poor. A majority of their residents have some of the disabilities which tend to go with age—blindness, deafness, strokes, dementia etc (tho' others are very intact, and have entered in order to have company and community . . . and [perhaps] a religious structure). The Little Sisters Homes, in the States, and elsewhere (the Order was founded in France, rapidly spread to England—Dickens wrote about them in the 1840's) originally had something akin to the Victorian architecture and great spaces of older State Hospitals—tho' these fine, high-ceilinged buildings have been largely abandoned now (as being "fire-traps," etc, too expensive to renovate, easier to demolish).

In England I would visit the Highlands Hospital—built in the 1880's as a fever hospital, covering many acres, and with low, isolated huts (as well as a main building) for hundreds, even thousands, of patients with infectious diseases. It was turned over for postencephalitic patients in the 1920's (when I visited it with De Niro in 1989–90—he was "studying" then to "be" a postencephalitic—there were only a handful of patients left in the once overflowing Wards, and the place was shut down three years later). And my (schizophrenic) brother was often a patient at Colney Hatch/Friern Barnet Hospital—the largest "Lunatic Asylum" in England (with more than 3000 patients at one point)—this too was closed in the 1990's. I used to visit him there.

Not to forget—Caenwood/Kenwood—the beautiful Convalescent Home on Hampstead Heath, in London, where I spent six weeks as a patient in 1974 (and which I wrote about in *A Leg to Stand On*). This seemed to me an "asylum" in the very best sense of the word—very different from the sense of "Total Institution" (used by Goffman, Rothman etc). Given the stereotypes of snake-pit, etc. one wants to bring out—as did some of the employees you spoke to—the *good* (and crucial) aspects of Asylums, as well as the dreadful ones.

I am rambling—and this is the point. That I am not sure how much my own scattered, rambling reminiscences can be cohered into a Foreword for your book (tho' I started having a go at writing one, and will continue)—especially as so much of my experience has been in Asylums for the Poor and Elderly, the Neurologically Disabled, the Convalescent, etc. rather than Lunatic Asylums. But give me your thoughts!*

Again, I am continually taken aback by the beauty and eloquence of your pictures . . . and now your Introduction matches these.

Best,
Oliver

* OS did write a foreword, which also appeared as an essay in *The New York Review of Books*, "The Lost Virtues of the Asylum."

To Robert John Aumann

COUSIN, MATHEMATICIAN

SEPTEMBER 11, 2008

2 HORATIO ST., NEW YORK

Dear Robert John,

Many thanks for your postcard. I am glad you liked my little piece on Darwin,* and I instantly accessed your (quite wonderful!) interview with Sergiu Hart,† as you suggested.

It gives such a rich picture of your intellectual life and trajectory (trajectories)—and has set me thinking in various directions, especially about games-theory in relation to natural selection (it seems obvious once this is mentioned, both being optimizing devices/strategies), the coordination of neurons, etc. It has made me rethink the whole notion of "rationality," and the unifying power of a Center for rationality.‡ And there is an equally rich picture of your emotional and family and social and collegial life—complemented by the splendid photos of you and colleagues, you and family, etc.

And specifically—this, perhaps, was a main reason for your suggesting it—the interview is fascinating—and moving—for your exploration of the questions "You are a deeply religious man. How does it fit in with a rational view of the world? How do you fit together science and religion?"

This is something I have always been tempted to bring up with you, but always steered away from.

I often worry on this score, for (especially here, in the States) there are passionate attempts to replace natural selection/evolution by Creationism and "Intelligent Design" etc, to replace natural models of reality by supernatural ones. This is bound to lead to conflict and contradiction—and worse. What you bring out beautifully, and with all sorts of examples (intellectual, moral, aesthetic etc) is that a rational religion does not seek models of reality, but has quite other concerns—and in this sense is "orthogonal" to science.

---

* "Darwin and the Meaning of Flowers."

† Hart, a student of Aumann's at the Hebrew University in Jerusalem, published this interview in the journal *Macroeconomic Dynamics.*

‡ Aumann, a Nobel laureate for his work in game theory, was a founder of the Center for Rationality at the Hebrew University.

How, even in a purely aesthetic sphere (tho' is anything "purely" aesthetic?), like playing the piano (which I have started doing for hours a day, since engaging a piano teacher, my first in more than sixty years), science cannot say all that much, or needs to be very modest and circumspect (I worry about the grandiosity of "neuroesthetics," "neuroethics," etc) and about the over-inclusiveness of E. O. Wilson's "consilience," which (it seems to me) aims to scientize everything in human life, everything in the world. (Perhaps this is a perennial temptation for some scientists, not excluding very great ones.)

The notion of "self-commitment" seems central to you—but this is quite different from a "categorical imperative," because it depends on choice, freedom of will (and this, as you say, God leaves to us). I especially loved your descriptions of the Sabbath, observing it, feeling it, in its fullest sense—not just a legalistic observation, but one which can only spring from a true religious feeling.

I fear I do not feel this myself, perhaps because in the mildly observant family in which I grew up, Shabbos was relatively dilute, somewhat attenuated; but I see how it could be felt—(and perhaps was felt by you, in a much more orthodox family, from the start)—and it makes me wistful.

I have to think on these matters, which I have not really done—I hesitate to write, but feel that a letter like this, rambling and disconnected as it is, is better than none.

With warmest good wishes,
Love,
Oliver

PS The next time you are in New York I hope I can bring my friend Ralph Siegel to dinner (you met him, briefly, at a birthday of mine). Ralph is a very good neuroscientist, but has also become quite orthodox (and Talmud-loving) in the last few years—and this he attributes to an Erev Shabbat he spent with my brother Marcus and his family in Sydney. For him, then, the sense of the Sabbath's beauty, and specialness, served as a path to religion.

To John Horgan

SCIENCE JOURNALIST, AUTHOR
OF *THE END OF SCIENCE*

DECEMBER 8, 2008
2 HORATIO ST., NEW YORK

Dear John,

Thanks for your Dec. 5 email. I enjoyed the event and think we made a good "team"!*

On the matter of "stage fright" I am somewhat ignorant, despite (or because) I suffer from it myself—perhaps less than I did, but nonetheless significantly. That is to say, I find my heart is racing a bit, my palms sweat, and my fingers and toes get icy. But within a minute or two of actually facing the audience, I (nearly always) feel different—and start to enjoy myself. Though people sometimes recommend beta-blockers [. . .] for the anxiety and the autonomic effects, I have never taken these—partly because I feel that this sort of tension, unpleasant though it is, is (for me, at least) a prerequisite of performing well. I do, however, need to be alone, or with a supportive or congenial presence, for half an hour or so before any talk or "performance." I can't bear moving straight from a social situation to a performance one. My sentiments are exactly those of (the younger) Bragg:†

> A good lecture is a *tour de force;* a good lecturer should be keyed up to a high pitch of nervous tension before it and limp and exhausted after it . . . If a sensitive lecturer is to give of his best, he must be left in peace for a period before the lecture starts. It is the refinement of cruelty to expect him to be social.

And I find I need—as perhaps we all do—a strong basic structure, along with the freedom to improvise at any point.

Best,
Oliver

* Horgan had invited OS to an onstage interview at Stevens Institute of Technology.

† Lawrence Bragg and his father, William Henry Bragg, jointly received a Nobel Prize for their work on X-ray crystallography.

To Nick Younes

MARCH 3, 2010

2 HORATIO ST., NEW YORK

Dear Nick,

How nice to get a *typewritten* letter—an extinct species, these days—and to visualize you with a Selectric on your lap.

Anne* sent us a detailed account of your incredible blizzard, and how she feared a tree would fall on the house. Many years ago—the day after reunification, when Kate and I went to Leipzig (which, unchanged from 1940, looked as if it were frozen in a time-warp), it started to snow, and snow, and snow . . . and we were deliciously marooned there for almost a week, and felt very grateful to the snow. A similar feeling was expressed by Auden, in his poem "Thank You, Fog."

I can imagine that with a computer you are "plugged in" 24/7—continual (but also dangerous) distraction—I am besieged by e-mail, tho' Kate filters 90% of them out. I see how one can never feel alone with one of those infernal machines going, and, sometimes, one needs to (have you read Anthony Storr's book *Solitude*?).

In regard to reading and writing I am the opposite to you. I can write non-stop (at rare, happy times) for hours—once, travelling to London, I was so absorbed that I noticed nothing external to me between Kennedy and Heathrow; as we were touching down I said to the stewardess "You didn't serve me any dinner!" She said she had, but that I had eaten "absentmindedly," and without pausing in my writing. I tend to find reading, on the other hand, continually interrupted by my own thoughts—autochthonously, or related to what I am reading—

You seem (if I do not misunderstand you) to speak of gathering facts, metaphors, etc as material for some future book, theory or synthesis. Surely they arouse associations, are in some sense integrated, *as* you incorporate them. But, in another sense (I imagine it is the same with all thinking people) they may be put away in some mental attic or basement until their time comes.

* Anne Mathews-Younes, Nick's mother, who had also become a friend of OS's.

I have to stop now—I have been standing* at my desk for the last 8 hours—I look forward to *seeing* you (and will read some Euripides).

Love,
Oliver

To Revella Levin

PSYCHOTHERAPIST

JUNE 1, 2010

2 HORATIO ST., NEW YORK

Dear Revella,

Thank you for your nice letter—your letters are always full of fire and interest.

I added a little to my "Asylum" piece (for the NY Rev. Books); and yes, I went to Geel, too briefly, and also enclose a copy of my foreword to a fine book by Eugeen Roosens (*Geel Revisited*). (And I have heard of similar places, going back centuries, in Morocco and Japan.) I wonder if my schizophrenic brother could have had a better life with community support, instead of being doped (and, in the 1990s, shocked—insulin coma) out of his mind the whole time. (I once tried Haldol myself, and felt like a zombie.)

I say a little about the two-facedness of "tranquillizers" in my *Mars* book—copy enclosed—but it seems to me that the matter is very complex and individual, and "need" must be balanced against "tolerance." There can be cruel and tragic dilemmas, as with my Awakenings patients and l-DOPA.

This has been a rather difficult six months for me. A total knee replacement tipped me (perhaps literally, because of the asymmetry) into an excruciating sciatica, and I had to have a laminectomy. Between these two operations I was effectively immobilized and housebound (unable to sit, so couldn't travel or go out).† Things are

* At this time, OS was plagued by sciatica and could not tolerate sitting.

† Since OS was unable to sit, either to eat or to write, we constructed a standing "desk" on his kitchen counter, using sturdy volumes of *The Oxford English Dictionary*.

easing now, and I am getting slowly back to a more normal life, which includes *swimming.* (I find this easier, and safer, than walking.) Three days before the knee surgery I had a hemorrhage in my right (melanoma) eye, and lost the sight in it completely. I am due to have a vitrectomy next week, and perhaps that will restore a little vision.

I *would* have been depressed by this accumulation of problems had I not been able to *write,* and *this* has kept me going. I have completed one book, *The Mind's Eye,* about visual perception and some of its vicissitudes, and ¾ written another (*Seeing Things*)* on visual hallucinations.

Since you are kind enough to ask, I enclose the (uncorrected) proof of my own visual story from *Mind's Eye.* The others are more upbeat—tales of accommodation, even transcendence, in face of adversity.

My best,
Oliver

To Revella Levin

JULY 2, 2010
2 HORATIO ST., NEW YORK

Dear Val,

A very short letter, because I have had two eye operations in the past three weeks, and reading is very difficult at this point.

I was most interested in your letter where you speak of Luria and the (so brief!) welcoming of psychoanalysis in the USSR. Luria himself speaks of his excitement when, as a nineteen-year-old, he got a letter from Freud.

I started to read your discussion of "Bathsheba," found it fascinating, but was/am not up to finishing it till my eyes are better.

I am not sure that I have ever spoken of schizophrenia as "curable" or "incurable"—these are not useful words here. I think that many people with schizophrenia can be "helped," whether by analysis,

* Later retitled *Hallucinations.*

supportive therapy, drugs, living in therapeutic communities, or all of these (as described by my gifted schizophrenic namesake Elyn Saks in her remarkable memoir *The Center Cannot Hold;* Elyn has given our more rigidly biological psychiatry colleagues something to think about). You yourself speak carefully, modestly, about moving people in schizophrenic bind and torments to . . . to the sort of neurosis we all have to live with. As yet, I have never really explored the schizophrenias—I am only a neurologist! I enclose a current piece.

My best,
Oliver

To Revella Levin

JULY 9, 2010 (77 TODAY!) [NOT SENT]
2 HORATIO ST., NEW YORK

Dear Val,

Thanks for your latest letter—I do like the image of (my) producing champagne from pickles.

Among other birthday mail (and e-mail) I got a letter from Jane Goodall whom I love and admire (how can one not?), and this made me think of the passion and steadfastness of her commitment to (the understanding, the lives, and now the almost-desperate plight of) chimpanzees—see her July 8 article (attached)—and her "advocacy" of/for them, for all imperiled species and life-forms.

I cannot help contrasting my 50+ years as a physician with her 50 years as a biologist/naturalist. Do I have any commitment comparable to hers? I could say that I do, that I have a commitment to the well-being of my patients, and of those who suffer, are tormented, are ignored, or misunderstood, or marginalization, etc. generally—and that I write about them with sympathy and force, so that their situations may be more widely known and understood. In this sense all I have written is a form of "advocacy," and this continues in the forthcoming book—perhaps, especially, in my piece on the face-blind (it too will be in the *New Yorker* soon), where I emphasize that this can afflict 2–3% of the population with such severity that they have great difficulty recognizing anyone (even, sometimes, their own reflection in a mirror), and may become quite socially withdrawn or otherwise limited in their relationships as a result. This is not generally known: it will be after my article appears. In this sense I will be an "advocate" for the face-blind.

On the question of reading: the problem is not just that one eye is blind, but that the other eye has moderately dense cataracts, and I fear to have it operated on, not having (so to speak) an eye to spare.

My best,
Oliver

To David Remnick

EDITOR IN CHIEF, *THE NEW YORKER*

MARCH 1, 2012
2 HORATIO ST., NEW YORK

Dear David,

I was very pleased and *thrilled*—and reassured!—by your response to *Hallucinations* (for at my age there are always the fears of declining powers and senile repetition). (Do you know a fabulous and very funny book called *Losing It* by William Miller?)

I never intended an entire book on hallucinations, only the first piece (on Charles Bonnet syndrome), but it grew and grew, almost against my will. I certainly didn't intend the "Altered States" chapter, but was persuaded, when I was in hospital last year, and unable to write, to reminisce and tell stories of my drug-taking days; so the personal parts of this chapter were *told* (to a friend who transcribed them), not written—but, conceivably, this made them more alive. I am glad you have selected this chapter, but a little nervous too, since I am "outing" myself here as a (former) drug-taker. I hope the first half of the piece is not too drastically edited—I think one needs some of the historical-cultural background—tho' I can see for myself how a couple of thousand words can be cut—and John Bennet has always been a marvellous and sensitive editor to me. The book will not come out till after the Election in November. Again, all my thanks for liking and taking this piece, as so many before—

My best,
Oliver

To Pam Belluck

HEALTH AND SCIENCE WRITER,
THE NEW YORK TIMES

SEPTEMBER 9, 2012
2 HORATIO ST., NEW YORK

Dear Pam Belluck,

I am reading *Island Practice* with great enjoyment—I think you capture the flavor of Nantucket, and of a very original doctor, perfectly.

Reading it brings back a vivid memory from 1979. I too love islands, and the previous year I had spent six weeks in Gore Bay, Manitoulin,* swimming, writing, wandering all over the island—when I returned in '79 the islanders had a proposition: You seem to love us here in Manitoulin, they said, and we love you. Our doctor, a New Zealander, has just retired. How would you feel about staying on here, and being our doctor? (or words to this effect). When I hesitated, they added, "The Province of Ontario would provide you with a house—and it's a good life here, as you've seen for yourself." I allowed myself to phantasy being an island doctor—island practice—for a couple of days, and then expressed my regrets. But your beautiful book brings all this back to me, and wondering how life would have been had I stayed on as their doctor.

I sense your own deep feeling for Nantucket, and Lepore, and all the other "characters" in your book.

With thanks and best wishes,
Oliver Sacks

* OS has the timing a year off here; his second visit to Manitoulin was in 1980.

To Marc Bekoff

BIOLOGIST

SEPTEMBER 27, 2012

2 HORATIO ST., NEW YORK

Dear Marc,

This will be a preliminary and (I think!) fairly short letter. I do not use a computer and I can no longer see my typing—so I hope my scribble is legible.

I enjoyed meeting you at the Institute* last Friday, and today find your wonderful package, with *The Emotional Lives of Animals* which I will try to read over the weekend (but reading is much slower now that I have to use a magnifying glass), and your charming, illustrated books for younger readers (I have sometimes wondered about doing something similar with Chemistry—a children's book of the Elements, with stories about their discovery and properties to make them seem exciting and *almost* [alive].

(Before I forget: do you know Alexandra Horowitz' book *Inside of a Dog*? She was, very briefly, a student of mine.)

I am very glad you enclosed Jane's† marvellous article—so incisive, so passionate, so *right*. I am (ardently!) with her on every point save one—the existence of God. What is important is that I (or anyone else) *can* be with her on every point, even if agnostic or atheistic. Darwin, I suspect (with his own passionate sense of continuity and connectedness) would have been.

For some reason I myself have written more about plants than about animals—I enclose copies of *Island* and of *Oaxaca Journal*, plus a piece on Darwin and flowers. I have in mind a (companion) piece—or pieces—on what especially distinguishes animals and plants: Animal Minds and Memories (and, if you will, "Souls"). A starting point for me is (Darwin's discussion of) the mental/emotional states of worms. I love Jane's story of herself—as a baby, taking earthworms into her bed—I will (again) be bringing pots of soil crawling with earthworms into my kitchen (I did this in '09 when I had an impulse to write about them). Darwin found that their "favorite" vegetables

* Institute for the Humanities at NYU.

† Goodall, who worked closely with Bekoff.

were onions, radishes and cabbage. These are my favorites too—which exemplifies the continuity of life!

So far I have written about insectivore plants (which Darwin calls "not only wonderful plants but sagacious animals"), sea-anemones, regular jellyfish, "box" jellyfish, sea-slugs (which Eric Kandel has studied so brilliantly) and cephalopods. [. . .] Now I have to get onto vertebrates—if I'm up to it.

Again, all my thanks—and sorry for an overlong "brief" letter.

Oliver

To Gay Sacks and family

OCTOBER 30, 2012

2 HORATIO ST., NEW YORK

Dear Gay, Carla, Elliot and All,*

I hope you have all received copies of *Hallucinations* from Picador. There was going to be a jolly publication party tonight, but that—like almost everything in NYC at the moment—is cancelled (hopefully just postponed).†

Thank you for messages (which Kate tells me of)—it looks like the loss of power, and everything related to electrical power (such as running water), will last for several days. I have never seen (no one has) NYC brought to her knees in this way. None of us, I think, quite took "Sandy" seriously—tho' I was fearful when I saw, on satellite pictures, that it was close to 1000 miles in diameter.

There have been a number of disasters or near-disasters—and no doubt there will be many to come. A frightening one was the failure of back-up power in NYU hospital—hundreds of patients, including people on respirators etc. and preemies in incubation, had to be taken elsewhere. I fear a substantial death-toll among them.

I can't help thinking that our total dependence on "mains" electricity

* Even after Marcus Sacks's death in 2004, OS remained close to his widow, Gay, as well as to her children and their own families.

† Much of Lower Manhattan had been flooded by Hurricane Sandy.

poses a huge potential danger like that of plant monocultures—potatoes, bananas, rubber, vines. William Cobbett* worried about this more than 2 centuries ago, was worried about Big City Life, and felt every cottage should have its own source of power. We need *vast* back-up batteries as a start.

My book comes out next Tuesday, Election Day—I can't help wondering how this business will affect both Election and book.

Again thanks for your messages, and all my love,
Oliver

THOUGH HE WAS BY NOW in a relationship with Bill Hayes, OS resisted coming out publicly as a gay man (until the publication of his autobiography, in May 2015, only a few months before his death). So it had been especially important for him to hear, throughout his life, from other gay people living open lives. One such person was Bob Buscombe, a friend from OS's San Francisco days.

To Bob Buscombe

MAY 14, 2013
2 HORATIO ST., NEW YORK

Dear Bob,

What a sweet and *happy* letter—out of the blue. It brings you and Spaulding Avenue back, as if yesterday, and yet it was more than fifty years ago (1960–61, I think—I didn't come to the States till the summer of '60). It was a colorful time. But I think we have both been lucky enough to live and age well, and gratefully—and old age has its colors, colorfulness too (and what a nice, handsome, laughing picture of you). [. . .]

I don't think of my life as "dazzling," but as an attempt to explore and understand things—human and otherwise. I have and have had

* Cobbett was a wide-ranging and influential writer who advocated self-sufficiency in his 1822 book, *Cottage Economy.*

many good friends, but never a "partner" (tho' I have a feeling that things here are changing). I am moved that *you* have had a partner for 50 years—and are planting your garden together for the Spring (it has been a wonderful Spring in the East, and my favorite quince trees are in full blossom). (I enclose a couple of tiny fern pieces.) Otherwise, I am slightly blind, slightly deaf, slightly lame, and—like you—enjoy a swim daily. (I enclose too a little swim piece.) I will look you up if I find myself in LV, and you look me up if you are in NY.

Lovely to hear from you—
Affectionately,
Oliver

To Louis Breger

PSYCHOTHERAPIST

JULY 1, 2013
2 HORATIO ST., NEW YORK

Dear Lou,

Thank you very much, first, for your very friendly and thoughtful letter of May 9. [. . .]

It and you have been in my mind this past weekend, since after starting your book* I have gone back to it—and read with great pleasure, and admiration, to the very end.

The fact that you show such sympathy and insight into the traumatized and neurotic Freud puts you in the right position to understand and criticize some of his later aberrations—and it made me read with an attention I can't extend to the "Freud bashers."

In my 47½ years with Shengold the "Oedipus complex" has rarely, if ever, been mentioned. But he has always brought things back to the (multiple) traumas of evacuation and school of my years between 6–10. Again, over-and-above (intellectual) insights, what has been

* Breger was the author of two books on Freud: *Freud: Darkness in the Midst of Vision—an Analytical Biography* (2000) and *A Dream of Undying Fame* (2009).

crucial to me is the sense of his empathy and warmth and support and the feeling, the knowledge, that he *likes* me. We sit and face each other—no couch (I do not know how it is with his other patients). There is little talk of "resistance," but much about "transference." It must, as you bring out, have been similar with *some* of Freud's patients, those who could bring out his *best* qualities as a human being and a therapist. It is odd, as you bring out, that he did not write (or hardly wrote) about his successful (and "happy") cases but more about his failures (and, implicitly, the obstinate theoretical biases behind these). [. . .]

With warm regards,
Oliver

*To Björk**

MUSICIAN

DECEMBER 8, 2013
2 HORATIO ST., NEW YORK

Dear Bjork,

I am so sorry—Billy and I will *not* be able to come to Iceland, to *you*, for the New Year's Eve (we had booked hotels and everything)—but we will *certainly* come later in the year.

I got the DVD of you and David Attenborough (and even some of me)—I was tickled and pleased when you called me "the David Attenborough of the Brain."

It has just started snowing in New York, and there is a crisp feeling of winter, and Christmas, in the air.

You must already have had lots of snow in Iceland—at least in the North.

We will be thinking of you on New Year's Eve, and remembering

* Björk, after reading *Musicophilia,* asked for a meeting with OS—she was incubating her *Biophilia* project at the time. This unlikely pair hit it off, and OS and Bill later visited her a couple of times in Iceland.

the wonderful evening with you, you all, last New Year's Eve—the great bonfire on the Beach, and everyone moving towards the Cathedral as midnight approached, and hugging and kissing and laughing and crying—the intensest expression of *being alive.*

Love from us both,
Oliver

PS: Hope you can read my writing!

To Gerald Edelman

DECEMBER 17, 2013
2 HORATIO ST., NEW YORK

Dear Gerry,

It is too long since we corresponded, but I often think of you, and hope that you are recovered after the difficult months earlier this year. I was very sorry I could not come in person to the special evening in July—but I was preoccupied with (80th) birthday visitors from England etc.

I have returned to writing about *in*vertebrate nervous systems and behaviours—jellyfish, worms, *Aplysia,* insects and—my favorite cephalopods. By chance I met J. Z. Young in 1949—he was my examiner in a school exam, and one of the questions was about Octopus (I ignored the others); then again, in 1960, when he gave the lectures which were elaborated into *A Model of the Brain;** and now I am reading his (1995) article (in *Cephalopod Neurobiology*) on "Multiple matrices in the memory system of Octopus."

Altho' the nomenclature is different, he speaks in terms of populations, of learning, categorizing, in terms of (salience-based) "selection," of "recurrent fibres," and of what is in effect "degeneracy" *i.e.* of processes which sound akin to neural Darwinism. I was intrigued that he had been moved to this sort of "population" thinking by his studies of Octopus learning (and brains) and could not help wondering whether

* Young's 1964 book.

the two of you—your TNGS and further theories are much more fully supported, as well as intellectually elaborated—had any contact.

There are all sorts of difficulties—Theodore Bullock was a pioneer in contending with some of these, and I remember writing to him in the early 1950's—and perhaps we will never be able to get the behavioural and physiological data we need with Octopus—but I am intrigued that your son David has been attracted to questions of cephalopod awareness, attention (I avoid saying "consciousness") etc—and ways of investigating them and their neural correlates. I hope to visit him when his lab is set up in Bennington.

I hope I haven't bent your ear too much with this letter, Gerry. I'd love to hear from you, and to visit you again.

My warmest good wishes,
Oliver

To Robert C. Stein

CORRESPONDENT

FEBRUARY 16, 2014
2 HORATIO ST., NEW YORK

Dear fellow-boy-chemist-and-octogenarian,

Thank you for your charming letter. I am especially glad that my *Uncle Tungsten* [. . .] had resonances for a *real* chemist like you.

I have, however, an opposite, at least different, philosophy from "Get rid of all those books and papers." While I am happy to give some books away—I am very glad that I have copies of so many of the letters received and written over the decades—for example, the many, and vivid, and deeply felt letters which passed between my parents and myself when I came to America in 1960. Where memories are often feeble and capricious, these have really allowed me to *revisit* (and write about) that part of my life—and to realize how insufficiently I appreciated the loving qualities of my parents at that special time. Having and reading and writing about these letters—more than 50 years later—seems to fill in and "complete" my sense of

To David Brody

CORRESPONDENT

NOVEMBER 21, 2014

2 HORATIO ST., NEW YORK

Dear fellow-atheist and skeptic,

Many thanks for your interesting letter. You describe a very special sort of "anticipatory" dream such as (I and) many people have in relation to nocturnal myoclonus—the sudden jerk of a limb, or the whole body, which can jolt one awake.

One friend of mine would always imagine that the TV set on a table at the foot of his bed was toppling off the table—he had to wake nightly to save it. The imagined leap was the dream "rationalization" of the myoclonus. For myself, the "anticipatory" (or, as I prefer to say, "rationalizing" or "explaining" or "contexting") dream can take all sorts of forms, but all involve a sudden (imagined) movement—falling into a ditch, whatever—I suspect that such dreams can be fashioned in a (split-) second between the first unperceived, unconscious, "intimation" of the jerk and its full realization. It might be possible to test this with patient video-EEG recordings (over many nights, for it only happens with most people, occasionally).

I read Dunne's book* as a boy—I think it (ingenious!) nonsense. I am no believer in "clairvoyance"—dreams or otherwise.

Thanks again for writing,
Oliver Sacks

To Revella Levin

JANUARY 13, 2015

2 HORATIO ST., NEW YORK

Dear Val,

Thanks for your good letter of Dec 7 (and an earlier one, from September).

* *An Experiment with Time* (1927) by J. W. Dunne, about precognitive dreams.

No, I haven't forgotten you—you are not at all forgettable!—nor have I been ill; but I have been totally absorbed in writing an Autobiography—something much more open and personal than anything I have written; and (I hope) at least moderately honest and not too self-serving. Writing it seems to me to have some affinities with being in analysis—a sort of self-analysis, but one embedded (as perhaps all analysis should be) in one's passions, intellectual and emotional; one's loves (and, yes, hates); all the events and accidents which interweave with whatever central thrust one's life has had—I find myself dropping into the past tense—and thinking sometimes I am writing about someone else, someone who is oddly familiar (in many ways), a stranger in others, but who, on the whole, I have come to like—more so than when I started.

Who knows (at our age) what the future holds? I wish us both a happy and productive 2015.

Love,
Oliver

16

Gratitude

2015

OVER THE CHRISTMAS BREAK in 2014, OS took off with Billy for a few days in the Caribbean; he had just delivered the final manuscript for *On the Move* to his publishers. He felt a little unwell while he was away; he thought he might have gallstones. Less than a month later, he was diagnosed with metastatic cancer—the ocular melanoma he had lived with for nine years had spread to his liver. He knew this would be fatal, even before his doctor delivered the bad news: he had eight months to live, sixteen at the outside.

It was immediately apparent how OS would handle this: he made a list of new priorities ("have fun, even be silly" was one of six directives to himself), and aside from time with those he loved most, writing, as always, was the most urgent and the most rewarding. In the months he had left, we—Billy, OS, and myself, along with some helpers—worked nearly around the clock, soaking up maximum time together and completing an astonishing number and range of essays. We were focused and, in a way, joyful. Time was precious, and we used it well. He sketched out a final plan for a posthumous collection, *The River of Consciousness*. We pulled out some articles that had been awaiting completion, and he started a few more. Altogether, 2015 would see the publication of nine new essays in *The New York Times, The New Yorker*, and *The New York Review of Books*.

The first order of business, however, even as OS was heading back from that first doctor's appointment, was to get word to Dan Frank, his editor at Knopf, to ask whether the publication of *On the Move* could be expedited, so that OS could live to see its reception. (Knopf agreed almost immediately to publish the book in early May instead of September, as they had originally planned.)

A few days later, OS wrote, almost in a single sitting, an essay called "My Own Life," conveying his gratitude for a life well lived, and saying, "It is up to me now to choose how to live out the months that remain to me. I have to live in the richest, deepest, most productive way I can." He put the essay away for a few weeks, awaiting the right time.

OS decided, too, that he wanted someone to record him reading passages of *On the Move* in his own voice—it would be his last testament as well as his coming out as a gay man.*

There was one more thing. Before he announced his diagnosis to the world, he had to tell his closest friends. He did this almost entirely in his favorite medium: letters, many dozens of them.

To Anna and Joseph Horovitz

FEBRUARY 5, 2015
2 HORATIO ST., NEW YORK

Dearest Anna and Joe,

I have some sad news. At the beginning of the year it was discovered that I had metastases (in the liver) from my eye melanoma. These things rarely metastasize, and I had nine good (and productive) years before my luck ran out. There are no really good treatments for this sort of cancer, but some treatments have a chance of slowing, though not completely stopping its advance.

I hope that in the time remaining I can write—I write, Joe, as you compose, and nothing gives me (or others) greater pleasure. I have several nearly-completed books which I would like to complete, even tho' I may/will not see their publication. I am exceedingly glad that I *did* compile an autobiography (just) before this hit me—it will be published here on May 1. [. . .]

Work apart, I hope that once a treatment is over (it will occupy and

* This project quickly became much more than a reading of selections from the book, expanding to many hours of OS reminiscing on camera to the documentary filmmaker Ric Burns. Burns's film, *Oliver Sacks: His Own Life,* was released in 2019.

debilitate me a little from mid Feb to mid April) I will be able to enjoy life as well, go out, see my friends—and, sometimes, say "Farewell." I hope then to come to England, provisionally, in May.

You have been both so important in my life—you are my favorite people!—and I look forward so much to seeing you again in the Spring.

Your clarinet pieces, Joe, have been on my bedside CD, since you gave them to me, and come fresh and delightful (and witty!) whenever I play them. My piano-teacher, Faine, adores them, and I encouraged her to write to you.

I find, as David did when he had his cancer, playing the piano a great solace, and for me, as for him, it is nearly always Bach now.

All my love to you both,
Oliver

To Robert B. Silvers

FEBRUARY 8, 2015
2 HORATIO ST., NEW YORK

Dear Bob,

The *Review* will soon be receiving an advance reading copy of my autobiography (*On the Move*), basically the uncorrected MS of 12/14. The book itself—with some corrections, revisions etc—is to be published on May 1.

Knopf advanced the publication-date from (its original) in September, because I have a serious medical problem now, and don't know what shape I will be in, or if I will be alive, in September.

Last month I was found to have metastases in the liver from my eye cancer. These ocular melanomas rarely metastasize, and I am very grateful to have had nine good (and productive) years before it decided to spread. There are no radical treatments for this, only some palliative treatments which may slow its spread. I feel well at the moment, but start one of the treatments (which may have some debilitating effects) next week. *This* week I will be making a video-reading of selections from the book, perhaps ad-libbing or improvising a bit—Ric Burns will edit this.

I was anxious to do this, to provide images as well as words, while I was still in good shape.

I may (or may not) (try to) publish a "final" piece about getting the news, facing death, doing (especially writing) what I can in the time I have left—it will quote Hume's final piece ("My Own Life") which he wrote, in a single day, when he knew he was mortally ill. "My Own Life" is honest, brave, realistic, as Hume always was—and I will use this title.

But at this point I want only to inform a few friends (and relatives)—and I have come very much to think of you as a friend, no less than an editor (the most wonderful of editors!) over the forty-plus years we have known each other. You have been an important and special figure for me over the years—and have inspired or suggested much that I might not have written about otherwise. I think especially of *Seeing Voices*.

This letter is to you, for you, and I write in confidence.

My best,
Oliver

*To Nicholas Naylor-Leyland**

FRIEND AND SWIMMING COMPANION

FEBRUARY 15, 2015
2 HORATIO ST., NEW YORK

Dear Nicholas,

I'm afraid I have some sad news to share. The cancer (melanoma) I had in my right eye, which was not supposed to metastasize, has done so—to my liver. This was discovered last month, after I had had some dark urine (I thought it was just a mild gallbladder attack).

* OS first met Naylor-Leyland on a trip to Curaçao organized by their swimming coach, Doug Stern. OS returned almost every year after that, often with Naylor-Leyland or another friend.

Treatments are only palliative; it can retard or reduce the metastases for a while—30% of my liver is now occupied by them—maybe extend survival for a few months, but that is it.

I am, at least, extremely glad this didn't happen earlier, and that I was granted nine good, productive years, to enjoy life, see friends and family, travel (and being me) *write,* and I hope I can continue to do so for at least some of the time that remains to me. I am especially glad I was able to complete my autobiography (*On the Move*), and that the publishers, working 24/7, will be bringing it out in a couple of months. It has a couple of *your* photos in it—the ones you took when we were in Curaçao—so you are listed among the photo-credits.

Things like this make one view one's life—and death—in a different way. I have a great sense of gratitude for all the good things, and above all, people there have been in my life—and you (since that morning we met in Curaçao, in '96) are among them. I have so many good (and often funny) memories of times spent with you—how, sharing a cabin on one of Doug's weekends, you kept turning up the heat, and I kept turning it down.

I have to go into hospital tomorrow, for "embolization" of (one half of) the liver—the emboli, hopefully, will reduce its load of metastases (they are very sensitive to oxygen lack)—and if this goes well, and I can tolerate it (the metastases release cytokines, inflammatory agents etc, as they necrose) I will have the other ½ embolized in a month or so. I feel fine now—have been swimming every day, and making a film with Ric Burns—but may be knocked out a bit by the embolizations.

But, with luck, I should bounce back, for a time, be able to enjoy life, to write, and (most important!) to see friends, those whom I love, and who have loved me. Not least—*you.*

All my love,
Oliver

To Jerome Bruner

FEBRUARY 15, 2015
2 HORATIO ST., NEW YORK

Dear Jerry,

It was inspiring to see you in such shape and spirits on your 99th birthday, and at the quiet lunch (with just you, Kate, and me) a couple of weeks later. And I was deeply grateful for your (2014) inscription in my (1983) copy of your autobiography *In Search of Mind.*

But now I have some sad news. When I told you, at the start of '06, that I had developed a cancer (melanoma) in the eye you said "Shit!"—which was exactly how I felt. These things rarely metastasize, and I had nine good and productive years—seeing my friends, travelling, writing—but now ("Shit!!!" redoubled) the thing has metastasized, to my liver. They are not too treatable, and I am forced to think of life—and death—in a new way.

I am especially sensible of all the good and generous things in my life—not least your wonderful review of *A Leg to Stand On* when it came out here in '84. I had been a bit depressed by a (nasty) review of it in England, and *your* review completely transformed my mood, my image of myself, and stimulated me to write nearly a dozen new pieces, in short order—pieces which became the basis of *The Man Who Mistook His Wife for a Hat.* You lifted me out of the abyss; and you have been such a good friend (and, implicitly, a mentor) ever since. [. . .]

I can only hope that I can preserve the same good energy in the time that remains to me—enjoying life, seeing my friends, and (of course) *writing* to the last.

All my love—and thanks dear Jerry,
Oliver

ON FEBRUARY 19, *THE NEW YORK TIMES* published OS's essay "My Own Life," which announced to the world that he was dying. By February 21, the *Times* had received such an outpouring of comments from readers that the editors decided to publish a selection of those as a feature article. And OS himself was deluged with thousands of letters and emails, many from fans, others from colleagues and friends. He read virtually all and replied to a huge number of them—especially, of course, those from friends and colleagues.

To V. S. Ramachandran

FEBRUARY 24, 2015
2 HORATIO ST., NEW YORK

Dear Rama,

I was very touched by your letter, especially your references to Crick and Richard Gregory, your feeling that such men should be immortal, and that you still hear their voices in your head. I felt that also about both of them—Edelman too—and have written something about them in my autobiography. I am amazed and moved—greatly honored too—that you include *me* in this pantheon of influences.

I am sorry to hear of the rough time *you* have been having with exaggeration of benign familial tremor. One relative of mine with this finds that a stiff drink b.i.d.* makes all the difference; another, a non-drinker, takes Xyrem. (I'm sure you have sampled both.)

I was in hospital for a week after my (right hepatic) embolizations last Monday, and still rather weak and in some pain. When these symptoms die down, I will have the left lobe done. I can't look beyond that point. I've never been one for metaphysics or mysticism, but will get Schrodinger's book (I know only his *What is Life?*).

You wd be welcome here any time.

Love,
Oliver

* Twice a day.

To Faine Wright

PIANO TEACHER

MARCH 8, 2015
2 HORATIO ST., NEW YORK

Dear Faine,

It was such a pleasure seeing you again last Wednesday—I am glad you are getting your energy back, sad you still have the pains.

What a pleasure it is to walk! Yesterday, with Billy, I went to a Fern meeting* in the NY Bot. Gdn—an anniversary: I first went to one in March 1993, though this one *may* be my last visit (I hope not!) there was a sense of farewell, and some tears all round (it is very rare for me to tear up, tho' I think I may *need* to, sometimes) and after the meeting—quite a long walk (for me), a mile perhaps, *in* the NYBG, especially beautiful with new fallen snow.

I have been practicing the Sarabande (from the 4th English Suite)—minor keys call to me, now, more than major ones—and trying to keep my fingers nimble with the C-minor Prelude. But I have to go into hospital on Wednesday for a second embolization—the first one, happily, worked brilliantly, and there was no hint of any metastases *outside* the liver. So I must defer a lesson to the following week. [. . .]

Love,
Oliver

* OS routinely attended the monthly meetings of the New York chapter of the American Fern Society, a group of amateur fern lovers. In *Oaxaca Journal,* he chronicled a trip to Mexico with some members of the group.

To Michael Jellinek

CHILD PSYCHIATRIST AND FORMER STUDENT

MARCH 20, 2015

2 HORATIO ST., NEW YORK

Dear Michael,

I was warmed—and delighted—by your letter: no letters move me more than those from former students, and I have received many in the last few weeks, all reminding me of those few years (1967–1973) when I met/taught medical students from Einstein. [. . .]

And *very* moved—I had forgotten (till your letter reminded me) that strange, front living-room in my Bronx Park East house, which had a single, enormous bright green carpet, tubs of ferns against the window—and, as you say, no furniture. And you, vital young people (tho' I was only, I suppose, 10–12 years older than you), sitting on the floor, munching, and talking about everything under the sun. I hope I gave *you*—and your fellow-students—a chance to talk too (I have a tendency to monologue, reminiscence, storytelling, maybe fabulation, myself).

I am happy to know that that brief contact 40 years ago had some influence on you, and that you have had a gratifying career as a child psychiatrist.

My warmest good wishes—and thanks,
Oliver

To Richard Aldrich

HISTORIAN*

APRIL 10, 2015
2 HORATIO ST., NEW YORK

Dear Richard,

I was very moved and, yes, comforted by your warm, generous letter—the "Warwick experience," so multifarious, meant a lot to me, though so brief. Being "historically-minded" (rather than "cutting edge") I especially enjoyed my time with you and the Institute of Advanced Study—and being able to speak about (and exemplify) the role of narrative, and having (Hawthorne's phrase haunts me) "an intercourse with the world."

I am currently (now that I have had two procedures to reduce the "melanoma load" in my liver, albeit temporarily) feeling better, and had quite a long swim today.

What you say about a (sort of) posthumous life in the memory of friends and others (and the friends who are around me now) *is* a real source of comfort, along with the feeling that (in some difficult to categorize way) I may have done something worthwhile with my life, and loved and been loved.

I was intrigued, taken aback, by your thought that we have been "spectacularly lucky" in living at this time, while everyone seems to put such a negative spin on everything (global warming, the Middle East, cybercrime, etc.). But certainly, for better or worse, we have seen more, *deep* changes in our lifetime than any previous generation. I hope my (not-*too*-egocentric) autobiography will give (as thru a peephole: an odd, idiosyncratic one) a (sort of) picture of our time, as well as of me. Dear Richard, thank you again for a very special letter.

Say "Hi" for me to people who remember me at Warwick.

Oliver

---

* Aldrich, a professor of international security, hosted OS during his visits to the University of Warwick, in England, where OS was a visiting professor during the last few years of his life.

AT THE END OF APRIL, we celebrated the publication of *On the Move* with publishers and friends before a final trip to the United Kingdom in May to say farewell to family and friends and old familiar places. We—OS, Billy, and KE—detoured to see Chesil Beach and the fossils of Lyme Regis, and visited a primate rescue center in Dorset, where OS communed with baby orangutans.

To Nick Lane

BIOCHEMIST, AUTHOR OF *LIFE ASCENDING*

MAY 13, 2015

2 HORATIO ST., NEW YORK

Dear Nick,

Some years ago you sent me *Life Ascending*, and it has been a bedside book ever since, one I pick up, and learn from, get inspired by, again and again.

And now your (astounding!) new book—*The Vital Question*. I have read many books and articles on "The Origin of Life," but yours has a depth, a detail, a logic—an experimental support (insofar as laboratory reactors can simulate alkaline vents etc)—and an *intellectual exuberance*—which is enthralling.

I picked it up when I was in London last week (after reading a review in *New Scientist*)—and have been reading it slowly (slowly because it is challenging in the best sense; and because I am half-blind now), and am only up to Chapter 4—slowly, but with great excitement.

When Crick was approaching his end, with complete equanimity, and working, literally, to his last hours, his only lament was that he would not see what scientific advance would be achieved by 2030—and now I am dying I have just this feeling. But your book is giving me a special joy—because it is not only a superb synthesis of the last half century, but a sort of peephole into the next half-century, and shows what a golden age of science we are in. This (curiously) is a great comfort to me—at this stage—as well as a joy.

Congratulations on your work—I have read (some of) your papers over the years, and often meant to write to you—and now *The Vital*

Question lets me write, and express my admiration both for your (and others') work and your presentation of it.

With best wishes,
Oliver

To Richard Wagner

FORMER NEIGHBOR OF THOM GUNN'S

JUNE 30, 2015
2 HORATIO ST., NEW YORK

Dear Dr. Wagner (can I say Richard?),

I was greatly interested, greatly *moved*, by your letter—your courage in being honest and forthright, at a time and on a subject* bound, sooner or later, to cause your ejection from the priesthood. In another few years, perhaps, with Pope Francis at the helm, these last bastions of Catholic bigotry may have fallen.

I like to think of you as living across the street when I visited Thom, and glad to know that *he* appreciated you and your work. I still miss him deeply—there were not too many people with whom I could be entirely open—and I like to think that his ghost is pleased that my title comes from his poems. (I find it a huge relief being open now to all and sundry—I am so glad I completed my book *before* I became ill.)

And what a liberation, an affirmation, for us all that the Supreme Court voted as it did.† I suspect that Ruth Bader Ginsburg, quite ill now, stayed on to ensure the 5/4 decision.

Thanks for your letter, and my very best wishes,
Oliver

---

* Homosexuality.

† A few days earlier, on June 26, 2015, the U.S. Supreme Court's decision in *Obergefell v. Hodges* upheld the right of two people of the same sex to marry.

To Robert John Aumann

JULY 29, 2015
2 HORATIO ST., NEW YORK

Dear Robert John,

Many thanks for your letter—I fear I was too dismissive-seeming or gnomic in my piece.* I fully agree that understanding consciousness (and its necessary connection with brain processes) is THE last frontier in Science, and, as you say, unlike any other problem. I am not sure whether the word "mechanistic" helps—I am not wedded to it, and get impatient when, say, colleagues (like Zeki or Ramachandran) give me neurological/mechanistic theories of beauty.

But I am equally, or more, impatient with any Cartesian dichotomies ("Mind"/Matter; Res cogitans/res extensa etc).

I think one *may* need to expand the word "consciousness" to include sentience, feeling, etc. which are surely well-developed not only in mammals, but many (?all) invertebrates—octopus etc. Certainly "Nature" has no problem giving animals an "inner" side, sentience, consciousness, etc.—even if "higher" forms of consciousness—autobiographic, self-consciousness etc. are (more or less) confined to humans (with perhaps a little in apes, dolphins etc).

The subject of (embodied) consciousness was largely avoided by neuroscientists—*taboo*—until, in 1979, Crick made a clarion-call (in *Scientific American*) for the necessity and legitimacy of its scientific investigation—and was devoted, to his death in '04, to finding (what he called) "neural correlates" of consciousness (especially visual consciousness). He wrote many papers and some books on the subject, and Christof Koch carries on the work.

To my mind Edelman was a deeper thinker here than Crick, and proposed various models and theories of it from 1978 to *his* death in 2014.

Both were a bit irked by the philosopher Colin McGinn (*The Problem of Consciousness*, etc) who said there would always be a gap between neuroscience, however advanced, and (an understanding

* "My Periodic Table."

of) Consciousness. He felt that we suffered from a "cognitive closure," a limitation which would prevent us ever thinking in the appropriate terms. He was sometimes called a "mysterian." If I say "I do not see it as a problem at all," it is because I see C (or its precursors—sentience etc) as a property of living entities (*living*—I am not a pan-psychist). (But I agree we don't have the least inkling of how/what happens—that we are *nowhere*.)

I do not know how long my own Consciousness will last. My Hebrew names are Eliahu Z'ave, do not know my mother's (she was Muriel Elsie Landau). My father was Schmuel Eliezer.

Love +++
Oliver

Billy says "hello," Kate too.

To Atul Gawande

JULY 30, 2015
2 HORATIO ST., NEW YORK

Dear Atul,

I get a great many letters, most thoughtful and heartwarming, but yours was very special, and meant a great deal to me.

(I am sorry I did not cite your book in *On the Move*, perhaps in the "Little Sisters" Section (pp 223–5), where I cited Victoria Sweet's book—do you know her/it?) I am so glad I was able to complete OTM before I became ill, and with no thought that I would get ill.

I like your comment about the continuing importance of, or maybe the current return to, elements etc.* I am currently writing a piece on Eyes—all sorts from those of jellyfish and scallops and jumping spiders and octopus to *our* (vertebrate) eyes, and was delighted to find a transcript of a talk on this which I gave to our

* Gawande had written a note appreciating the connection between OS's love for the chemical elements in his childhood and a *New York Times* essay he had just published about turning eighty (in 2013) and mercury, element 80.

"Field Club" at school in 1950, very nice to discover some constancies in one's life. I am also trying to write something about the (deadly) effects of Social Media, when they absorb people, to the exclusion of everything else, throughout their waking hours—there is a terrifying long-short story by E. M. Forster called "The Machine Stops," written c 1900—do you know it? It is incredible that he foresaw such possibilities then.*

But I do not know if I can complete the pieces—I fear I am losing ground fast (I had to have 2 liters of ascites† tapped a couple of days ago—and I fear it is reaccumulating fast).

When I think or write in pessimistic (even apocalyptic) terms I am pulled back by the image of people like *you*—good, powerful articulate people, who can do something to reverse the evil trends in Medicine, or Society generally, and guide us to a saner course. It is people like you—good scientists, good physicians etc—who give me hope for the future—a hope one needs very much when one's own life is close to the end, and the negativities of life (so well put by the Pope in his Encyclical or Martin Rees in *Our Final Hour*) seem to darken the horizon.

Thank you again for your very good, deep letter—I too hope we can connect again.

Oliver

* OS was working on an essay about Forster's piece, "Life Continues," which would be published posthumously.

† A medical term for the condition that occurs when excess fluid collects in the abdomen, usually as a result of liver failure.

To Robert John Aumann

AUGUST 9, 2015
2 HORATIO ST., NEW YORK

Dearest Robert John,

You have been much in my mind lately—I have been thinking of how our 20 years' friendship has been so deeply important and moving for me.

Visiting you, and the "cousinhood," in Israel last year was, in a way, a critical experience for me, enjoying the embrace of family in a way which I had not known since childhood. In particular, the Erev Shabbat meal and prayers and songs with you and your family were a high point of my experience (Billy's, too). Also crucial, and moving me at the deepest levels, was your 2004 interview, in which, among many other things, you spoke of the beauty of the Sabbath for you, its stoppage of time, its remoteness from all daily concerns.

These thoughts have incited me to write a little essay ("Sabbath"), which I enclose. After some minor revisions, especially towards the end, I hope I can persuade the New York Times to publish it. It will constitute, so to speak, "my last words."

I am losing ground fast, I am very weak, but fortunately fairly comfortable. Unless the second immunotherapy infusion tomorrow can retard the rapid spread of metastases, I doubt if I shall outlast this month. Please give my love and warmest greetings to your family, and to all the cousins in Israel. Thank you for everything.

Love,
Oliver

PS I met a Rabbi today. [. . .] We talked freely of everything, and said the *Shema* together. (I will keep a copy by my bedside so I can become more fluent.) He suggested coming again and going over some Talmud with me. He seems learned and sensitive—I think I will do this.

He wondered too if I might not say, with him, the Mourner's Kaddish for my parents (I never did).

To Gay Sacks

AUGUST 9, 2015

2 HORATIO ST., NEW YORK

Dearest Gay,

My cancer has been rushing ahead like a tidal wave, and nothing seems able to slow it down. I am very weak now, though not too uncomfortable, but I am moving towards terminal care and doubt if I will outlast this month.

Although I know you and Carla and Jonny* wondered about advancing your visit here, I am in general averse now to visits and phone calls, although I am happy to receive emails or letters.

I've been writing a piece ("Sabbath"), of which I enclose an uncorrected copy—I hope that with some revision it will be acceptable to the Times this week; and, as you see, this has memories of Friday evenings, Erev Shabbat, and their peacefulness and specialness during my growing up days, and then I shift to how special it was to see all of you celebrate the Shabbat in Sydney, and how this actually "converted" Ralph Siegel.

I feel now that I have done my work, at least done what I can, and that the time that remains to me should be a sort of peaceful, end-of-life Sabbath, and I do have a sense of serenity and acquiescence, as I think Marcus did in his final days.

I know that my illness has been especially painful for you, for all of you, to see, because it so resembles Marcus's illness in a way, as I myself resemble him in many ways. So it will be a sort of double bereavement for you all. I am sorry, Elliot,† that I did not answer your lovely letter of some months ago.

Coming to Australia [thirty] years ago, returning when I could, and now staying in close touch with you, has been an important part of my family life—perhaps the most important part.

There are so many good memories, Gay, of us exploring together—going to Tasmania, going to the Barrier Reef, going to the Daintree Forest, and not least going up to Hamelin Pond and Shark Bay, in the

* Gay and Marcus's daughter and son-in-law. In February, as soon as they heard of OS's diagnosis, Gay, Carla, and John had flown to New York to visit for a week.

† Gay and Marcus's son.

North West, and seeing the stromatolites growing there as they have been doing for three billion years. I don't have the strength to write separately to Carla and John or to Elliot and Katya or India (who sounds like a very remarkable young woman), and must just send my farewell, gratitude, and deepest love to all of you,

Oliver

To Dan-Eric Nilsson

ZOOLOGIST

AUGUST 10, 2015

2 HORATIO ST., NEW YORK

Dear Dr. Nilsson,

I am a clinical neurologist with some background in marine biology—and a special interest in animal eyes. I have vivid memories of seeing the amazing eyes (rhopalia) of *Charybdaea marsupialis* in the Natural History Museum (in London) when I was a boy, dissecting cephalopod eyes, etc.

For some reason this early interest has come back to my mind, strongly, and I want to tell you how much pleasure your many papers (over the last 35 years) have given me—as has your book *Animal Eyes.* It excited me to think of the creative energy and enthusiasm you have poured into your work, its *adventurousness*—and how tenaciously you have stayed with (this wonderful) subject. I tried to write a little myself, but I am mortally ill now, and have to put such projects aside. But I want to say "THANK YOU" before I go.

Oliver Sacks

To Robert B. Silvers

AUGUST 15, 2015
2 HORATIO ST., NEW YORK

Dear Bob,

My cancer is advancing at great speed, and I do not know if I will last the month.

I would like to put together a (shortish) book to be called *The River of Consciousness and other essays.* Most of these—the two "Darwin" essays (flowers, and earthworms); the fallibility of memory; and "Scotoma"—were edited and published by you (the only others are "Speed" and my three "final" *NY Times* pieces).

I would like to dedicate this book of essays to you, in gratitude for all you have meant to me over the years.

I hope this will be agreeable with you.

With my love,
Oliver

To Dan Frank

AUGUST 15, 2015
2 HORATIO ST., NEW YORK

Dear Dan,

I am going down fast, and do not know how long I can hope to retain consciousness and coherence.

Of my many scattered works—clinical, personal, essays etc, some published some not—I would like, first, to put together (they are all published, and in pretty good shape) a modest (c 40,000 word) book to be entitled *The River of Consciousness and other Essays*. The others being the two "Darwin" pieces "Flowering Plants" and "Earthworms etc," "The Fallibility of Memory" and "Scotoma," all of which were published by Bob Silvers—and to dedicate the book to him. The other "major" essay is "Speed" (which was published in the *New Yorker*); and my three 2015 Op-Ed pieces for the *NY Times* (c another 4000 words in all); "Scotoma" (maybe just use its subtitle "Forgetting and

Neglect in Science") is ?12–15,000 words, including many footnotes—a bit meandering, but the only general essay I have written on the history of Science.

I have drafted a brief (still rough) Preface about the writing of these pieces, along with Acknowledgements and Dedication (I have also written to Bob Silvers tonight, asking if my dedicating the book to him would be acceptable/agreeable).

Let us talk—*soon*—because time is running out.

Love,
Oliver

OS CONTINUED TO WORK ON other essays during his last two weeks, and to make notes about books to be published posthumously. He died at home on August 30, 2015.

Acknowledgments

Life with Oliver Sacks was never boring, and simply to keep up with his thoughts and output occupied far more than a forty-hour week. I am immensely grateful to the talented and devoted people who joined me in the effort to keep his affairs running smoothly, acted as wise sounding boards, cheerfully pursued all manner of strange research requests, vetted his incoming and outgoing correspondence (lest he unwittingly commit himself to a series of lectures or mistakenly consign a royalty check to the wastebin), copied and filed said correspondence, and did much, much more. As only they will fully appreciate, the job of assisting Oliver was endless, and might entail driving him to see patients, driving him to see his own doctors, listening to stories, making patients and "subjects" feel at home, taking him to the pool, swimming with him, ordering various fishy delights from Russ & Daughters, chocolate typewriters from Li-Lac, or books from Three Lives, Googling some topic on the smartphones that Oliver liked to call "answer boxes" (his own phone was rudimentary), or explaining why he might like to meet with someone called "Björk"—to say nothing of traveling with him, organizing his lectures, soothing his nerves, ordering swimming goggles, microwaving a bagel for exactly thirty-three seconds, and fetching endless cups of very hot tea (Lapsang souchong with a hint of Darjeeling, specially blended by McNulty's). We laughed a great deal, ate tons of sushi, drank genever or Belgian beer with Oliver, and made sure he got where he needed to go, both literally and metaphorically. We forayed to shuttle launches, mineral mines, and chemical factories. This present book, like so many of his works, would have been impossible without the unflagging labors of Sheryl Carter, Hailey Wojcik, and Hallie Parker—all of whom Oliver adored. Yolanda Rueda, similarly adored, ensured for many years that Oliver's home and bookshelves would be kept sparkling, that his fridge was always full of Jell-O, fish, tabouli, and other delicacies. She kept us all smiling, and many of the correspondents represented in this book partook of her tea and digestive biscuits.

Diana Beck and Juan Martinez also contributed to the daily running of our lives, as did a cadre of others, from Oliver's swim coach to his piano teacher to the many health care professionals who looked after him. He appreciated each of them.

To the multitude of readers who have been inspired to pick up a pen or open a computer to write to Oliver, much gratitude. Your reactions (both positive and negative), suggestions, ideas, and stories continue to make a very real difference to us and to other readers.

Bill Hayes, Oliver's partner in later life, has been a source of inspiration, wisdom, and sanity from the day he walked into Oliver's world. Bill kindly read several drafts of this book and provided crucial support and feedback to me in too many ways to count.

My own deepest gratitude and appreciation go to my husband, Allen Furbeck, and our son, Kai Furbeck, who shared their lives and minds (and me) with Oliver, and accompanied us on many travels both literary and actual. To have two such extraordinary humans in my life amazes me every day.

I could not have begun to approach this volume of letters without the time and talents of two people. Bill Morgan, archivist extraordinaire, devoted several years during the late 1990s and early 2000s to organizing the vast jumble of Oliver's archive and creating a massive and detailed catalog. This enabled Oliver to retrieve letters and sources to remind himself of his early life as he wrote his 2015 memoir, *On the Move*. It also yielded nearly seventy bankers boxes of correspondence.

To boil down this enormous collection of letters to and from Oliver, I was supremely lucky to have Ben Kravitz, a talented writer and editor who spent several years helping me shape this book. Ben organized and reorganized material over and over again; he made a first selection and subsequent ones; he learned to decipher Oliver's handwriting and transcribed hundreds of handwritten letters. He read and reread many drafts and added thoughtful suggestions, corrections, and opinions ranging from grammatical subjects to philosophical observations. Without Ben, this book could never have been completed. I owe him a huge debt of gratitude.

The people who work at Alfred A. Knopf and its Vintage Books imprint are legendary for their passion and talents (including the best sales reps in the business, as I found in my first job as a bookseller long ago). Deb Garrison, my editor, provided wisdom and enthusiasm and stunned me with her ability to recall the smallest details after repeated readings of unreasonably large drafts. Bonnie Thompson, who copyedited this book (as well as *Musicophilia* and *Hallucinations*), has clarified and focused infelicitous words and saved me from numerous gaffes and errors, whether factual, grammatical, logical, or moral. Kathy Hourigan helmed Knopf's production of this book and many of its predecessors; working with her over the past quarter century has been a joy. Thanks also to Ellen Feldman, Zuleima Ugalde, and the Knopf and Vintage design, promotion, and marketing teams, and to Chip Kidd for yet another perfect jacket design.

Laura J. Snyder generously shared her knowledge of Oliver's early life to clarify certain details (I hope I have gotten them right). John McCaskey generously lent his talents to designing a custom database for the Sacks correspondence.

Melvin Erpelding graciously spent time with me reminiscing about his friendship with Oliver.

Mackenzie Kristofco made excellent suggestions to the final draft, and chased down many reference materials along the way. His contributions to organizing the Oliver Sacks archive continue.

Many others helped point me in the right direction, clarifying time lines, facts, and ideas. From the extended Sacks family, Gay Sacks, Anna Horovitz, Alexa Gardner, Tony Gardner, and Joan Caplin reminded me of family history and relationships. Many others I called upon were great friends or colleagues of Oliver's, among them Orrin Devinsky, Jacqui Graham, Mark Homonoff, Andrew Lees, Keith McNally, Edward Mendelson, Rachel Miller, Tom Miller, William Miller, Walter Parkes, Tobias Picker, Nick Rodman, Ingrid Rodman-Holmes, Steve Silberman, Paul Theroux, Concetta Tomaino, and Jeff Towns. Uwe Naumann helped me chase down obsolete German words. I also thank Catherine Barnett, Douglas Braaten at the New

York Academy of Sciences, Sarah Chalfant, Marie Cyprien, Molly Friedrich, Michael Hackenberg, Charles Harris at *Natural History* magazine, Jonathan Kurtis, Lance Lee, Michael Nott, and Jim Poyser for their help.

Dan Frank, Oliver's editor at Knopf for two decades, guided me in this project, as he had with so many of Oliver's earlier books. He generously devoted precious hours, when he had few remaining, to meticulously reviewing letters and sharing his thoughts, giving me a framework for moving forward. Though he did not live to see the finished manuscript, Dan's spirit, his keen sense of humor, and his love of words hover over this book, which is dedicated to his memory.

Selected Bibliography

Books by Oliver Sacks with year of first publication:

Migraine (1970)
Awakenings (1973)
A Leg to Stand On (1984)
The Man Who Mistook His Wife for a Hat (1985)
Seeing Voices: A Journey into the World of the Deaf (1989)
An Anthropologist on Mars (1995)
The Island of the Colorblind (1996)
Uncle Tungsten (2001)
Oaxaca Journal (2002)
Musicophilia (2007)
The Mind's Eye (2010)
Hallucinations (2012)
On the Move (2015)
Gratitude (2015)
The River of Consciousness (2017)
Everything in Its Place (2019)

Essays by OS referred to in this volume, with first publication information followed by, in most cases, the title of the book in which a version (usually updated, sometimes largely rewritten) is included. For a complete list of publications, see oliversacks.com.

"The Aging Brain"
Everything in Its Place

"Altered States"
The New Yorker, August 27, 2012
Revised version in *Hallucinations*

"An Anthropologist on Mars"
The New Yorker, December 27, 1993
Revised version in *An Anthropologist on Mars*

"The Autist Artist"
The New York Review of Books, April 25, 1985
Revised version in *The Man Who Mistook His Wife for a Hat*

"The Axonal Dystrophies" (with W. Jann Brown)
Bulletin of Los Angeles Neurological Society, January 1965

"The Case of Anna H."
The New Yorker, October 7, 2002
Revised version, "Sight Reading," in *The Mind's Eye*

"The Case of the Colorblind Painter" (with Robert Wasserman)
The New York Review of Books, November 19, 1987
Revised version in *An Anthropologist on Mars*

"Chaos and Awakenings"
Appendix to the 1990 edition of *Awakenings*

"Credit to a Pioneer [Étienne-Jules Marey]"
The New York Times Science section, letter to the editor, June 17, 2003

"Cupid's Disease"
The Man Who Mistook His Wife for a Hat

"Darwin and the Meaning of Flowers"
The New York Review of Books, November 20, 2008
Revised version in *The River of Consciousness*

"Dear Mr. A. . . ."
A chapter in *W. H. Auden: A Tribute,* edited by Stephen Spender, 1975
Revised version in *On the Move*

"The Disembodied Lady"
The Man Who Mistook His Wife for a Hat

"The Divine Curse"
Life, September 1988
Revised version included in "Travels with Lowell" in *Everything in Its Place*

"The Dog Beneath the Skin"
The Man Who Mistook His Wife for a Hat

"Early Work on Elephant Gait Not to Be Forgotten"
Letter to the editor, *Nature,* May 15, 2003
Expanded version, "The Elephant's Gait," in *Everything in Its Place*

"Face-Blind"
The New Yorker, August 30, 2010
Revised version in *The Mind's Eye*

"The Fallibility of Memory"
The River of Consciousness

"The Great Awakening"
The Listener, October 26, 1972

"Hard Times for Curious Minds"
The New York Times, May 13, 1999

"Humphry Davy: The Poet of Chemistry"
The New York Review of Books, November 4, 1993
Revised version in *Everything in Its Place*

"In the River of Consciousness"
The New York Review of Books, January 15, 2004
Revised version in *The River of Consciousness*

"The Landscape of His Dreams"
The New Yorker, July 27, 1992
Revised version in *An Anthropologist on Mars*

"The Last Hippie"
The New York Review of Books, March 26, 1992
Revised version in *An Anthropologist on Mars*

"The Leg"
London Review of Books, June 17–30, 1982
Revised version in *A Leg to Stand On*

"Life Continues"
Everything in Its Place

"The Lost Mariner"
The New York Review of Books, February 16, 1984
The Man Who Mistook His Wife for a Hat

"The Lost Virtues of the Asylum"
The New York Review of Books, September 24, 2009
Revised version in *Everything in Its Place*

"The Man Who Mistook His Wife for a Hat"
London Review of Books, May 19, 1983
The Man Who Mistook His Wife for a Hat

"A Matter of Identity"
The Man Who Mistook His Wife for a Hat

"The Mental Life of Plants and Worms, Among Others"
The New York Review of Books, April 24, 2014
Revised version, "Sentience: The Mental Lives of Plants and Worms," in *The River of Consciousness*

"The Mind of A. R. Luria"
The Listener, June 28, 1973

"The Mind's Eye"
The New Yorker, July 28, 2003
Revised version in *The Mind's Eye*

"Murder"
The Man Who Mistook His Wife for a Hat

"Musical Ears"
London Review of Books, May 3, 1984
Revised as "Reminiscence" in *The Man Who Mistook His Wife for a Hat*

"Music and the Brain"
Foreword to *Clinical Applications of Music in Neurologic Rehabilitation,* edited by Concetta M. Tomaino, 1998

"My Own Life"
The New York Review of Books, February 19, 2015
Gratitude

"My Periodic Table"
The New York Times, July 24, 2015
Gratitude

"Mysteries of the Deaf"
The New York Review of Books, March 27, 1986
Revised version in *Seeing Voices*

"Neurology and the Soul"
The New York Review of Books, November 22, 1990

"Obituary: Professor A. R. Luria: Pioneer Brain Specialist"
The Times (London), September 5, 1977

"A Passage to India"
The Man Who Mistook His Wife for a Hat

"The Possessed"
The Man Who Mistook His Wife for a Hat

"Prodigies"
The New Yorker, January 9, 1995
Revised version in *An Anthropologist on Mars*

"Recalled to Life"
The New Yorker, October 31, 2005
Revised version in *The Mind's Eye*

"Remembering Francis Crick"
The New York Review of Books, March 24, 2005
Revised version in *On the Move*

"The Revolution of the Deaf"
The New York Review of Books, June 2, 1988
Revised version in *Seeing Voices*

"Sabbath"
The New York Times, August 14, 2015
Gratitude

"Scotoma"
In *Hidden Histories of Science,* edited by Robert B. Silvers, 1995
Revised version in *The River of Consciousness*

"Speak, Memory"
The New York Review of Books, February 21, 2013
Revised version, "The Fallibility of Memory," in *The River of Consciousness*

"Speed"
The New Yorker, August 23, 2004
Revised version in *The River of Consciousness*

"Stereo Sue"
The New Yorker, June 11, 2006
Revised version in *The Mind's Eye*

"A Surgeon's Life"
The New Yorker, March 16, 1992
Revised version in *An Anthropologist on Mars*

"To See and Not See"
The New Yorker, May 10, 1993
Revised version in *An Anthropologist on Mars*

"Travel Happy (1961)"
Antaeus, Autumn 1988
Abridged version, as part of "San Francisco," in *On the Move*

"Travels with Lowell"
Everything in Its Place

"The Twins"
The New York Review of Books, February 28, 1985
The Man Who Mistook His Wife for a Hat

"Water Babies"
The New Yorker, May 26, 1997
Everything in Its Place

"Witty Ticcy Ray"
London Review of Books, March 19, 1981
The Man Who Mistook His Wife for a Hat

"Yes, Father-Sister"
The Man Who Mistook His Wife for a Hat

List of Correspondents

The beginning of each letter to a given correspondent is indicated by boldface page numbers in the Index. People with abbreviated or omitted last names have been de-identified for privacy.

Aldrich, Richard
Alvarez, Al
Auden, W. H.
Aumann, Robert John
Bandel, Andrea
Barnett, Catherine
Barry, Susan R.
Beasley, Barbara
Bekoff, Marc
Belluck, Pam
Bennet, John
Best, Nancy
Bird, Seymour
Björk
Bogdan, Robert
Bonnard, Augusta
Borden, Bill
Böttiger, Lars Erik
Breger, Louis
Brenner, A. K.
Brody, David
Bruner, Jerome
Buscombe, Bob
Cahill, Kevin
Campbell, Donald
Case, R. E.
Chopra, Deepak
Cohen, Joanne
Cohen, Kalman
Cole, Jonathan
Crick, Francis
Cunningham, Jean
Dahl, Tom
Dallas, Duncan
Damasio, Antonio
Daroff, Robert B.
De Luccia, Josephine
Dennett, Daniel
Devinsky, Orrin
Diamond, Jared
Donald, Merlin
Draper, Dave
Dyson, Freeman
Eban, Abba
Edelman, Gerald
Edgar, Kate
Edwards, Peter
Erpelding, Mel
Fahn, Stanley
Fite, George L.
Fox, Orlan
Franck, Vanya
Frank, Dan
Friel, Brian
Frumkes, Lewis
Gardner, Howard
Gawande, Atul
Gelfman, Jane
Genevieve, Mother
Godwin, Dr.
Goldberg, Elkhonon
Goodacre, David
Goodall, Jane
Gould, Stephen Jay

Greene, Raymond
Gregory, Richard
Gunn, Thom
Haycraft, Colin
Hayes, Bill
Henryk-Gutt, Rita
Herrick, Michael
Herz, Rachel
Hobson, Mr.
Hoffmann, Roald
Horgan, John
Horovitz, Anna and Joe
Hughes, James
Hymson, T. J.
Ivins, Marsha
James, Harriett
Jamison, Kay Redfield
Jellinek, Michael
Jennifer F.
John Z.
Katzman, Robert
Keller, Evelyn Fox
Kermode, Frank
Kiell, Paul J.
Klein, Georg
Koch, Christof
Koven, Barnet
Kurtis, Jonathan
Landau, Helena
Lane, Nick
Lee, David
Lee, Lance
Levi-Montalcini, Rita
Levin, Revella
Lindenbaum, Richard
Listener, The, eds.
Luria, A. R.
Maguire, John
Marie Geraldine Therese, Sister
Mathias, Jane
McCormick, Kenneth
McEwen, Dr.
Messeloff, Charles
Miller, Jonathan
Miller, Mrs.
Miller, Rachel
Moorhead, Hugh S.
Mountcastle, Vernon
Mueller, Jonathan
Murphey, Tim
Nascimento, Rodrigo Delfino
Näslund, Görel Kristina
Naylor-Leland, Nicholas
New York State Office of Mental Health
Nicholls, Henry
Nightingale, Benedict
Nilsson, Dan-Eric
Osmond, Humphry
Palmer, Orrin
Papavisiliou, Paul S.
Parkes, Walter
Payne, Christoper
Pearl, Bill
Perrier, Larry
Picker, Tobias
Pinchbeck, Daniel
Pinter, Harold
Poyser, Jim
Presedo, Emilio
Pyle, Eleanor
Ramachandran, V. S.
Rapin, Isabelle
Remnick, David
Riggs, Austin
Riley, Charles A.
Rimland, Bernard
Rodman, F. Robert
Ronza, Marina Molino
Rose, Augustus S.
Rose, Innes
Roth, Emma
Rutherford, Susan
Sacks, Elsie
Sacks, Gay
Sacks, Marcus
Sacks, Michael
Sacks, Samuel
Salter, Leon J.
Samuel, Alison
Schaller, Susan
Seymour L.
Shapiro, Harvey
Shatan, Chaim F.
Shearer, Rhonda
Shubin, Neil
Silberling, Edwyn
Silvers, Robert B.
Singer, Peter
Snellgrove, Marin
Social and Health Issues Review, eds.
Sontag, Susan
Southam, Brian
Stein, Robert C.
Stent, Gunther

Stephan, Karl D.
Stern, Gerald
Stokoe, William
Stone, Sharon
Stow, Cyrus J.
Strawson, Galen
Swales, Peter
Swann, William
Tanaka, Gayle
Theroux, Paul
Tighe, Lillian
Tunberg, William
Turner, Paul
Vincze, Jenö
Wagner, Richard
Warvarovsky, Mike
Wasserman, Robert
Waymouth, Barbara
Weir, Peter
Wendy
Weschler, Lawrence
Whitzman, Stephen
Williams, Robin
Wilson, Frank
Wiltshire, Stephen
Wolpert, Lewis
Wright, Faine
Younes, Nick

Index

Boldface numbers indicate the first page of a letter to a correspondent.

Illustration Credits

Unless otherwise specified, all photographs are from the collection of the Oliver Sacks Foundation.

The photograph of Thom Gunn, Hyde Park, London, 1959, is by Rollie McKenna. © The Rosalie Thorne McKenna Foundation, courtesy Center for Creative Photography, the University of Arizona Foundation.

The photograph of Robert Rodman is from the collection of Ingrid Rodman-Holmes and used with her kind permission.

The photograph of Picador Books authors is by Julian Calder, used by permission.

The photograph of Marsha Ivins in microgravity is courtesy of NASA.

The photograph of OS with Jonathan Miller, as well as the photograph of OS with umbrella, are by Tom Miller. © Tom Miller, used by permission.

A NOTE ABOUT THE AUTHOR

Dr. Oliver Sacks, referred to by *The New York Times* as "the poet laureate of medicine," spent more than fifty years working as a neurologist and writing books about the neurological predicaments and conditions of his patients, including *Musicophilia, The Man Who Mistook His Wife for a Hat,* and *Awakenings.* Over the years, he received many awards, including honors from the John Simon Guggenheim Memorial Foundation, the National Science Foundation, the American Academy of Arts and Letters, the American Academy of Arts and Sciences, and the Royal College of Physicians. He was a frequent contributor to *The New Yorker, The New York Review of Books,* and the *London Review of Books.* His work has been translated into dozens of languages and has inspired generations of physicians, artists, and readers around the world.

oliversacks.com

A NOTE ABOUT THE EDITOR

For more than three decades, Kate Edgar worked with Dr. Sacks as editor, researcher, assistant, and friend. She is executive director of the Oliver Sacks Foundation, which seeks to preserve and extend narrative medicine through the works of Sacks and others, and to support the destigmatization of diverse mental and neurological conditions.

A NOTE ON THE TYPE

This book was set in Minion, a typeface produced by the Adobe Corporation specifically for the Macintosh personal computer and released in 1990. Designed by Robert Slimbach, Minion combines the classic characteristics of old-style faces with the full complement of weights required for modern typesetting.

Composed by North Market Street Graphics
Lancaster, Pennsylvania

Printed and bound by LSC Communications
Harrisonburg, Virginia

Book design by Pei Loi Koay